D1579156

Analog and Digital Electronics

A First Co

621·381

Revised Se

Analog and Digital Electronics

A FIRST COURSE

Revised Second Edition

Peter H. Beards

NAPIER UNIVERSITY, EDINBURGH, UK

Prentice Hall

London New York Toronto Sydney Tokyo Singapore
Madrid Mexico City Munich

First published 1987; Second edition published 1991

This revised second edition published 1996 by
Prentice Hall Europe
Campus 400, Maylands Avenue
Hertfordshire HP2 7EZ
A division of
Simon & Schuster International Group

Typeset in 10/12 pt Times
by MCS Ltd, Salisbury

Printed and bound in Great Britain by
T. J. Press (Padstow) Ltd.

Library of Congress Cataloging-in-Publication Data

Beards, Peter H., 1932–
    Analog and digital electronics : a first course / Peter H. Beards.
        p. cm.
    Includes bibliographical references (p.) and index.
    ISBN 0-13-571753-1
        1. Electronics. 2. Analog electronic systems. 3. Digital
electronics. I. Title.
TK7816.B43   1990
621.381—dc20   90-7563
                                                                   CIP

British Library Cataloguing in Publication Data

Beards, P. H. (Peter Henry) *1932–*
    Analog and digital electronics.—2nd. ed.
    1. Electronics
    I. Title
    537.5

    ISBN 0-13-571753-1 pbk

1 2 3 4 5   00 99 98 97 96

For Josephine

Contents

Preface to the Revised Second Edition

Part of the motivation for this revision has arisen from requests from lecturers who use the book. Several have asked for an introduction to PSpice. Simulation of electronic circuits has been prominent in the development of electronics in recent years, but the cost of software and the machines to run it have been prohibitive for many educational institutions. However, PSpice is a public domain analog simulator which can be run on IBM PCs, so many departments use it to give their students an introduction to simulation. PSpice is applicable to analog circuits containing either discrete components or op amps, so rather than introduce it into one chapter it has been covered as an appendix and referenced to several chapters in the existing text.

Other additions to the second edition include some more applications of op amps and, by request, the 555 timer. This was originally excluded because I thought its popularity was declining, but it appears to be one of the most durable of ICs.

P. H. B.

Preface to the Second Edition

In writing a book of this type one of the problems an author has to face is that of selection or, to put it another way, of what to leave out. Although thermionic valves are still used in certain applications they do not feature significantly in the mainstream of electronic engineering and therefore may be omitted, perhaps with regret – although this may be nostalgia. During the early years of transistor development solid-state physics occupied a large proportion of broadly based textbooks but since the vast majority of electronic engineers work in applications rather than semiconductor fabrication, this emphasis is no longer defensible. Even within the semiconductor fabrication industry many engineers will be employed on circuit design and layout rather than device development.

The stimulus for this book arose from the difficulty of trying to find a text which covered, in broad terms, the foundation work of electronics for an electrical/electronic engineering degree course. There are many excellent books covering parts of this field. Electronic circuits are particularly well provided for, with a number of books giving comprehensive coverage of both analog and digital circuits. Indeed, for the group of students on which this text is focused the depth of treatment is frequently excessive. However, electronic circuit books usually omit everything but the most cursory treatment of combinational and sequential logic which is a topic normally well developed by students in their pre-final years. Equally, the subject of digital systems has stimulated a plentiful supply of books which are perfectly adequate in their coverage of both binary logic and its realization in electronic circuits, but the nature of the subject excludes analog electronics.

By limiting the depth of treatment to that associated with pre-final year level, I believe it is possible to provide a foundation in analog and digital electronics suitable for degree and diploma courses within the confines of a single textbook. The range of topics which can be covered provides a suitable context for the inclusion of analog/digital interfacing and also for an introduction to the operation and application of microprocessors, nowadays commonplace in the earlier years of electrical engineering courses.

As noted, semiconductor theory has deliberately been minimized and has largely been confined to Chapter 1 which concentrates on the operation of diodes and bipolar junction transistors (BJTs). Chapter 2 covers the principles of amplification and establishes techniques for biasing. Chapter 3 introduces the concept of the small-signal equivalent circuit, i.e. the concept of modeling for electronic circuit analysis and design. This leads into the frequency response of BJT amplifiers and to both medium- and high-frequency equivalent circuits.

Chapter 4 covers the operation of the field effect transistor (FET) and its rather more important derivative, the MOSFET. Again, bias and equivalent circuits are described. Chapter 5 is an introduction to the conceptually and analytically difficult subject of negative feedback (NFB). It establishes the benefits and costs of NFB and evaluates the properties of a representative range of both BJT and FET amplifiers with various feedback configurations. Chapters 6 and 7 describe power amplifiers and a representative selection of analog circuits, some of which form the basic building blocks of integrated circuits.

Probably the most important development in analog electronics after the invention of transistors was the integrated circuit operational amplifier (op amp). Chapter 8 describes the properties of op amps and Chapter 9 deals with a selection of op amp applications.

Waveform generation can be separated into sinusoidal and nonsinusoidal circuits, and these are the subjects of Chapters 10 and 11. Undoubtedly electronic engineering was revolutionized by the development of electronic logic. Chapters 12 to 17 form an introduction to some of the more important features of this subject, Chapter 18 deals with the analog/digital interface, and Chapters 20 and 21 with the operation and application of microprocessors. In summary, the remaining chapters are as follows. Chapter 12 is an introduction of binary logic and logic functions. From here combinational logic and methods of minimizing logic functions are developed with particular emphasis on mapping. Chapter 13 describes some of the more important applications of combinational logic. Chapter 14 introduces sequential logic through a description of the various types of flip flop. Flip flops (sometimes called bistables) are the basic components of asynchronous and synchronous counters and shift registers all of which are described in this chapter. Chapter 15 is a general treatment of the analysis and design of asynchronous sequential circuits with both gates and flip flops. For the most part the treatment is limited to circuits with one secondary variable although an example of a two-secondary design is included. The chapter also includes a description of the primary plane method of asynchronous sequential circuit design. It is important for the digital system designer to understand the properties of commercially available hardware. Chapter 16 describes the important properties of digital integrated circuits and the operation of the more important digital IC families. Chapter 17 is concerned with electronic memories, including RAMs, ROMs, PROMs, PLAs and PALs. Chapter 18 describes various types of digital-to-analog and analog-to-digital converter and includes handshaking. The justification for leaving diode applications until the latter stages of the book is that a coherent and wide-ranging treatment of the subject draws widely on material from preceding chapters. Topics included in Chapter 19 are rectifiers, regulated power supplies including the switching regulator, clipping, clamping and detection.

Chapters 20 to 22 form an introduction to microprocessors for engineers. In the first edition this work was confined to 8-bit machines using the Motorola M6800 as the working example. In this edition the microprocessor work has been supplemented by an additional chapter which describes a 16-bit machine, The Intel 8088.

Chapter 20 begins with the main components of a microprocessor system but the main emphasis is on application and covers addressing modes, an instruction set, branching and subroutines, with specimen programs in assembly language. Chapter 21 is devoted to microprocessor input/output (I/O) by describing, in some detail, the Motorola peripheral interface adapter. It includes an example of interfacing a microprocessor to an analog device via a PIA and A/D converter. More powerful microprocessors are now commonplace. The Intel 8088 is described as a working example of a 16-bit machine in a new Chapter 22. Again the emphasis is on engineering examples, in this case using the parallel peripheral interface as a device for effecting input/output applications. A glossary of abbreviations and symbols used is included at the beginning of the book.

An Instructor's Manual for this new edition is available for adoptors of this text.

ACKNOWLEDGEMENTS

In preparing this text I have received help and advice from many of my colleagues in the Electrical and Electronic Engineering Department at Napier Polytechnic. I am especially grateful to Abid Almaini, David Binnie, Bill Buchanan, Frank Greig, Maureen Hume, Eion Johnston, George Lauder, Ron McHugh (now ex-colleague), Bill Pryde, John Sharp and Barbara Urquhart. I am also indebted to Peter Williams, Director of MEDC, Paisley, for many helpful comments on the first draft of the book. The following companies have been generous in allowing me to use some of their data sheets and diagrams:

Fairchild Semiconductor Ltd
Intel
Motorola Semiconductors
National Semiconductor
RS Components Ltd
Tektronix UK Ltd
STC Components Ltd

I should also like to record by thanks to SCOTVEC for permission to use some of their examination questions and to my wife Josephine for help with the proof reading.

P. H. B.

Glossary of Abbreviations and Symbols

Acc	Accumulator	dB	Decibel
ADC	Analog-to-digital converter	dB(m)	Decibel level relative to 1 mW
A_f	Gain with negative feedback	DIL	Dual in line
A_i	Current gain	DTL	Diode transistor logic
A_p	Power gain	DVM	Digital voltmeter
A_v	Voltage gain	EAROM	Electrically alterable read only memory
A_{vf}	Voltage gain with negative feedback	ECL	Emitter coupled logic
A_{vh}	Voltage gain at high frequency	EEPROM	Electrically erasable and programmable read only memory
A_{vo}	Voltage gain at medium frequency		
A_{vof}	Voltage gain at medium frequency with negative feedback	EPROM	Erasable and programmable read only memory
		f_β	Beta cutoff or 3 dB down frequency of a transistor
A_{vol}	Open-loop voltage gain		
ALU	Arithmetic/logic unit	FET	Field effect transistor
BCD	Binary coded decimal	FF	Flip flop (or bistable)
BJT	Bipolar junction transistor	FPLA	Field programmable logic array
BSI	British Standards Institution	f_c	Closed loop cutoff frequency
CAD	Computer-aided design	f_{co}	Open loop cutoff frequency
CB	Common base	g_{fs}	Mutual conductance or transconductance (FET)
CC	Common collector		
C_c	Collector capacitance	g_m	Mutual conductance or transconductance (generally)
CD	Common drain		
C_{ds}	Capacitance drain–source	HF	High frequency
CDI	Collector diffusion isolation	h_{FB}	DC current gain of a common base BJT
C_e	Diffusion capacitance of a BJT		
CE	Common emitter	h_{FE}	DC current gain of a common emitter BJT
CF	Change function		
C_{gd}	Capacitance gate–drain	h_{fe}	Small-signal current gain of a common emitter BJT
C_{gs}	Capacitance gate–source		
Ck	Clock	h_{ie}	Small-signal input resistance of a common emitter BJT
CMRR	Common mode rejection ratio		
CMOS	Complementary metal oxide semiconductor	h_{oe}	Small signal output conductance of a BJT
CPU	Central processing unit	h_{re}	Small-signal reverse voltage transfer ratio of a BJT
Cr	Clear		
DAC	Digital-to-analog converter	$\parallel$	In parallel with

IC	Integrated circuit	PISO	Parallel in, serial out	
I_{CBO}	Collector leakage current of a BJT with emitter open circuit	Pk	Peak	
I_{CEO}	Collector leakage current of a BJT with base open circuit	Pk–Pk	Peak to peak	
		PLA	Programmable logic array	
I_{DSS}	Drain saturation current of an FET	PMOS	P channel MOSFET	
		PROM	Programmable read only memory	
I^2L (or IIL)	Integrated injection logic	PSRR	Power supply rejection ratio	
I_{os}	Input offset current of an op amp	RAM	Random access memory	
		$r_{bb'}$	Base spreading resistance	All of hybrid
I/O	Input/output			
IR	Instruction register	$r_{b'e}$	Input resistance	Π model
IGFET	Insulated gate field effect transistor (usually MOSFET)	$r_{b'c}$	Feedback resistance	of BJT
		r_{ce}	Output resistance	
IEEE	Institute of Electrical and Electronic Engineers	r_{ds}	Drain resistance of FET	
		R_i	Input resistance	
JFET	Junction field effect transistor (usually just FET)	R_{if}	Input resistance with NFB	
		RMS	Root mean square	
K map	Karnaugh map	R_o	Output resistance	
KCL	Kirchhoff's Current Law	R_{of}	Output resistance with NFB	
KVL	Kirchhoff's Voltage Law	ROM	Read only memory	
LF	Low frequency	R_{TH}	Thermistor resistance	
ln	Log to base ε	RTL	Resistor transistor logic	
LSI	Large-scale integration	$\theta_{JA(JC)}$	Thermal resistance junction to ambient (or junction to case)	
LSB	Least significant byte or bit depending on context	SIPO	Serial in, parallel out	
MF	Medium (or mid) frequency	SISO	Serial in, serial out	
MNOSFET	Metal nitride oxide semiconductor field effect transistor	SSI	Small-scale integration	
		t_{PD}	Propagation delay time	
MOSFET	Metal oxide semiconductor field effect transistor	t_{PHL}	Time for high to low transition	
		t_{PLH}	Time for low to high transition	
M/S	Mark space ratio	t_r	Rise time	
MSB	Most significant byte or bit depending on context	TTL	Transistor transistor logic	
		UL	Unit load	
MSI	Medium-scale integration	ULA	Uncommitted logic array	
MUX	Multiplexer	$V_{CE(sat)}$	Collector emitter saturation voltage of a BJT	
M/V	Multivibrator			
NFB	Negative feedback	VMOS	Vertical MOSFET	
NMH	High state noise margin	V_{IH}	High input voltage	
NML	Low state noise margin	V_{IL}	Low input voltage	All of a
NMOS	N channel MOSFET	V_{OH}	High output voltage	digital IC
Op Amp	Operational amplifier	V_{OL}	Low output voltage	
PAL	Programmable array logic	V_{os}	Input offset voltage of an op amp	
PC	Program counter			
PCB	Printed circuit board	V_P	Pinch-off voltage of a FET	
PIA	Peripheral interface adapter	V_z	Breakdown voltage of a zener diode.	
PIPO	Parallel in, parallel out			

1

An Introduction to Semiconductor Devices

Chapter Objectives
1.1   Semiconductor Crystals – Silicon
1.2   Thermally Generated Hole/Electron Pairs
1.3   Current Flow in an Intrinsic Crystal
1.4   Doping
1.5   pn Junction
1.6   pn Junction in Equilibrium
1.7   pn Junction Diode I/V Characteristics
1.8   Reverse Breakdown Voltage of a pn Diode
1.9   Bipolar Junction Transistor (BJT)
1.10   Current Conventions and BJT Symbols
1.11   Common Emitter (CE) Characteristics
1.12   Common Base Configuration
1.13   Gallium Arsenide (GaAs)
        Summary
        Revision Questions

The purpose of this chapter is to explain the basic mechanism of semiconductor devices and to deduce their characteristics. Particular emphasis will be given to the operation of pn junctions because they are fundamental to most semiconductor devices. At this stage the only transistor to be described will be the bipolar junction transistor (BJT); field effect devices will be covered in Chapter 4. Having read the chapter, it is important that the reader remembers the operating characteristics of devices rather than the physical mechanism that produces them.

1.1 SEMICONDUCTOR CRYSTALS – SILICON

Semiconductors are materials with conductivities between those classed as insulators and those classed as conductors. The most common substance used in semiconductor (solid state) devices is silicon (Si). At one time germanium was more common

1

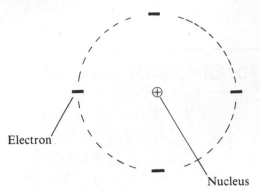

Electron

Nucleus

Fig. 1.1   Bohr Model of Silicon Atom with Valence Electrons only Shown

than silicon, but currently it is little used. A semiconductor material which would appear to have considerable potential for device manufacture is gallium arsenide (GaAs).

Figure 1.1 shows the well-known Bohr model of an atom of silicon but, for simplicity, with the inner orbital electrons omitted. It will be noted that the outer orbit has four electrons.† The positive charge held by the nucleus exactly balances

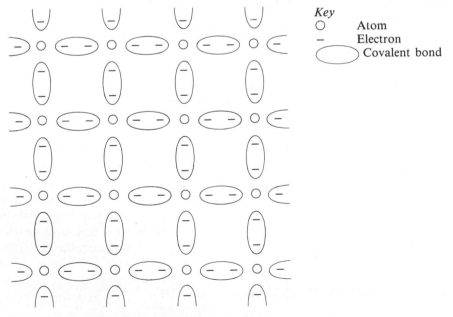

Key
○   Atom
–   Electron
⬭   Covalent bond

Fig. 1.2   Intrinsic Silicon Crystal Lattice

† This is also the case for germanium and, in general, the theory covered in this chapter, which is written with reference to silicon, applies equally well to germanium.

the total negative charge on the electrons leaving the atom with zero net charge — an important point.

Figure 1.2 is an attempt to depict, in two dimensions, the structure of a silicon crystal. Each of the outer orbit electrons — the valence electrons — of an atom acts as if it pairs up, i.e. forms a bond, with a valence electron of an adjacent atom. Consequently, each atom behaves as if it has 8 valence electrons, 4 of its own and one from each of 4 neighboring atoms. A lattice structure of silicon atoms formed in this way is called an intrinsic silicon crystal. The structure does not give rise to any free electrons for conduction. At all points in the crystal the net charge is zero.

1.2 THERMALLY GENERATED HOLE/ELECTRON PAIRS

Even at room temperature (often taken as 25 °C) the heat energy added to an intrinsic crystal makes the lattice structure vibrate. The additional energy is not distributed evenly, with the effect that the lattice breaks down at some points and as a result some electrons break out of the bonding arrangement and become free for electrical conduction. Being electrons they are, of course, negative charge carriers.

Holes

Removing an electron from a point in the lattice entails moving a negative charge which therefore destroys the charge neutrality: once an electron has been removed

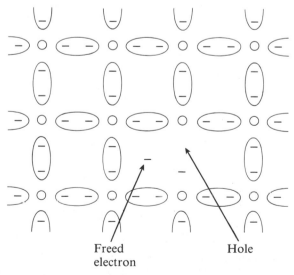

Freed electron   Hole

Fig. 1.3   A Thermally Generated Hole/Electron Pair

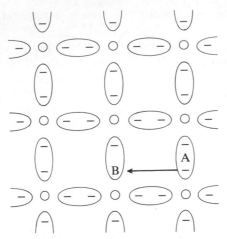

Fig. 1.4 Electron Moving from A to B but Positive Charge, i.e. Hole, Moves from B to A, which is therefore Current Direction

the crystal becomes positively charged at that point. Such a point of positive charge in a crystal is called a *hole*. A freed electron and the consequent hole are shown in Fig. 1.3. It is evident that free electrons and holes are thermally generated in pairs.

Since the lattice vibrates it is quite possible for an electron from an atom near to one where a hole has been formed to escape from its position in the lattice and occupy the hole as illustrated in Fig. 1.4. In that case, although the particle which has moved is the electron, i.e. from A to B, the point in the crystal carrying a positive charge has moved with the hole, i.e. from B to A. Electric current involves the movement of charge and in a very real sense holes are positive charge carriers.

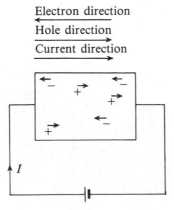

Fig. 1.5 Current Direction in an Intrinsic Crystal

1.3 CURRENT FLOW IN AN INTRINSIC CRYSTAL

At normal temperatures current will flow through an intrinsic crystal if a dc supply is connected across it as in Fig. 1.5. Electrons, being negative, will be attracted towards the positive side of the supply while holes will move in the opposite direction. However, current direction is defined, by convention, as the direction of movement of positive charge and therefore both holes and electrons contribute to current flowing from right to left through the crystal in Fig. 1.5, i.e. in the hole direction. As temperature increases so does the rate at which hole/electron pairs are generated. Therefore conductivity increases with temperature, and this characteristic is qualitatively different from that of conductors where the opposite is the case.

1.4 DOPING

Although heat increase at normal operating temperatures does increase the conductivity of silicon, much larger increases are effected by doping silicon with selected impurities. Equally important is the fact that we can choose to increase conductivity either by increasing the number of free electrons or by increasing the number of holes.

n Type Silicon

By introducing into the silicon crystal lattice structure, during its formation, a small proportion of atoms with five valence electrons – Fig. 1.6 shows one added at A

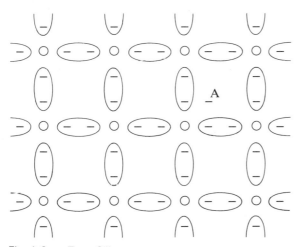

Fig. 1.6   n Type Silicon

– an electron surplus to the requirements of forming the lattice is added for each impurity atom. These electrons are not tied into the lattice and are therefore available for conduction irrespective of the operating temperature. Because all of the additional charge carriers produced in this way are, being electrons, negative, the material is said to be n type silicon. It must be stressed, however, that each impurity atom has zero net charge, so n type silicon also has zero net charge – the phrase 'n type' simply denotes that the additional current carriers are negative. An impurity element commonly used for this purpose is phosphorus; others are arsenic and antimony. These impurity elements are sometimes called donors.

p Type Silicon

Impurity atoms with just three valence electrons each can also be introduced in the formation of an otherwise pure silicon crystal. The effect is as shown at B in Fig. 1.7, where a hole is formed where there is an impurity atom. Since holes are positive charge carriers the material is called p type silicon although, once again, charge neutrality is maintained. Boron, gallium and indium are all p type impurities, generally known as acceptors.

Doping Concentration

Although the introduction of impurities does significantly increase the conductivity of silicon – far more than the increase attributable to thermal generation of charge carriers at normal operating temperatures – typically, the proportion of impurity atoms to silicon atoms is of the order of 1 in 10^6. There are good reasons for choosing the doping concentration with care, as we shall see when we consider the bipolar junction transistor.

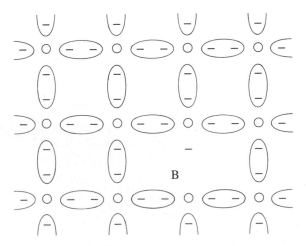

Fig. 1.7   p Type Silicon

Majority and Minority Carriers

Even in a doped crystal the process of generating hole/electron pairs thermally continues but in, say, an n type crystal there will clearly be more electrons than holes and vice versa in a p type crystal. It is sometimes useful to use the terminology *majority* and *minority* carriers. Thus, in p type silicon, holes are majority carriers and electrons minority carriers. In n type silicon it is the other way round.

1.5 pn JUNCTION

A simple way of representing impurity atoms is shown in Fig. 1.8. For example, a hole in p type silicon is represented by an unringed + which indicates that the positive charge is mobile. If it is removed it leaves the ringed, i.e. immobile, electron ⊖ and charge neutrality is destroyed because this point in the crystal is now negatively charged. (This will be so because the departure of a hole is caused by the arrival of an electron.) Similarly, an n type impurity has a mobile (unringed) negative charge − and an immobile (ringed) positive charge ⊕.

A pn junction, which could be formed by diffusing p type impurities into one end of an n type crystal, is shown in Fig. 1.9. Only impurity atoms are included and the situation depicted is as the junction might appear at the instant of its formation. Although there is a junction between the two types of silicon it is nevertheless

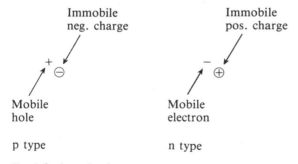

Fig. 1.8   Impurity Atoms

Fig. 1.9   pn Junction with No Recombination

important to think of the whole structure as a single crystal lattice. Therefore free electrons from the n type silicon will move into holes in the p type silicon to produce the situation shown in Fig. 1.10. Here on the p side of the junction there is a layer of negative charges because the holes have gone, and on the n side there is a layer of positive charges because the free electrons have gone. Since this layer around the junction is now depleted of free charge carriers it is called the *depletion layer*. This layer is, essentially then, an insulator with surplus negative charge on the p side and surplus positive charge on the n side. It is therefore a charged capacitance with a pd of V_j – the junction (or contact) voltage across it (Fig. 1.11). The magnitude of V_j is a few tenths of a volt. Once the depletion layer has been created, the potential difference across it forms a barrier to the further passage of holes from the p to the n region and to electrons from the n to the p region.

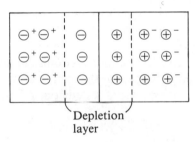

Fig. 1.10   pn Junction after Recombination at the Junction

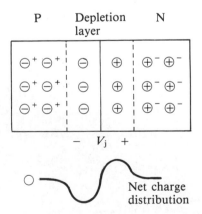

Fig. 1.11   Distribution of Charge and Voltage across the Junction as a Consequence of the Formation of the Depletion Layer

1.6   pn JUNCTION IN EQUILIBRIUM

To conclude that once the depletion layer has formed there will be no more move-ment of charge carriers across the junction would be an oversimplification because we must not forget thermally generated minority carriers in each region. The polarity of V_j is such as to assist their passage across the junction so there will be a continuous flow of holes from the n to the p region and of electrons from the p to the n region. This constitutes a current flow from the n to the p region. With no external connections to the pn junction the net current flow must be zero. We can account for this by noting that some majority carriers do acquire sufficient energy to jump the potential barrier at the junction and thus counterbalance the minority carrier current, and maintain the junction in equilibrium. Both of these currents rise with temperature:

(a) minority carrier generation simply does, as noted, increase with temperature, and

(b) more majority carriers gain enough energy to surmount the junction voltage as temperature rises.

The movement of the different types of charge carrier across the junction is il-lustrated in Fig. 1.12.

1.7   pn JUNCTION DIODE I/V CHARACTERISTICS

A diode is a device which allows significant current to flow in one direction only. This property is useful for many purposes such as converting ac to dc. A variety of diode circuits are described in Chapter 19.

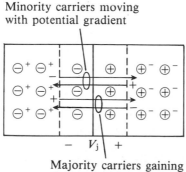

Fig. 1.12   pn Junction in Equilibrium

Reverse Bias

Once the foregoing description of a pn junction in equilibrium is assimilated its operation as a diode is easy to understand. A dc supply connected as in Fig. 1.13 draws majority carriers away from the junction and therefore widens the depletion layer. It also increases the voltage across the junction (check by going round the loop and noting that it increases V_j) and virtually precludes any majority carrier current at all. Minority carriers still cross the junction but since the polarity of V_j is such as to sweep all minority carriers in the vicinity of the junction across it, this current will not increase with supply voltage V once the majority carrier current has been eliminated. However, an increase in temperature increases the generation of minority carriers so the current I is temperature dependent. This current is called the reverse saturation current I_0 and for a low-power silicon diode at room temperature its magnitude will be about 20 nA. (For germanium I_0 would be about 1 μA.)

Because conduction is negligible the polarity of voltage, or bias as it is usually known, is called reverse and the I and V polarities are reckoned to be negative. Figure 1.14 is a typical reverse I/V characteristic.

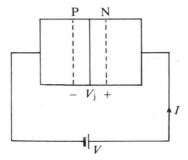

Fig. 1.13   Reverse biased pn Junction

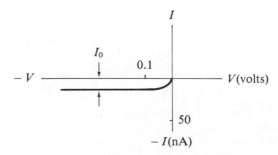

Fig. 1.14 I/V Characteristic of Reverse Biased pn Junction

Forward Bias

The effect of forward bias (Fig. 1.15) is to push majority carriers towards the junction, thus narrowing the depletion layer, and to lower the junction voltage. Majority carriers now cross the junction in increasing numbers as V rises. Figure 1.16 is a typical forward I/V characteristic for a silicon diode. Its shape shows that I is virtually negligible until V reaches V_C – the cut-in voltage.† Typically V_C for a silicon diode is 0.6 V. By 0.7 V significant conduction has been established and from then on the I/V characteristic may be considered linear. It also rises quite steeply and for many applications it is acceptable to use the approximation that when conducting,

$$V = V_{D(ON)} = \text{constant}$$

at, say 0.7 V or 0.8 V. For other purposes the slope of the linear part of the characteristic may have to be taken into account and it will correspond to a resistance of the order of 10 Ω.

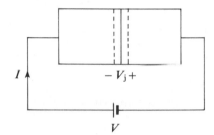

Fig. 1.15   Forward Biased pn Junction

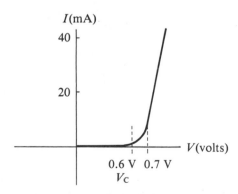

Fig. 1.16   *I/V* Characteristic of Forward Biased pn Junction

---

† Theoretically, the complete characteristic for a silicon diode covering both forward and reverse characteristics is a continuous exponential function: $I = I_0(e^{kV} - 1)$ where k is temperature dependent but is about 20 (volt^{-1}) at room temperature. However, the effect of diode resistance makes the characteristic virtually linear once a useful forward current is established.

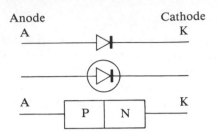

Fig. 1.17   Diode Symbol with Corresponding pn Diagram

Diode Symbol

The accepted circuit symbols for a diode are shown in Figure 1.17 with the corresponding terminals marked on the pn diagram below it. The ringed component symbolizes a discrete (free standing) diode; unringed the diode is part of an integrated circuit. This differentiation is also used with transistor symbols. Conduction occurs when $V_{AK} \geqslant V_C$.

Piecewise Linear *I/V* Characteristic of a pn Diode

Our interest in diodes is in using them as circuit elements, and for many applications it is satisfactory to approximate a diode I/V characteristic to the two sections shown in Fig. 1.18. These are an OFF section where $V < V_{D(ON)}$ and I is negligible, and an ON section where $V \geqslant V_{D(ON)}$ and I increases linearly as V rises above $V_{D(ON)}$.

This section of the characteristic is that of a resistance R_d where:

$$R_d = \frac{V - V_{D(ON)}}{I} \tag{1.1}$$

A diode can therefore be represented as an open circuit (or very high resistance) if $V < V_{D(ON)}$ and as a voltage $V_{D(ON)}$ and a resistance R_d when $V \geqslant V_{D(ON)}$ (Fig.

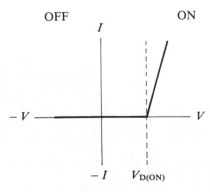

Fig. 1.18   Useful Approximation to pn Diode *I/V* Characteristic

$V_{D(ON)}$

Fig. 1.19 Equivalent Circuit of pn Diode when ON

1.19). When using the approximation that V is constant at $V_{D(ON)}$ when the diode conducts, then $R_d = 0$ and the ON section of the characteristic will be vertical.

1.8 REVERSE BREAKDOWN VOLTAGE OF A pn DIODE

If the reverse voltage V is increased beyond a certain value then I increases rapidly (Fig. 1.20) and the diode is said to have broken down. There are two mechanisms which can cause breakdown. One is called *avalanching*. It has been noted that under reverse bias conditions minority carriers pass through the depletion layer and as V increases so does their velocity. In passing through the depletion layer they are likely to collide with atoms in the lattice and dislodge electrons. These in turn are accelerated by the electric field across the depletion layer so the dislodging process builds up and a large current flows.

The other mechanism is the zener effect. When the doping concentration is high the depletion layer will be narrow so the electric field across the layer will be high and when V reaches a certain value the field will force electrons out of the lattice structure.

For both breakdown mechanisms the voltage at which breakdown occurs is called the zener voltage. In most diodes, when breakdown occurs, the lattice structure is irreparably damaged unless the reverse current is limited. For these devices a quantity called *peak inverse voltage* (PIV) is specified which must not be exceeded for safe working. However, there is also a class of diodes called *zener*

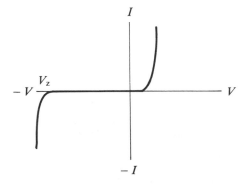

Fig. 1.20 *I/V* Characteristic of pn Diode Extended to Reverse Breakdown

diodes which are designed to use the breakdown (zener) voltage V_Z as a reliable reference voltage. They can be used as the reference source in circuits designed to produce an accurate source of current (Chapter 7) or voltage (Chapter 19) or as a means of protecting components susceptible to damage from excessive voltage (Chapter 4). Such diodes can use either zener or avalanching breakdown, the latter usually being for higher voltages.

1.9 BIPOLAR JUNCTION TRANSISTOR (BJT)

A bipolar junction transistor (BJT) can be connected up in such a manner that it amplifies current: this property can be used in conjunction with passive components to amplify voltage, as we shall see in Chapter 2. BJTs can also be used as high-speed switches in applications that will be described in the chapters on digital electronics (Chapter 11 onwards).

A (BJT) consists of either a narrow layer of p type silicon sandwiched between two n layers – the npn transistor shown in Figure 1.21, or an n layer sandwiched between two p layers, the pnp transistor. The three layers are called emitter (E), base (B) and collector (C).

An important feature of the BJT is that the emitter is much more heavily doped than the base. This is illustrated in Fig. 1.21 where the mobile charge carriers are indicated by plus or minus signs as appropriate. With the B–E junction forward biassed, as it is in Fig. 1.22 by a dc supply V_{BB}, which in this circuit is the base–emitter voltage V_{BE}, then most of the current will consist of electrons crossing from the emitter into the base where they become minority carriers.

At this point the width of the base, i.e. the distance from emitter to collector, becomes an important factor. In fact it is much narrower than Fig. 1.22 would suggest and for a high frequency or for a fast switching transistor the base width would be of micron dimensions. This means that once these electrons have been injected into the base where they are, as noted, minority carriers, they are in the

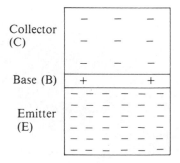

Fig. 1.21   npn Bipolar Junction Transistor

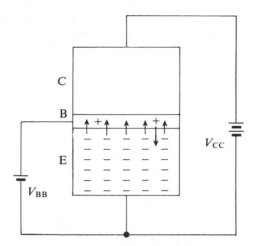

Fig. 1.22 With B–E Junction Forward Biased, Most of the Charge Carriers Crossing the Junction will be Electrons Entering the Base from the Transmitter

vicinity of the B–C junction. With the positive side of V_{CC} on the collector this junction is reverse biased so most of these minority carriers (electrons) are swept across the B–C junction into the collector. Since the collector acquires excess negative charge in this way, collector current I_C flows, as shown, from V_{CC} into the collector (Fig. 1.23).

There are two factors which cause the number of charge carriers crossing from base into collector to be less than the number crossing the emitter–base junction. First the base is a p region, albeit a lightly doped one, so holes will pass from base

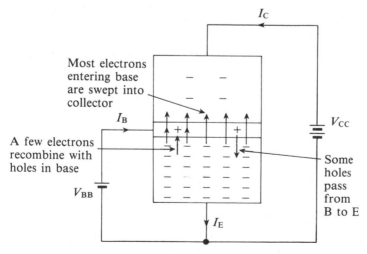

Fig. 1.23 Movement of Charge Carriers in npn BJT

to emitter in the usual way for a forward biased pn junction. Second, some of the electrons entering the base will combine (a process called recombination) with holes before coming under the influence of the B–C voltage. Both of these factors (illustrated in Fig. 1.23) make the base negative so a small current I_B flows into the base.

By definition,

$$I_C = h_{FE} I_B \tag{1.2}$$

where h_{FE} is the static, forward, common emitter, current transfer ratio or, more simply, the CE dc current gain. To some extent this is an approximation because I_C does not wholly consist of the useful component, $h_{FE} I_B$. There is also a leakage current flowing into the collector due to thermally generated current carriers in the BJT. This current is designated I_{CEO}, the current flowing from collector to emitter with the base open circuit (thus ensuring that $I_B = 0$). This current is clearly temperature dependent and in good circuit design it can be made negligible, as we shall see in Section 2.10. If I_{CEO} is negligible then equation (1.2) is acceptable for use in circuit design. Otherwise,

$$I_C = h_{FE} I_B + I_{CEO} \tag{1.3}$$

Sometimes β is used instead of h_{FE}.

1.10   CURRENT CONVENTIONS AND BJT SYMBOLS

BJT symbols are used in Figs. 1.24a and b for npn and pnp transistors connected in the common emitter configuration. The phrase 'common emitter' means that the emitter terminal is common to both input and output voltages, a matter to be elaborated in Chapter 2. The arrow on the npn transistor indicates that current normally flows out of the emitter while in the pnp case the arrow indicates emitter current normally flowing inwards. This text will continue to concentrate on circuits

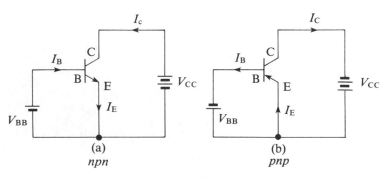

Fig. 1.24   BJT Symbols with Normal Voltage Polarities and Current Directions

using npn transistors, but to work out an explanation of pnp transistor operation simply reverse all voltage and current directions and exchange the roles of holes and electrons. There are circuits which benefit from the availability of the two types of BJT, and some of these will be described.

In the npn case of Fig. 1.24a, then using Kirchhoff's current law (KCL),

$$I_E = I_C + I_B \tag{1.4}$$

and with the dc supplies connected as shown the currents actually will flow in the directions indicated on the diagram. Thus, if $I_C = 1$ mA and $I_B = 0.01$ mA, then $I_E = 1.01$ mA. Classical network theory conventions hold that currents should always be labeled as flowing into the network. Using this convention the equation would be written

$$I_E + I_B + I_C = 0 \tag{1.5}$$

and using the same example I_E would be -1.01 mA. In the interest of understanding concepts of electronic circuits this book will label currents in the direction that they actually flow, e.g. as in Fig. 1.24. Where this conflicts with a definition, the discrepancy will be explained.

1.11 COMMON EMITTER (CE) CHARACTERISTICS

Input (Base–Emitter) Characteristics I_B vs. V_{BE}

It is necessary for npn transistor action that V_{CE} is positive. Fixing V_{CE} at, say, 10 V (Fig. 1.25a) then we can plot I_B vs. V_{BE} for normal operation. Since the B–E region is a pn junction the shape of the characteristic (Fig. 1.25b) is, as expected, similar to that of a pn diode. However, since most of the current carriers crossing the B–E

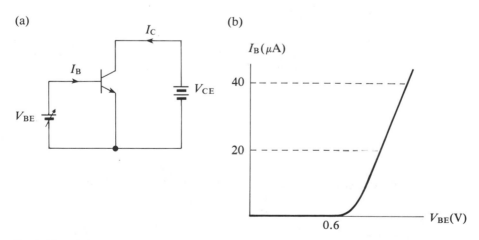

Fig. 1.25   (a) CE Configuration (b) CE Input Characteristic

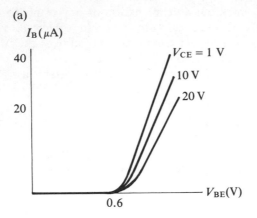

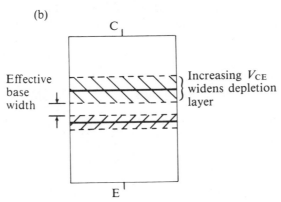

Fig. 1.26   Increasing V_{CE} widens B–C Depletion Layer, Reducing Effective Base Width and Reducing I_B

junction go to the collector then base current values will be much smaller than for a diode. In fact they will be reduced by a factor of h_{FE}.

Changing the value of V_{CE} does have a slight effect, as shown in the complete set of characteristics of Fig. 1.26a: I_B actually falls as V_{CE} rises for a given value of V_{BE}. This occurs because of the Early effect, which is illustrated in Fig. 1.26b. It shows the B–E and C–B depletion layers. The important one is the C–B depletion layer because, as V_{CE} rises, the C–B reverse bias increases which causes the depletion layer to widen and, in particular, to extend deeper into the base. This reduces the effective base width. Therefore fewer electrons are lost in recombination with holes while moving across the base so there is less negative charge to neutralize by base current. The effect of V_{CE} on input characteristics is, however, negligible in most applications and we shall often simply use a single characteristic of I_B vs. V_{BE}, such as the one drawn in Fig. 1.25b, to characterize common emitter input performance.

Output (Collector) Characteristics I_C vs. V_{CE}

The output characteristics of Fig. 1.27b are graphs of I_C vs. V_{CE} for various values of I_B, i.e. I_B is a parameter (Fig. 1.27a). Beginning with $I_B = 0$, the only current flowing through the transistor is I_{CEO}. This is usually negligible. It is shown in Fig. 1.27b as a horizontal line almost coincident with the V_{CE} axis. Fixing I_B at, say, 10 μA then at low values of V_{CE} (below, say, 0.2 V), current carriers in the base are not efficiently collected and the transistor is said to be *saturated*. In these circumstances,

$$I_B > \frac{I_C}{h_{FE}} \tag{1.6}$$

As V_{CE} rises, I_C rises and then levels out. The 'leveling-out' voltage is called $V_{CE(SAT)}$ and is usually between 0.2 V and 1.0 V.

Once $V_{CE(SAT)}$ is exceeded the transistor is no longer saturated and is said to be in its *active region*. Now

$$I_B = \frac{I_C}{h_{FE}}$$

or, more appropriately,

$$I_C = h_{FE} I_B.$$

Increasing I_B in 10 μA increments and plotting I_C vs. V_{CE} results in the family of

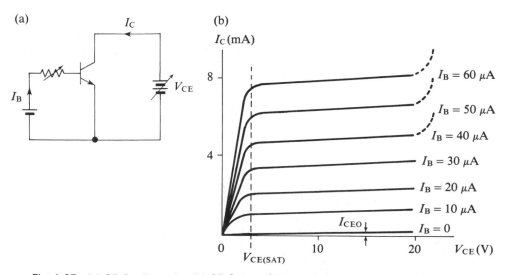

Fig. 1.27 (a) CE Configuration (b) CE Output Characteristics

output characteristics of Fig. 1.27b. Features to note are:

(a) In the active region I_C is almost independent of V_{CE} and almost totally dependent on I_B.

(b) The slight increase in I_C with V_{CE} is due to the Early effect because increasing V_{CE} widens the B–C depletion layer, which reduces the effective base width and increases the efficiency of collection of electrons from base into collector.

(c) Beyond a certain value of V_{CE} there is a rapid rise in I_C (dotted in Fig. 1.27b). This occurs because the reverse biased B–C junction breaks down, which damages the transistor. A value of $V_{CE(max)}$ is normally quoted for transistors which must not be exceeded.

(d) The value of $V_{CE(SAT)}$ does not significantly increase with I_B.

Maximum Ratings

The following are maximum ratings for a Texas 2N 3708 and are typical for a general purpose npn transistor:

Collector–base voltage	30 V
Collector–emitter voltage (base open circuit)	30 V
Emitter–base voltage	−6 V
Collector current	30 mA
Continuous device dissipation at 25 °C in free air (to be derated by 2.5 mW/°C up to 125 °C)	250 mW

1.12 COMMON BASE CONFIGURATION

When the configuration shown in Fig. 1.28 operates as an amplifier the base will be common to input and output voltages. The dc current gain of the circuit is

$$h_{FB} = \frac{I_C}{I_E} \tag{1.7}$$

which is clearly less than unity, usually between 0.97 and 0.995. (On data sheets h_{FB}

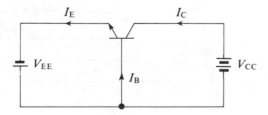

Fig. 1.28   CB Configuration

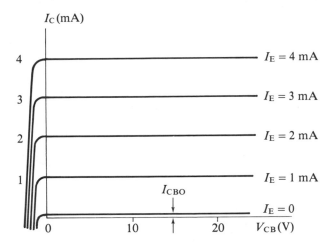

Fig. 1.29   CB Output Characteristics

is often quoted as a negative number because conventional network theory requires I_E to be labeled as an inward flowing current which will therefore be negative in this instance.) Sometimes α is used instead of h_{FB}.

With no current gain the common base circuit is clearly less useful than the common emitter circuit but in its favor it should be observed from the characteristics of Fig. 1.29 that the dependence of I_C on V_{CB} in the CB circuit is even less than the dependence of I_C on V_{CE} in the CE circuit.

Note also that even when $V_{CB} = 0$, I_C has leveled out. The leakage current between collector and base with emitter open circuit (Fig. 1.30) is labeled I_{CBO} (sometimes I_{CO}). By a similar reasoning to the one which produced equation (1.3), we can say that

$$I_C = h_{FB}I_E + I_{CBO} \tag{1.8}$$

where h_{FB} is the common base dc current gain.

Then $I_C = h_{FB}(I_C + I_B) + I_{CBO}$

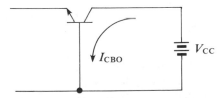

Fig. 1.30   Leakage Current I_{CBO} in CB Configuration

from which

$$I_C = \frac{h_{FB}}{1 - h_{FB}} I_B + \frac{I_{CBO}}{1 - h_{FB}} \qquad (1.9)$$

Comparing this with equation (1.3),

$$\frac{h_{FB}}{1 - h_{FB}} = h_{FE} \qquad (1.10)$$

and

$$I_{CEO} = \frac{I_{CBO}}{1 - h_{FB}}$$

If $\quad h_{FB} = 0.99$

$$h_{FE} \doteqdot 100$$

$$I_{CEO} = 100 \, I_{CBO}$$

$$I_{CEO} \doteqdot h_{FE} I_{CBO} \qquad (1.11)$$

Transistor manufacturers usually quote I_{CBO} as being of the order of 100 nA at 25 °C for a lower-power, general-purpose transistor. It does, however, tend to double for every 10 °C increase in temperature. Since I_{CEO} will be $I_{CBO}/(1 - h_{FB}) \doteqdot h_{FE} I_{CBO}$, temperature stability is an important consideration, which will be covered in Chapter 2.

1.13 GALLIUM ARSENIDE (GaAs)

For many years the potential of this material for the manufacture of high frequency, high speed semiconductors was known before useful devices were actually made. Many of the difficulties have been overcome and GaAs devices are readily available. The most common device is the MESFET (metal-semiconductor FET) which is probably now dominant in microwave amplifiers. Operating frequencies around 20 GHz are commonplace; 60 GHz is quite feasible. GaAs is also used in the manufacture of fast switching transistors for high bit rate digital circuits.

Gallium (Ga) is a group III element (3 electrons in its outer shell) and arsenic a group V element. A GaAs crystal is made up of Ga and As pairs so that it is essentially the same as an intrinsic Si crystal. However, free electrons have a higher mobility in GaAs than in Si and this enables GaAs devices to be operated at higher frequencies than Si devices. Also, electrons generally have a higher mobility than holes, so n type GaAs devices are more common than p type.

GaAs can be doped with a number of different elements (e.g. sulphur and selenium) which facilitates the production of materials with a wide range of conductivities: high conductivity materials are useful for transistors with low power dissipation, while low conductivity materials are ideal for integrated circuit substrates. These two factors hold out the attractive prospect for semiconductor manufacturers of integrated circuits consisting of large numbers of transistors capable of operating at high speeds.

SUMMARY

The following points should be remembered:

1. The general shape of silicon diode I/V characteristics, and the polarity of and approximate values of conducting and nonconducting voltages.
2. Polarity and typical values of V_{BE} and V_{CE} for transistor operation.
3. The general shape of transistor input and output characteristics.
4. The meaning of saturation region and active region of operation.
5. The definition $I_C = h_{FE} I_B$ and the conditions under which it holds.
6. $I_{CEO} \doteq h_{FE} I_{CBO}$, and doubles for every $10\,^{\circ}C$ rise in temperature.

REVISION QUESTIONS

1.1 Justify the statement that in p type silicon current flow occurs by the movement of positive charge although the only moving particles are electrons.
1.2 Justify the statement that n type silicon has zero net charge.
1.3 Explain why there is a potential difference across a pn junction without any external supplies.
1.4 Describe the movement of majority and minority carriers in a pn diode in equilibrium, i.e. with no externally applied voltage.
1.5 Describe the movement of carriers across a pn junction (a) when forward biased, (b) when reverse biased.
1.6 Justify qualitatively the I/V characteristic of a pn diode.
1.7 Describe the avalanche and zener breakdown mechanisms of a pn diode.
1.8 Why is the emitter of a BJT more heavily doped than the base?
1.9 How can a substantial current flow in the collector of a BJT if the base–collector junction is reverse biased?
1.10 Deduce the general shape of the collector characteristics of a BJT and indicate their salient features.
1.11 Explain the meaning of the terms 'saturation region' and 'active region' of a BJT.
1.12 Under what conditions does the relationship $I_C = h_{FE} I_B$ hold?

2

The Common Emitter Amplifier – Biasing

Chapter Objectives

2.1   Notation for Labeling Electrical Quantities
2.2   Control of Collector Current
2.3   BJT as a Voltage Amplifier
2.4   Simplified Equivalent Circuit of CE Amplifiers
2.5   Graphical Assessment of Voltage Gain – Load lines
2.6   Confirmation of Load Line Result by Equivalent Circuit
2.7   Safe Working Limits
2.8   Biasing a Common Emitter Amplifier
2.9   Calculation of Quiescent Condition in an Emitter Bias Circuit Taking h_{FE} and I_B into Account
2.10   Control of Operating Point Stability by Choice of Bias Components
2.11   Coupling and Decoupling Capacitors – The Separation of AC and DC in an Amplifier
        Summary
        Revision Questions

The aims of this chapter are:

(a) to explain the means by which a BJT can be used to amplify an alternating voltage;
(b) to show how voltage gain can be estimated graphically;
(c) to explain the need to bias a BJT for linear amplification and to describe methods of biasing.

2.1   NOTATION FOR LABELING ELECTRICAL QUANTITIES

Before beginning the main material of this chapter it would be as well to deal with the BSI/IEEE standard for labeling voltage, current and power quantities in elec-

24

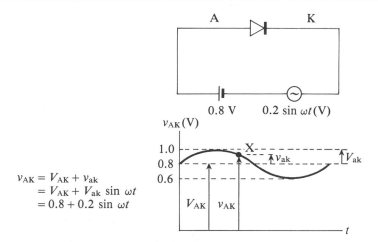

$$v_{AK} = V_{AK} + v_{ak}$$
$$= V_{AK} + V_{ak} \sin \omega t$$
$$= 0.8 + 0.2 \sin \omega t$$

Fig. 2.1   Diagram Illustrating BSI/IEEE Standard for Labeling Electrical Quantities

tronic circuits. Figure 2.1 shows an alternating voltage of 0.2 V peak superimposed on a direct voltage of 0.8 V applied across a diode. This is accompanied by a graph of the voltage on A wrt K and the parts of the standard relevant to this work may be explained with reference to this graph as follows:

(a) Upper case first letters with upper case subscripts are for dc and quiescent values. Thus, the quiescent voltage (i.e. the voltage when the varying component is zero) across the diode is $V_{AK} = 0.8$ V. We shall often use $V_{AK(Q)}$ to emphasize that it is a quiescent value. A supply voltage carries a double subscript so it could be written that $V_{AK(Q)}$ is supplied from a dc source V_{AA}.

(b) Upper case first letters with lower case subscripts are for peak and rms values of ac quantities. For Fig. 2.1 then, $V_p = 0.2$ V or $V_{p-p} = 0.4$ V. We shall often use rms values of signal components for electronic circuits, e.g. V_{ak} or V_s.

(c) Lower case first letters with lower case subscripts represent the instantaneous value of varying components only. In Fig. 2.1 at X, $v_{ak} = 0.1$ V.

(d) Lower case first letters and upper case subscripts are for instantaneous total values. In Fig. 2.1 at X, $v_{AK} = 0.9$ V. It follows that the graph is, as labeled, a graph of v_{AK} vs. t.

The same standard applies for current (I or i) and power (P or p).

2.2   CONTROL OF COLLECTOR CURRENT

Adding v_s, an alternating signal to be amplified, in series with V_{BB}, is the only addition to the basic CE configuration (Fig. 2.2) described in Section 1.11. As it stands the circuit will not amplify v_s but it does include some important concepts

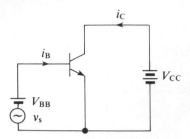

Fig. 2.2   Common Emitter Circuit with Signal Input Voltage v_s

which are used in the amplifying process. Assuming that v_s is a voltage sine wave,

$$v_s = V_p \sin \omega t,$$

then the voltage waveform applied between base and emitter, v_{BE}, consists of v_s superimposed on V_{BB} as shown in Fig. 2.3.

One condition for linear amplification, which means that the ac components of input and output should be identical waveforms – in this case sine waves – is that V_{BB} must be chosen so that v_s does not reduce v_{BE} to a level where it is operating on the nonlinear part of the input characteristic. This is illustrated in Fig. 2.4 where V_{BB} (i.e. $V_{BE(Q)}$) = 0.8 V and $V_p = 0.05$ V. Consequently v_{BE} oscillates between 0.75 V and 0.85 V and it is not until v_{BE} falls below 0.75 V that the input characteristic becomes nonlinear. In these circumstances the ac component of i_B is a sine wave oscillating between 10 μA and 40 μA around a quiescent base current $I_{B(Q)} = 25$ μA. Note that v_{BE} and i_B are in phase, as emphasized in the v_{BE} and i_B waveforms of Fig. 2.5.

Two factors could make i_B become nonlinear. First, if the peak value of the input signal V_p rises, say to 0.15 V, the negative going part of v_{BE} goes into the nonlinear region of the input characteristic so the negative-going half-cycle of i_B is flattened off (Fig. 2.6). Second, even if V_p remains at 0.05 V but V_{BB} is reduced, say to 0.75 V, then again the negative half-cycle of i_B will be flattened (Fig. 2.7).

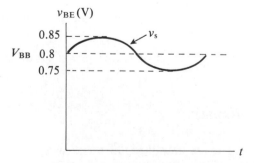

Fig. 2.3   Base–Emitter Voltage Waveform v_{BE} Consisting of Sine Wave of 0.05 V Peak Superimposed on Direct Voltage of 0.8 V

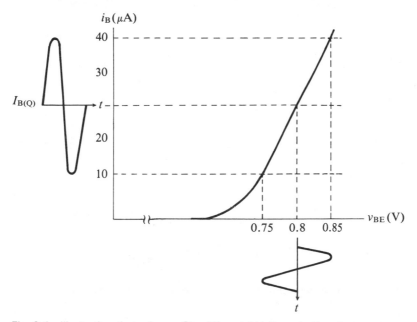

Fig. 2.4 Illustrating that when a Sine Wave of Voltage is Superimposed on a Bias which Puts It on to the Linear Part of the Characteristic it will Produce Sinusoidal Variations of the Base Current

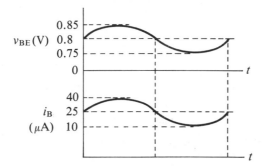

Fig. 2.5 Signal Components of v_{BE} and i_B are in Phase

This leads us to an important concept in analog electronics – that of *bias*. To achieve a linear waveform of base current it is clearly necessary for the input signal which is to be amplified, v_s, to be *biased* by V_{BB} on to the linear section of the input characteristic. Essentially, although the input signal v_s may have positive and negative half-cycles, by being biased, it makes v_{BE} alternately more and less positive, which in turn makes i_B alternately more and less positive.

Before finishing with the input circuit it is worth noting that the linear part of the input characteristic, being a graph of current vs. voltage, can be represented by

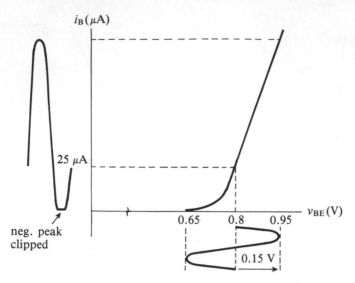

Fig. 2.6   Negative Half-cycle of i_B Clipped when Amplitude of v_s is Excessive

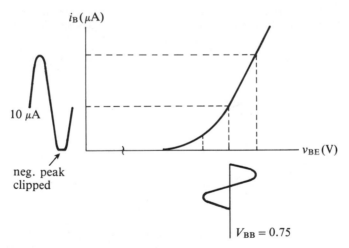

Fig. 2.7   Negative Half-cycle of i_B Clipped when Bias Voltage V_{BB} is too Low

a conductance or resistance. Traditionally the latter is used and the resistance h_{ie} is given by,

$$h_{ie} = \frac{\Delta v_{BE}}{\Delta i_B} \tag{2.1}$$

$$= \frac{0.85 \text{ V} - 0.75 \text{ V}}{40 \ \mu\text{A} - 10 \ \mu\text{A}} = 3.3 \text{ k}\Omega$$

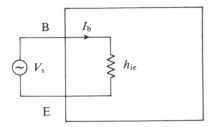

Fig. 2.8   Network Representing the Effective Input to a BJT for Small Signals when Biased for Linear Operation

Assuming that an input signal of rms voltage V_s operates only on the linear part of the characteristic, the rms value of the signal component of base current I_b resulting from this will be

$$I_b = \frac{V_s}{h_{ie}} \qquad (2.2)$$

and as far as small-signal ac conditions are concerned the input between base and emitter behaves like the simple network of Fig. 2.8. The idea here is that the effect of a dc source to bias the BJT into linear operation is accepted but the dc source is then ignored and the network simply represents the small-signal performance of the input circuit.

DC and AC Current Gains h_{FE} and h_{fe}

It was noted in Chapter 1 that $I_C = h_{FE} I_B$ so taking $h_{FE} = 80$ then, for Fig. 2.2, when $I_{B(Q)} = 25\ \mu A$, $I_{C(Q)} = 2$ mA. An important parameter (to be dealt with more fully in Chapter 3) is the small-signal current gain h_{fe}. This is the ratio of small *changes* in collector current to the small *changes* in base current which produce them under linear operating conditions. The parameter h_{fe} is defined as

$$h_{fe} = \frac{\Delta i_C}{\Delta i_B}\bigg|_{V_{CE} = \text{constant}} \qquad (2.3)$$

Assuming that the BJT in Fig. 2.2 has $h_{fe} = 100$ then for every 1 μA *change* in i_B, i_C *changes* by 100 μA. Therefore, developing the calculations already made with respect to Fig. 2.2, the waveform of i_C will be as shown in Fig. 2.9. For example, as i_B (Fig. 2.5) increases from 25 μA to 40 μA, i.e. by 15 μA, i_C increases by h_{fe} times as much, i.e. by 1.5 mA.

The network of Fig. 2.8 representing the input of a BJT operating under linear conditions can now be extended as in Fig. 2.10. Again, having established linear operation (a) by biasing v_s on to the linear part of the input characteristic, and (b) by holding v_{CE} positive with the dc supply V_{CC} (Fig. 2.2), our main interest is in the changing component of the collector current. As v_{BE} increases, i_B increases by Δi_B to produce $\Delta i_C = h_{fe} \Delta i_B$. This current increase flows from V_{CC} *into* the collector out

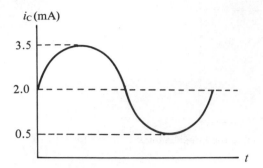

Fig. 2.9   Collector Current Waveform for Fig. 2.2

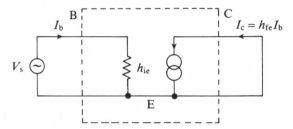

Fig. 2.10   Equivalent of Input and Output Circuits of BJT Operating under Small-signal Linear Conditions

of the emitter and back to V_{CC}. Again, omitting the dc supplies once linear operation is attained, Fig. 2.10 represents the small-signal ac performance of the transistor. The rms component of the ac components of base and collector currents are indicated on the diagram by I_b and I_c.

2.3   BJT AS A VOLTAGE AMPLIFIER

In Fig. 2.2, v_{CE} is constant so there is no ac output voltage. In Fig. 2.11 a load resistor R_L has been added in series with the collector so that any change in i_C will produce a varying voltage across R_L. Assuming the same voltage and current values that have been worked out before and repeated in Figs. 2.12a, b and c, it is now possible to see that Δv_{CE}, which is the circuit output, will be an amplified version of $v_s = \Delta v_{BE}$ given a suitable value of R_L. Incidentally, the term 'common emitter', which was introduced in Chapter 1, now becomes more meaningful because we can see that input and output are common to the emitter.

Let $R_L = 2.7 \text{ k}\Omega$ and let $V_{CC} = 12$ V. Now we shall calculate v_{CE} for the quiescent, maximum and minimum values of i_C. Referring to the CE amplifier of

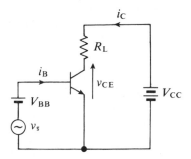

Fig. 2.11   Common Emitter Amplifier

Fig. 2.11 then, at all times,

$$v_{CE} = V_{CC} - i_C R_L \tag{2.4}$$

and for the values quoted

$$v_{CE} = 12 - i_C \times 2.7$$

(i_C in mA). For the quiescent condition

$$i_C = I_{C(Q)} = 2 \text{ mA}$$

Therefore $v_{CE} = V_{CE(Q)} = 12 - 2 \times 2.7 = 6.6$ V
For $i_C = 3.5$ mA

$$v_{CE} = 12 - 3.5 \times 2.7 = 2.55 \text{ V}$$

For $i_C = 0.5$ mA

$$v_{CE} = 12 - 0.5 \times 2.7 = 10.65 \text{ V}$$

From these calculations the waveform of v_{CE} has been drawn at Fig. 2.12d. It is an inverted version of the other waveforms, i.e. it is in antiphase with the input v_s. This is because any increase in i_C increases the volt drop across R_L but reduces the output v_{CE}, which is V_{CC} minus this volt drop (equation (2.4). The important point is that the peak-to-peak value of Δv_{CE} is $10.65 - 2.55 = 8.1$ V, whereas $\Delta v_{BE} = v_s = 0.1$ V peak to peak, so there is an amplification of 81 times. This could equally well have been expressed in peak or rms values. To account for the phase inversion it is better to calculate A_v, the voltage amplification or gain for corresponding values of Δv_{BE} and Δv_{CE}. Therefore,

$$A_v = \frac{\Delta v_{CE}}{\Delta v_{BE}} = \frac{10.65 - 2.55}{0.75 - 0.85} = -81 \text{ times.}$$

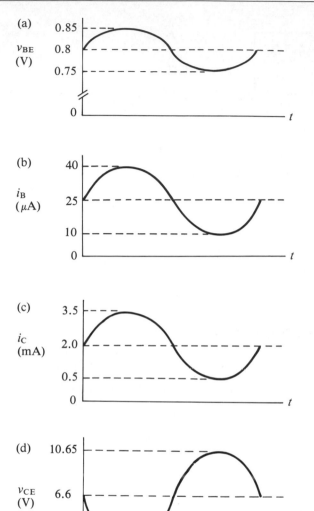

Fig. 2.12   Waveforms for the CE Amplifier of Fig. 2.11. Note Phase Relationships of Four Waveforms

2.4   SIMPLIFIED EQUIVALENT CIRCUIT OF CE AMPLIFIERS

The simple equivalent circuit of Fig. 2.10 can now be extended. Again, quiescent values are of no concern so neglecting V_{CC} in Fig. 2.11, R_L is connected from C to E to complete the representation of the CE amplifier as the linear network of Fig.

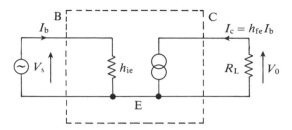

Fig. 2.13   Simplified Equivalent Circuit of CE Amplifier Operating under Small-signal Linear Conditions

2.13. The rms value of the output will be $V_0 = V_{ce}$. To calculate A_v,

$$I_b = \frac{V_s}{h_{ie}}$$

$$V_0 = V_{ce} = - I_c R_L = - h_{fe} I_b R_L = - h_{fe} \frac{V_s}{h_{ie}} R_L$$

Therefore $A_v = \dfrac{V_0}{V_s} = \dfrac{- h_{fe} R_L}{h_{ie}}$ (2.5)

Substituting the values already used,

$$A_v = \frac{-100 \times 2.7}{3.3} = -81$$

as before.

The equivalent circuit of Fig. 2.13 is not an accurate representation of a single-stage CE amplifier. This is mainly because the definition of h_{fe} specifies that v_{CE} must be constant: clearly this is not the case; if it were there would not be any output voltage. However, the effect of v_{CE} on h_{fe} is not large and for most purposes Fig. 2.13 is acceptable for amplifier calculations. The subject will be dealt with more fully in Chapter 3.

2.5   GRAPHICAL ASSESSMENT OF VOLTAGE GAIN – LOAD LINES

For a CE amplifier (Fig. 2.14) with collector load R_L and supply voltage V_{CC}, the relationship between v_{CE} and i_C is, as previously noted,

$$v_{CE} = V_{CC} - i_C R_L$$ (2.6)

and it is linear. This relationship is expressed graphically by the load line of Fig. 2.15. Two points easily calculated from equation (2.6) are

at A,
$$i_C = 0 \qquad \text{therefore } v_{CE} = V_{CC};$$

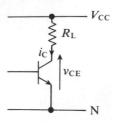

Fig. 2.14   CE Amplifier with Collector Load and Supply Only Shown

at B,

$$v_{CE} = 0 \qquad \text{therefore} \quad i_C = \frac{V_{CC}}{R_L}$$

These two points are marked on the axes of Fig. 2.15 and joined to form the *load line*. This line is the locus of the only pairs of values of i_C and v_{CE} which can exist for given values of V_{CC} and R_L. It should be noted that the line depends *only* on V_{CC} and R_L and is totally independent of the transistor characteristics. In fact, if the BJT in Fig. 2.14 were to be replaced by any other device, e.g. a diode, a field effect transistor or even a resistor, then the load line would be the same; we would simply alter the current and voltage suffixes.

However, we can include the BJT collector characteristics on the same i_C vs. v_{CE} axes (Fig. 2.16). The load line is now a locus of all (but nevertheless the only) sets of i_B, i_C and v_{CE} that can exist for a particular BJT connected in a CE amplifier with certain values of R_L and V_{CC}. Including the BJT characteristics does indicate that certain points on the load line cannot be realized. For example, point B (Fig. 2.15) where $v_{CE} = 0$ V cannot be realized because transistor action requires that $v_{CE} > V_{CE(SAT)}$; the smallest value of v_{CE} and the largest value of i_C will be at point C in Fig. 2.16.

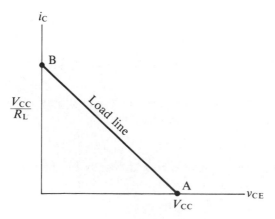

Fig. 2.15   Load Line for Fig. 2.14

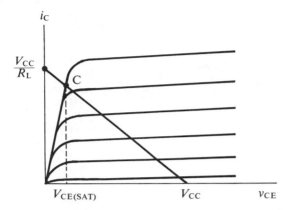

Fig. 2.16 Load Line Superimposed on a Set of CE Collector Characteristics

It is probably true to say that the load line is more of an educational tool used to illustrate certain aspects of transistor amplifier operation than a practical technique. The following example is intended to increase the reader's insight into the process of voltage amplification by looking at the type of example used in Section 2.3 in a different manner.

□ **Example 2.1**

For the circuit of Fig. 2.17 find the voltage amplification A_v using the BJT characteristics of Fig. 2.18.
A good method is to work backwards from the ouput characteristics to the input. First draw the load line.

Point A is simply $i_C = 0$; $v_{CE} = V_{CC} = 12$ V.

Point B is $v_{CE} = 0$; $i_C = \dfrac{V_{CC}}{R_L} \doteq 3.6$ mA.

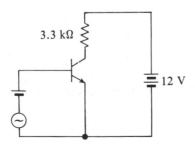

Fig. 2.17 Circuit for Example 2.1

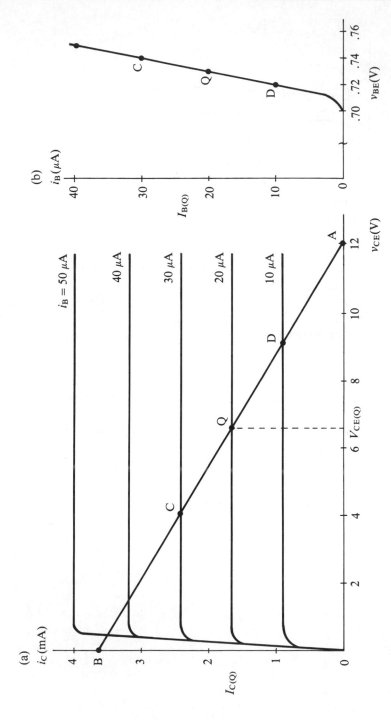

Fig. 2.18   (a) Load Line for $R_L = 3.3\ k\Omega$; $V_{CC} = 12\ V$
(b) Input Characteristic

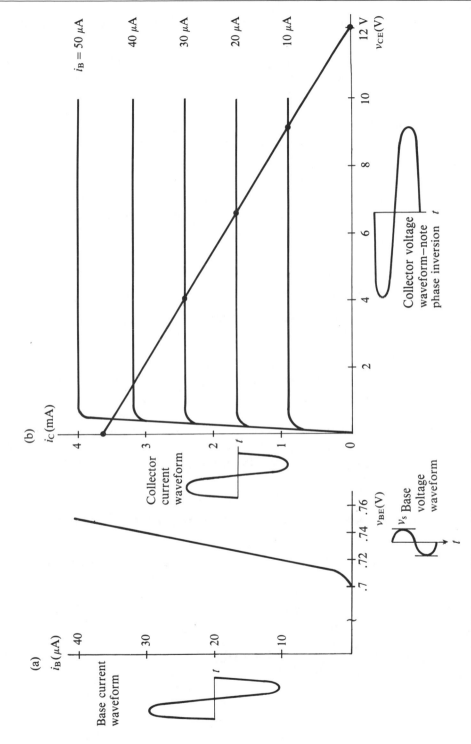

Fig. 2.19 (a) Base Voltage and Current Waveforms Superimposed on Input Characteristic
(b) Collector Current and Voltage Waveforms Superimposed on Output Characteristics

The load line is drawn by joining A and B. Then we can choose three points on the load line as follows:

(1) The quiescent or operating point Q roughly in the middle of the load line. In Fig. 2.18a the $i_B = 20\ \mu A$ characteristic is suitable, which gives, reading off the characteristic, $I_{C(Q)} = 1.65$ mA, $V_{CE(Q)} = 6.5$ V.

Then take equal increments of base current, say 10 μA, above and below $I_{B(Q)}$.

(2) At point C, $i_B = 30\ \mu A$ and $i_C = 2.4$ mA; $v_{CE} = 4$ V.
(3) At point D, $i_B = 10\ \mu A$ and $i_C = 0.9\ \mu A$; $v_{CE} = 9$ V.

In each case check that the readings fit the equation

$$v_{CE} = 12 - i_C \times 3.3.$$

These values of base current can now be transposed to the input characteristic (Fig. 2.18b) and for the three points we have

$$I_{B(Q)} = 20\ \mu A;\ V_{BE(Q)} = 0.73\ V$$
$$i_B\ \ \ = 30\ \mu A;\ v_{BE}\ \ \ = 0.74\ V$$
$$i_B\ \ \ = 10\ \mu A;\ v_{BE}\ \ \ = 0.72\ V$$

From these figures the voltage gain A_v can be calculated as follows:

$$A_v = \frac{\Delta v_{CE}}{\Delta v_{BE}} = \frac{9 - 4}{0.72 - 0.74} = \frac{5}{-0.02} = -250 \text{ times}$$

Redrawing Fig. 2.18 but now beginning with the input characteristic as in Fig. 2.19a, one can see all of the waveforms deduced in Section 2.3 at a glance. Superimposing a sine wave on the v_{BE} axis oscillating between 0.72 V and 0.74 V gives a similar waveform on the i_B axis. The i_B values are transposed to the output characteristics and lead to the waveforms of i_C and v_{CE} drawn in Fig. 2.19b. Working through from v_{be} to v_{ce} the phase inversion is also evident.   □

2.6   CONFIRMATION OF LOAD LINE RESULT BY EQUIVALENT CIRCUIT

Graphical methods and small-signal equivalent circuit techniques of finding voltage gain have a common basis in theory so their validity can be tested by the consistency of the results they give. Thus for the previous example we can estimate h_{ie} and h_{fe} from the characteristics and calculate A_v.

From Fig. 2.18b,

$$h_{ie} = \frac{\Delta v_{BE}}{\Delta i_B} = \frac{0.74\ V - 0.72\ V}{30\ \mu A - 10\ \mu A} = 1\ k\Omega$$

From Fig. 2.18a,

$$h_{fe} = \frac{\Delta i_C}{\Delta i_B}\bigg|_{V_{CE} = const}$$

At $V_{CE(Q)} = 6.5$ V,

$$h_{fe} = \frac{2.4 \text{ mA} - 0.9 \text{ mA}}{30 \text{ } \mu\text{A} - 10 \text{ } \mu\text{A}} = 75$$

These figures are obtained by drawing a vertical line through $V_{CE} = 6.5$ V and reading off values of i_B and the corresponding values of i_C. Using equation (2.5) again,

$$A_v = \frac{-h_{fe}R_L}{h_{ie}} = \frac{-75 \times 3300}{1000}$$

$$= -247$$

Bearing in mind that equation (2.5) is an approximation and that graphical errors such as line thickness affect accuracy, this result is remarkably close to the graphical estimate for A_v of -250. In practice, 10% difference is as good as can be expected.

2.7 SAFE WORKING LIMITS

Another use of the load line is in choosing supply voltage and load resistor values which avoid exceeding the maximum safe power dissipation specified for a particular BJT. For example, suppose a BJT is quoted as having a maximum con-

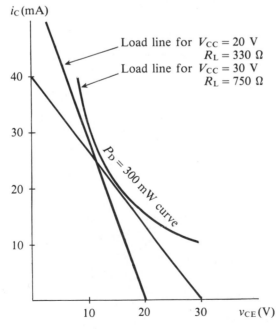

Fig. 2.20 Curve for Maximum Power Dissipation in BJT

tinuous power dissipation $P_{D(max)} = 300$ mW. Power dissipation in a BJT is given by $v_{CE} \times i_C$. A hyperbola representing $v_{CE} \times i_C = 300$ mW has been drawn in Fig. 2.20. To avoid excessive power dissipation the circuit load line must always be below this hyperbola. For example, if $V_{CC} = 30$ V then drawing a load line which is tangential to the maximum dissipation curve gives $i_C = 40$ mA at $v_{CE} = 0$ V. Therefore the load line represents a load resistance of $(30 \text{ V}/40 \text{ mA}) = 750 \ \Omega$. This is the minimum permissible load resistance value if $V_{CC} = 30$ V and $P_{D(max)} = 300$ mW. Using $V_{CC} = 20$ V would permit load resistance values down to 330 Ω. Safe working limits are not, however, confined to maximum power dissipation; there are also values of i_C and v_{CE} which must not be exceeded.

☐   **Example 2.2**

Given that

$$V_{CE(max)} = 25 \text{ V}$$
$$I_{C(max)} \ = 40 \text{ mA}$$
$$P_{D(max)} \ = 200 \text{ mW}$$

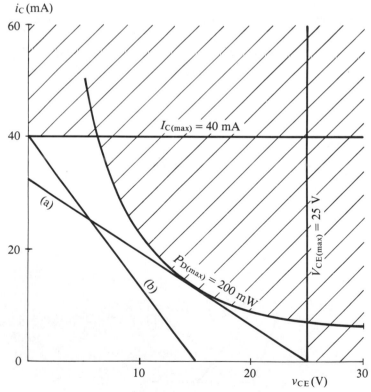

Fig. 2.21   Graph for Example 2.2

Choose minimum values of R_L for safe working given that (a) $V_{CC} = 25$ V, (b) $V_{CC} = 15$ V.

First draw on the i_C vs. v_{CE} axes of Fig. 2.21 three lines:

a horizontal line at $I_{C(max)} = 40$ mA
a vertical line at $V_{CE(max)} = 25$ V
a hyperbola for $P_{D(max)} = 200$ mW

Then shade the area for all values of i_C vs. v_{CE} in excess of any of these three values. Now the load line must not enter the shaded area.

(a) For $V_{CC} = 25$ V the tangent to the $P_{D(max)}$ curve cuts the i_C axis at 32 mA, which is less than $I_{C(max)}$, so the load resistance represented by this line is permissible and will be

$$R_L = \frac{25 \text{ V}}{32 \text{ mA}} = 780 \ \Omega.$$

(b) For $V_{CC} = 15$ V the line tangential to the $P_{D(max)}$ curve cuts the i_C curve at about 54 mA which is greater than $I_{C(max)}$ so the resistance represented by this line cannot be used. Instead draw the load line from $v_{CE} = 15$ V to $i_C - 40$ mA, which is well below the $P_{D(max)}$ curve and therefore acceptable on all three counts. The minimum resistor value is

$$\frac{15 \text{ V}}{40 \text{ mA}} = 375 \ \Omega. \qquad \square$$

Resistor Tolerances

Unless some special purpose warrants the high cost of close tolerance components, the designer will have to use standard components. In choosing resistors this usually means 10% preferred values. For the last example the nearest preferred values above 780 Ω and 375 Ω are 820 Ω and 390 Ω, but if the actual component randomly chosen is 10% less then in each case the actual value will be below the minimum permissible safe value. Therefore the minimum 10% preferred values should be 1 kΩ and 470 Ω.

2.8   BIASING A COMMON EMITTER AMPLIFIER

For linear operation it is necessary, as we have seen, for the BJT of a CE amplifier to be biassed on to the linear sections of its input and output characteristics. A frequent requirement is that $V_{CE(Q)}$ (Fig. 2.22) should be approximately $V_{CC}/2$ although this is by no means always the case. Referring to the idealized output characteristics and load line of Fig. 2.23, the quiescent or operating point Q has been chosen so that $V_{CE(Q)} = V_{CC}/2$. As the input signal v_s increases, so does the peak-to-peak value of v_{CE} and, as the waveform drawn below the v_{CE} axis shows, it can

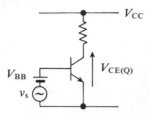

Fig. 2.22 $V_{CE(Q)}$ is the Quiescent Value of v_{CE} when $v_s = 0$

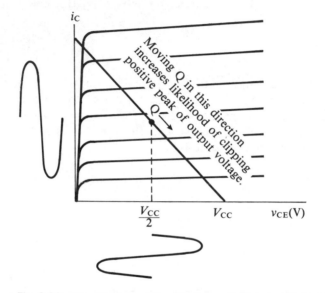

Fig. 2.23   Bias Point Q set for Maximum Undistorted Output Voltage Swing

approach V_{CC} before clipping, i.e. flattening, of the waveform peak occurs. This is called the condition of maximum undistorted output signal. With practical transistors it will be somewhat less than V_{CC}. If the bias is shifted then one of the output waveform peaks will be clipped before the peak-to-peak value of v_{CE} approaches V_{CC}. For example, if $V_{CE(Q)} = 0.75\ V_{CC}$ then the positive peak of v_{CE} will clip when, at best, the peak-to-peak output voltage is $V_{CC}/2$.

In Fig. 2.22 the quiescent values of base and collector current are controlled by V_{BB}. Provision of two dc supplies is inconvenient and costly, and in this section we shall design circuits which eliminate the need for V_{BB} and operate from a single supply V_{CC}. The provision of bias to establish suitable quiescent values is a dc matter, so reference will be made to quantities such as I_B and V_{CE} – it will be understood that quiescent values are being referred to.

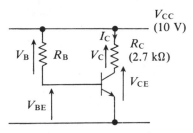

Fig. 2.24   Base Bias Circuit

Base Bias

The components of Fig. 2.24 can be designed to provide self bias and a good way to describe the circuit is to approach it from a design point of view.

☐   **Example 2.3**

Given $V_{CC} = 10$ V and $R_C = 2.7$ kΩ, choose a value of R_B which will make $V_{CE} = 6$ V.

BJT data: $h_{FE} = 50$, $V_{BE(ON)} = 0.7$ V.

As with many of these problems a good approach is to work backwards from the output. With $V_{CC} = 10$ V and $V_{CE} = 6$ V, then $V_C = 4$ V.
Therefore

$$I_C = \frac{V_C}{R_C} = \frac{4 \text{ V}}{2.7 \text{ k}\Omega} = 1.5 \text{ mA}.$$

In the active region of the BJT,

$$I_C = h_{FE} I_B$$

Therefore, to produce a certain value of I_C, in this case, 1.48 mA, the value of I_B must be,

$$I_B = \frac{1.5 \text{ mA}}{50} = 30 \text{ } \mu A$$

Once the BJT is conducting V_{BE} only changes by a few tenths of a volt at the most, so making the approximation that $V_{BE(ON)}$ is constant at 0.7 V,

$$V_B = V_{CC} - V_{BE(ON)} = 9.3 \text{ V}.$$

Therefore, to make $I_B = 29.6 \text{ } \mu A$,

$$R_B = \frac{9.3 \text{ V}}{30 \text{ } \mu A} = 310 \text{ k}\Omega. \qquad\qquad ☐$$

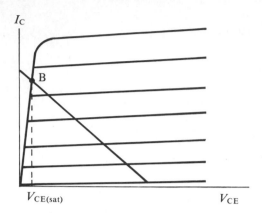

Fig. 2.25   Thermal Runaway Causes I_c to Rise until the BJT is Saturated at Point B

DISADVANTAGES OF THE BASE BIAS CIRCUIT

Dependence of Operating Point on h_{FE}:   It is important to realise that, in Fig. 2.24, once R_B has been chosen, the value of I_B will be fixed whatever the value of h_{FE}. The value of I_C will therefore be $h_{FE}I_B$. Reference to BJT data sheets usually shows that a particular type of BJT will have an h_{FE} between certain minimum and maximum values, typically 50 and 200. Returning to the previous example, if the BJT had an $h_{FE} = 100$, then with the same value of R_B, $I_B = 30\ \mu A$ as before but $I_C = 3.0\ mA$, which is twice the current required to make $V_{CE} = 6$ V. Now, $V_C = 8$ V and $V_{CE} = 2$ V, which is approaching $V_{CE(SAT)}$. If $h_{FE} \geqslant 125$ the BJT will saturate and therefore the circuit will be useless for linear amplification.

Thermal Instability: In Section 1.9 it was noted that a component of I_C was the leakage current I_{CEO}. Repeating equation (1.3),

$$I_C = h_{FE}I_B + I_{CEO}$$

It was also noted that I_{CEO} is temperature dependent but could be made negligible with good circuit design. However, in the base bias circuit of Fig. 2.24, when the circuit is switched on current flows through the BJT so its temperature rises. This causes I_{CEO} to increase, which in turn causes temperature to rise, i.e. the action is cumulative. Eventually I_C may rise to the point B on the load line of Fig. 2.25 where the BJT has become saturated. This process is called 'thermal runaway'. One consequence is that there can be no negative-going excursion of v_{CE} and another may be damage to the BJT through overheating. Germanium transistors are particularly prone to thermal runaway.

Emitter Bias

Figure 2.26 is designed to reduce the dependence of the operating point on h_{FE} and to stabilize the operating point against temperature change, thereby minimizing the possibility of thermal runaway. As an aid to understanding the operation of this

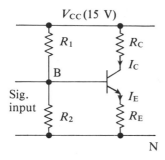

Fig. 2.26   Emitter Bias Circuit

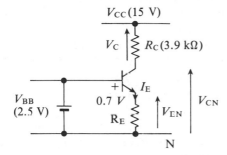

Fig. 2.27   Explanatory Form of Emitter Bias Circuit

circuit, consider Fig. 2.27. Taking $V_{BE(ON)} = 0.7$ V, $V_{EN} = V_{BB} - 0.7$

Therefore $I_E = \dfrac{V_{BB} - 0.7}{R_E}$

Note (1) that I_E is independent of h_{FE}, and (2) that what the circuit does is to fix I_E; hence the name *emitter bias*.

Now if $h_{FE} > 50$, $I_C \doteqdot I_E$

(If $h_{FE} = 50$ the error is 2% and falls as h_{FE} rises.)

The principle then, is that having established I_E, I_B takes up a value $I_E/(1 + h_{FE})$ and as $I_C \doteqdot I_E$ its value is predictable and independent of h_{FE}. In the base bias circuit, I_B is established and $I_C = h_{FE} I_B$, which is unpredictable.

☐   **Example 2.4**

Given the figures in brackets in Fig. 2.27, choose a value of R_E to make $V_{CN} = 9$ V.

For this, $V_C = 6$ V

Therefore $I_C = \dfrac{6\text{ V}}{3.9\text{ k}\Omega} = 1.54$ mA.

$V_{EN} = V_{BB} - V_{BE} = 2.5 - 0.7 = 1.8$ V.

$I_E \doteqdot I_C = 1.54$ mA

Therefore $R_E = \dfrac{V_{EN}}{I_E} = \dfrac{1.8\text{ V}}{1.54\text{ mA}} \doteqdot 1.2$ kΩ.

Note that $V_{CE} = V_{CC} - V_C - V_{EN} = 7.2$ V   □

Another advantage of emitter bias is that the circuit is thermally stable. As temperature rises when the circuit is switched on, I_C and I_E also tend to rise, which increases V_{EN}. But $V_{BN} = V_{BB}$ is fixed so $V_{BE} = V_{BN} - V_{EN}$ tends to fall. Reference to an input characteristic shows that once the BJT is conducting, I_B changes quite sharply for small changes in V_{BE} (we have taken V_{BE} as being virtually constant at 0.7 V for design purposes). Therefore if V_{BE} does tend to fall it will produce a corresponding fall in I_B and therefore I_C, which counteracts the original temperature increase. This mechanism is a form of negative feedback which will be dealt with in Chapter 5.

The circuit of Fig. 2.27 is, however, unsatisfactory, not only in requiring a second dc supply but also because V_{BB} short-circuits the input signal. In Fig. 2.26 the resistors R_1 and R_2 potentially divide V_{CC} so that the voltage across R_2, V_{BN}, simulates the effect of V_{BB} without short-circuiting the input signal. If we were to repeat the previous problem but use Fig. 2.26, then clearly there are an infinite number of pairs of values of R_1 and R_2 which will make $V_{BN} = 2.5$ V, and herein lies a difficult design problem. In Fig. 2.28, which is the same circuit as Fig. 2.26 with currents marked, it can be seen that I_B flows through R_1 but not R_2. Making R_1 and R_2 small makes I_1 and I_2 large and the value of I_B will be relatively insignificant. Now once I_E and I_C are fixed, then $I_B = I_C/h_{FE}$, but since h_{FE} is unpredictable we cannot know what will be the value of I_B. However, if I_B is insignificant com-

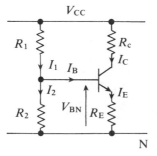

Fig. 2.28   Emitter Bias Circuit with Currents Marked. $I_2 = I_1 - I_B$. Therefore V_{BN} Depends

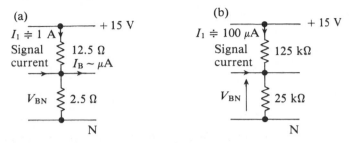

Fig. 2.29 (a) I_1 is so Large Compared with I_B that V_{BN} is Virtually Independent of I_B. However, Signal Current is Shunted by Resistors
(b) I_1 is Comparable with I_B, therefore V_{BN} is Dependent on I_B. Shunting Effect on Signal Current is Negligible

pared with I_1 and I_2 it does not matter, and V_{BN} will be predictable. Further, V_{BN} will be held constant however the BJT currents change. This will hold V_{EN} constant and maintain thermal stability. The point is illustrated by the extremely low and impracticable values of Fig. 2.29a where I_B will be in microamps, so V_{BN} will be predictable and constant as required.

Apart from the current drain on V_{CC}, small values of R_1 and R_2 divert input signal current away from the BJT base, so on these grounds large values of R_1 and R_2 are preferable. However, the value of I_B, which is dependent on the unpredictable h_{FE}, is comparable with I_1 and I_2 given resistors of the order of those shown in Fig. 2.29b, so V_{BN} will be unpredictable and so will the operating point. Further, as I_C and I_E rise with temperature, I_B falls because of minority carrier current flowing into the collector and out of the base. This will make V_{BN} rise as V_{EN} rises, so there is no fall in V_{BE} to offset the increase in leakage current and the circuit will not be thermally stable. Clearly the designer is faced with compromise in choosing resistor values. At this stage we shall consider a nonrigorous but effective method of designing an emitter bias circuit and look at a more rigorous design at the end of the chapter.

Nonrigorous Design Procedure

☐ **Example 2.5**

For Fig. 2.30 choose values of R_1, R_2 and R_E to make $V_{CE} = 7$ V.
Transistor data:

$h_{FE(min)} = 50$; $h_{FE(max)} = 150$; $V_{BE(ON)} = 0.8$ V.

We can begin by suggesting that when the design is complete $V_{EN} = 0.1\ V_{CC}$. Whatever value is chosen for V_{EN} it must be deducted from V_{CC} in determining the voltage available for output voltage swing. A deduction of 10% of the supply, in this case 1.5 V, is usually acceptable.

With $V_{EN} = 1.5$ V and $V_{CE} = 7$ V, $V_C = 6.5$ V.

Therefore $I_C = \dfrac{6.5}{2.2} \div 3 \text{ mA} \div I_E$

Therefore $R_E = \dfrac{V_{EN}}{I_E} = \dfrac{1.5 \text{ V}}{3 \text{ mA}} = 500 \ \Omega.$

To make $V_{EN} = 1.5 \text{ V},$

$$V_{BN} = V_{BE} + V_{EN} = 0.8 + 1.5 = 2.3 \text{ V}.$$

Taking h_{FE} to be midway between the minimum and maximum value, i.e. $h_{FE} = 100$, then typically,

$$I_B = \dfrac{I_C}{h_{FE}} = 30 \ \mu A$$

As noted, I_1 and I_2 must be significantly larger than I_B to ensure reasonable stability and predictability of operating point.

Therefore make $I_1 = 10 I_B = 300 \ \mu A$

Then $\qquad\qquad\quad I_2 = \ \ 9 I_B = 270 \ \mu A$

$$R_2 = \dfrac{V_{BN}}{I_2} = \dfrac{2.3 \text{ V}}{270 \ \mu A} = 8.5 \text{ k}\Omega$$

$$V_1 = V_{CC} - V_{BN} = 12.7 \text{ V}$$

$$R_1 = \dfrac{V_1}{I_1} = \dfrac{12.7 \text{ V}}{300 \ \mu A} = 42.3 \text{ k}\Omega.$$

Using 10% preferred values,

$$R_E = 470 \ \Omega, \ R_1 = 39 \text{ k}\Omega, \ R_2 = 8.2 \text{ k}\Omega. \qquad\qquad\qquad \Box$$

Without further investigation at this stage the results seem to be acceptable. The shunting effect of R_1 and R_2 on the input signal will be considered in Chapter 3, but being larger than a typical value of h_{ie} most of the signal current will go into the

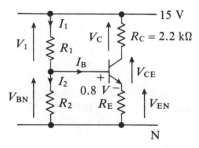

Fig. 2.30   Circuit for Example 2.5

base as required. The current drain from V_{CC} into R_1 at 300 μA is only one-tenth of I_C so it is not excessive, but at ten times I_B it gives reasonable stability and predictability. The error in V_{CE} by having transistors with either $h_{FE(min)}$ or $h_{FE(max)}$ will be calculated in Section 2.9. First we will complete this section by looking at an approximate but frequently used method of estimating the quiescent condition.

Approximate Method of Estimating Quiescent Condition in an Emitter Bias Circuit

☐ **Example 2.6**

Find V_{CN} in Fig. 2.31.

It cannot be assumed that the designer has made $V_{EN} = 0.1\ V_{CC}$. However, since we only seek an approximate solution we can assume that h_{FE} is high enough for I_B to be ignored and therefore treat R_1 and R_2 as a simple potential divider.

Then $V_{BN} = \dfrac{10}{10 + 47} \times 10\ \text{V} = 1.7\ \text{V}.$

Assume $V_{BE} = 0.7\ \text{V},$

$$V_{EN} = V_{BN} - V_{BE} = 1.0\ \text{V}$$

Then $I_C \doteqdot I_E = \dfrac{V_{EN}}{R_E} = \dfrac{1.0\ \text{V}}{680\ \Omega} = 1.5\ \text{mA}$

$$V_C = I_C \times R_C = 1.5 \times 2.7 = 4.1\ \text{V}$$

Therefore $V_{CN} = 5.9\ \text{V}$ ☐

2.9 CALCULATION OF QUIESCENT CONDITION IN AN EMITTER BIAS CIRCUIT TAKING h_{FE} AND I_B INTO ACCOUNT

The following example is a more accurate method of finding the quiescent condition than the one used in the last section. However, it is necessary to know h_{FE}.

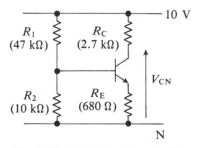

Fig. 2.31   Circuit for Example 2.6

☐   **Example 2.7**

Given the 10% preferred values calculated for the circuit of Fig. 2.30, it would be instructive to see how V_{CE} varies by using transistors with from minimum to maximum values of h_{FE}. This circuit is repeated in Fig. 2.32 with resistor values.
(a) $h_{FE} = h_{FE(min)} = 50$.

The problem could be solved by using Kirchhoff's laws to find I_1 and I_B in Fig. 2.32. A better method is to redraw the circuit as in Fig. 2.33 and replace that part of the circuit to the left of B and N by its Thevenin equivalent (Fig. 2.34) in which

$V_T = V_{BN(OC)}$ (i.e. V_{BN} with circuit to the right of B and N disconnected)

$$V_T = \frac{8.2}{8.2 + 39} \times 15 = 2.61 \text{ V}$$

R_B = resistance to the left of B and N with V_{CC} replaced by a short circuit. This puts 8.2 kΩ and 39 kΩ in parallel.

$$R_B = \frac{8.2 \times 39}{47.2} = 6.8 \text{ k}\Omega.$$

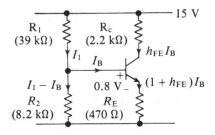

Fig. 2.32   Circuit for Example 2.7

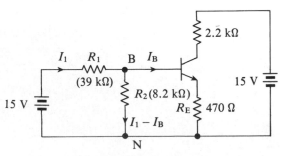

Fig. 2.33   Figure 2.32 Redrawn to Indicate how $R_1 R_2$ Potential Divider Chain may be Redrawn

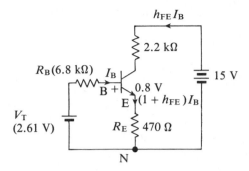

Fig. 2.34   Fig. 2.32 with $R_1 R_2$ Replaced by Thevenin Equivalent

Now apply KVL (Kirchhoff's Voltage Law) around loop V_T–R_B–B–E–R_E–N.

$$V_T = R_B I_B + V_{BE} + R_E (1 + h_{FE}) I_B \tag{2.7}$$

$2.61 = 6.8\ I_B + 0.8 + 0.47 \times 51\ I_B$

$1.81 = 30.8\ I_B$

$I_B\ = 59\ \mu A$

$I_E\ = 51 I_B = 3.0\ \text{mA};\qquad V_{EN} = I_E R_E - 1.4\ V$

$I_C\ = 2.95\ \text{mA};\qquad\qquad V_C = I_C R_C = 6.5\ V$

$V_{CE} = V_{CC} - V_{EN} - V_C = 7.1\ V$

(b) To calculate V_{CE} when $h_{FE} = 150$, simply use this value in equation (2.7).

Then $2.61 = 6.8\ I_B + 0.8 + 0.47 \times 151\ I_B$,

giving $\qquad I_B = 23\ \mu A$

$\qquad\qquad V_{EN} = 1.6\ V$

$\qquad\qquad V_C = 7.6\ V$

$\qquad\qquad V_{CE} = 5.8\ V$

Summarizing, as h_{FE} increases from 50 to 150, V_{CE} falls from 7.1 V to 5.8 V. $\square$

2.10   CONTROL OF OPERATING POINT STABILITY BY CHOICE OF BIAS COMPONENTS

The stabilization of quiescent conditions against temperature change has been described qualitatively in Section 2.8. In equation (1.11) it was noted that

$$I_{CEO} \doteqdot h_{FE} I_{CBO}$$

and that for a silicon transistor I_{CBO} is reckoned to double for every 10 °C increase in temperature. Therefore, taking the worst case for a 2N3709,

$$I_{CBO(max)} = 100 \text{ nA at } 25\,^{\circ}\text{C}$$
$$h_{FE(max)} = 165$$

Then at $55\,^{\circ}\text{C}$, I_{CBO} will have increased eightfold and

$$I_{CEO} = 165 \times 8 \times 100 \text{ nA} \doteqdot 0.13 \text{ mA}$$

For many applications this may be acceptable but the emitter bias circuit does reduce it by a factor which depends on the choice of R_E and R_B (i.e. R_1 and R_2 in parallel).

Repeating equation (1.9) we have

$$I_C = \frac{h_{FB}}{1 - h_{FB}} I_B + \frac{I_{CBO}}{1 - h_{FB}}$$

and using equations (1.10) and (1.11),

$$I_C \doteqdot h_{FE} I_B + h_{FE} I_{CBO} \tag{2.8}$$

Observing Fig. 2.34 we can rewrite equation (2.7),

$$V_T = I_B R_B + V_{BE} + (I_C + I_B) R_E \tag{2.9}$$

From equation (2.8),

$$I_B = \frac{I_C}{h_{FE}} - I_{CBO}$$

Substituting for I_B in equation (2.9) and rearranging,

$$V_T - V_{BE} = I_C \left(\frac{R_B + R_E}{h_{FE}} + R_E \right) - I_{CBO}(R_B + R_E)$$

Taking $V_T - V_{BE}$ to be constant,

$$\frac{dI_C}{dI_{CBO}} = \frac{R_B + R_E}{(R_B + R_E)/h_{FE} + R_E} \tag{2.10}$$

which is called the stability factor S.

Two extreme cases for S are:

(a) $R_B = 0$

$$S = \frac{h_{FE}}{1 + h_{FE}} \doteqdot 1$$

This is ideal because I_C is only increased by I_{CBO} and not by $h_{FE} I_{CBO}$. It is not practicable, however, because, as we have seen, R_1 and R_2 must be high enough to avoid shunting the signal current.

(b) $R_E = 0$

$$S = h_{FE}$$

This is the worst case because the circuit offers no stabilization against temperature increase.

□ **Example 2.8**

Transistor data for Fig. 2.35:

$$V_{BE(ON)} = 0.8 \text{ V}$$

$$I_{CBO} \text{ at } 25\,^\circ\text{C} = 100 \text{ nA}$$

$$h_{FE} = 100$$

It is required that $V_{CN} = 7$ V and that I_C changes by no more than 10 μA above the design value. Choose values of R_1, R_2 and R_E.

With temperature increasing from 25 $^\circ$C to 55 $^\circ$C, I_{CBO} increases eightfold, i.e. from 100 nA to 800 nA. Therefore $\delta I_{CBO} = 700$ nA

Therefore $\dfrac{\mathrm{d}I_C}{\mathrm{d}I_{CBO}} = \dfrac{10}{0.7} \doteqdot 14$

Substituting in equation (2.10) and rearranging gives

$$R_B \doteqdot 15 \, R_E$$

$$I_C = \frac{4.5 \text{ V}}{2.2 \text{ k}\Omega} \doteqdot 2 \text{ mA}$$

Therefore $I_B = 20 \; \mu$A

At this stage we can suggest a value for R_E and proceed, iteratively if necessary. Using $V_{EN} = 0.1 \, V_{CC} = 1.2$ V

$$R_E = \frac{1.2 \text{ V}}{2 \text{ mA}} = 600 \; \Omega$$

Therefore $R_B = 15 \, R_E = 9$ kΩ.

Therefore $\dfrac{R_1 R_2}{R_1 + R_2} = 9$ kΩ. $\hfill (2.11)$

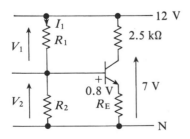

Fig. 2.35   Circuit for Example 2.8

Now $V_1 = 10 \text{ V} = I_1 R_1$

Therefore $I_1 = \dfrac{10}{R_1}$

and $V_2 = 2 \text{ V} = (I_1 - I_B)R_2$ where $I_B = 0.02 \text{ mA}$.

Therefore $2 = (I_1 - 0.02)R_2 = \left(\dfrac{10}{R_1} - 0.02\right)R_2$ (2.12)

Solving equations (2.11) and (2.12) gives

$R_1 = 11 \text{ k}\Omega$

$R_2 = 50 \text{ k}\Omega$

If these values produce too large a shunting effect then R_E can be increased, but since I_E is already fixed by the circuit requirements then the penalty is a reduced output voltage swing for a given value of V_{CC}. □

2.11 COUPLING AND DECOUPLING CAPACITORS – THE SEPARATION OF AC AND DC IN AN AMPLIFIER

The emitter bias circuit of Fig. 2.26 has been supplemented in Fig. 2.36 by three capacitors. There is also a resistance R_L which represents an external load such as the input resistance to another stage of amplification or some transducer. In a low-power amplifier R_2 will be several kilohms. Without C_1, the source resistance R_s shunts R_2 and reduces its effective value. Further, the source may have a dc component which is not to be amplified. In either case the bias conditions of the BJT could be altered. Separating the source and the amplifier by C_1 allows ac to pass from source to amplifier but blocks dc. The reactance of C_1 should be considerably less than the amplifier input resistance (a matter to be explained in Chapter 3), to avoid losses.

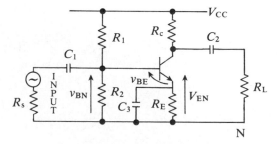

Fig. 2.36 Coupling and Decoupling Capacitors

C_2 has a similar function. If only the signal component of the output voltage is to be fed to R_L then the quiescent component of v_{CN} must be blocked. C_2 will pass ac without significant loss if carefully chosen, but will block dc. The function of C_3 is more difficult to understand. As the instantaneous value of the input signal rises, i_C and i_E will also increase and therefore V_{EN} will rise. Now

$$v_{BE} = v_{BN} - V_{EN}$$

so v_{BE} will increase by less than the input signal. This phenomenon, called negative feedback, will be dealt with more fully in Chapter 5. However, the consequence is that the rise in i_C will not be as large as expected so R_E causes a gain reduction. To prevent this R_E is shunted by C_3 which holds V_{EN} constant. In effect, if chosen to have a low reactance at the lowest required frequency, it acts as an ac short-circuit so that there will be no signal component superimposed on V_{EN}. C_3 is called a decoupling capacitor.

SUMMARY

1. For linear amplification dc bias is required so that the input signal only causes variation of v_{BE} and i_B over the linear section of the BJT input characteristic.
2. Signal components of v_{BE}, i_B and i_C are in phase.
3. Amplification of an input signal is achieved by using it to control variations of i_C through an external load resistance.
4. The resulting output voltage is in antiphase with the input voltage.
5. A circuit can be drawn which represents the BJT when biased for linear operation so that reasonable approximations can be made for gain calculations.
6. The CE amplifier load line diagram is a useful aid to understanding the mechanism of voltage amplification.
7. There are safe working limits for various BJT electrical quantities which must not be exceeded.
8. Base bias of a BJT has two drawbacks.
9. Emitter bias circuits can be designed to overcome these drawbacks.
10. The choice of components for emitter bias circuits is a compromise; the factors involved are predictability of quiescent conditions, thermal stability, shunting of signal current and current drain.
11. Coupling and decoupling capacitors are used to reduce the conflict between quiescent requirements and signal requirements.

REVISION QUESTIONS

2.1 Calculate V_{CE} in Fig. Q.2.1.

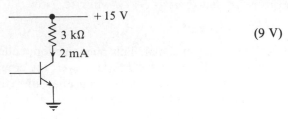

(9 V)

Fig. Q.2.1

2.2 In Fig. Q.2.2, $I_B = 20 \, \mu A$; $h_{FE} = 80$. Calculate I_C and V_{CE}.

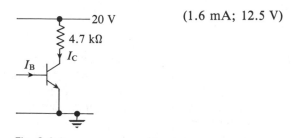

(1.6 mA; 12.5 V)

Fig. Q.2.2

2.3 In Fig. Q.2.3, $v_s = 0.1 \sin \omega t$ (V); $I_{B(Q)} = 80 \, \mu A$; $h_{ie} = 2 \, k\Omega$.
   (a) Plot i_B and calculate its maximum and minimum values.
   (b) If $I_{C(Q)} = 3$ mA and $h_{fe} = 50$, plot i_C and v_{CE} and calculate maximum and minimum values.
   (c) Calculate A_v.

(130 μA; 30 μA; 5.5 mA; 0.5 mA; 0.1 V; 9.1 V; -45)

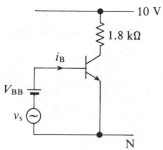

Fig. Q.2.3

2.4 For Fig. Q.2.4:
  (a) Estimate h_{ie}.
  (b) Plot i_B, indicating $I_{B(Q)}$ and extreme values of i_B.
  (c) Given $I_{C(Q)} = 2$ mA and $h_{fe} = 80$, plot i_C and v_{CE} and estimate A_v.
  (d) Evaluate $h_{fe}R_L/h_{ie}$.

(Estimated answers: (a) 3.3 kΩ (b) 41 μA; 11 μA (c) 3.2 mA; 0.8 mA; 15.5 V; 2.1 V; -134)

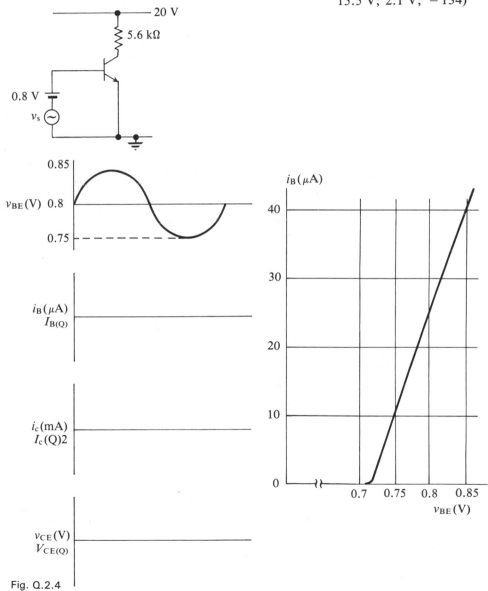

Fig. Q.2.4

2.5 With axes of i_C against v_{CE}, draw the following load lines when the supply voltage V_{CC} is:
(a) 10 V (b) 15 V; for load resistances (i) 10 kΩ, (ii) 5 kΩ (iii) 1 kΩ.

2.6 On axes of I_C against V_{CE}, indicate the limits of the operating region for a transistor whose absolute maximum ratings are:

Maximum power dissipation at 20 °C = 200 mW,
Maximum collector current   = 30 mA,
Maximum collector voltages   = 30 V.

2.7 A BJT with collector characteristics shown in Fig. Q.2.7 has $V_{CE(max)} = 15$ V; $I_{C(max)} = 10$ mA; $P_{D(max)} = 20$ mW. Choose a suitable pair of values for V_{CC} and load resistance (10% tolerance) for safe working.

(14 V; 3.3 kΩ)

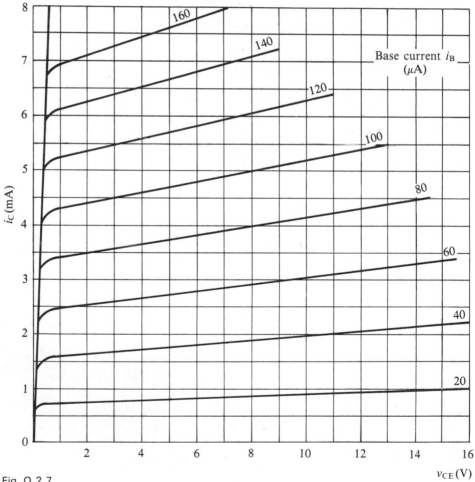

Fig. Q.2.7

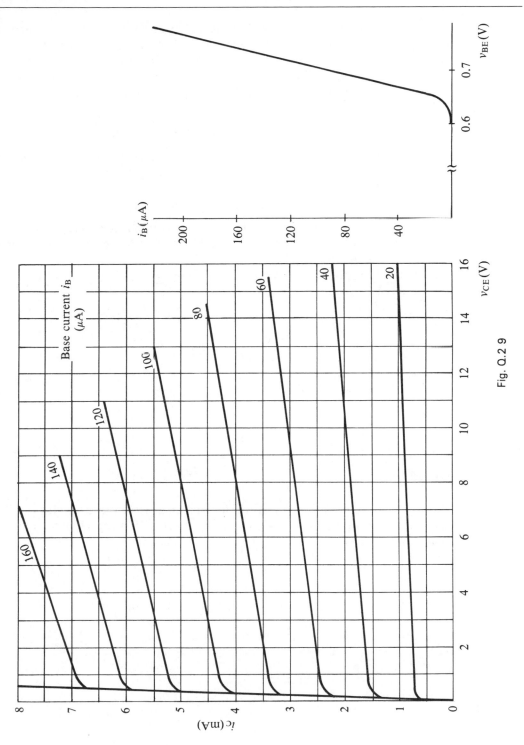

Fig. Q.2 9

2.8   A transistor whose collector characteristics are given in Fig. Q.2.7 is used in a single-stage amplifier having a resistive load of 3.5 kΩ. The supply voltage is 14 V and quiescent base current is 40 μA. Determine the quiescent collector current and voltage and also the changes in the collector current and voltage for an increase in base current of 20 μA and hence estimate the current gain of the stage.

$$(1.8 \text{ mA}; \ 7.7 \text{ V}; \ 0.9 \text{ mA}; \ -3.1 \text{ V}; \ 45)$$

2.9   A single-stage CE amplifier with $V_{CC} = 15$ V and load resistance of 2 kΩ uses a BJT with the characteristics of Fig. Q.2.9. THe circuit is biased so that the quiescent base current is 100 μA. If a signal voltage $v_s = 25 \sin \omega t$ (mV) is applied to the input, plot on a common time axis v_{BE}, i_B, i_C and v_{CE}.

2.10 (a) For the circuit shown in Fig. Q.2.10 it is required that $V_{CE(Q)} = 6$ V and $I_{C(Q)} = 2.5$ mA. Determine the values of R_B and R_L for a transistor with $h_{FE} = 100$. Assume $V_{BE} = 0.8$ V.
       (b) If the transistor is replaced by one having $h_{FE} = 200$, determine the new value of $V_{CE(Q)}$.

$$(368 \text{ k}\Omega; \ 1.6 \text{ k}\Omega; \ 2 \text{ V})$$

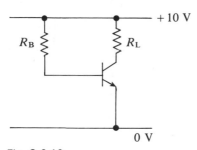

Fig. Q.2.10

2.11 The current gain for the transistor in Fig. Q.2.11 ranges from 40 to 100. Calculate the variation in $V_{CE(Q)}$ that is possible for the circuit. Assume $V_{BE} = 0.7$ V.

$$(8.1 \text{ V to } 0 \text{ V})$$

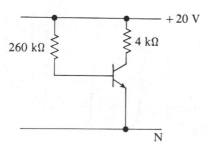

Fig. Q.2.11

2.12 For the circuit in Fig. Q.2.12, the resistor values are as follows:

$R_1 = 120$ kΩ, $R_2 = 22$ kΩ, $R_3 = 4$ kΩ, $R_4 = 1$ kΩ.

Using the same transistor as for Q.2.11, determine the variations in collector current and the emitter and collector voltages that are possible. Assume $V_{BE} = 0.7$ V.

(I_C: 1.6 mA to 2.0 mA; V_{CN}: 13.6 V to 12 V; V_{EN}: 1.65 V to 2 V)

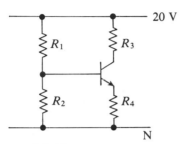

Fig. Q.2.12

2.13 Calculate approximate values of $V_{BN(Q)}$, $V_{EN(Q)}$, $I_{C(Q)}$ and $V_{CN(Q)}$ for the circuit of Fig. Q.2.13. ($V_{BE(ON)} = 0.7$ V)

(2.4 V; 1.7 V; 1.7 mA; 7.4 V)

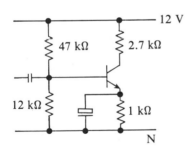

Fig. Q.2.13

2.14 Given that $h_{FE} = 50$ and $V_{BE(ON)} = 0.7$ V for the BJT in Fig. Q.2.14, calculate accurate and 10% preferred resistor values to give $V_{CN(Q)} = 9$ V.

($R_1 = 50$ kΩ, use 47 kΩ; $R_2 = 9.6$ kΩ, use 10 kΩ; $R_e = 1.17$ kΩ, use 1.2 kΩ)

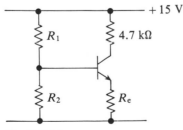

Fig. Q.2.14

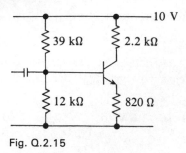

Fig. Q.2.15

2.15 Calculate $V_{CN(Q)}$ accurately in Fig. Q.2.15 when (a) $h_{FE} = 40$, (b) $h_{FE} = 100$. ($V_{BE(ON)} = 0.8$ V)

(6.8 V; 6.3 V)

3

BJT Equivalent Circuits

Chapter Objectives
3.1   General Concepts and Definition
3.2   The Decibel
3.3   The h (Hybrid) Parameter Small-signal Equivalent Circuit of a BJT
3.4   Relationship Between h Parameters and Input and Output Characteristics
3.5   Variation of h Parameters
3.6   Amplifier Analysis Using the Hybrid (h) Equivalent Circuit
3.7   Cascaded Stages
3.8   An Amplifier Calculation Including the Effect of h_{re}
3.9   Frequency Response of an Amplifier
3.10   Simplified Hybrid II circuit for the Analysis of BJT Amplifiers at MF
        Summary
        Revision Questions

The aim of this chapter is to demonstrate that when a BJT is operating under linear conditions it can be represented by a simple model – a linear network – which facilitates the calculation of voltage, current and power gain and input and output resistance. These linear networks, better known as small-signal equivalent circuits, enable us to calculate component values in order to meet design specifications without having to resort to graphical methods. However, it should be noted that the process is restricted to small-signal operation and is not valid, for example, in power amplifier calculations, as we shall see in Chapter 6.

3.1   GENERAL CONCEPTS AND DEFINITIONS

Voltage, Current and Power Gain

A properly biased amplifier can, as far as small ac signals are concerned, be treated as a two-port network (Fig. 3.1). There will be input voltage V_i and input current

Fig. 3.1 Two-port Network. Terminals Marked for BJT in CE Mode

I_i to the network, which will feed, to an external load R_L, output voltage V_o and output current I_o. Invariably amplifiers do not have four terminals but three, one of which is common to input and output. This is the case with the CE amplifier where the emitter is common and the base is the other input terminal and the collector the other output terminal. B, C and E have been marked on Fig. 3.1 as an example. For this network,

$$\text{Voltage gain } A_v = \frac{V_o}{V_i}$$

$$\text{Current gain } A_i = \frac{I_o}{I_i}$$

$$\text{Power gain } A_p = \frac{P_o}{P_i} = \frac{V_o I_o}{V_i I_i} = A_v A_i$$

In this book, small-signal symbols such as V_i, I_o, etc., represent rms values of sinusoidal quantities unless otherwise specified. As such they can also be treated as phasors so that if, for example, A_v is positive it indicates that V_o and V_i are in phase, and if negative that they are in antiphase.

Input and Output Resistance

Generally the input terminals of an amplifier will appear as an impedance to an ac source. Unless we are dealing with very low or high frequencies this impedance will simply approximate to a resistance – the amplifier input resistance R_i. Its actual composition may be due to several components such as the BJT itself and bias resistors. For a given input voltage V_i (Fig. 3.2),

$$I_i = \frac{V_i}{R_i}$$

An important but conceptually more difficult property is output resistance R_o. Any network or part of a network which is linear and which can be brought out to a pair of terminals can, invoking Thevenin's theorem, be replaced by a constant-voltage source (constant rms ac in this case) and a resistance. In Fig. 3.2 these are E and R_o respectively. R_o limits the output current because the maximum signal

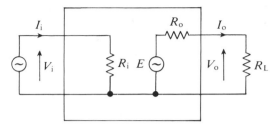

Fig. 3.2   Illustration of Input Resistance R_i and Output Resistance R_o of an Amplifier

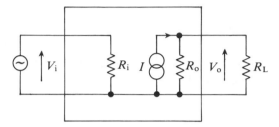

Fig. 3.3   Amplifier Output Represented as a Norton Equivalent Circuit

current that the amplifier can deliver flows when its output is short-circuited (i.e. $R_L = 0$) and will be

$$I_{o(max)} = \frac{E}{R_o}$$

If the output had been drawn as a Norton equivalent (Fig. 3.3) then the circuit would be a constant alternating current generator I in parallel with R_o. $V_{o(max)}$ would occur with the output open-circuited ($R_L = \infty$). Then

$$V_{o(max)} = IR_o$$

Maximum power will be delivered by the amplifier to the load R_L when $R_L = R_o$. Using Fig. 3.2, $P_o = I_o^2 R_L$

$$P_{o(max)} = \left(\frac{E}{2R_o}\right)^2 R_o = \frac{E^2}{4R_o} \tag{3.1}$$

Clearly if R_o can be reduced by circuit design then $P_{o(max)}$ will rise.

3.2   THE DECIBEL

Apart from the fact that aural sensitivity is logarithmic, it is easier to work out the gain (or loss) of a system if the gain of each part is calculated as a logarithmic value: the overall gain is then found by addition. The basic unit – the bel – is inconveniently large so we invariably use the decibel (bel ÷ 10) – the dB – which is defined as

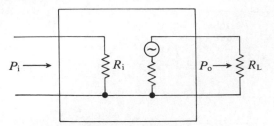

Fig. 3.4 Gain in dB is $10 \log_{10}(P_o/P_i)$

a power ratio.

$$\text{Gain in dB for Fig. 3.4} = 10 \log_{10} \frac{P_o}{P_i} \qquad (3.2)$$

If we have a loss, i.e. $P_o < P_i$, then the result is negative. The overall gain of a network of cascaded elements is found by adding the gains and subtracting the losses. It is usually less convenient to measure power than voltage, so if we measure V_i and V_o for Fig. 3.5 we have:

$$\text{Gain in dB} = 10 \log_{10} \frac{P_o}{P_i}$$

$$= 10 \log_{10} \frac{V_o^2/R_L}{V_i^2/R_i}$$

$$= 20 \log_{10} \frac{V_o}{V_i} - 10 \log_{10} \frac{R_L}{R_i} \qquad (3.3)$$

Thus, measuring voltages and knowing resistances yields power gain. However, in many cases the resistance part is ignored and it is customary to quote voltage gain as:

$$A_v = 20 \log_{10} \frac{V_o}{V_i} \quad \text{dB} \qquad (3.4)$$

and current gain as:

$$A_i = 20 \log_{10} \frac{I_o}{I_i} \quad \text{dB} \qquad (3.5)$$

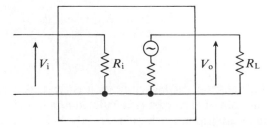

Fig. 3.5 Power Gain of Network in dB is $20 \log_{10}(V_o/V_i) - 10 \log_{10}(R_L/R_i)$

☐ **Example 3.1**

If voltage gain is 500, $A_v = 20 \log_{10} 500 = 54$ dB
If gain is subject to fluctuations of ± 6 dB,
then $6 = 20 \log_{10} dA_v$

$dA_v = 2$

Then max. gain $= 500 \times 2 = 1000$
    min. gain $= 500 \div 2 = 250$
Check: $20 \log_{10} 1000 = 60$ dB   (i.e. $54 + 6$)
        $20 \log_{10} \ 250 = 48$ dB   (i.e. $54 - 6$)   ☐

The preceding definitions are for ratios but the dB can be used as a measure of power or voltage relative to certain standard levels.

(a) Power – in telecommunications the milliwatt standard is commonly used where,

$$1 \text{ mW} = 0 \text{ dB(m)}$$

thus $200 \text{ mW} = 10 \log_{10} \dfrac{200}{1} = +26 \text{ dB(m)}$

$0.2 \text{ mW} = 10 \log_{10} \dfrac{0.2}{1} = -10 \log_{10} 5 = -7 \text{ dB (m)}$

(b) Voltage – 600 Ω is a standard terminating resistance in telecommunications and 1 mW dissipated in 600 Ω gives 0.775 V, which is defined as 0 dB(V).

Thus $5V = 20 \log_{10} \dfrac{5}{0.775} = +16.2 \text{ dB(V)}$

3.3   THE *h* (HYBRID) PARAMETER SMALL-SIGNAL EQUIVALENT CIRCUIT OF A BJT

It is now necessary to develop a two-port network (the contents of the box in Fig. 3.6) which is equivalent to the CE-connected BJT in Fig. 3.7, which is assumed to be biased for linear operation. The four terminal quantities in general use for two-port networks, i.e V_i, I_i, V_o and I_o, will be V_{be}, I_b, V_{ce} and I_c† for the BJT.

In using two-port network theory to develop an equivalent circuit, any two of these quantities can be defined as dependent variables in terms of the other two as independent variables. The most common choice for the BJT is to make V_{be} and I_c the dependent variables, which gives rise to the following important equations:

$$V_{be} = h_{ie}I_b + h_{re}V_{ce} \tag{3.6}$$

$$I_c = h_{fe}I_b + h_{oe}V_{ce} \tag{3.7}$$

† Standard definitions will show I_o as flowing *out* of the network, i.e. opposite in direction to I_c.

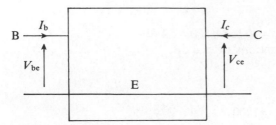

Fig. 3.6 Two-port Network with Terminal Voltages and Currents which Assume that the Network is a BJT in CE Configuration

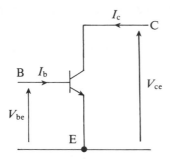

Fig. 3.7 Contents of Two-port Network Must Represent BJT Biased for Linear Operation

The four h parameters are properties of the BJT and enable it to be represented as a linear network. For any type of BJT, values of the h parameters will be quoted and they are all that needs to be known about the BJT to calculate such quantities as voltage and current gain of amplifiers.

Figure 3.8 is the equivalent circuit which can be used to represent Fig. 3.7. The circuit follows from equations (3.6) and (3.7). Taking first the left-hand section of the circuit and using KVL (Kirchhoff's Voltage Law), the voltage V_{be} is equal to the voltage across h_{ie} ($h_{ie}I_b$) plus the value of the voltage source $h_{re}V_{ce}$. Using KCL (Kirchhoff's Current Law) for the right-hand part of the circuit, the collector current I_c consists of the value of the current source $h_{fe}I_b$ plus the current in the conductance h_{oe}, which will be $h_{oe}V_{ce}$.

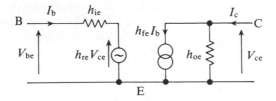

Fig. 3.8 Hybrid Equivalent Circuit of BJT in CE Configuration

The four h parameters can be described as follows:

h_{ie} is the input base–emitter resistance – measured in ohms – typically 500 Ω to
   5 kΩ

h_{re} is the reverse voltage transfer ratio and is unitless – typically 10^{-4} to 10^{-3}.

h_{fe} is the forward current transfer ratio and is unitless – typically 50 to 500.

h_{oe} is the collector–emitter output conductance – measured in siemens – typically
   1 μS to 100 μS.

Each h parameter has a double subscript. The first letter follows from the description and the second, an 'e' in each case, indicates that the parameter refers to CE operation. There are common base and common collector h parameters, e.g. h_{ib}, h_{fc}, etc. However they are much less important than the CE h parameters and manufacturers do not always quote them. If they are ever required there are conversion formulae for deriving them from CE parameters.

Precise definitions of the h parameters can be obtained from equations (3.6) and (3.7) by making the independent variables zero one at a time. Then if $V_{ce} = 0$ in both (3.6) and (3.7),

$$V_{be} = h_{ie}I_b. \quad \text{Therefore } h_{ie} = \left.\frac{V_{be}}{I_b}\right|_{V_{ce}=0} \tag{3.8}$$

$$I_c = h_{fe}I_b. \quad \text{Therefore } h_{fe} = \left.\frac{I_c}{I_b}\right|_{V_{ce}=0} \tag{3.9}$$

Alternatively, when $I_b = 0$,

$$V_{be} = h_{re}V_{ce}. \quad \text{Therefore } h_{re} = \left.\frac{V_{be}}{V_{ce}}\right|_{I_b=0} \tag{3.10}$$

$$I_c = h_{oe}V_{ce}. \quad \text{Therefore } h_{oe} = \left.\frac{I_c}{V_{ce}}\right|_{I_b=0} \tag{3.11}$$

3.4   RELATIONSHIP BETWEEN h PARAMETERS AND INPUT AND OUTPUT CHARACTERISTICS

The input and output characteristics describe BJT performance graphically while the h parameters represent BJT performance, under linear operating conditions, numerically. However, if both are valid then we should be able to define the h parameters from the linear sections of the characteristics.

(a) h_{ie} – in equation (3.8): making $V_{ce} = 0$, i.e. no alternating output voltage, is the same as making v_{CE} constant. Referring now to a set of input characteristics (Fig. 3.9), this is the same as choosing just one characteristic, say $v_{CE} = 10$ V.

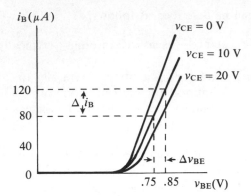

Fig. 3.9   Calculating h_{ie}

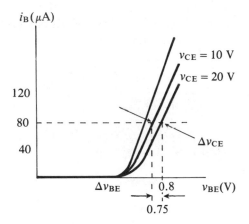

Fig. 3.10   Calculating h_{re}

Then taking the linear section of this characteristic,

$V_{be} = \Delta v_{BE}$ and

$I_b = \Delta i_B$ so

$$h_{ie} = \frac{\Delta v_{BE}}{\Delta i_B}\bigg|_{v_{CE}=10\text{ V}} = \frac{0.1\text{ V}}{40\ \mu\text{A}} = 2.5\text{ k}\Omega.$$

(b) h_{re} – it has been noted in Section 1.11 that the input characteristics vary with V_{CE}. Therefore, if i_B is held constant, altering v_{CE} will alter v_{BE}. This is shown in Fig. 3.10 where i_B is held at 40 μA and as v_{CE} is increased from 10 V to 20 V, v_{BE} increases by 0.05 V.

$$h_{re} = \frac{V_{be}}{V_{ce}}\bigg|_{I_b=0} = \frac{\Delta v_{BE}}{\Delta v_{CE}}\bigg|_{i_B=80\ \mu\text{A}} = \frac{0.05}{10} = 5 \times 10^{-3}$$

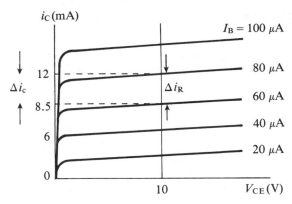

Fig. 3.11   Calculating h_{fe}

h_{re} is more likely to be about one-tenth of this value. It is the least important of the *h* parameters and is neglected in the vast majority of applications.

(c) h_{fe} – holding v_{CE} constant at 10 V in the output characteristic of Fig. 3.11, then h_{fe} – the increase of i_C caused by the increase of i_B – can be estimated.

$$h_{fe} = \frac{I_c}{I_b}\bigg|_{V_{ce}=0} = \frac{\Delta i_C}{\Delta i_B}\bigg|_{v_{CE}=10\text{ V}} = \frac{3.5\text{ mA}}{20\ \mu A} = 175$$

(d) h_{oe} – holding i_B constant at 60 μA in Fig. 3.12, it can be seen that i_C does increase slightly as V_{CE} increases.

$$h_{oe} = \frac{I_c}{V_{ce}}\bigg|_{I_b=0} = \frac{\Delta i_C}{\Delta v_{CE}}\bigg|_{i_B=60\ \mu A} = \frac{0.2\text{ mA}}{10\text{ V}} = 20\ \mu S.$$

This is a second-order effect and although usually more significant than h_{re}, it is often omitted from calculations.

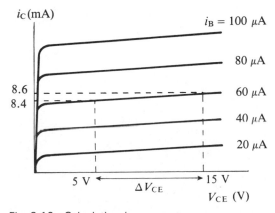

Fig. 3.12   Calculating h_{oe}

3.5   VARIATION OF *h* PARAMETERS

Manufacturers usually quote maximum, minimum and typical values of h parameters under certain test conditions. The conditions usually quoted are I_C, V_{CE}

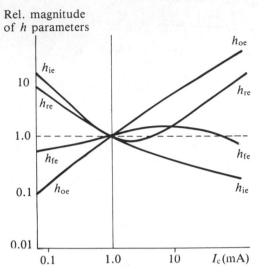

Fig. 3.13   Variation of h Parameters with I_C

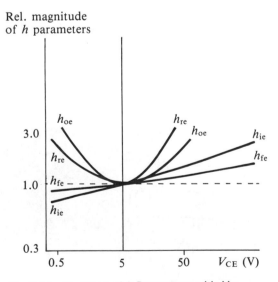

Fig. 3.14   Variation of h Parameters with V_{CE}

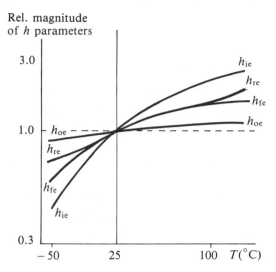

Fig. 3.15   Variation of *h* Parameters with Free Air Temperature

and free air temperature, T. For a low-power BJT these typically would be:

$$
\begin{aligned}
I_C &= 1 \text{ mA} \\
V_{CE} &= 5 \text{ V} \\
T &= 25\,^{\circ}\text{C}
\end{aligned}
$$

Variation of any of these quantities causes the *h* parameters to vary and Figs 3.13–3.15 illustrate the manner of these variations on curves normalized to the values quoted above. For example, if a BJT has $h_{ie} = 2$ kΩ at $I_C = 1$ mA then at $I_C = 20$ mA, it has to be multiplied by 0.05 i.e. h_{ie} falls to 100 Ω. The curves also show that at extremes of operating points the approximations made by ignoring h_{oe} and h_{re} may not be valid. For example, at high collector currents both of these parameters increase quite significantly.

3.6   AMPLIFIER ANALYSIS USING THE HYBRID (*h*) EQUIVALENT CIRCUIT

Evaluation of gain by load line is cumbersome and limited. As circuits become more complicated and as the need arises to find quantities other than gain, e.g. input and output resistance, then the load line is discarded in favor of the equivalent circuit as long as linear conditions obtain. It is first necessary to draw the equivalent circuit and then to use standard network theory for analysis. In the following examples, h_{re} will be considered negligible but, for completeness, a single example including h_{re} will be worked out at the end of this section.

☐ **Example 3.2**

Using Fig. 3.16 as an example, find A_v given

$h_{ie} = 1.5 \text{ k}\Omega$; $h_{fe} = 70$; $h_{oe} = 25 \text{ }\mu\text{S}$; $h_{re} = 0$.

To draw the equivalent circuit the following procedure is recommended:

(a) Assume that all capacitors are ac short-circuits unless the capacitor values are given and that there is a specific requirement to take into account their effect on the circuit performance.

(b) Assume that all dc power supplies are ac short-circuits. This is justifiable because we are analyzing the ac performance of the circuit only, and any pair of terminals at a fixed voltage, e.g. V_{CC} and N, cannot have an alternating voltage superimposed on them. Therefore, dc power supplies are ac short-circuits.

(c) Draw the h parameter model of the BJT, labeling the terminals B, C and E.

(d) Add the other components to the BJT equivalent circuit noting (a) and (b).

The resulting circuit, which is equivalent to Fig. 3.16, is given in Fig. 3.17. It is correct to show R_B being connected between B and E because the top end of R_B in Fig. 3.16 is connected to V_{CC}. But as (b) above makes V_{CC} to N a short-circuit in the equivalent circuit, then the top end of R_B is effectively connected to E in the equivalent circuit. Similarly, the end of R_C connected to V_{CC} will also go to E in Fig. 3.17. As the internal resistance of the source V_s is zero, then R_B will not

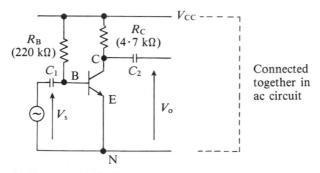

Fig. 3.16   Circuit for Example 3.2

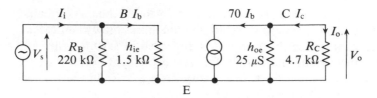

Fig. 3.17   Equivalent Circuit of Fig. 3.16

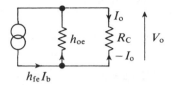

Fig. 3.18 In Output Circuit $h_{fe}I_b$ Splits Between R_C and h_{oe}

affect I_b, so

$$I_b = \frac{V_s}{h_{ie}}$$

The only source of signal current in the output is $h_{fe}I_b$ which divides between R_c and h_{oe} in proportion to their conductances (Fig. 3.18).

Convention dictates that the direction of I_o is *defined* as shown. However, since the current in R_C is part of $h_{fe}I_b$ then I_o will be negative. Therefore,

$$I_o = -\frac{1/R_C}{1/R_C + h_{oe}}\, h_{fe}I_b$$

$$I_o = -\frac{1}{1 + h_{oe}R_C}\, h_{fe}\, \frac{V_s}{h_{ie}}$$

$$V_o = I_o R_C$$

Therefore, $\dfrac{V_o}{V_s} = -\dfrac{h_{fe}}{h_{ie}}\dfrac{R_C}{1 + h_{oe}R_C}$ \hfill (3.12)

$$= \frac{-219}{1 + 0.118} = -186$$

NOTES
1. The negative sign indicates that V_o and V_s are in antiphase.
2. Comparing equation (3.12) with equation (2.5), note that the gain is reduced by $(1 + h_{oe}R_C)$ because h_{oe} has been taken into account. The term $R_C/(1 + h_{oe}R_C)$ is the resistance of $R_C \parallel h_{oe}$ so equation (2.5) is still valid provided that the value of the resistance in the numerator of the equation is the total resistance between collector and emitter. In this example the reduction, which gives the more accurate result, is 12%. The approximation becomes more accurate by the extent to which R_C is less than $1/h_{oe}$. The voltage gain in dB will be

$$A_v(\text{dB}) = 20\log_{10}186$$

$$= 45.3\ \text{dB.} \hfill \square$$

☐ **Example 3.3**

In Fig. 3.19 a finite source resistance $R_s = 1$ kΩ has been included in the circuit. For Fig. 3.20 find V_o, I_o and P_o and hence find A_v, A_i and A_p. Also find R_i.

Transistor data $h_{ie} = 5$ kΩ; $h_{fe} = 50$; $h_{oe} = 100$ μS; $h_{re} = 0$.

The identification of actual quantities such as V_o is not ambiguous, but to calculate gain may involve finding either V_o/V_{be} or V_o/V_s. In this case it is quite clear that R_s belongs to the source so $A_v = V_o/V_{be}$. Also, the input resistance R_i is the resistance looking into the amplifier at the B–E terminals.

From the equivalent circuit of Fig.3.20 it is clear that,

$$R_i = R_B \parallel h_{ie}$$

$$= \frac{100 \times 5}{105} = 4.76 \text{ kΩ}$$

$$V_{be} = \frac{R_i}{R_i + R_S} V_s$$

$$= \frac{4.76}{4.76 + 1} \times 1 \text{ mV} = 0.83 \text{ mV}$$

Note that 0.17 mV of V_s has been lost across R_s.

$$I_b = \frac{V_{be}}{h_{ie}} = \frac{0.83 \text{ mV}}{5 \text{ kΩ}} = 0.17 \text{ μA}$$

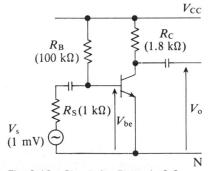

Fig. 3.19   Circuit for Example 3.3

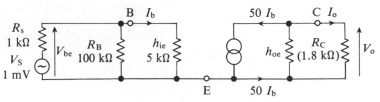

Fig. 3.20   Equivalent Circuit for Fig. 3.19

Let the total load across C and $E = R_T$

Then $R_T = R_c \parallel h_{oe} = 1.52$ kΩ.

Then $V_o = -h_{fe} I_b \times R_T$

$$= -50 \times 0.17 \times 10^{-6} \times 1.53 \times 10^3 = 13 \text{ mV}$$

Then $A_v = \dfrac{V_o}{V_{be}} = \dfrac{-13}{0.83} = -15.6$

The input current I_i is the total current supplied by the source to the amplifier and is therefore the base current plus the current in R_B. This will be the current into R_i.

Therefore, $\quad I_i = \dfrac{V_{be}}{R_i} = \dfrac{0.83 \text{ mV}}{4.76 \text{ k}\Omega} = 0.174 \ \mu\text{A}.$

The output current is the current supplied to the 1.8 kΩ load resistance.

Therefore, $\quad I_o = \dfrac{V_o}{R_c} = \dfrac{-13 \text{ mV}}{1.8 \text{ k}\Omega} = -7.2 \ \mu\text{A}$

Therefore, $\quad A_i = \dfrac{I_o}{I_i} = \dfrac{-7.2}{0.174} = -41.3$

$$P_i = I_i^2 R_i$$
$$= 0.174^2 \times 10^{-12} \times 4.76 \times 10^3 \text{ W}$$
$$= 0.144 \times 10^{-9} \text{ W}$$

$$P_o = I_o^2 R_c$$
$$= 7.2^2 \times 10^{-12} \times 1.8 \times 10^3 \text{ W}$$
$$= 93.3 \times 10^{-9} \text{ W}$$

Therefore,

$$A_p = \frac{93.3}{0.144} = 648$$

Also, $A_p = A_v A_i = 15.6 \times 41.3 = 644$

These should be exactly equal; the error occurs in rounding figures.

$A_v(\text{dB}) = 20 \log_{10} 15.6 = 23.9 \text{ dB}$

$A_i(\text{dB}) = 20 \log_{10} 41.3 = 32.3 \text{ dB}$

$A_p(\text{dB}) = 10 \log_{10} 644 = 28.1 \text{ dB}$ □

□ **Example 3.4**

For the amplifier of Fig. 3.21 calculate the ratios V_o/V_s and I_o/I_i and also find the output resistance of the amplifier considered as the circuit feeding R_L.

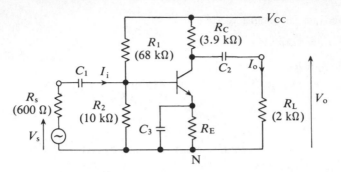

Fig. 3.21 Example 3.4 – an Amplifier with Emitter Bias and External Load

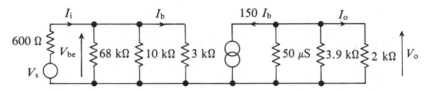

Fig. 3.22 Equivalent Circuit of Fig. 3.21

Transistor data:

$h_{fe} = 150$; $h_{ie} = 3$ kΩ; $h_{oe} = 50$ μS; $h_{re} = 0$.

In the equivalent circuit (Fig. 3.22) the resistors R_1 and R_2 will be in parallel because V_{CC} is connected to N. For the same reason, R_C and R_L are in parallel. Further, as C_3 acts as an ac short-circuit across R_E then the emitter is connected to N.

Input resistance $R_i = 68$ kΩ ‖ 10 kΩ ‖ 3 kΩ
$$= 2.23 \text{ k}\Omega$$

Therefore $I_i = \dfrac{V_s}{R_s + R_i} = \dfrac{V_s}{2.83 \times 10^3}$

$$V_{be} = \frac{R_i}{R_s + R_i} V_s = \frac{2.23}{2.83} V_s = 0.79 \ V_s$$

$$I_b = \frac{0.79 \ V_s}{h_{ie}} = \frac{0.79 \ V_s}{3 \times 10^3}$$

Total conductance between collector and emitter $= G_{ce}$

$$G_{ce} = h_{oe} + \frac{1}{R_C} + \frac{1}{R_L} = 806 \ \mu S$$

$$V_o = \frac{-h_{fe} I_b}{G_{ce}} = \frac{150 \times 0.79 \ V_s}{806 \times 10^{-6} \times 3 \times 10^3} = -49 V_s$$

Therefore $\dfrac{V_o}{V_s} = -49$

$$I_o = \frac{V_o}{R_L} = \frac{V_o}{2 \times 10^3}$$

$$\frac{I_o}{I_i} = \frac{V_o/(2 \times 10^3)}{V_s/(2.83 \times 10^3)}$$

$$= \frac{V_o}{V_s} \times \frac{2.83}{2} = -49 \times 1.41 = -69.$$

The output resistance R_o is the resistance looking back from R_L into the amplifier and is therefore $R_C \parallel h_{oe} = 3.3$ kΩ. □

In previous examples, if R_C is regarded as the load fed from the transistor, then $R_o = 1/h_{oe}$.

3.7 CASCADED STAGES

Transistors can be cascaded to provide extra gain in an arrangement such as Fig. 3.23. The capacitor C_2 separates the dc conditions of the two stages so that V_{CN1} and V_{BN2} are independent of one another. However, assuming that all capacitors are virtually short-circuits at the lowest signal frequency of interest, then the signal output of Q_1, V_{cn1}, becomes V_{bn2}. This method of coupling is called resistor–capacitor or, more usually, RC, coupling.

The overall gain of the amplifier cannot be calculated by simply taking the gain of each stage and multiplying them (or adding if they are in dB) unless in calculating the gain of the first stage, the loading effect of the input resistance to the second stage is taken into account. Very often the best method is to draw the complete equivalent circuit and work through it.

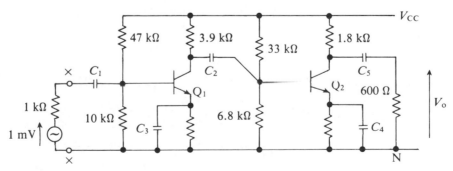

Fig. 3.23 Cascaded (RC Coupled) Stages

☐ **Example 3.5**

For Fig. 3.23 calculate V_o and the power gain in dB from the input terminals X–X to the 600 Ω load.
Transistor data (for both transistors):

$$h_{ie} = 2 \text{ k}\Omega; \; h_{fe} = 100; \; h_{oe} = 50 \; \mu\text{S}; \; h_{re} = 0.$$

In the equivalent circuit of Fig. 3.24 it will be observed that h_{oe} would be, in both stages, shunted by resistors considerably less than $1/h_{oe}$ (20 kΩ) so the error introduced by neglecting it is small.

$$R_i = 47 \text{ k}\Omega \parallel 10 \text{ k}\Omega \parallel 2 \text{ k}\Omega = 1.6 \text{ k}\Omega$$

$$I_i = \frac{V_s}{R_s + R_i} = \frac{1 \text{ mV}}{2.6 \text{ k}\Omega} = 0.38 \; \mu\text{A}.$$

$$P_i = I_i^2 R_i = 0.23 \times 10^{-9} \text{ W}$$

$$V_{be} = \frac{R_i}{R_i + R_s} \times V_s = \frac{1.6}{2.6} \times 1 \text{ mV} = 0.62 \text{ mV}$$

Therefore $I_{b1} = \dfrac{V_{be}}{h_{ie}} = \dfrac{0.62 \text{ mV}}{2 \text{ k}\Omega} = 0.31 \; \mu\text{A}$

Total conductance across collector and emitter of $Q_1 = G_{ce1}$

$$G_{ce1} = \frac{1}{5.6 \text{ k}\Omega} + \frac{1}{33 \text{ k}\Omega} + \frac{1}{6.8 \text{ k}\Omega} + \frac{1}{2 \text{ k}\Omega} = 0.86 \text{ mS}$$

Therefore $V_{be2} = -\dfrac{h_{fe} I_{b1}}{G_{ce1}} = -\dfrac{100 \times 0.31 \times 10^{-6}}{0.86 \times 10^{-3}} \text{ V}$

$$= -36 \text{ mV}$$

$$I_{b2} = \frac{V_{be2}}{h_{ie}} = \frac{-36 \times 10^{-3}}{2 \times 10^3} = -18 \; \mu\text{A}$$

Total resistance between collector and emitter of $Q_2 = R_{ce2}$

$$R_{ce2} = 1.8 \text{ k}\Omega \parallel 600 \; \Omega = 450 \; \Omega$$

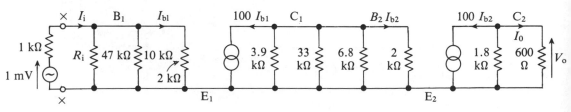

Fig. 3.24   Equivalent Circuit of Fig. 3.23 (h_{oe} omitted)

Therefore $V_o = -h_{fe}I_{b2}R_{ce2}$

$$= 100 \times 18 \times 10^{-6} \times 450 \text{ V}$$

$$= 0.81 \text{ V}$$

$$P_o = \frac{V_o^2}{R_L} = \frac{0.81^2}{600} = 1.1 \text{ mW}$$

Therefore $A_p = \dfrac{P_o}{P_i} = \dfrac{1.1 \times 10^{-3}}{0.23 \times 10^{-9}} = 4.8 \times 10^6$

$$A_p(\text{dB}) = 10 \log_{10} A_p = 67 \text{ dB} \qquad\qquad \square$$

One of the aims of the designer should be to couple as efficiently as possible the current $h_{fe}I_{b1}$ from Q_1 into the base of Q_2 as I_{b2}. In this example $h_{fe}I_{b1}$ is 31 μA and I_{b2} is 18 μA. More efficient coupling would be achieved by increasing the collector load of Q_1 and the bias chain of Q_2 but, as noted in Chapter 2, several factors have to be taken into account in choosing these resistors.

3.8 AN AMPLIFIER CALCULATION INCLUDING THE EFFECT OF h_{re}

$\square$ **Example 3.6**

Find A_v for the circuit of Fig. 3.25.
Transistor data:

$$h_{ie} = 2.5 \text{ k}\Omega; \; h_{fe} = 100; \; h_{oe} = 20 \; \mu\text{S}; \; h_{re} = 2 \times 10^{-4}.$$

It is clear from the equivalent circuit, Fig. 3.26, that R_B (47 k$\Omega \parallel$ 10 kΩ) does not affect I_b, and therefore A_v, but that $h_{re}V_o$ does.

$$V_o = -100 \; I_b R_{ce}$$

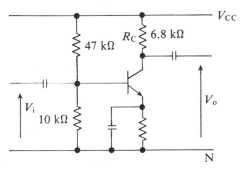

Fig. 3.25   Circuit for Example 3.6

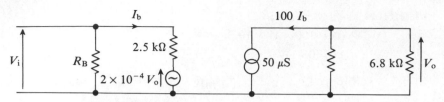

Fig. 3.26   Equivalent Circuit of Fig. 3.25

where $R_{ce} = R_C \parallel h_{oe} = 6.8\ \text{k}\Omega \parallel 20\ \mu\text{S} = 6.0\ \text{k}\Omega$.

Therefore $V_o = 100\ I_b \times 6\ \text{V}$ (I_b in mA)

Using KVL for the input current,

$$V_i = 2.5\ I_b + 2 \times 10^{-4}\ V_o$$
$$= 2.5\ I_b - 2 \times 10^{-4} \times 100\ I_b \times 6$$
$$= I_b(2.5 - 0.12) = 2.38\ I_b.$$

This is the only time the effect of h_{re} appears in the analysis and its effect is to increase I_b by 0.12 in 2.5, i.e. by approximately 5%.

$$V_o = -h_{fe}I_bR_{ce} = -100 \times \frac{V_i}{2.38} \times 6$$

$$\frac{V_o}{V_i} = A_v = -252$$

Neglecting h_{re},

$$A_v = \frac{h_{fe}R_{ce}}{h_{ie}} = -240$$

As the gain at -252 is high it maximizes the effect of h_{re}, but even so for most purposes it is negligible. $\square$

3.9   FREQUENCY RESPONSE OF AN AMPLIFIER

Unless amplifiers are required to have an uneven frequency response, e.g. for filtering or for special effects, they tend to have a falling gain at low and high frequencies (LF and HF) with a flat response over a substantial frequency band usually termed medium frequencies (MF). At LF the falling response is usually due to capacitors in series with the signal path, i.e. coupling and decoupling capacitors. At HF it will be due to the fact that the BJT itself will have reduced gain, or to stray capacitance in leads, etc.: these effects can be represented by capacitance in parallel with the signal path.

3.9.1 LF Response

In Fig. 3.27, which could represent a single-stage amplifier such as Fig. 3.19, a capacitor C_1 couples a source to an input R_i. At MF, C_1 acts as a short circuit. At LF the reactance of the capacitor, X_{C_1} will become significant. As frequency f falls, X_{C_1} rises, so input current I_i, and therefore V_i, fall.

At MF, $I_i = \dfrac{V_s}{R_1}$ where $R_1 = R_s + R_i$

At LF, $I_i = \dfrac{V_s}{R_1 - jX_{C_1}}$

and as $V_i = I_i R_i$, then V_o and therefore the overall gain V_o/V_s fall. It should be noted that the ratio V_o/V_i is not affected by this coupling but as V_o falls then the effective gain as measured by V_o/V_s falls.

Let $A_{vo} = $ MF gain.

Then as $V_o \propto I_i$

$$V_o \propto \frac{V_s}{R_1}$$

Let $A_{vl} = $ LF gain

then $V_o \propto \dfrac{V_s}{R_1 - jX_{C_1}}$

and $\dfrac{A_{vl}}{A_{vo}} = \dfrac{R_1}{R_1 - jX_{C_1}} = \dfrac{1}{1 - jX_{C_1}/R_1}$

i.e. $\dfrac{A_{vl}}{A_{vo}} = \dfrac{1}{1 - j/2\pi f C_1 R_1}$ (3.13)

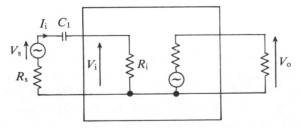

Fig. 3.27   A Single Coupling Capacitor Causing Loss at LF

Defining a particular frequency $f_1 = \dfrac{1}{2\pi C_1 R_1}$,

$$\frac{A_{vl}}{A_{vo}} = \frac{1}{1 - jf_1/f} \qquad\qquad (3.14)$$

$$\left|\frac{A_{vl}}{A_{vo}}\right| = \frac{1}{\sqrt{1 + (f_1/f)^2}} \qquad\qquad (3.15)$$

Note that as f falls $\left|\dfrac{A_{vl}}{A_{vo}}\right|$ falls.

$$\text{Arg }\frac{A_{vl}}{A_{vo}} = \tan^{-1}\frac{f_1}{f} \qquad\qquad (3.16)$$

At $f = f_1$,

$$\frac{A_{vl}}{A_{vo}} = \frac{1}{1 - j} \text{ and}$$

$$\left|\frac{A_{vl}}{A_{vo}}\right| = \frac{1}{\sqrt{2}} \text{ and}$$

$$\text{Arg }\frac{A_{vl}}{A_{vo}} = +45°.$$

Calculating the fall in gain in dB represented by $|A_{vl}/A_{vo}|$ at this frequency,

$$20 \log_{10}\frac{1}{\sqrt{2}} = -3$$

i.e. at f_1 the gain has fallen by 3 dB.

The frequency at which the LF gain is 3 dB down on the MF gain is known as the lower cutoff frequency (sometimes break or corner frequency).

Bode Diagrams

Plotting $|A_{vl}/A_{vo}|$ and $\text{Arg}(A_{vl}/A_{vo})$ against f gives the two graphs of Fig. 3.28, which is called a Bode diagram. As f falls below f_1 then f_1/f becomes greater than 1 and eventually equation (3.14) can be approximated to

$$\frac{A_{vl}}{A_{vo}} = \frac{1}{-jf_1/f} = j\frac{f}{f_1}$$

At frequencies where this approximation is valid, $|A_{vl}/A_{vo}|$ is proportional to f. Thus if f is halved, gain is halved, i.e. gain falls by 6 dB. A halving of frequency is a reduction of 1 octave, and gain is said to fall by 6 dB/octave. A reduction in frequency by a factor of 10 – a decade – reduces gain by a factor of 10, i.e. gain falls by 20 dB/decade. Also, at frequencies where $A_{vl}/A_{vo} = jf/f_1$, then $\text{Arg}(A_{vl}/A_{vo}) = +90°$.

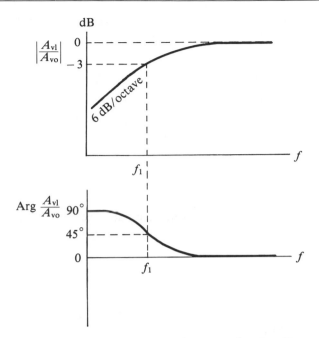

Fig. 3.28   Bode Diagram for LF Response of an Amplifier with a Single Coupling Capacitor

Note that in equation (3.13), R_1 is the total series resistance in the circuit, i.e. it includes the source resistance R_s.

A similar LF response will be caused by an RC coupling between stages. If h_{oe} had been included in Fig. 3.24, then the relevant part of the circuit is as shown in Fig. 3.29 where R_C is the collector load of Q_1, and R_B the parallel equivalent of the bias chain for Q_2. Then reducing the section to the left of C to its Thevenin equivalent, it will be seen that the total effective series resistance in series with C is

$$R_c \| \frac{1}{h_{oe}} \text{ plus } R_B \| h_{ie}$$

Now the cutoff frequency and LF response can be calculated as before. In the circuit

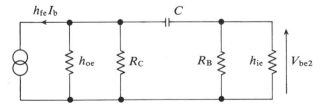

Fig. 3.29   Equivalent Circuit for Calculating the Loss caused by an RC Coupling Between Two CE Stages

of Fig. 3.23 there is another capacitor coupling to the 600 Ω load. These effects are additive and to find the total LF loss, simply add the three separate losses in dB.

LF Loss Due to Decoupling Capacitor

Another source of LF loss occurs because of the increasing reactance of decoupling capacitors, e.g. C_3 in Fig. 3.21, as frequency falls. Once it can no longer be regarded as a short-circuit, then gain falls and, as we shall see in Chapter 5, it eventually falls to

$$A_v \doteq -\frac{R_C}{R_E}$$

Having developed the analysis for LF response it can readily be used as a design tool, for example, in the choice of coupling capacitor necessary to ensure a maximum lower cutoff frequency.

☐ **Example 3.7**

For Fig. 3.21 choose a value of C_1 to make the lower cutoff frequency 50 Hz.
As calculated in Example 3.4,

$$R_s + R_i = 2.83 \text{ k}\Omega$$

Lower cutoff frequency $f_1 = \dfrac{1}{2\pi C_1 R_1}$

from which $C_1 = 1.1 \ \mu\text{F}.$ ☐

3.9.2   HF Response

Before investigating the causes of falling gain at HF it is generally true to say that, for analytical purposes, they can be represented by shunt capacitance. Using again

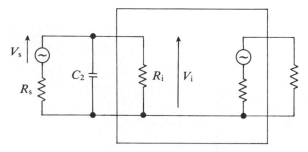

Fig. 3.30   Shunt Capacitance Causing HF Loss

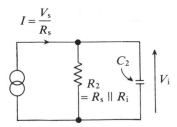

$$I = \frac{V_s}{R_s}$$

Fig. 3.31 Norton Equivalent of Input Section of Fig. 3.30

the example of an amplifier input (Fig. 3.30), the coupling capacitor has negligible effect at HF and is therefore omitted, but shunt capacitance C_2 is included instead. Converting the input to its Norton equivalent (Fig. 3.31), $R_2 = R_s \parallel R_i$ and $I = V_s/R_s$.

At MF, $V_i = IR_2$

At HF, $V_i = I \dfrac{R_2/j\omega C_2}{1/(j\omega C_2) + R_2} = \dfrac{IR_2}{1 + j\omega C_2 R_2}$

Then $\dfrac{A_{vh}}{A_{vo}} = \dfrac{1}{1 + j\omega C_2 R_2}$ (3.17)

where A_{vh} = HF gain.

Defining $f_2 = \dfrac{1}{2\pi C_2 R_2}$

$$\frac{A_{vh}}{A_{vo}} = \frac{1}{1 + jf/f_2}$$ (3.18)

Following the LF response analysis,

$\left| \dfrac{A_{vh}}{A_{vo}} \right|$ will be 3 dB down at $f = f_2$

and Arg $\dfrac{A_{vh}}{A_{vo}} = -45°$ at $f = f_2$

As f increases above f_2, $|A_{vh}/A_v|$ will fall at approaching 6 dB/octave and $\mathrm{Arg}(A_{vh}/A_{vo})$ will be $-90°$.

The complete Bode diagram from LF to HF is shown in Fig. 3.32.

The bandwidth is defined as the frequency band between the cutoff frequencies f_1 and f_2, i.e. bandwidth $= f_2 - f_1$. In practice, for the type of amplifier under discussion, f_1 will be a few Hz or tens of Hz, whereas f_2 will be in kHz or even MHz, so effectively, the bandwidth is f_2. Note that the value of R_2 for Figs 3.30 and 3.31 is the total parallel resistance, i.e. it includes the source resistance R_s. For the RC

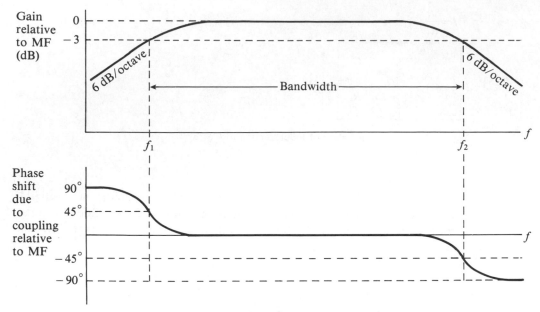

Fig. 3.32 Bode Diagram of an Amplifier with a Single Capacitance Coupling and a Single Shunt Capacitance

coupled stages of Fig. 3.29,

$$R_2 = R_c \parallel \frac{1}{h_{oe}} \parallel R_B \parallel h_{ie}.$$

To determine the high frequency response of an amplifier it is necessary to quantify C_2 and R_2. For this it is necessary to study the behavior of transistors at HF.

3.9.3 Hybrid Π Model of a BJT for HF Analysis

Losses at HF are more complicated than at LF because the performance of the BJT itself must be taken into account. However the HF analysis leading to equations (3.17) and (3.18) is appropriate and in the end the problem is reduced to one of finding values of C_2 and R_2.

The h parameter model contains no frequency-dependent components and is therefore unsatisfactory at HF where a fall in current gain relative to MF is easily demonstrated. The most commonly used HF model is the hybrid Π circuit shown in Fig. 3.33. The components can be justified as follows. Base spreading resistance $r_{bb'}$ represents the resistance presented to charge carriers passing from emitter to base. It separates the terminals B and B'. Only voltage applied between B' and E produces output current. It is typically 50 Ω and is often neglected with little consequent error. Input resistance $r_{b'e}$ represents charge which has to be replaced in the

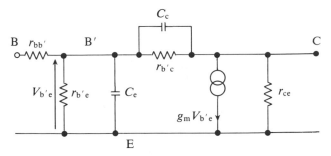

Fig. 3.33 Hybrid Π Model for HF Analysis of a BJT

base. It is therefore equivalent to h_{ie} and is around 1 kΩ. Diffusion capacitance C_e − as v_{BE} rises and then falls with an input signal, so there will be an increase followed by a decrease of the concentration of charge carriers entering the base. However, there is an inevitable time lag between voltage and concentration of carriers associated with their finite mobility. This lag is represented by C_e which will be of the order of 100 pF. Collector capacitance C_c is simply the capacitance of the reverse biased collector base junction. Although much less than C_e at around 4 pF, its effect can be amplified, as will be demonstrated by analysis in the section on the Miller effect. Feedback resistance $r_{b'c}$ represents the voltage feedback from collector to base due to the Early effect. It has the same effect as h_{re}, is around 1 MΩ, and is usually neglected. Collector−emitter resistance r_{ce} represents the slight increase in collector current with collector−emitter voltage and is equivalent to h_{oe}. At around 50 kΩ it can either be neglected or combined with the load resistance. Transconductance g_m essentially represents the BJT as a device which produces collector current I_c in proportion to base emitter voltage $V_{b'e}$. As with the h parameter model, it is defined in terms of constant v_{CE}, i.e. there is an ac short-circuit across the output terminals C-E.

Then $I_c = g_m V_{b'e}$

whereas the h parameter model gives

$$I_c = h_{fe} I_b$$

Therefore $g_m V_{b'e} = h_{fe} I_b$ \hfill (3.19)

Beta Cutoff Frequency f_β and Unity Gain Frequency f_T

The parameters f_β and f_T are defined in terms of current gain values with the output short-circuited. Under these conditions (Fig. 3.34), C_c and $r_{b'c}$ are in parallel with C_e and $r_{b'e}$ and have values such that they can be neglected. At MF then, the circuit is reduced to Fig. 3.35, i.e. C_e may be neglected, and from equation (3.19)

$$g_m V_{b'e} = h_{fe} I_b = \frac{h_{fe} V_{b'e}}{r_{b'e}}$$

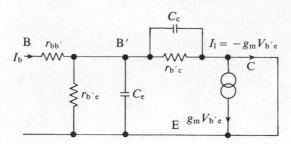

Fig. 3.34   With Hybrid Π Model Output Short-circuited, C_c and $r_{b'c}$ are in Parallel with $r_{b'e}$ and C_e

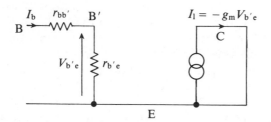

Fig. 3.35   Hybrid Π Circuit with Output Short-circuited at MF

Therefore $g_m = \dfrac{h_{fe}}{r_{b'e}}$ $\hspace{4cm}$ (3.20)

At HF, C_e is now included (Fig. 3.36) and current gain A_i is given by:

$$A_i = \frac{I_1}{I_b} = \frac{-g_m V_{b'e}}{I_b} = -\frac{g_m}{1/r_{b'e} + j\omega C_e}$$

Substituting from equation (3.20),

$$A_i = \frac{-h_{fe}}{1 + j\omega C_e r_{b'e}} = \frac{-h_{fe}}{1 + jf/f_\beta} \hspace{3cm} (3.21)$$

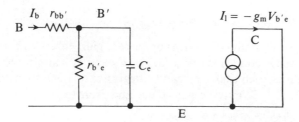

Fig. 3.36   Hybrid Π Circuit with Output Short-circuited at HF

Recognizing the form of equation (3.18), f_β gives the 3 dB down or beta cutoff frequency.

When A_i falls to unity at $f = f_T$,

$$1 = \frac{-h_{fe}}{1 + jf_T/f_B}$$

but as $h_{fe} \gg 1$

$$\left| \frac{f_T}{f_\beta} \right| = h_{fe}$$

i.e. $\quad f_T = h_{fe}f_\beta$

This is called the unity gain frequency or the short-circuit 'current gain × bandwidth' product. Usually it is the h parameters for a BJT that are given, but since these can be converted into hybrid Π values then f_β and f_T can be calculated if C_e is known.

☐ **Example 3.8**

For a certain BJT, $h_{fe} = 80$; $h_{ie} = 2$ kΩ; $C_e = 100$ pF.

Find f_β and f_T.

$$r_{b'e} = h_{ie} = 2 \text{ k}\Omega$$

$$f_\beta = \frac{1}{2\pi C_e r_{b'e}} = 0.8 \text{ MHz}$$

$$f_T = h_{fe}f_\beta = 64 \text{ MHz}$$ ☐

Hybrid Π Circuit with Resistive Load

Consider the simplified hybrid Π circuit of Fig. 3.37. If it is fed from a current source I_b, $r_{bb'}$ has no effect on $V_{b'e}$. Being equivalent to h_{re}, $r_{b'c}$ can be neglected and r_{ce} can be included with R_L or neglected if $R_L \ll r_{ce}$.

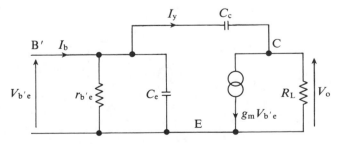

Fig. 3.37   Hybrid Π Circuit with Load Resistance

At MF,

$$V_o = -g_m V_{b'e} R_L.$$

i.e. $A_v = \dfrac{V_o}{V_{b'e}} = -g_m R_L.$

HF Response – Miller Effect

Once the BJT has a resistive load and becomes a voltage amplifier, the analysis can no longer be conducted by taking C_c as being in parallel with C_e as it was when calculating the short-circuit current gain. Connecting any passive component between the input and output terminals of a voltage amplifier – in this case connecting C_c between B' and C – causes the effect of that component to be magnified. Referring again to Fig. 3.37 and considering voltages wrt E,

Voltage at B' $= V_{b'e}$

Voltage at C $= -A_v V_{b'e}$

Therefore voltage across $C_c = V_{b'e}(1 + A_v)$

and $I_y = V_{b'e}(1 + A_v)j\omega C$

It follows that for a given value of input voltage $V_{b'e}$, the extra input current produced because of C_c is as if a capacitor $(1 + A_v)C_c$ were connected across the input terminals. For example, if $C_e = 100$ pF, $C_c = 4$ pF and $A_v = 100$, then the effective capacitance across the input is 100 pF + 404 pF, i.e. C_c has the greater effect. This amplification of the effect of a passive component is called Miller effect. It is an example of shunt connected voltage negative feedback to be discussed in Chapter 5. It is sometimes deliberately exploited, as we shall see in Sections 9.6 and 11.7.

To determine the HF response now refer to Fig. 3.38 in which $C_i = C_e + (1 + A_v)C_c$. V_o will still be $-g_m V_{b'e} R_L$. Therefore finding the frequency response is a matter of determining $V_{b'e}$. Applying Norton to Fig. 3.38 converts the input circuit to the form in Fig. 3.39, where

$$I_s = \frac{V_s}{R_s} \text{ and } R_i = R_s \parallel r_{b'e}$$

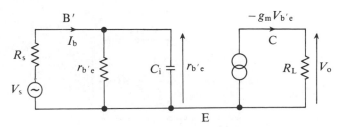

Fig. 3.38   Hybrid Π Circuit with $C_i = C_e + (1 + A_v)C_c$

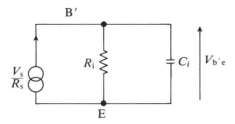

Fig. 3.39　Norton Equivalent of Input Side of Fig. 3.38

Then $V_{b'e} = \dfrac{V_s}{R_s} \dfrac{1}{1/R_i + j\omega C_i} = \dfrac{V_s}{\dfrac{R_s (1 + j\omega C_i R_i)}{R_i}}$

Comparing this with the general equations (3.17) and (3.18) for HF response, A_{vh} will be 3 dB down at $f = f_2$, where

$$\omega C_i R_i = 1$$

i.e.　$f_2 = \dfrac{1}{2\pi C_i R_i}$

☐　**Example 3.9**

A CE amplifier is fed from a source of internal resistance 1 kΩ and drives a collector load of 2.2 kΩ. The transistor data are:

$$h_{ie} = 2 \text{ k}\Omega; \; h_{fe} = 120; \; h_{oe} = 20 \; \mu S; \; C_e = 100 \text{ pF}; \; C_c = 3 \text{ pF}.$$

Find the midband gain and the upper cutoff frequency.

At MF, $A_v = -g_m R = \dfrac{-h_{fe}}{h_{ie}} R = -210$　　$[R = R_L \parallel h_{oe}]$

$R_i = R_s \parallel r_{b'e} = R_s \parallel h_{ie} = 667 \; \Omega.$

$C_i = C_e + (1 + A_v)C_c = 100 + 221 \times 3 = 763 \text{ pF}.$

$f_2 = \dfrac{1}{2\pi R_i C_i} = 312 \text{ kHz.}$　　　　　　　　　　　　　　☐

3.10　SIMPLIFIED HYBRID Π CIRCUIT FOR THE ANALYSIS OF BJT AMPLIFIERS AT MF

An approximation for MF voltage gain we have used several times is

$$A_v = \dfrac{-h_{fe}R_L}{h_{ie}}$$

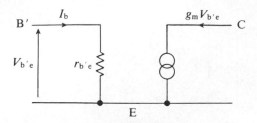

Fig. 3.40   Simplified Hybrid Π Model at MF

From previous comments on the inconsistency of h parameters it might be supposed, for example, given that h_{fe} lies between 50 and 200, then A_v could not be predicted within a maximum to minimum range of $4:1$. However, the physical mechanism of BJTs shows that, for given operating conditions, a BJT with a large h_{fe} will also have a large h_{ie}. Therefore the approximation for A_v given above might be more predictable than suggested by the simple application of h parameters.

When the hybrid Π is shorn of the less important resistances and, for MF analysis, all capacitors, we have the simplified version of Fig. 3.40. It has the same configuration as the simplified h parameter model ($h_{oe} = 0; h_{re} = 0$) in which

$$h_{ie} = r_{b'e}$$

$$h_{fe} I_b = g_m V_{b'e}$$

and as previously demonstrated,

$$\frac{h_{fe} V_{b'e}}{h_{ie}} = g_m V_{b'e}$$

Therefore $g_m = \dfrac{h_{fe}}{h_{ie}}$

It can be shown from semiconductor theory that g_m depends only on $I_{C(Q)}$ and absolute temperature T. It is given by

$$g_m = \frac{I_{C(Q)}}{V_T}$$

where $V_T = \dfrac{T}{11\,000} = 26$ mV at $25\,^{\circ}$C.

Therefore $g_m = \dfrac{I_{C(Q)}}{26\text{ mV}}$

For example, at $I_{C(Q)} = 10$ mA, $T = 25\,^{\circ}$C

$$g_m = \frac{10\text{ mA}}{26\text{ mV}} = 380\text{ mS.}$$

☐ **Example 3.10**

Given $V_{BE(ON)} = 0.8$ V in Fig. 3.41, calculate A_v at 25 °C.

$$V_{BN} \doteq \frac{10}{10 + 39} \times 12 \text{ V} = 2.45 \text{ V}$$

$$V_{EN} = 2.45 - 0.8 = 1.65 \text{ V}$$

$$I_{C(Q)} \doteq I_{E(Q)} = \frac{1.65 \text{ V}}{1 \text{ k}\Omega} = 1.65 \text{ mA}$$

$$g_m = \frac{I_{C(Q)}}{V_T} = \frac{1.65 \text{ mA}}{26 \text{ mV}} = 63 \text{ mS}$$

$$A_v = - g_m R_L = - 63 \times 2.7 = - 170$$

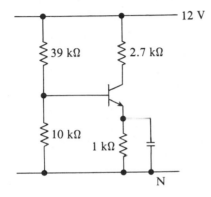

Fig. 3.41 Circuit for Example 3.10 ☐

SUMMARY

Having read this chapter students should:

1. Remember the hybrid (h) equivalent circuit of a BJT operating under linear conditions (Fig. 3.8).
2. Be able to draw the equivalent circuit of an amplifier incorporating the h equivalent circuit.
3. Be able to analyze amplifiers under small-signal conditions using the equivalent circuit.
4. Remember and be able to use decibel notation.
5. Remember the hybrid Π circuit and be able to apply it for the analysis of BJT amplifiers at HF.
6. Be able to calculate upper and lower cutoff frequencies.
7. Be able to use a simplified version of the hybrid Π model for analysis of BJT circuits at MF.

REVISION QUESTIONS

In all cases assume the transistor is biased for linear operation.

3.1   Find V_{ce} in Fig. Q.3.1 given

$h_{ie} = 2$ kΩ; $h_{fe} = 80$; $h_{oe} = 0$; $h_{re} = 0$.

$(-1.2$ V$)$

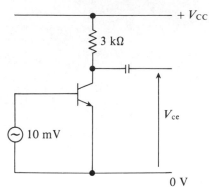

Fig. Q.3.1

3.2   With the same transistor as in Q.3.1, find V_{ce} in Fig. Q.3.2.

$(-0.68$ V$)$

3.3   Repeat Q.3.2 but with $h_{oe} = 100$ μS.

$(-0.53$ V$)$

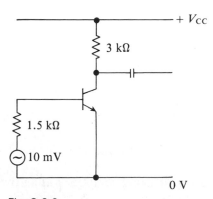

Fig. Q.3.2

3.4   Given $h_{ie} = 3$ kΩ; $h_{fe} = 150$; $h_{oe} = 50$ μS; $h_{re} = 0$, find, in Fig. Q.3.4,

$$\frac{I_c}{I_b}; \frac{V_{ce}}{V_{be}}; A_p = \frac{\text{Power in } 4.7 \text{ k}\Omega}{\text{Power into transistor}}$$

$(120; \; -190; \; 23\,000)$

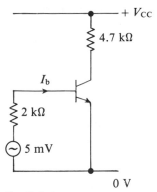

Fig. Q.3.4

3.5   If $h_{oe} = 0; h_{re} = 0$, prove that for a CE amplifier,

$$A_v = \frac{h_{fe}}{h_{ie}} R, \text{ where } R = \text{collector load.}$$

3.6   The hybrid parameters of a transistor in common emitter connection are $h_{ie} = 1200\ \Omega$, $h_{fe} = 45$, $h_{oe} = 30\ \mu S$ and $h_{re} = 0$. The transistor is connected between a source of internal resistance $1500\ \Omega$ and a load of resistance $10\ k\Omega$. Calculate:
(a) the current gain I_c/I_b
(b) the voltage gain V_{ce}/V_{be}
(c) the corresponding power gain.

$$((a)\ -34.6;\ (b)\ -288;\ (c)\ 9.96 \times 10^3)$$

3.7   The simplified equivalent circuit of a transistor amplifier stage is shown in Fig. Q.3.7. Derive expressions for:
(a) the voltage gain V_o/V_1
(b) the current gain I_o/I_1
Calculate the values of (a) and (b) above if the transistor has the following parameters:

$$h_{ie} = 1\ k\Omega;\ h_{fe} = 40;\ h_{oe} = 50\ \mu S.$$

$$((a)\ -160;\ (b)\ -32)$$

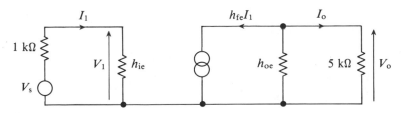

Fig. Q.3.7

3.8   A transistor used in a common emitter circuit has the following parameters: $h_{ie} = 1.2$ kΩ; $h_{fe} = 40$; $h_{oe} = 50$ μS; $h_{re} = 0$. The magnitude of the voltage source is 50 mV and its internal resistance is 600 Ω. Calculate:
(a) the input current;
(b) the current in a 5000 Ω collector load;
(c) the voltage gain of the circuit in dB;
(d) the power gain of the circuit in dB.

                                       ((a) 27.8 μA; (b) 0.89 mA; (c) 42.5 dB; (d) 36.3 dB)

3.9   A transistor is used in a common emitter circuit and delivers an output of 6 V rms across a 5000 Ω load. It has the following parameters:

       $h_{ie} = 1$ kΩ; $h_{fe} = 50$; $h_{oe} = 50$ μS; $h_{re} = 0$. Calculate:
(a) the input current;
(b) the current gain of the stage;
(c) the magnitude of the voltage source if its internal resistance is 500 Ω;
(d) the power gain in dB.

                                     ((a) 30 μA; (b) -40; (c) 45 mV; (d) 39 dB)

3.10   In the simplified transistor amplifier circuit shown in Fig. Q.3.10, the hybrid parameters are $h_{ie} = 800$ Ω, $h_{fe} = 47$, $h_{oe} = 80$ μS, $h_{re} = 0$. The input sinusoidal signal current is 0.1 mA rms. Draw the small-signal equivalent circuit and hence calculate:
(a) the collector signal current;
(b) the current in the 1.5 kΩ output resistor, I_o;
(c) the output voltage, V_o;
(d) the overall voltage amplification in dB.

                             ((a) 3.06 mA; (b) -2.35 mA; (c) -3.53 V; (d) 35.8 dB)

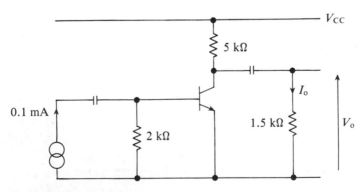

Fig. Q.3.10

3.11   The BJT of Fig. Q.3.11 has h parameters in the ranges shown.
$h_{fe}(\text{min}) = 50$
$h_{fe}(\text{max}) = 250$

$h_{ie}(min) = 1\ k\Omega$
$h_{ie}(max) = 5\ k\Omega$
$h_{oe} = 25\ \mu S$
$h_{re} = 0$

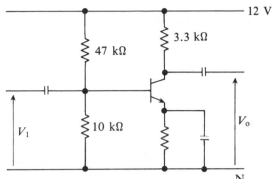

Fig. Q.3.11

The amplifier output is connected to an external resistive load of 2 kΩ. Calculate the maximum and minimum voltage gain in dBs. All capacitors have negligible reactance at the frequency under consideration.

(49.6 dB; 21.7 dB)

3.12 For the circuit shown in Fig. Q.3.12, the transistor has the following small-signal hybrid parameters:

$h_{ie} = 1.5\ k\Omega,\ h_{fe} = 100,\ h_{oe} = 0.1\ mS,\ h_{re} = 0.$

The effect of the capacitors may be assumed to be negligible. Draw the small-signal equivalent circuit and hence determine:
(a) the voltage gain V_0/V_1;
(b) the current gain I_0/I_1;
(c) the power gain of the stage;
(d) the output voltage V_o.

((a) −76; (b) −53.9; (c) 4096; (d) 42.6 mV)

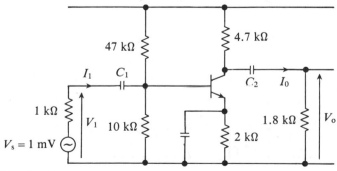

Fig. Q.3.12

3.13   Defining overall voltage gain for Fig. Q.3.12 as V_o/V_s, find a value of C_1 to make the lower cutoff frequency 10 Hz and C_2 to make the lower cutoff frequency 40 Hz.

$$(7 \ \mu F; \ 0.8 \ \mu F)$$

3.14   For Fig. Q.3.14 calculate V_o. Data for both transistors:

$h_{ie} = 2 \ k\Omega, \ h_{fe} = 100, \ h_{oe} = 25 \ \mu S, \ h_{re} = 0.$

$$(0.21 \ V)$$

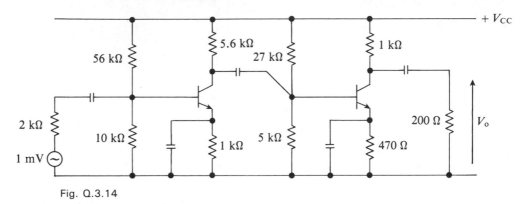

Fig. Q.3.14

3.15   Given that a BJT has $h_{fe} = 50, \ h_{ie} = 1 \ k\Omega, \ C_e = 80 \ pF$, find its beta cutoff and unity gain frequencies.

$$(2 \ MHz; \ 100 \ MHz)$$

3.16   For Fig. Q.3.16 calculate the effective source resistance by evaluating the Thevenin equivalent of the circuit feeding the BJT input. Find the midband gain and using the effective source resistance find the bandwidth, given: $h_{ie} = 1.5 \ k\Omega, \ h_{fe} = 100, \ h_{oe} = 50 \ \mu S, \ C_e = 70 \ pF, \ C_c = 2 \ pF.$

$$(-220; \ 0.4 \ MHz)$$

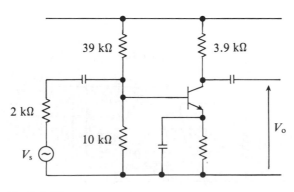

Fig. Q.3.16

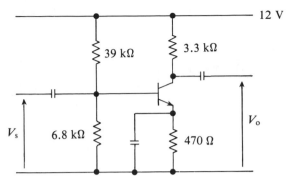

Fig. Q.3.17

3.17 Using the LF hybrid Π equivalent circuit, calculate the voltage gain of Fig. Q.3.17. Assume normal temperature operation, i.e. $25\,^\circ$C.

(-292)

4

Field Effect Transistors

Chapter Objectives

4.1   FET Properties
4.2   FET Operation
4.3   Pinch-off Voltage V_P and Drain Saturation Current I_{DSS}
4.4   Self-bias in a Common Source Amplifier
4.5   FET Small-signal Equivalent Circuit
4.6   MOSFET Operation
4.7   VMOS
        Summary
        Revision Questions

In this chapter there will be a description of the various types of field effect transistor (FET) and an explanation of their characteristics. Many of the principles of amplifiers already explained in the context of BJTs also apply to FETs: this enables biassing and equivalent circuits to be covered as well as a description of the devices themselves within a single chapter. The term FET is now normally used as an abbreviation for junction field effect transistor – the abbreviations JFET and JUGFET having declined in popularity. A more recent device, the metal oxide semiconductor field effect transistor, is invariably abbreviated to MOSFET (occasionally to MOST) to distinguish it from the FET. Although a derivative of the FET there are now more MOSFETs than any other type of transistor.

4.1   FET PROPERTIES

All FETs are unipolar devices although the word 'unipolar' is not used in naming them. It signifies that, unlike the BJT where the movement of both holes and electrons is essential to its operation, a FET uses only one type of carrier in its signal current path. The relative merits of FETs can be summarized as follows.

Advantages
1. Very high input resistance – several megohms for an FET and thousands of megohms for a MOSFET compared with a kilohm or so for a BJT.
2. Less noisy – the passage of charge carriers across a pn junction is a somewhat random process and is therefore inherently noisy. Since collector current in a BJT is produced by charge carriers crossing two pn junctions, the additional noise will be much greater than the corresponding current in an FET where charge carriers do not have to cross any pn junctions at all.
3. Crucially for the development of the high packing density essential to the development of such integrated circuits as microprocessors is the fact that a MOSFET requires very little area on an integrated circuit (IC). This is the reason behind the introductory remark concerning the preponderance of MOSFETs. For example, with very large scale integration (VLSI), ICs containing a million MOSFETs can now be made.
4. FETs are thermally more stable than BJTs.

Disadvantage
At present the BJT retains an advantage where high speed switching and high-power handling capacity are required and in particular, where simultaneously high speed and high power are required. However, FETs and MOSFETs are gaining ground on BJTs. In this context there will be a description of the VMOSFET at the end of the chapter.

4.2   FET OPERATION

Figure 4.1 can be used to describe the operation of an FET although the normal construction is somewhat different. Its basic component is a bar of n type silicon with p regions diffused at the sides. As such it is called an n channel FET. As with the BJT, the complementary pattern is also available – a p type bar with n regions diffused at the sides – a p channel FET. The metal contacts at the ends of the bar are

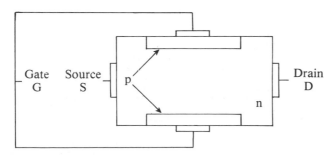

Fig. 4.1   n Channel FET

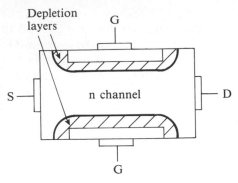

Fig. 4.2   Depletion Layers in an FET

called source S and drain D, and are analogous to emitter and collector in the BJT. The p regions are commoned to a terminal called the gate G (analogous to the BJT base).

Depletion layers will form around the pn junctions (Section 1.5) and in the absence of external supply voltages the depletion layers extend into the bar as shown in Fig. 4.2. Between source and drain then, there is a clear path for conduction to take place in an uninterrupted channel of n type silicon. To investigate the I/V characteristics of this channel the gate is now connected to the source (Fig. 4.3), i.e. $V_{GS} = 0$ V and a dc source is connected between drain and source so that V_{DS} is positive. As soon as V_{DS} is increased, I_D flows and initially increases with V_{DS} in a linear manner. This current consists of electrons flowing from S to D, i.e. conventional current flowing in at D and out at S. With V_{GS} held at 0 V and V_{DS} being positive, the n channel is positive wrt the p gate so the junctions are reverse biased. Therefore none of the current I_D entering the drain goes to the gate, but instead it all comes out of the source. For FETs then, in normal operation, I_D and I_S are equal

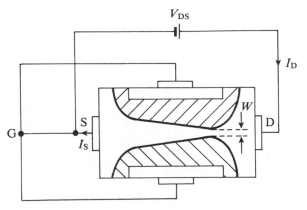

Fig. 4.3   Narrowing of Channel at Drain end of n Channel FET when V_{DS} is Positive

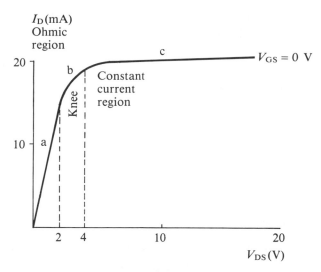

Fig. 4.4 FET Drain Characteristic when $V_{GS} = 0$ V

and it is only necessary to remember that the current flowing out of the source is equal to I_D.

With current now flowing through the channel from D to S the whole channel (n) is now positive wrt the gate (p) but more so at the drain end. This causes the depletion layer to widen near the drain, as shown in Fig. 4.3, an effect which eventually narrows the effective channel width W to the extent that, with increasing V_{DS} there is very little increase in I_D. The effect of these observations can be summarized in the common source I_D/V_{DS} characteristic for $V_{GS} = 0$ V (Fig. 4.4). The characteristic has three identifiable regions:

(a) an ohmic region, where $I_D \propto V_{DS}$ until the 'channel narrowing' effect operates;
(b) a 'knee' – typically between 2 V and 4 V, and therefore higher and less well pronounced than for a BJT, which is a disadvantage of a FET; and
(c) an almost constant-current region where V_{DS} has little effect on I_D. There is also a breakdown region (not shown) where, when V_{DS} is increased above a quoted maximum for an FET, I_D increases rapidly, and the device will be damaged.

A FET only becomes useful if I_D can be controlled by an input signal. In Fig. 4.5, V_{GS} has been made negative. This makes the pn junctions reverse biased and therefore widens the depletion layers before increasing V_{DS} from 0 V. The consequence is to reach the 'knee' and constant-current regions of the drain characteristic at a lower value of V_{DS} than when $V_{GS} = 0$ V. Figure 4.6 shows a set of drain characteristics for various values of V_{GS}. They are similar to the BJT collector characteristics except in one important feature – the output current I_D for the FET is controlled by gate–source *voltage* whereas, for the BJT, I_C is controlled by base *current*: by contrast with a BJT, a FET is clearly *voltage controlled*.

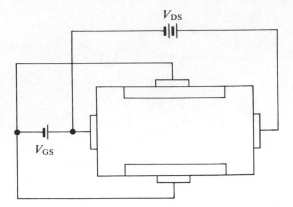

Fig. 4.5   n Channel FET with V_{GS} Negative

Figure 4.7a is the same circuit as Fig. 4.5 with an approved symbol for the n channel FET replacing physical representation. The inward pointing arrow on the gate distinguishes it from the p channel FET symbol in Fig. 4.7b. On this latter diagram it will be noted that all polarities have been reversed. There are other symbols in use for FETs, notably one in which the gate is connected to the middle of the bar. With this symbol drain and source are indistinguishable: this does not

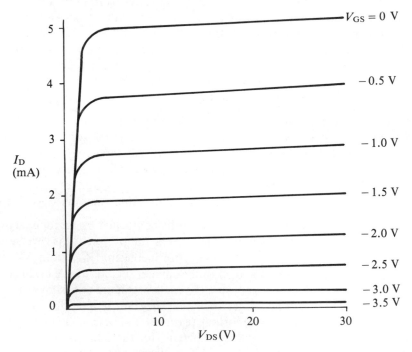

Fig. 4.6   FET Drain Characteristics

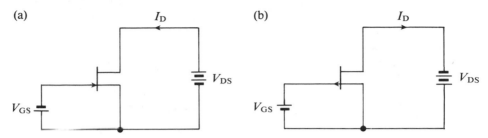

Fig. 4.7   n and p FET Symbols with Normal Voltage and Current Polarities (a) n Channel FET (b) p Channel FET

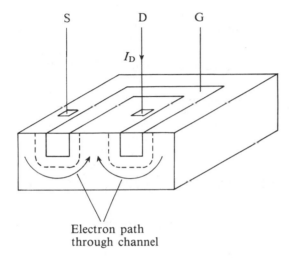

Electron path
through channel

Fig. 4.8   FET Geometry Illustrating Drain and Source as Being Dissimilar

seem unreasonable referring back to Fig. 4.1 which depicts the FET as having a symmetrical geometry. However, Fig. 4.8 is a truer representation of a FET in which there is clearly no drain-to-source symmetry. Even so, the explanation of the formation of a channel whose effective width is depletion layer controlled stands.

4.3   PINCH-OFF VOLTAGE V_P AND DRAIN SATURATION CURRENT I_{DSS}

Referring to the drain characteristics of Fig. 4.6 it may be inferred that any significant reduction of V_{GS} below -3.5 V would prevent the flow of any drain current at all. At such a voltage the depletion layers close the channel completely – it is said to be *pinched off*. By decreasing V_{GS} until this effect occurs, i.e. I_D falls to zero at some positive value of V_{DS}, we find V_P, the pinch-off voltage.

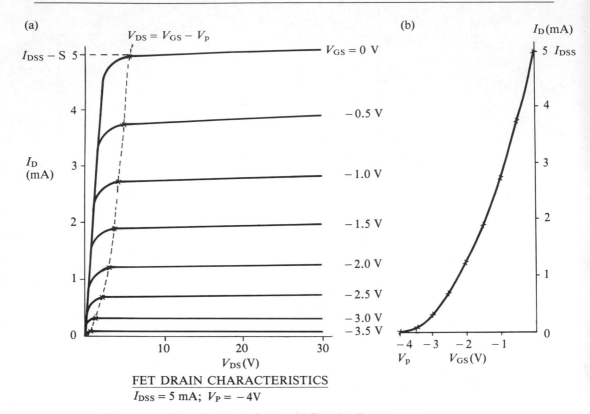

FET DRAIN CHARACTERISTICS
$I_{DSS} = 5$ mA; $V_P = -4$V

Fig. 4.9   (a) Drain Saturation Current (b) Transfer Characteristic

Redrawing the previous set of drain characteristics in Fig. 4.9a, one can also note on the $V_{GS} = 0$ V characteristic the value of I_D at which the characteristic has just leveled off to its 'constant current' region. This current is the *drain saturation current* I_{DSS}. For each characteristic, the value of V_{DS} at which the constant current region begins is given by

$$V_{DS} = V_{GS} - V_P \tag{4.1}$$

For an n channel FET both quantities on the right-hand side of this equation are negative, so for the $V_{GS} = -1$ V characteristic, then taking $V_P = -4$ V the constant-current region begins at

$$V_{DS} = -1 \text{ V} - (-4 \text{ V}) = +3 \text{ V}$$

These points are joined together by a dotted line in Fig. 4.9a.

FET Transfer Characteristic I_D vs. V_{DS}

Provided the FET is being operated to the right of the dotted line in Fig. 4.9a — which it must be for linear operation — then V_{DS} has little effect on I_D, which is therefore almost completely controlled by V_{GS}. The term 'constant current' implies independence of V_{DS} but not of V_{GS}. Therefore, drawing a vertical line in Fig. 4.9a at, say, $V_{DS} = 10$ V, we can read off the values of V_{GS} required to produce corresponding values of I_D: these have been plotted on the transfer characteristic of Fig. 4.9b.

Three features should be noted on this characteristic:

(a) By definition it crosses the I_D axis ($V_{GS} = 0$ V) at I_{DSS}.
(b) By definition it crosses the V_{GS} axis ($I_D = 0$ mA) at V_p.
(c) It is not linear.

This last feature could have been deduced from the drain characteristics which are clearly less evenly spaced than the collector characteristics of a BJT. The nature of this nonlinearity of the transfer characteristic can be derived from a quantitative study of semiconductor physics. Here the transfer characteristic is simply quoted as

$$I_D = I_{DSS}\left(1 - \frac{V_{GS}}{V_P}\right)^2 \tag{4.2}$$

Unlike the BJT there are no input characteristics for FETs simply because there is no input (i.e. gate) current. The load line technique can be used to illustrate the operation of a common source (CS) amplifier and to establish required quiescent conditions.

☐ **Example 4.1**

For the circuit of Fig. 4.10 the FET has the drain characteristics of Fig. 4.11.

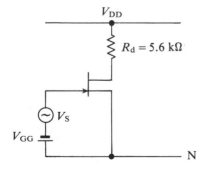

Fig. 4.10   Circuit for Example 4.1

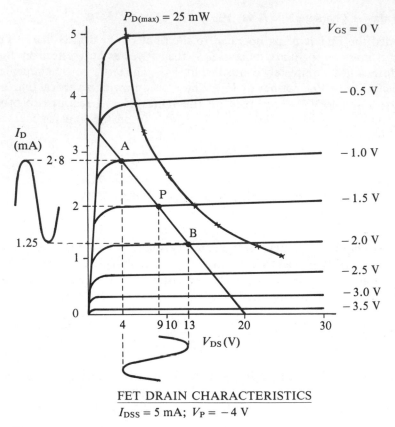

FET DRAIN CHARACTERISTICS
$I_{DSS} = 5$ mA; $V_P = -4$ V

Fig. 4.11   Drain Characteristics for Example 4.1

Maximum ratings for the FET are:

$$V_{DS(max)} = 30 \text{ V}$$
$$I_{D(max)} = 10 \text{ mA}$$
$$P_{D(max)} = 25 \text{ mW}$$

Choose a suitable value of V_{DD}, a suitable operating point, and estimate the voltage gain A_v.

First plot on the drain characteristics of Fig. 4.11 the maximum power dissipation curve,

$$P_D = V_{DS} \times I_D = 25 \text{ mW}$$

The load line equation for the circuit corresponds, in form, to equation (2.4) established for the BJT, i.e.,

$v_{DS} = V_{DD} - i_D R_d$ from which,

$$i_D = \frac{V_{DD}}{R_d} - \frac{v_{DS}}{R_d} \tag{4.3}$$

To satisfy equation (4.3) the load line must have a slope of $-1/5.6$ kΩ. To satisfy the FET maximum ratings it must, at all points, be below the maximum power dissipation curve and also below $V_{DS(max)} = 30$ V and $I_{D(max)} = 10$ mA. The load line drawn in Fig. 4.11 satisfies all of these conditions. It cuts V_{DS} at 20 V so this will be the value of the supply voltage V_{DD}. To select an operating point, working from the characteristic $V_{GS} = -1.5$ V allows v_{GS} to be varied by at least 0.5 V in each polarity without nonlinearity, i.e. without going out of the constant-current region of the characteristics when v_{GS} is increasing and without approaching pinch-off when v_{GS} is decreasing. With the operating point P at $V_{GS(Q)} = -1.5$ V, then $I_{D(Q)} = 2$ mA and $V_{DS(Q)} \doteq 9$ V. For an input signal $v_s = 0.5 \sin \omega t$ (V), the amplifier is fluctuating between points A and B which gives the following limits for i_D and v_{DS}:

At A, $i_D = 2.8$ mA; $v_{DS} \doteq 4$ V

At B, $i_D = 1.25$ mA; $v_{DS} = 13$ V

Therefore $A_v = \dfrac{\Delta v_{DS}}{\Delta v_{GS}} = \dfrac{13\text{ V} - 4\text{ V}}{-2\text{ V} - (-1\text{ V})} = -9$ $\qquad\square$

This example highlights the similarities between FET and BJT amplifiers: biassing the device to the most linear parts of the characteristics and control of output current through an external load resistor resulting in an output voltage signal in antiphase with the input signal. In the case of the FET, however, there is no current or power gain or, if preferred, A_i and A_p are infinite. Again the use of drain characteristics is less useful for gain calculation than the use of equivalent circuits. The FET small-signal equivalent circuit will be described in Section 4.5.

4.4 SELF-BIAS IN A COMMON SOURCE AMPLIFIER

Figure 4.12 is the simplest self-bias circuit for a CS amplifier. The function of R_g is to hold the gate voltage V_{GN} at 0 V. In effect, the gate is an open circuit so if, for example, electrons were to build up on it, its voltage would become negative and eventually pinch off the FET. The resistor R_g allows charge to leak away from the gate to ground. As the leakage current through R_g will be of the order of nano-amps, a value as high as 1 MΩ will develop a gate-to-ground voltage V_{GN} of less than 0.1 V. Thus with any value of R_g up to 1 MΩ we can take it that $V_{GN} = 0$ V.

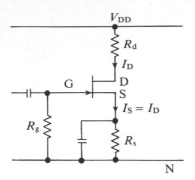

Fig. 4.12   Self-bias in a CS Amplifier

As gate current is of the order of nanoamps it can be neglected. As noted, I_D and I_S are virtually equal so that when I_D flows, a voltage $I_D R_s$ will be developed across R_s which makes the source positive, i.e.

$$V_{SN} = I_D R_s.$$

Since $V_{GN} = 0$ V then $V_{GS} = -I_D R_s$, or

$$R_s = \frac{-V_{GS}}{I_D}.$$

There are no absolute voltage requirements, only relative ones, and by making the gate 0 V wrt N and the source positive then V_{GS} will indeed be negative. Referring back to Example 4.1, the quiescent requirement is:

$$I_{D(Q)} = 2 \text{ mA}; \ V_{GS(Q)} = -1.5 \text{ V}$$

This will be achieved by making

$$R_s = \frac{1.5 \text{ V}}{2 \text{ mA}} = 750 \ \Omega.$$

Connection of a capacitor across R_s prevents loss of gain by feedback. One error introduced by this self-bias circuit is that with the source held positive, in the example by 1.5 V, the effective supply voltage V_{DD} is reduced by 1.5 V. However, the error is small and if required, can be allowed for by drawing the load line from V_{DD} to V_{SN}, i.e. from 18.5 V.

Use of Transfer Characteristic to Determine Quiescent Condition

Given the requirement to achieve a certain operating point, the transfer characteristic can be used to find a suitable value of the source resistor R_s. This can be illustrated by the following example.

□ **Example 4.2**

In Fig. 4.13 it is required that $V_{DN(Q)} = 12$ V. The FET data are $I_{DSS} = 6$ mA; $V_P = -4$ V. Choose values for R_s and R_g.

Voltage across $R_d = V_{DD} - V_{DN(Q)} = 8$ V

Therefore $I_{D(Q)} = \dfrac{8 \text{ V}}{5.6 \text{ k}\Omega} = 1.4$ mA

The transfer characteristic can be obtained using equation (4.2) and the values of I_{DSS} and V_p given. It is drawn in Fig. 4.14.

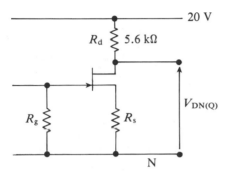

Fig. 4.13   Circuit for Example 4.2

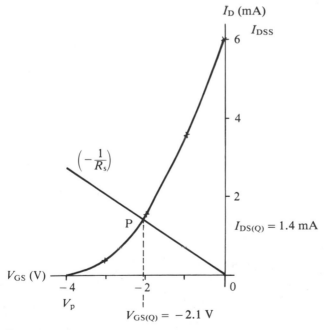

Fig. 4.14   Transfer Characteristic and Load Line for Example 4.2

Marking $I_{D(Q)} = 1.4$ mA, it can be seen that

$V_{GS(Q)} = -2.1$ V

Therefore $R_s = \dfrac{-V_{GS(Q)}}{I_{D(Q)}} = 1.5$ kΩ.

A suitable value for R_g would be 470 kΩ in the absence of any specific data about gate current or any specific requirement for input resistance. □

The line joining the origin and the working point P in Fig. 4.14 is given by

$-V_{GS} = I_D R_s$

As with all load lines it is totally independent of the FET and simply expresses the fact that if $R_s = 1.5$ kΩ, then

$-V_{GS} = I_D \times 1.5$ kΩ.

The transfer characteristic, on the other hand, is exclusively a function of the FET. As they coexist in the same circuit, the point of intersection is the operating point.

Numerical Application of Transfer Characteristic to Determine Quiescent Condition

If I_{DSS} and V_p of an FET are known, then using the transfer characteristic equation (4.2), calculation concerning the quiescent condition can be carried out without using graphs.

□ **Example 4.3**

Confirm the results obtained in Example 4.2 by using the transfer characteristic equation.

As before, $I_{D(Q)} = 1.4$ mA

$$I_D = I_{DSS}\left(1 - \frac{V_{GS}}{V_P}\right)^2 \qquad \text{(equation (4.2) repeated)}$$

$$1.4 = 6\left(1 - \frac{V_{GS}}{-4}\right)^2$$

from which

$V_{GS} = -\ 2.08$ V

Therefore $R_s = \dfrac{2.08}{1.4} \doteqdot 1.5$ kΩ (as before) □

Bias Circuit to Accommodate Spread of I_{DSS} and V_P

As with BJTs, manufacturers of FETs do not give precise values for I_{DSS} and V_P

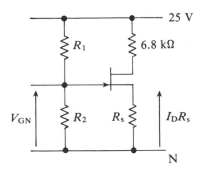

Fig. 4.15 Circuit for Example 4.4

for a particular type of FET, but rather quote maximum and minimum values. For some FET types the spread can be extremely wide and sometimes they are coded by colored dots into bands of I_{DSS} and V_P. The designer can choose either to take average values and use the techniques described or decide on the tolerance to be permitted and design a circuit which guarantees an acceptable result irrespective of whether I_{DSS} and V_P are maximum or minimum. A circuit which can be designed to accommodate a spread of I_{DSS} and V_P is shown in Fig. 4.15. Its action, which is quite different from the apparently similar BJT emitter bias circuit, will be described by the following example.

☐ **Example 4.4**

In Fig. 4.15 it is required that $V_{DN(Q)}$ lies between 12 V and 15 V. Choose suitable values of R_1, R_2 and R_s given the following FET data:

$$I_{DSS(max)} = 6 \text{ mA}; \; V_P = -5 \text{ V}$$
$$I_{DSS(min)} = 3 \text{ mA}; \; V_P = -3 \text{ V}$$

First calculate $I_{D(Q)\,(max)}$ and $I_{D(Q)\,(min)}$.

When $V_{DS(Q)} = 12$ V, $I_{D(Q)\,(max)} = \dfrac{13 \text{ V}}{6.8 \text{ k}\Omega} = 1.9 \text{ mA}$

When $V_{DS(Q)} = 15$ V, $I_{D(Q)\,(min)} = \dfrac{10 \text{ V}}{6.8 \text{ k}\Omega} = 1.5 \text{ mA}$

Then plot the two transfer characteristics, one for $I_{DSS} = 6$ mA; $V_P = -5$ V (A), and one for $I_{DSS} = 3$ mA; $V_P = -3$ V (B) (Fig. 4.16).

It is clear that $I_{D(Q)}$ will be maximum for a FET with curve A and minimum for a FET with curve B. Then mark $I_{D(Q)\,(max)}$ at P on curve A and $I_{D(Q)\,(min)}$ at Q on curve B. The operating point must lie on the line joining P and Q and will be within the permissible tolerance. Extended, the line cuts the V_{GS} axis at $+3.6$ V.

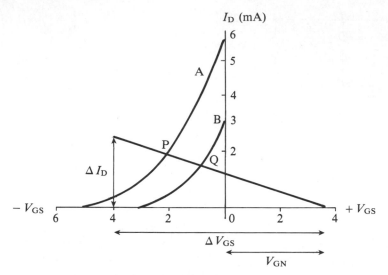

Fig. 4.16   Transfer Characteristics for Example 4.4

Referring to Fig. 4.15, the voltage across $R_2 = V_{GN}$.

Therefore $V_{GS} = V_{GN} - I_D R_s$

This is the equation of the line joining P, Q and the V_{GS} axis.

When $I_D = 0$, $V_{GS} = V_{GN} = +3.6$ V.

The slope of the line is

$$\frac{\Delta I_D}{\Delta V_{GS}} = \frac{2.5 \text{ mA}}{-7.6 \text{ V}} \doteq \frac{-1}{3 \text{ k}\Omega}$$

Therefore $R_s = 3$ kΩ (use nearest preferred value = 3.3 kΩ).
Now it is only necessary to choose R_1 and R_2 to make $V_{GN} = +3.6$ V, which is simple, as $I_G = 0$. They should be high values if R_{in} is required to be high, but not above 1 MΩ if V_{GN} is not to be significantly in error due to the few nanoamps of I_G.

$$V_{GN} = \frac{R_2}{R_1 + R_2} \, 25 = 3.6 \text{ V}$$

Therefore $R_1 \doteq 6\, R_2$

If $R_2 = 82$ kΩ; $R_1 = 490$ kΩ (use 470 kΩ).   □

FET as a Voltage-controlled Resistance

In Section 4.2 it was noted that the drain characteristics of an FET are virtually linear at low values of V_{DS} and are therefore said to be 'ohmic'. Figure 4.17 is a set

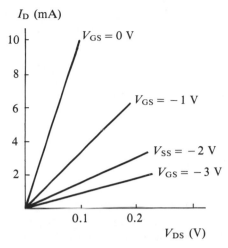

Fig. 4.17 Ohmic Region of FET Drain Characteristics

of FET drain characteristics which have only been drawn for low values of V_{DS} so that the ohmic region is pronounced. It shows that not only are the characteristics linear, but also that their slope changes with V_{GS}. Therefore, between drain and source, an FET behaves like a resistance,

$$r_{DS} - \frac{V_{DS}}{I_D}$$

whose value rises as V_{GS} becomes more negative. Figure 4.18 is a graph of r_{DS} vs. V_{GS} extracted from Fig. 4.17.

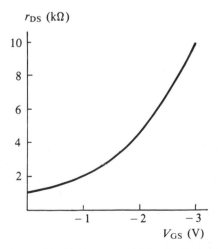

Fig. 4.18 Resistance vs. V_{GS} for FET Operated in Ohmic Region of Drain Characteristics

A FET operated as a voltage-controlled resistance (VCR) is useful as a control component. For example, where it is required to control amplifier gain so that the signal output is constant, a VCR can be used as one of the gain-determining components of the amplifier, and when the output rises the VCR value is altered to reduce the gain. This is an example of automatic gain control (AGC) which is commonly used in radio receivers to maintain output constant as radio signal strength varies.

4.5   FET SMALL-SIGNAL EQUIVALENT CIRCUIT

Having developed techniques for establishing linear operating conditions, and having used a load line to demonstrate a graphical approach to finding voltage gain, the next step is to develop an equivalent circuit to represent an FET operating under small-signal linear conditions for the analysis of FET amplifiers. Since an FET takes no input (gate) current, the equivalent circuit between gate and source is simply an open circuit. It is therefore only necessary to develop an output (drain–source) equivalent circuit which gives the signal drain current I_d as a function of signal input voltage V_{gs} (i.e. the signal voltage connected across the gate–source open circuit) and the output signal voltage V_{ds}. An equation, then, from which the drain–source equivalent circuit can be developed is

$$I_d = g_{fs}V_{gs} + g_{ds}V_{ds}$$

but more often, probably because of long-standing practice inherited from valve equivalent circuits,

$$I_d = g_{fs}V_{gs} + \frac{1}{r_{ds}}V_{ds} \tag{4.4}$$

The two FET small-signal parameters contained in this equation are

g_{fs} – the transconductance or mutual conductance (sometimes labeled g_m)

and

$$r_{ds}\left(=\frac{1}{g_{ds}}\right) - \text{the drain resistance.}$$

Equation (4.4) is realized by the circuit of Fig. 4.19. It includes the gate as an isolated terminal. However, when an external signal is connected between G and S, then this voltage, V_{gs}, will largely determine the drain signal current I_d.

To define, first, g_{fs}, make $V_{ds} = 0$, i.e. $v_{DS} =$ constant. Then from equation (4.4),

$$g_{fs} = \frac{I_d}{V_{gs}}\bigg|_{V_{ds}=0}$$

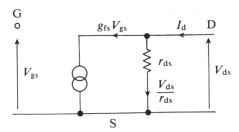

Fig. 4.19 Small-signal Equivalent Circuit

Holding v_{DS} constant at 15 V on the drain characteristics of Fig. 4.20,

$$g_{fs} = \frac{(3.95 - 2.85)\ \text{mA}}{(-0.5 - (-1))\ \text{V}} = 2.2\ \text{mS}.$$

To find r_{ds}, make $V_{gs} = 0$ in equation (4.4), i.e. $v_{GS} = $ constant. Then,

$$r_{ds} = \frac{V_{ds}}{I_d}\bigg|_{V_{gs}=0}$$

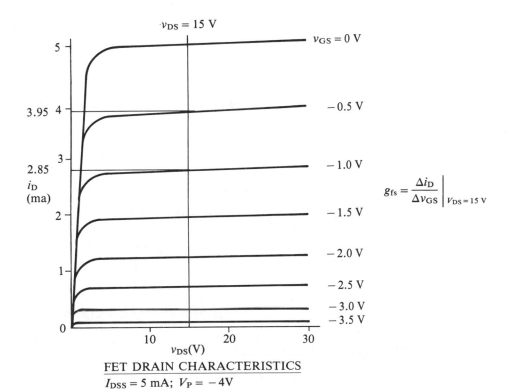

FET DRAIN CHARACTERISTICS

$I_{DSS} = 5\ \text{mA};\ V_P = -4\text{V}$

Fig. 4.20

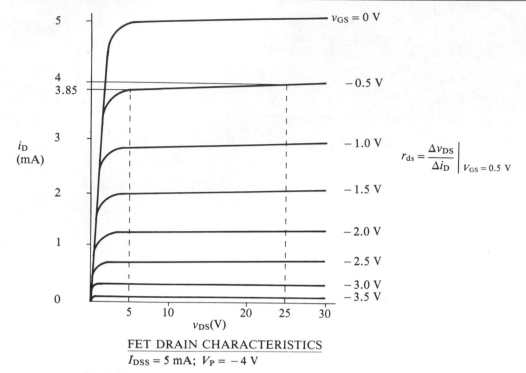

$$r_{ds} = \frac{\Delta v_{DS}}{\Delta i_D} \bigg|_{V_{GS} = 0.5\ V}$$

FET DRAIN CHARACTERISTICS
$I_{DSS} = 5\ mA;\ V_P = -4\ V$

Fig. 4.21

Holding v_{GS} constant at $-0.5\ V$ on the drain characteristics of Fig. 4.21,

$$r_{ds} = \frac{(25 - 5)\ V}{(4.0 - 3.85)\ mA} = 130\ k\Omega.$$

Analyzing FET Amplifier Equivalent Circuits

The rules established for drawing BJT amplifier equivalent circuits apply equally to FETs, i.e. treat dc power supplies and capacitors as short-circuits at MF and then connect all external components to the FET terminals.

☐ **Example 4.5**

Develop an expression for the voltage gain of the CS amplifier of Fig. 4.22 and hence determine V_o.

FET data: $g_{fs} = 5\ mS;\ r_{ds} = 50\ k\Omega.$

The equivalent circuit is shown in Fig. 4.23.

The only current source in the output circuit is $g_{fs}V_{gs}$ which flows into $R_d \parallel r_{ds}$.

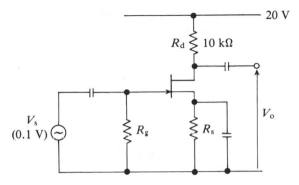

Fig. 4.22   Circuit for Example 4.5

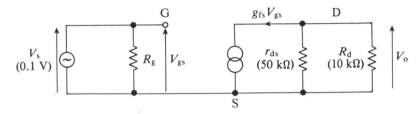

Fig. 4.23   Equivalent Circuit of Fig. 4.22

Therefore $V_o = -g_{fs}V_{gs} \times \dfrac{R_d r_{ds}}{R_d + r_{ds}}$

In this particular circuit $V_{gs} = V_s$

Therefore $A_v = \dfrac{V_o}{V_s} = \dfrac{-g_{fs}R_d r_{ds}}{R_d + r_{ds}}$ (4.5)

$\qquad = -g_{fs}R$

where $R = R_d \parallel r_{ds}$ □

In Fig. 4.23

$\qquad R = \dfrac{50 \times 10}{60} = 8.3 \text{ k}\Omega.$

$\qquad A_v = -5 \times 10^{-3} \times 8.3 \times 10^3 = -41.5$

Therefore $V_o = -4.15$ V

A useful approximation when $r_{ds} \gg R_d$ is that

$\qquad R \doteqdot R_d$

and $A_v = -g_{fs}R_d$ (4.6)

LIBRARY
LAUDER COLLEGE

In the last example this approximation would give a 20% error, but with $R_d = 5$ kΩ and $r_{ds} = 100$ kΩ the error would only be 5%. Frequently manufacturers do not quote a value of r_{ds} on FET data sheets, the assumption being that for most applications it is high enough to be ignored.

Introducing an external load resistance and a source with finite output resistance leads to slightly more complicated problems.

☐ **Example 4.6**

For Fig. 4.24 calculate the voltage gain (V_o/V_s) in dB and choose a value for C_1 which gives an LF cutoff of 20 Hz, and a value for C_2 which gives an LF cutoff of 50 Hz.

FET data:

$g_{fs} = 6$ mS,

$r_{ds} = 60$ kΩ.

The equivalent circuit is drawn in Fig. 4.25 and includes C_1 and C_2 whose values have to be calculated.

Let $R_B = 120$ kΩ ‖ 220 kΩ = 78 kΩ.

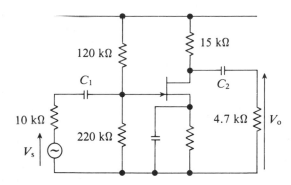

Fig. 4.24 Circuit for Example 4.6

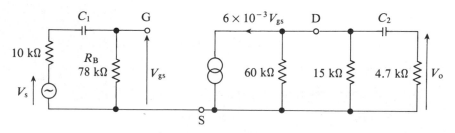

Fig. 4.25 Equivalent Circuit of Fig. 4.24

At MF,

$$V_{gs} = \frac{78}{78 + 10} \, V_s = 0.88 \, V_s.$$

$$R_{ds} = 60 \text{ k}\Omega \parallel 15 \text{ k}\Omega \parallel 4.7 \text{ k}\Omega$$

$$G_{ds} = \frac{1}{R_{ds}} = 16.7 \, \mu S + 67 \, \mu S + 213 \, \mu S = 297 \, \mu S$$

$$V_o = - g_{fs} V_{gs} R_{ds} = - g_{fs} \times 0.88 \, V_s \frac{1}{G_{ds}}$$

$$\frac{V_o}{V_s} = \frac{-6 \times 10^{-3} \times 0.88}{297 \times 10^{-6}} = -17.7$$

Therefore A_v (dB) $= 20 \log_{10} 17.7 = 25.0$ dB.

To find the value of C_1 to give LF cutoff at 20 Hz, refer to Fig. 4.26. Then $X_{C_1} = 88$ kΩ at 20 Hz.

$$\text{i.e. } 88 \times 10^3 = \frac{1}{2\pi f C_1}$$

$$C_1 = 0.09 \, \mu F$$

To find the value of C_2 to give LF cutoff at 50 Hz refer to Fig. 4.27

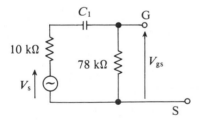

Fig. 4.26   Circuit for Finding Value of C_1 to give LF Cutoff at 20 Hz

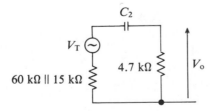

Fig. 4.27   Circuit for Finding Value of C_2 to give LF Cutoff at 50 Hz

Then $X_{C_2} = (60 \text{ k}\Omega \parallel 15 \text{ k}\Omega) + 4.7 \text{ k}\Omega$ at 50 Hz

$= 16.7 \text{ k}\Omega$

$C_2 = 0.19 \, \mu\text{F}.$ □

4.6   MOSFET OPERATION

The letters MOSFET stand for metal-oxide-semiconductor field effect transistor. There are two basic types – enhancement and depletion, with p and n versions of each basic type. An n channel enhancement MOSFET is shown in Fig. 4.28. Two heavily doped n regions, which form the drain and source, have been diffused into a p type substrate. On top of the substrate is a very thin but highly insulating layer of SiO_2 with slots to give access to drain and source. Aluminum is then added in three sections as shown. The two outer sections are the contact points to drain and source. The middle section covers the section of substrate between drain and source and is the gate. Being insulated from the substrate ensures (a) that whatever the polarity between gate and substrate, the gate takes no current, and (b) that the input resistance is very high. In addition to the drain, source and gate terminals associated with the FET, the MOSFET also has the substrate which is usually internally con-

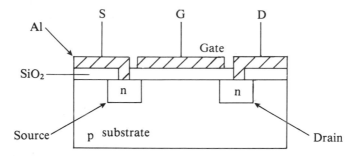

Fig. 4.28   n Channel Enhancement MOSFET

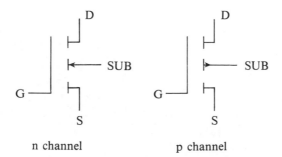

Fig. 4.29   Enhancement MOSFET Symbols

nected to the source. A p channel MOSFET is formed by interchanging p and n materials. Symbols for n and p channel enhancement MOSFETs are given in Fig. 4.29.

n Channel Enhancement MOSFET Operation

The gate, insulating layer, and substrate form a capacitor. When source and substrate are earthed, and also the gate is earthed, i.e. $V_{GS} - 0$ V, the capacitor is discharged. With n type source and drain separated by p substrate there is no path for current to flow when V_{DS} is made positive. When V_{GS} is made positive the gate is positively charged and corresponding negative charges, i.e. electrons, are attracted into the region of substrate between drain and source as shown in Fig. 4.30. These electrons will either be minority carriers from the substrate or majority carriers from

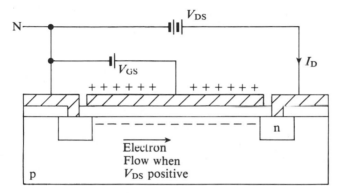

Fig. 4.30 Formation of a Channel of electrons Between Drain and Source by making V_{GS} Positive

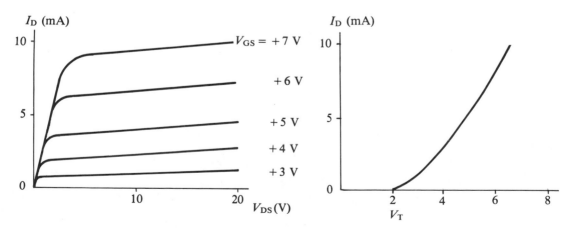

Fig. 4.31 Drain and Transfer Characteristics of an n Channel Enhancement MOSFET

drain and source. They will form a continuous channel of electrons between drain and source so that when V_{DS} is made positive I_D flows as illustrated in Fig. 4.30.

n Channel Enhancement MOSFET Characteristics

Drain and transfer characteristics of an n channel MOSFET are shown in Fig. 4.31. Referring first to the transfer characteristic, it is not until V_{GS} has reached V_T – the threshold voltage – that there will be enough electrons in the channel to produce any significant drain current. Once V_{GS} exceeds V_T then, as the drain characteristics show, it only requires V_{DS} to reach a relatively small value (say 1 V to 3 V) before I_D becomes almost independent of V_{DS}.

Biassing an n Channel Enhancement MOSFET

Unlike the n channel FET, V_{GS} for the n channel enhancement MOSFET must be positive for use as an amplifier. This can be achieved by the circuit of Fig. 4.32 which must be designed so that

$$V_{GN} > V_{SN}$$

Indeed, the circuit can be designed by omitting R_s and connecting source to ground.

An alternative circuit is shown in Fig. 4.33 where R_1 makes the gate voltage equal to the drain voltage. Since the gate takes no current, R_1 can be a very high resistance, i.e. several megohms. To determine the operating point consider the transfer characteristic of Fig. 4.34. Since $V_{GS} = V_{DS}$ then the axes are the same as for the drain characteristics, so the load line for Fig. 4.33, i.e. $V_{DD} = 15$ V, $R_d = 5$ kΩ, can be drawn in as shown. Then the point of intersection between the transfer characteristic and the load line is the only possible point of coexistence and will be the operating point (6 V, 1.8 mA).

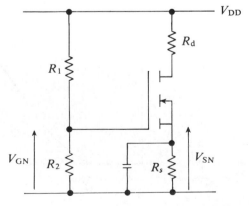

Fig. 4.32   n Channel Enhancement MOSFET Self-bias Circuit

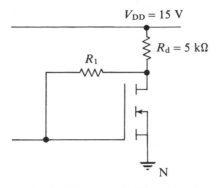

Fig. 4.33   Alternative Self-bias Circuit

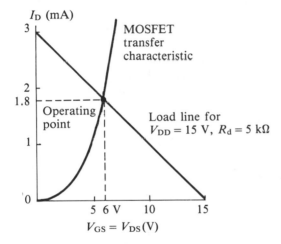

Fig. 4.34   Diagram to Find Operating Point for Fig. 4.31

n Channel Depletion MOSFET

The difference between n channel depletion and enhancement MOSFETs is that the depletion device is manufactured with a lightly doped n channel between drain and source, as shown in Fig. 4.35, so that even when $V_{GS} = 0$ V, drain current will flow when V_{DS} is made positive. (For the p channel depletion MOSFET the channel will be lightly p doped.) This enables the device to be operated in two modes. First, if V_{GS} is made negative then the channel conductivity will fall because it will be depleted of electrons. Alternatively, if V_{GS} is made positive, conductivity will rise because the device operates in the same way as the enhancement MOSFET. Drain and transfer characteristics are as shown in Fig. 4.36. The n channel depletion MOSFET can therefore be operated with positive, negative or zero bias. Positive

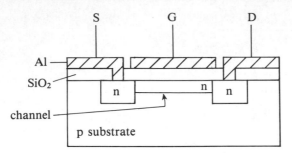

Fig. 4.35   n Channel Depletion MOSFET

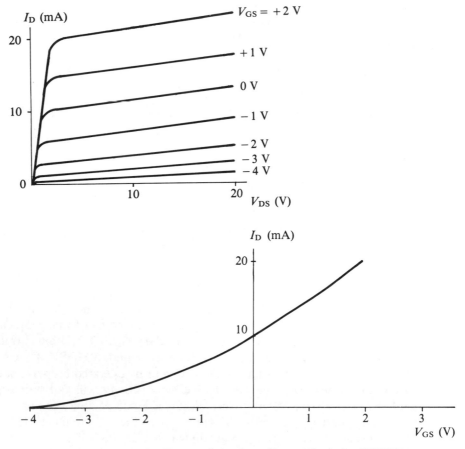

Fig. 4.36   Drain and Transfer Characteristics for n Channel Depletion MOSFET

bias can be provided as for the n channel enhancement MOSFET and negative bias as for the n channel FET.

MOSFET Gate Protection

The MOSFET SiO_2 layer is not only of a very high resistance but also extremely thin. It is therefore possible for charge to build up on the gate when it is unconnected, and for this charge to produce an electric field across the SiO_2 layer high enough to break it down irreparably. Some MOSFETs have a built-in zener diode between gate and source which limits the voltage between them to a safe level. MOSFETs without this internal protection are supplied with the leads clipped together, or embedded in some conducting material. The leads should not be separated until the MOSFET is connected into circuit.

MOSFET Equivalent Circuit

The MF small-signal equivalent circuit is the same as the FET equivalent circuit of Fig. 4.19.

FET HF EQUIVALENT CIRCUIT

To evaluate the performance of FETs and MOSFETs at HF, it is necessary to add the three capacitances shown in Fig. 4.37, i.e.

Capacitance between gate and source $- C_{gs}$
Capacitance between gate and drain $- C_{gd}$
Capacitance between drain and source $- C_{ds}$

As with the hybrid Π equivalent circuit of the BJT, C_{gd} is likely to cause the greatest HF loss, not because it may be larger than the other two, but because its impact is amplified by Miller effect (see Section 3.9). Using the approximation that

$$A_v = - g_{fs}R_d$$

where R_d = drain load resistance, then the effect of C_{gd} is to appear as a capacitance $(1 + g_{fs}R_d)C_{gd}$ between gate and source.

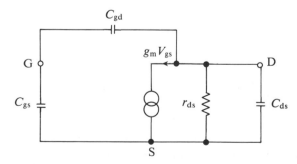

Fig. 4.37   HF Equivalent Circuit of FET

4.7 VMOS

As power amplifiers and as switches, FETs and MOSFETs have two disadvantages:

(a) the channel length is such that, when conducting, the resistance between drain and source is undesirably high.
(b) capacitance between electrodes reduces switching speed.

The VMOS FET (usually just VMOS) in which 'V' stands for 'vertical', is better in regard to both of these factors and is being increasingly used for both power and switching applications. In the VMOS of Fig. 4.38 the effective channel length is the depth of the p region as marked by L_c. When V_{GS} is positive electrons will be induced in this region and provide a path for current to flow 'vertically' from source through the channel and into the substrate which is the drain terminal.

Although the construction essentially forms two transistors, one each side of the V-shaped groove, they can be paralleled, thus halving the channel resistance. Further, the large drain area simplifies the task of making efficient contact with a heat sink.

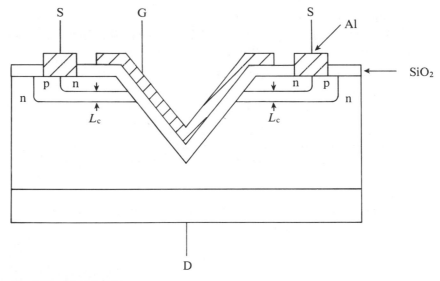

Fig. 4.38   VMOS FET

SUMMARY

The main points covered in this chapter are:

1. Descriptions of the operation of n channel FETs and both n channel depletion and n channel enhancement MOSFETs. From these descriptions it is possible to deduce the operation of the complementary (i.e. p channel) devices.

2. From the description of the devices typical drain and transfer characteristics have been drawn.
3. The transfer characteristic equation of a FET has been stated.
4. Self-bias circuits have been explained and the transfer characteristic equation has been used to determine quiescent conditions.
5. The small-signal equivalent circuit for all FET devices has been developed and used for the analysis and design of common source amplifiers.

REVISION QUESTIONS

4.1 For Fig. Q.4.1 determine $I_{D(Q)}$ and $V_{DS(Q)}$ using the characteristics of Fig. 4.6.

(2.8 mA; 11 V)

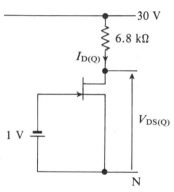

Fig. Q.4.1

4.2 For Fig. Q.4.2 it is required that $V_{DS(Q)} = 12$ V. Given $V_P = -5$ V, $I_{DSS} = 4$ mA, find $V_{GS(Q)}$.

(−1.8 V)

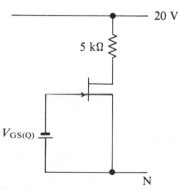

Fig. Q.4.2

4.3 To make $I_D = 3$ mA in Fig. Q.4.3 choose suitable values for R_s and R_g given $V_P = -4$ V, $I_{DSS} = 6$ mA.

(390 Ω; 470 kΩ)

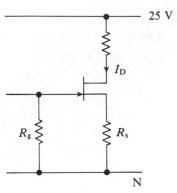

Fig. Q.4.3

4.4 What is the value of $I_{D(Q)}$ in Fig. Q.4.4 given that $I_{DSS} = 8$ mA, $V_p = -4$V? (*Hint*: write the transfer characteristic as $I_D = I_{DSS}\left(1 - \dfrac{-I_D \times 500}{V_P}\right)^2$, solve for I_D and take the practical value.)

(3.1 mA)

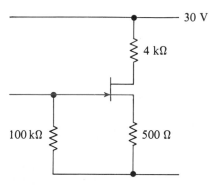

Fig. Q.4.4

4.5 For the circuit shown in Fig. Q.4.5, determine the small-signal voltage gain V_o/V_1. The FET parameters are $g_{fs} = 2$ mS, $r_{ds} = 100$ kΩ.

(−66)

4.6 For the circuit shown in Fig. Q.4.6, determine the small-signal output voltage V_o. The FET parameters are $g_{fs} = 2$ mS and $g_{ds} = 0.02$ mS.

(19.4 mV)

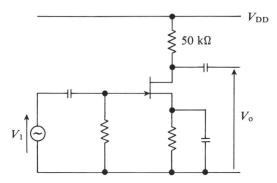

Fig. Q.4.5

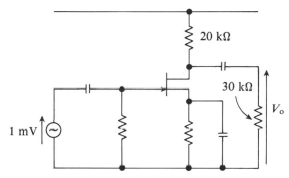

Fig. Q.4.6

4.7 It is required to bias the circuit of Fig. Q.4.7 so that the drain current
= 0.8 mA, $V_{DN(Q)} = 12$ V.
Determine the values of R_s and R_L for Fig. Q.4.7 given that:

$$V_P = -2 \text{ V}, \quad I_{DSS} = 1.65 \text{ mA}, \quad g_{fs} = 2 \text{ mS}, \quad r_{ds} \gg R_L.$$

Find the small-signal voltage gain.

(750 Ω; 15 kΩ; −30)

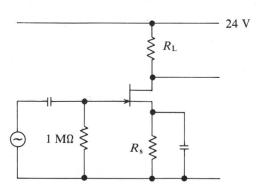

Fig. Q.4.7

4.8   The n-channel FET in Fig. Q.4.8 has the parameters $I_{DSS} = 1$ mA, $V_P = -1$ V. If the quiescent voltage at the drain is 10 V, find the value of R_1.

(2 kΩ)

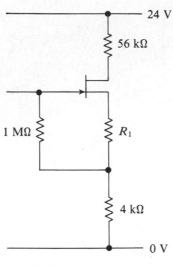

Fig. Q.4.8

4.9   An n channel FET has a drain load of 2.2 kΩ and dc supply of 20 V. It is required that $V_{DN(Q)}$ is between 10 V and 12 V. Calculate suitable resistor values.

FET data: $I_{DSS(max)} = 10$ mA, $V_p(max) = -6$ V, $I_{DSS(min)} = 5$ mA, $V_{P(min)} = -4$ V.

(1.6 kΩ; 100 kΩ; 33 kΩ)

4.10 The transfer characteristic of an n channel depletion MOSFET is drawn in Fig. Q.4.10. Using the circuit of Fig. 4.34, calculate the value of drain load

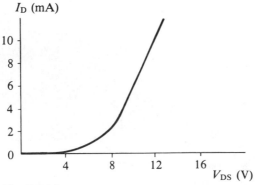

Fig. Q.4.10

to be used with a dc supply of 24 V to give $V_{DN(Q)} = 10$ V. Estimate g_{fs} at the working point by measuring the slope of the transfer characteristic and hence calculate the voltage gain in dB.

$$(2.4 \text{ k}\Omega; \ 14 \text{ dB})$$

4.11 Give the component values for a single stage common source amplifier which is required to have a voltage gain of 15 dB and a quiescent drain voltage $V_{DN(Q)} = 12$ V when fed from a 20 V dc supply.
FET data

$$g_{fs} = 3 \text{ mS} \qquad r_{ds} = 50 \text{ k}\Omega$$
$$I_{DSS} = 8 \text{ mA} \qquad V_P = -4 \text{ V}$$
$$(R_d = 1.9 \text{ k}\Omega; \qquad R_s = 270 \ \Omega)$$

5

Negative Feedback

Chapter Objectives

5.1   Advantages and Disadvantages of NFB
5.2   Effect of NFB on Gain
5.3   Sampling and Mixing Methods – Analysis of Various NFB Amplifiers
5.4   Use of NFB to Extend Bandwidth
5.5   Use of NFB to Reduce Harmonic Distortion
5.6   Use of NFB to Reduce Noise
5.7   Stability
        Summary
        Revision Questions

Conceptually, negative feedback (NFB) is one of the more demanding aspects of electronic engineering. However, it is the means by which the engineer exercises control over amplifier properties such as gain. Without NFB many amplifier properties are dominated by such factors as device performance which, as we have seen, are somewhat unpredictable. Analysis of NFB amplifiers can be extremely lengthy unless approximations are made. Provided these are done judiciously the sacrifice in accuracy is insignificant. Without approximations, analysis is sometimes so laborious that there is a danger of losing sight of the main points which should be brought out.

5.1   ADVANTAGES AND DISADVANTAGES OF NFB

In no particular order of importance the advantages of NFB are:

(a) Circuit performance can be made virtually independent of device parameters provided their range of values is known. For example, if h_{fe} for a BJT can be guaranteed to be greater than, say, 50, then voltage gain can be made virtually independent of h_{fe}.

136

(b) The stability of an amplifier's properties can be improved when some of its component values or power supply voltages change, e.g. due to aging effects or temperature change.

(c) Input resistance (R_i) and output resistance (R_o) of amplifiers can be fixed by circuit design. For example, the input resistance to a CE amplifier need not be dominated by h_{ie} (see Section 3.6) but can be made either much greater or much less than h_{ie} by the application of NFB.

(d) Amplifier frequency response can be extended at both LF and HF.

(e) Under certain conditions nonlinearity (percentage of harmonic distortion) and signal to noise ratio (S/N) of an amplifier can both be improved.

There are two disadvantages of NFB:

(a) All of the aforementioned advantages have to be 'paid for' by sacrificing gain. For this reason negative feedback is sometimes called 'degenerative' feedback. This is not such a serious cost as it might appear. For example, as we shall see when describing operational amplifiers (Chapter 8), large voltage gain ($A_v > 10^4$) is cheap.

(b) When used in large quantities NFB increases the likelihood of instability, i.e. the tendency to oscillate and produce an unwanted output. This is not an argument against using NFB but a warning that increased care with design and construction are necessary.

5.2   EFFECT OF NFB ON GAIN

Before proceeding to the general principles of NFB it may be helpful to illustrate the concept with a particular example. In describing the self-bias circuits of Chapters 2 and 4, the desirability of decoupling the emitter (CE amplifier) and the source (CS amplifier) by short-circuiting them, for signal frequencies, with a capacitor was justified on the grounds of avoiding gain reduction. In Fig. 5.1 the source-decoupling capacitor has been omitted. As noted in Section 4.5 the output voltage can be approximated to,

$$V_o \doteq -g_{fs}R_d V_{gs}$$

i.e. the output is proportional to V_{gs}. When R_s is decoupled, then since the decoupling capacitor effectively ties source to ground,

$$V_{gs} = V_s$$

where V_s is the signal to be amplified, and

$$A_v \doteq -g_{fs}R_d.$$

However, in Fig. 5.1, V_{gs} is not equal to V_s. As V_s increases, $I_s(= I_d)$ increases and

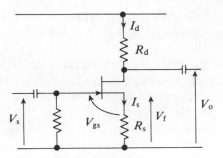

Fig. 5.1 NFB in a Common Source Amplifier

V_f increases. Therefore

$$V_{gs} = V_s - V_f \tag{5.1}$$

and $V_{gs} < V_s$.

In amplifying V_{gs} the action of the circuit does not change, i.e. $V_o = -g_{fs}R_d V_{gs}$, but V_{gs} has fallen. Examining Fig. 5.1 and equation (5.1) it can be seen that, instead of V_s appearing between G and S, the voltage V_f opposes V_s, i.e. it reduces the proportion of V_s appearing between G and S: it is said to be fed back in opposition to the input signal V_s. In effect, since it is V_s which is to be amplified, the effective gain A_{vf} is given by

$$A_{vf} = \frac{V_o}{V_s} \tag{5.2}$$

where A_{vf} is the gain with NFB, and is less than

$$A_v = \frac{V_o}{V_{gs}} \tag{5.3}$$

the gain without NFB. V_f is called the feedback voltage and the feedback is said to be negative because V_f is fed back in opposition to the input signal V_s and causes a reduction in output voltage for a given input signal.

Later we shall analyze NFB circuits such as Fig. 5.1, but first consider Fig. 5.2, which can be used to derive a fundamental feedback equation. Here the external load resistance is $R_1 + R_2$. With R_s decoupled,

$$V_{gs} = V_s - V_f$$

as before. The output voltage V_o is developed across R_1 and R_2 but a proportion β of V_o is fed back in opposition to V_s, where

$$\beta = \frac{-R_2}{R_1 + R_2}$$

(It is necessary for β to be negative because the phase inversion of the common source amplifier implies that A_v is negative.)

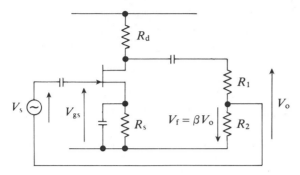

Fig. 5.2   Series Voltage NFB in a CS Amplifier

Then $V_f = \beta V_o = \beta A_v V_{gs}$

Therefore $V_{gs} = V_s - \beta V_o = V_s - \beta A_v V_{gs}$,

where $A_v = \dfrac{V_o}{V_{gs}}$ and is the voltage gain without NFB.

Then $A_{vf} = \dfrac{V_o}{V_s} = \dfrac{A_v V_{gs}}{V_{gs} + A_v V_{gs}}$

$$= \frac{A_v}{1 + \beta A_v} \tag{5.4}$$

So far, these introductory points have been made in terms of voltage. In Fig. 5.2 a proportion of the output *voltage*, βV_o is fed back in *series* with, but in opposition to, V_s, and as such is called series applied, voltage derived NFB, or usually, just series voltage NFB. However, the signal fed back can be derived from the output current and also, instead of feeding back a voltage to be connected in series with the input signal, a current can be fed back to be connected in shunt with the input current. Figure 5.3 embodies the ideas of NFB more generally than the series voltage NFB circuit of Fig. 5.2.

The various quantities S_s and so on represent signals which can either be voltages or currents. By *sampling* either output voltage or current the β network feeds back a signal $S_f = \beta S_o$. (In Fig. 5.2 this is the $R_1 R_2$ network.) Then S_f is *mixed*

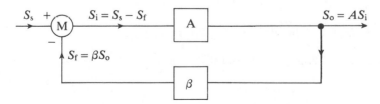

Fig. 5.3   General Model of an NFB Circuit

by the network M either in series or shunt with S_s so that $S_i = S_s - S_f$. (In Fig. 5.2 this is effected by connecting the bottom of V_s to the R_1R_2 junction rather than connecting it to the common ground rail.) In Fig. 5.3, A, the gain without the NFB, is

$$A = \frac{S_o}{S_i}$$

and is a current or voltage gain depending on the nature of S_i and S_o.

$$S_s = S_i + S_f$$
$$= S_i + \beta S_o$$
$$= S_i + \beta A S_i$$

Therefore $A_f = \dfrac{S_o}{S_s} = \dfrac{A S_i}{S_i + \beta A S_i}$

$$= \frac{A}{1 + \beta A} \tag{5.5}$$

which has the same form as equation (5.4) but is derived from more general considerations.

Applying equation (5.4) to an amplifier whose voltage gain without NFB is 100, then if $\beta = 0.1$,

$$A_{vf} = \frac{100}{1 + 0.1 \times 100} = 9.1$$

If the gain without NFB were -100 as, for example, in a single stage CE amplifier, then β must be negative to ensure that the feedback is negative, i.e.

$$A_{vf} = \frac{-100}{1 + (-0.1 \times -100)} = -9.1$$

The quantity $(1 + \beta A)$ is the factor by which the gain without NFB is reduced. Sometimes this is expressed as the amount of feedback in dBs. Referring back to the example where $A_v = 100$ and $\beta = 0.1$,

$$A_v(dB) = 20 \log_{10} 100 = 40 \text{ dB.}$$

$$A_{vf}(dB) = 20 \log_{10} 9.1 = 19.2 \text{ dB.}$$

Amount of feedback in $dB = 20 \log_{10}(1 + \beta A_v)$

$$= 20 \log_{10} 11 = 20.8 \text{ dB}$$

i.e. $A_v (dB) - (1 + \beta A_v) dB = A_{vf}(dB)$

$$40 - 20.8 \qquad\qquad = 19.2 \text{ dB}$$

Gain Stability

As noted, one of the advantages of NFB is that it reduces the dependence of gain on such factors as parameter spread, aging components, power supply fluctuations, and so forth. Equations (5.4) and (5.5) can be used to establish the principle of gain stabilization as follows. Suppose that $\beta A \gg 1$. Then

$$A_f = \frac{A}{1 + \beta A} \doteq \frac{A}{\beta A} = \frac{1}{\beta} \tag{5.6}$$

Since β is established by the passive components of the feedback network then gain depends only on these components and is independent of the amplifier itself, provided that its gain is large enough to ensure that $\beta A \gg 1$. In these circumstances such factors as parameter spread are of no consequence.

Example 5.1

An amplifier has a voltage gain $A_v = 1000$ and feedback factor $\beta = 0.1$. Then $\beta A_v = 100$ and

$$A_{vf} = \frac{1000}{1 + \beta A_v} \doteq \frac{1000}{100} = 10 = \frac{1}{\beta}$$

If now A_v falls to 500, $\beta A_v = 50$ which is still much greater than 1. Then

$$A_{vf} \doteq \frac{500}{50} = 10 = \frac{1}{\beta}$$

and the approximation of equation (5.6) holds. $\square$

Expressing the point about gain stability more generally, then using equation (5.5),

$$A_f = \frac{A}{1 + \beta A}$$

$$\frac{dA_f}{dA} = \frac{1}{(1 + \beta A)^2}$$

$$\frac{dA_f}{A_f} = \frac{1}{1 + \beta A} \frac{dA}{A} \tag{5.7}$$

In words, this equation means that change in gain (dA) as a proportion of total gain (A) is reduced by a factor $(1 + \beta A)$ when NFB is applied.

Example 5.2

Suppose the gain (voltage or current) of an amplifier without NFB is 1000 and can change by 20%, i.e.

$$A = 1000 \pm 200.$$

It is required to reduce the change to 2% by applying NFB. Then using equation (5.7),

$$0.02 = \frac{0.2}{1 + \beta 1000}$$

i.e. $\beta = 0.009$

and $A_f = \dfrac{1000}{1 + 0.009 \times 1000} = 100$

When A rises to 1200 then,

$$A_f = \frac{1200}{1 + 0.009 \times 1200} = 101.7$$

Equation (5.7) appears to predict that the result should be 102. The error occurs because equation (5.7) is derived by using differential calculus but is then applied to large changes (20%). However, the equation does provide a satisfactory basis for stability calculation. An objection, of course, is that gain has been reduced by a factor of 10, but using two stages with NFB, then with both at maximum gain (a worst case) the gain is 10 343, i.e. 3% up, as opposed to 20% without NFB.   □

5.3   SAMPLING AND MIXING METHODS – ANALYSIS OF VARIOUS NFB AMPLIFIERS

The choice of output quantity sampled – voltage or current – decides whether the output resistance R_o increases or decreases. The way this sample of the output is mixed with the input signal – series or shunt – decides whether the input resistance R_i increases or decreases. *Sampling controls R_o; mixing controls R_i.*

Identification of Sampling and Mixing

Figure 5.4a is a block diagram of a voltage derived, series applied, or simply, a voltage series, NFB amplifier. It is evident that Fig. 5.4b is a simple realization of series voltage NFB but in many circuits the type of NFB is not always so easy to identify. In this case the output voltage could be reduced to zero by short-circuiting the collector load resistance. This would have little effect on the signal current but clearly it would make $V_o = 0$ V and there would be no NFB. Therefore the feedback is voltage derived. The signal fed back is a voltage added in series with V_s rather than a current added in shunt with I_i so the feedback is series applied.

In Fig. 5.4b, V_s is the signal to be amplified and V_i the signal connected to the amplifier input terminals to be amplified by A_v rather than by the feedback gain A_{vf}.

Then $A_v = \dfrac{V_o}{V_i};$ $A_{vf} = \dfrac{V_o}{V_s}$

(a)

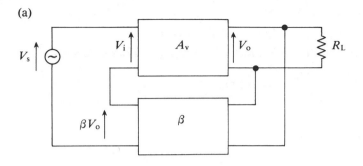

(b)

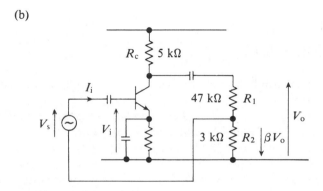

Fig. 5.4 (a) Block Diagram of Series Voltage NFB
(b) Series Voltage NFB in a CE Amplifier

and $V_i = V_s - \beta V_o$

from which we have previously established (equation (5.4)) that

$$A_{vf} = \frac{A_v}{1 + \beta A_v}$$

If there were no NFB then

$$R_i = \frac{V_i}{I_i}.$$

With NFB,

$$R_{if} = \frac{V_s}{I_i} = \frac{V_i + \beta A_v V_i}{I_i}$$

$$= R_i(1 + \beta A_v) \tag{5.8}$$

In other words, the effect of $A_v V_i$ – the feedback voltage – is to oppose the flow of input current I_i and increase input resistance by a factor $(1 + \beta A_v)$.

☐ **Example 5.3**

In Fig. 5.4b, given $h_{ie} = 2$ kΩ, $h_{fe} = 80$, find R_i.
Neglecting the loading effect of the $R_1 R_2$ feedback network,

$$A_v \div \frac{h_{fe} R_c}{h_{ie}} = \frac{-80 \times 5}{2} = -200$$

$$\beta = \frac{-3}{47 + 3} = -0.06$$

(The negative sign ensures that βA_v is positive for this circuit.)
Therefore $1 + \beta A_v = 13$

$$R_i = h_{ie} = 2 \text{ k}\Omega$$

and $R_{if} = h_{ie}(1 + \beta A_v) = 26$ kΩ.   ☐

It is worth noting that although there will be a corresponding fall in voltage gain A_{vf} of 13 times (22 dB) there will not necessarily be a corresponding fall in output voltage. Voltage gain $A_{vf} = V_o/V_s$, where V_s is the voltage across the input terminals. Consider, however, Figs. 5.5a and b, where the signal source to be amplified is an emf V_s with source resistance $R_s = 20$ kΩ. Then defining V_{in} as the voltage across the amplifier input resistance – without NFB 2 kΩ, with NFB, 26 kΩ – then application of NFB causes V_{in} to rise from 0.09 V_s to 0.56 V_s – a gain in input voltage of 15.8 dB. Therefore the overall loss of output voltage by using NFB is only 6.2 dB and this may be more than compensated for by the benefits listed in the introduction to this chapter.

Output Resistance

Before formally analyzing a series voltage NFB amplifier, the effect of this type of NFB on output resistance R_o can be understood qualitatively. Referring to Fig. 5.4a, which represents all series voltage NFB amplifiers, suppose that R_L were to be reduced in value. One effect is to make V_o fall because the increase in output current will cause an increase in the volt drop across the output resistance R_{of}; Fig. 5.7

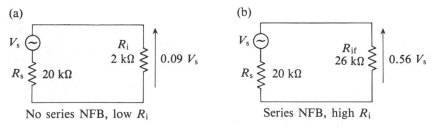

Fig. 5.5   Demonstrating that by Increasing R_i with Series NFB, Input Voltage to Amplifier Increases

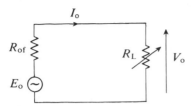

Fig. 5.6   If NFB Reduces R_{of} then V_o Will Tend to Remain Constant as R_L is Varied

illustrates the point. If V_o falls then βV_o – the feedback voltage in Fig. 5.4a – falls, so V_i rises and V_o follows it. In other words, the effect of the feedback is to sustain the output V_o with reduced load resistance – a characteristic of a circuit with low output resistance. In the limit, if $R_{of} = 0$, V_o remains constant whatever the value of R_L. This reduced output resistance is a consequence of the NFB being voltage derived and is independent of the method of mixing.

To confirm the general result that R_o increases with the application of series voltage NFB (shunt voltage NFB will be dealt with later in this section), note that in Fig. 5.7, A_v is the open circuit voltage gain without NFB, i.e. with $R_L = \infty$ and the β network disconnected, $V_o = A_v V_i$; R_o is the output resistance without NFB.

Then with load connected and NFB applied,

$$A_v V_i = V_o + I_o R_o$$

$$V_s = V_i + \beta V_o = \frac{V_o + I_o R_o}{A_v} + \beta V_o$$

Therefore $V_o = \dfrac{A_v V_s}{1 + \beta A_v} - \dfrac{I_o R_o}{1 + \beta A_v}$

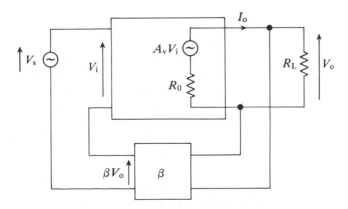

Fig. 5.7   Block Diagram to Determine Effect of Voltage NFB on Output Resistance

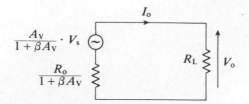

Fig. 5.8 Equivalent of Fig. 5.7 Output Circuit

From this the equivalent of the output circuit can be drawn (Fig. 5.8) which shows that the output resistance with NFB is

$$R_{of} = \frac{R_o}{1 + \beta A_v} \tag{5.9}$$

and is, as predicted, reduced.

There are other methods of calculating R_o, e.g. by taking the ratio of open-circuit voltage to short-circuit current or by suppressing V_s and calculating the current I supplied by a voltage V connected across the output terminals giving $R_o = V/I$.

☐ **Example 5.4**

Applying this last result to Fig. 5.4b, then without NFB (feedback link disconnected), R_o is the resistance looking back into the amplifier between C and N. Disregarding h_{oe},

$$R_o = R_L \parallel (R_1 + R_2) \doteqdot 4.5 \text{ k}\Omega.$$

With $\beta A = 13$

$$R_{of} = 321 \ \Omega. \qquad \qquad ☐$$

Figure 5.4b has the disadvantage that the input and output signals do not have a common terminal. If they were connected to a source and load, each with one terminal grounded, then R_2 would be short-circuited. Before describing a series voltage NFB amplifier which does not have this disadvantage, it is necessary to deal with a particular class of series voltage NFB amplifiers called followers.

Followers

BUFFERING

It is often necessary to connect between a source of high resistance and a load of much lower resistance a circuit called a buffer, which provides suitable loading conditions for the source and avoids a large voltage and power loss. For example,

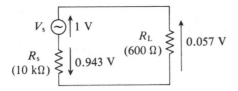

Fig. 5.9 Effect of Connecting a Source of High Resistance to a Load of Much Lower Resistance without Buffering

suppose we want to feed a 600 Ω load from a source of $V_s = 1$ V, $R_s = 10$ kΩ. If this is done directly as in Fig. 5.9 then $V_o = 57$ mV and the volt drop across R_s will be 943 mV, which is clearly undesirable. Further, if the source is only linear under certain load conditions then the direct connection might lead to distortion. A buffer circuit, of which the emitter follower now to be described is an example, is connected between source and load so that the actual resistance connected across the source is much higher than the load resistance itself.

EMITTER FOLLOWER ACTION

In the basic emitter follower circuit of Fig. 5.10, from which bias resistors have been omitted, as v_s makes v_{BN} more positive, i_C and i_E increase, so v_o goes more positive, i.e. the emitter follows the base. Therefore input and output are in phase. The circuit is sometimes called a common collector amplifier because, as we can see in the small-signal equivalent circuit (Fig. 5.11) the collector is common to input and output signals. Some manufacturers quote common collector h parameters – h_{fc}, h_{ic}, etc. – but they are rarely used because the circuit is readily analyzed with CE h parameters and the results give an immediate basis for comparison with, for example, the CE amplifier.

The follower action results in a voltage gain A_v which is positive but slightly less than unity. In feedback terms,

$$V_{be} = V_s - V_o$$

i.e. the whole of the output is fed back in series with the input V_s and $\beta = 1$: the

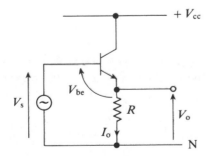

Fig. 5.10 Basic Emitter Follower

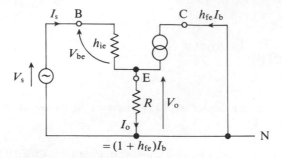

Fig. 5.11   Equivalent Circuit of Emitter Follower

circuit is an example of series voltage NFB. However, rather than treat the emitter follower as a feedback circuit, it is convenient simply to analyze Fig. 5.11 as a small-signal equivalent circuit in its own right.

In Fig. 5.11 h_{oe} and h_{re} have been omitted. If h_{oe} were included it would be in parallel with R so it is easily included if comparable with R.

Current gain $A_i = \dfrac{I_o}{I_s} = \dfrac{I_e}{I_b}$

$$= \frac{(1 + h_{fe})I_b}{I_b} = (1 + h_{fe}) \tag{5.10}$$

$$R_i = \frac{V_{bn}}{I_b} = \frac{h_{ie}I_b + (1 + h_{fe})I_bR}{I_b} \qquad (V_{bn} = V_s)$$

$$= h_{ie} + (1 + h_{fe})R \tag{5.11}$$

This will often be much higher than h_{ie} and can often be approximated to

$$R_i \doteqdot h_{fe}R \tag{5.12}$$

Voltage gain $A_v = \dfrac{V_o}{V_s} = \dfrac{I_oR}{I_sR_i}$

$$= \frac{A_iR}{R_i}$$

$$= \frac{(1 + h_{fe})R}{h_{ie} + (1 + h_{fe})R} \tag{5.13}$$

In most cases this is nearly 1, which is consistent with 100% voltage NFB.

Although the voltage gain is less than 1 there is a substantial power gain A_p.

$$A_p = \frac{P_o}{P_i} = \frac{V_oI_o}{V_sI_s} = A_vA_i \doteqdot A_i = (1 + h_{fe}) \tag{5.14}$$

OUTPUT RESISTANCE

In calculating R_o we must take into account R_s. Assume that R is the load and

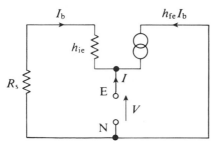

Fig. 5.12   Circuit for Calculating Emitter Follower Output Resistance

that R_o is the resistance one 'sees' looking back into the amplifier between E and N. Then using the technique of suppressing V_s and impressing a voltage V between E and N as in Fig. 5.12,

$$R_o = \frac{V}{I}$$

$$I = -(1 + h_{fe})I_b$$

$$V = -I_b(h_{ie} + R_s)$$

Then $R_o = \dfrac{h_{ie} + R_s}{1 + h_{fe}}$ \hfill (5.15)

$$\doteq \frac{h_{ie} + R_s}{h_{fe}} \tag{5.16}$$

If R is regarded as part of the amplifier ac coupled to some external load, then

$$R_{out} = R \parallel R_o$$

☐ **Example 5.5**

Using again the 1 V, 10 kΩ source and 600 Ω load, now buffer them with an emitter follower as in Fig. 5.13. Assume $h_{fe} = 100, h_{ie} = 1$ kΩ. From the equivalent circuit of Fig. 5.14,

$$V_s = 1 \; V = I_b(10 + 1) + 101 \; I_b \times 0.6 \qquad \text{(resistance in kΩ, current in mA)}$$

$$I_b = \frac{1}{71.6} \; \text{mA}$$

$$V_o = 101 \; I_b \times 0.6 = 0.85 \; V$$

which is obviously better then the 57 mV obtained without buffering. The result can be confirmed using equations (5.11) and (5.13), but it is often better simply to work from first principles by analyzing the equivalent circuit.

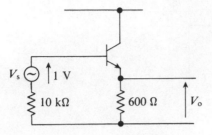

Fig. 5.13   Circuit for Example 5.5

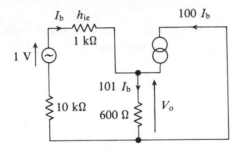

Fig. 5.14   Equivalent Circuit of Fig. 5.13   □

BANDWIDTH IMPROVEMENT

Another advantage of buffering is that it can extend the upper cutoff frequency. For example, if the 600 Ω load in the previous example is shunted by 100 pF then Fig. 5.15 is a Norton equivalent of the load being driven by a source of resistance R_n which will be 10 kΩ without the buffer and

$$\frac{R_s + h_{ie}}{1 + h_{fe}} = 109 \ \Omega$$

with the buffer. Then taking the upper cutoff frequency as $1/2\pi C R_T$ in each case, where $R_T = R_n \parallel 600$, the result is 2.8 MHz without buffering and 17.3 MHz with buffering.

BIASING

The simplest practical emitter follower, biased, with source and load ac

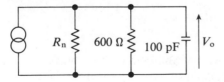

Fig. 5.15   Circuit to Demonstrate Increase in Bandwidth with Buffering

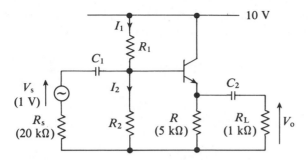

Fig. 5.16   Emitter Follower with Bias Resistors

coupled, is shown in Fig. 5.16. The emitter bias principles established in Chapter 2 apply in the choice of R_1 and R_2 but it may be necessary to choose higher values than in a CE amplifier if a priority is to maintain a high input resistance.

☐   **Example 5.6**

Choose values of R_1 and R_2 in Fig. 5.16 so that $V_{EN(Q)} = 5$ V and $R_i > 30$ kΩ. Calculate V_o and R_o looking back from R_L and the values of C_1 and C_2 so that each considered separately gives a lower cutoff frequency of 50 Hz.

Assume $h_{FE} = h_{fe} = 100$ and $h_{ie} = 2$ kΩ.

$$I_E = \frac{5\ V}{5\ k\Omega} = 1\ mA$$

$$I_B \doteqdot \frac{I_E}{h_{FE}} = 10\ \mu A$$

$$V_{BN} = V_{EN} + V_{BE} = 5 + 0.8 = 5.8\ V$$

Let $I_1 = 4I_B = 40\ \mu A$

$$R_1 = \frac{V_1}{I_1} = \frac{4.2\ V}{40\ \mu A} \doteqdot 100\ k\Omega$$

$$I_2 = I_1 - I_B = 3I_B = 30\ \mu A$$

$$R_2 = \frac{V_{BN}}{I_2} = \frac{5.8\ V}{30\ \mu A} \doteqdot 190\ k\Omega$$

Let input resistance looking into B and N be R_{is}.

Then $R_{is} \doteqdot h_{fe}R$

Where $R = 5\ k\Omega \parallel 1\ k\Omega = 0.83\ k\Omega$

$R_{is} = 83\ k\Omega$

$R_i = R_{is} \parallel R_i \parallel R_2 \doteqdot 37\ k\Omega$

If R_i had been less than the specified minimum (30 kΩ) then we could either make $I_i = 3I_B$ and repeat the calculations or use the bootstrap circuit of section 7.1.

$$V_{bn} = \frac{R_i}{R_s + R_i} \times 1 \text{ V} = 0.65 \text{ V}$$

Then using equation (5.13),

$$V_o = \frac{101 \times 0.83}{2 + 101 \times 0.83} V_{bn} = 0.63 \text{ V}$$

In calculating R_o, the effective value of source resistance R_s is 20 k$\Omega \parallel R_1 \parallel R_2 = 15.3$ kΩ

Then using equation (5.16),

$$R_o = \frac{15.3 + 2}{100} \parallel 5 \text{ k}\Omega = 173 \text{ }\Omega \parallel 5 \text{ k}\Omega \doteq 173 \text{ }\Omega.$$

To find C_1,

$$f = \frac{1}{2\pi C_1 R_3}$$

where $R_3 = R_s + R_i = 57$ kΩ

To find C_2

$$f = \frac{1}{2\pi C_2 R_4}$$

when $R_4 = R_o + R_L = 1173 \text{ }\Omega$

then $C_1 = 0.056 \text{ }\mu\text{F}$

$C_2 = 2.7 \text{ }\mu\text{F}$ $\square$

SOURCE FOLLOWER

A source follower (Fig. 5.17) has no advantage over a common source amplifier as regards input resistance because of the inherently high input resistance of FETs. However, as we shall see by analyzing the equivalent circuit (Fig. 5.18), the circuit does have the benefit of low output resistance. In this circuit the quiescent voltage between source and ground will have to be much larger than $V_{GS(Q)}$ so R_1 and R_2

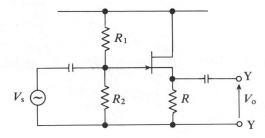

Fig. 5.17 Source Follower

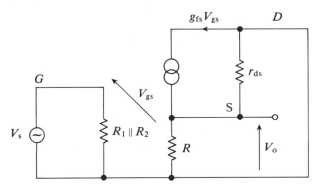

Fig. 5.18   Equivalent Circuit of Source Follower

must be chosen to establish a satisfactory working point. Note that r_{ds} is in parallel with R and can either be neglected or incorporated into R. Then,

$$V_o = g_{fs}V_{gs}R = g_{fs}(V_s - V_o)R$$

therefore $V_o + g_{fs}V_oR = g_{fs}V_sR$

and $A_v = \dfrac{V_o}{V_s} = \dfrac{g_{fs}R}{1 + g_{fs}R}$ (5.17)

It will be noted that this has the same form as $A/(1 + \beta A)$ – in this case $A = g_{fs}R$ and $\beta = 1$.

From equation (5.17),

$$V_o = V_s \frac{R}{1/g_{fs} + R}$$

This can be represented by the circuit of Fig. 5.19 and therefore,

$$R_o = \frac{1}{g_{fs}}$$ (5.18)

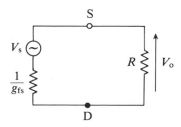

Fig. 5.19   Derivation of Equation (5.18)

☐ **Example 5.7**

In Fig. 5.17, $R = 2.7$ kΩ, $g_{fs} = 4$ mS and $r_{ds} = 50$ kΩ. Calculate A_v and R_o looking back from terminals Y–Y.

As $r_{ds} \gg R$, r_{ds} can be neglected.

Then using equation (5.17),

$$A_v = \frac{V_o}{V_s} = \frac{4 \times 10^{-3} \times 2.7 \times 10^3}{1 + (4 \times 10^{-3} \times 2.7 \times 10^3)} = 0.92$$

Output resistance looking back into the FET between source and drain is

$$\frac{1}{g_{fs}} = 250 \ \Omega.$$

Therefore $R_o = 250 \ \Omega \parallel 2.7$ kΩ

$$\doteqdot 230 \ \Omega \qquad\qquad\qquad\qquad\qquad ☐$$

Before returning to the problem of finding a series voltage NFB circuit with substantial voltage gain and a common input and output terminal, we shall first describe series current NFB.

Series Current NFB

Figure 5.20 shows schematically a series applied, current derived NFB amplifier. As implied by the name, the voltage fed back in series opposition to the input voltage is derived from the output current. In Fig. 5.21 the emitter resistor is not decoupled and therefore a voltage V_f is fed back in opposition to V_s so that,

$$V_{be} = V_s - V_f$$

On the face of it, the mode of NFB is the same as for the emitter follower but in this case the circuit output signal V_o is taken from the collector. If R_c is short-circuited, $V_o = 0$, but there will still be a signal current through R_e and therefore the

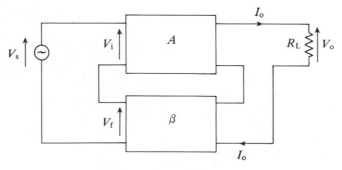

Fig. 5.20   Block Diagram of Series NFB

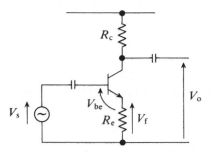

Fig. 5.21   Series Current NFB in a CE Amplifier

feedback signal V_f will be maintained. Therefore V_f is derived from the output *current*.

Analyzing the equivalent circuit (Fig. 5.22),

$$V_s = h_{ie}I_b + (1 + h_{fe})I_bR_e$$

Therefore $R_{if} = \dfrac{V_s}{I_b} = h_{ie} + (1 + h_{fe})R_e$

If R_e is comparable with h_{ie} and $h_{fe} \gg 1$, then

$$R_{if} \doteqdot h_{fe}R_e$$

as with the emitter follower.

$$A_{if} = \frac{I_c}{I_b} = h_{fe}$$

$$A_{vf} = \frac{I_cR_c}{I_bR_{if}} = \frac{A_iR_c}{R_{if}}$$

$$= \frac{h_{fe}I_bR_c}{(h_{ie} + [1 + h_{fe}]R_e)I_b}$$

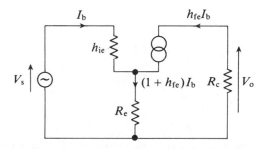

Fig. 5.22   Equivalent Circuit of Fig. 5.21

If $h_{ie} \ll (1 + h_{fe})R_e$, then

$$A_{vf} \doteq \frac{h_{fe}I_b R_c}{h_{fe}R_e I_b} = \frac{R_c}{R_e} \tag{5.19}$$

and A_{vf} is independent of the amplifier gain without NFB.

Output Resistance R_{of}

To give a qualitative explanation of the effect of current NFB on output resistance, a similar argument can be developed to the one given for voltage NFB. Any reduction in load resistance R_L (Fig. 5.23) tends to increase output current. Therefore the voltage fed back increases, which causes a reduction in voltage gain. Therefore, when load resistance is reduced, the fall in output voltage is greater than it would be without current NFB. This is characteristic of high output resistance.

Figure 5.24 is the equivalent circuit of Fig. 5.21 with h_{oe} included. Without h_{oe} the output resistance with or without NFB is infinite because we are looking back into a constant current source $h_{fe}I_b$. With h_{oe} and no NFB,

$$R_o = \frac{1}{h_{oe}}$$

To calculate R_{of}, suppress V_s, apply a voltage V across the output terminals and replace $h_{fe}I_b$ and h_{oe} by their Thevenin equivalent, as shown in Fig. 5.25.

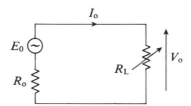

Fig. 5.23   If NFB Increases R_o Then Any Change in R_L Tends to Magnify the Change in V_o, which is Characteristic of High Output Resistance

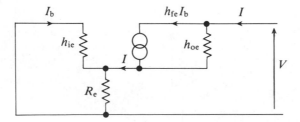

Fig. 5.24   Equivalent Circuit for Finding Output Resistance of CE Amplifier With Series Current NFB

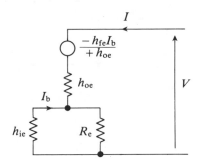

Fig. 5.25   Thevenin Equivalent Circuit of Fig. 5.24

Then $\qquad R_{of} = \dfrac{V}{I}$

$$-h_{ie}I_b = R_e(I_b + I)$$

and $\qquad I_b = -\dfrac{R_e I}{h_{ie} + R_e}$ $\hfill$ (A)

$$V + \frac{h_{fe}I_b}{h_{oe}} = \frac{I}{h_{oe}} - h_{ie}I_b$$

$$V = \frac{I}{h_{oe}} - I_b\left(h_{ie} + \frac{h_{fe}}{h_{oe}}\right) \hfill \text{(B)}$$

Substituting (A) in (B) and solving,

$$R_{of} = \frac{1}{h_{oe}} + \frac{R_e}{h_{ie} + R_e}\left(h_{ie} + \frac{h_{fe}}{h_{oe}}\right)$$

$$\doteqdot \frac{1}{h_{oe}} + \frac{R_e}{h_{ie} + R_e}\frac{h_{fe}}{h_{oe}} \hfill \text{(5.20)}$$

and the application of current NFB increases output resistance by the second term in equation (5.20).

Two-stage Series Voltage NFB Amplifier

Apart from the disadvantage of there being no common terminal for input and output signals, the application of NFB in Fig. 5.4b may result in its having insufficient gain for a particular purpose. In the accompanying block A_v may be realized by a multistage amplifier in which the extra stage(s) can be used to compensate for gain lost by NFB.

In Fig. 5.26 the output of Q_2, as well as being developed across the 3.9 kΩ load, is also developed across the 2.7 kΩ/47 Ω chain. As V_s goes positive, V_o goes positive and with the Q_1 emitter resistor decoupled, any increase in voltage across the 47 Ω

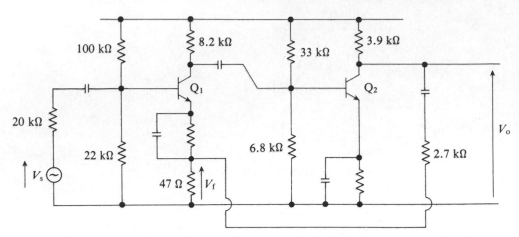

Fig. 5.26   Two-stage Amplifier with Series Voltage Feedback Over Two Stages and Series Current Feedback on the First Stage

resistor reduces V_{be1}. Therefore the feedback is negative. Short-circuiting the 3.9 kΩ resistor eliminates the feedback, which is therefore voltage derived.

$$V_{be1} = V_s - V_f$$

so the feedback is series applied.

☐   **Example 5.8**

Using the values in Fig. 5.26 and taking $h_{ie} = 2.5$ kΩ, $h_{fe} = 100$ for both transistors, calculate A_{vf}, R_{if} and R_{of}.

     Without NFB,

$$R_i \text{ (at } B_1) = h_{ie} + (1 + h_{fe})R_e = 7.2 \text{ kΩ}$$
$$R_o = 3.9 \text{ kΩ} \parallel (2.7 \text{ kΩ} + 47 \text{ Ω}) = 1.6 \text{ kΩ}$$

For stage 1 the effective collector load is R_1 where

$$R_1 = (8.2 \parallel 33 \parallel 6.8 \parallel 2.5) \text{ kΩ} = 1.43 \text{ kΩ}$$

The 47 Ω emitter resistor provides series current NFB.

Therefore $A_{v1} = \dfrac{-h_{fe}R_1}{h_{ie} + (1 + h_{fe})R_e} = -19.7$

For stage 2 the effective collector load R_2 is the same as R_o, i.e. $R_2 = 1.6$ kΩ.

Therefore $A_{v2} = \dfrac{-h_{fe}R_2}{h_{ie}} = -64$

Then $A_v = A_{v1}A_{v2} = 1261$

With NFB

Overall NFB is applied by dividing the output voltage between the 47 Ω and 2.7 kΩ resistors and feeding the voltage dropped across the 47 Ω resistor back in series with the input

Therefore $\beta = \dfrac{47}{2700 + 34} = 0.017$

Then $(1 + \beta A_v) = 22.4$

and R_{if} (at B_1) $= R_i(1 + \beta A_v) = 161$ kΩ

$$R_{if} = (122 \parallel 100 \parallel 22) \text{ k}\Omega = 16 \text{ k}\Omega$$

$$A_{vf} = \frac{A_v}{1 + \beta A_v} = 56$$

$$R_{of} = \frac{R_o}{1 + \beta A_v} = 71 \ \Omega$$

Note that A_{vf} is near to the value of $\dfrac{1}{\beta} - 59$.

However, it should be noted that R_{oe1} is dependent on h_{fe} so the voltage gain only becomes virtually independent of the transistor parameters if the feedback resistor (47 Ω) is much less than R_{oe1}. $\qquad\qquad\square$

Shunt Voltage NFB

In Fig. 5.27, as V_s increases, V_o decreases so I_f increases, and

$$I_b = I_s - I_f$$

thus demonstrating that the feedback is negative and shunt applied. If R_c is short-circuited there will be no voltage gain and no feedback current I_f, so the feedback is voltage derived. Referring back to Section 3.9 where Miller effect was described, it will be noted that R_f will produce a similar effect to C_c in Fig. 3.38. R_f

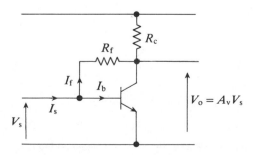

Fig. 5.27 CE Amplifier with Shunt Voltage NFB

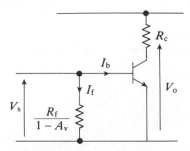

Fig. 5.28 R_f of Fig. 5.27 Replaced by Miller Resistance $R_f/(1 - A_v)$

will not appreciably affect the voltage gain because $I_b = V_s/h_{ie}$ and we can use the approximation that

$$A_{vf} = A_v = \frac{-h_{fe}R_c}{h_{ie}}$$

$$I_f = \frac{V_s - A_v V_s}{R_f} = \frac{V_s(1 - A_v)}{R_f} \tag{5.21}$$

where A_v is negative.

The input circuit, then, behaves as if a resistor $R_f/(1 - A_v)$ (Fig. 5.28) were connected between B and E. Then

$$R_{if} = \frac{R_f}{(1 - A_v)} \parallel h_{ie} \tag{5.22}$$

Although R_f does not affect the voltage gain defined as $A_v = V_o/V_{be}$, it will reduce V_o if the source has a finite resistance R_s as in Fig. 5.29. This occurs because the reduced input resistance causes a higher proportion of V_s to be dropped across R_s than without NFB. If A_v is large then $I_f > I_b$ and as V_s increases, I_f increases and the volt drop across R_s increases so that V_{be} will be negligible compared with V_s for a large value of R_s.

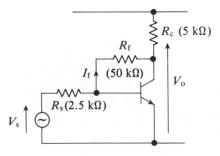

Fig. 5.29   Where Source Resistance is Finite, the Effect of Reduced Input Resistance will Cause a Fall in V_o

Then $I_f = \dfrac{V_s}{R_s}$

and $\quad V_o = -I_f R_f = -\dfrac{V_s}{R_s} R_f$

$$\frac{V_o}{V_s} = -\frac{R_f}{R_s} \tag{5.23}$$

This is the operational amplifier principle, which will be dealt with in Chapter 9.

R_f also improves bias stability but if it is not desirable for R_f to affect the dc conditions then a capacitor should be connected in series with it.

OUTPUT RESISTANCE

Using the technique of suppressing V_s and applying a voltage V across the output terminals (Fig. 5.30) then,

$$R_{of} = \frac{V}{I}$$

$$V = (I - h_{fe}I_b)R_f + h_{ie}I_b \tag{A}$$

$$h_{ie}I_b = [I - (1 + h_{fe})I_b]R_s$$

from which

$$I_b = \frac{IR_s}{h_{ie} + (1 + h_{fe})R_s} \tag{B}$$

Substituting for I_b in (A) leads to

$$R_{of} \doteqdot \frac{h_{ie}(R_f + R_s)}{h_{ie} + h_{fe}R_s} \tag{5.24}$$

R_c and h_{oe} will be in parallel with R_{of}.

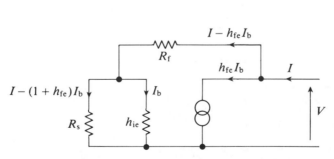

Fig. 5.30 Equivalent Circuit for Finding Output Resistance of Fig. 5.29

☐ **Example 5.9**

For Fig. 5.29 find $\dfrac{V_o}{V_{be}}$, $\dfrac{V_o}{V_s}$, R_{if} and R_{of}.

BJT data: $h_{fe} = 80$; $h_{ie} = 1.5 \text{ k}\Omega$.

$$A_v = \frac{V_o}{V_{be}} = \frac{-h_{fe}R_c}{h_{ie}} = -267$$

With a gain of this magnitude we can use the approximation

$$\frac{V_o}{V_s} = -\frac{R_f}{R_s} = -20.$$

Using equation (5.22),

$$R_{if} = \frac{50 \text{ k}\Omega}{268} \parallel 1.5 \text{ k}\Omega = 165 \ \Omega$$

Using equation (5.24),

$$R_{of} = \frac{1.5 \text{ k}\Omega(50 \text{ k}\Omega + 2.5 \text{ k}\Omega)}{1.5 \text{ k}\Omega + 80 \times 2.5 \text{ k}\Omega} = 391 \ \Omega$$

Output resistance $= R_c \parallel R_{of} = 362 \ \Omega$. ☐

☐ **Example 5.10**

For Fig. 5.31 calculate A_{vf}, A_{if}, R_{if} and R_{of} (Y–Y).
FET data: $g_{fs} = 5 \text{ mS}$; $r_{ds} = 100 \text{ k}\Omega$.
In this circuit $V_{gs} = V_i$
From the equivalent circuit of Fig. 5.32,

$$V_i = 150I_i + 3.2(I_i - g_{fs}V_i)$$
$$17V_i \doteqdot 153I_i$$

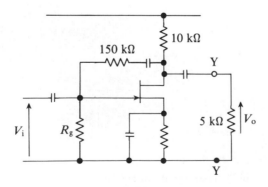

Fig. 5.31   Circuit for Example 5.10

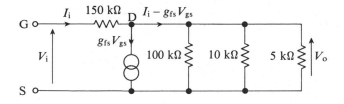

Fig. 5.32   Equivalent Circuit of Fig. 5.31

Therefore $R_{if} = \dfrac{V_i}{I_i} = 9 \text{ k}\Omega$

Current in $R_f = \dfrac{V_i}{9 \text{ k}\Omega}$

$$g_{fs} V_i = \frac{5}{10^3} V_i$$

Therefore current in R_f can be neglected and

$$A_{vf} = \frac{V_o}{V_i} = -g_{fs} \times 3.2 = -16$$

$$A_{if} = \frac{V_o/(5 \text{ k}\Omega)}{V_i/R_i} = \frac{-16 \times 9}{5} \doteqdot -29$$

To find R_{of} refer to Fig. 5.33 which includes a source resistance of 10 kΩ.

$$V_{gs} = (I - g_{fs} V_{gs}) R_s$$

$$V_{gs} = \frac{I R_s}{1 + g_{fs} R_s} \tag{A}$$

$$V = (I - g_{fs} V_{gs})(R_f + R_s) \tag{B}$$

Substituting (A) in (B) gives

$$V = I(R_f + R_s) - \frac{g_{fs} I R_s (R_f + R_s)}{1 + g_{fs} R_s}$$

$$\frac{V}{I} = R_{of} = \frac{R_f + R_s}{1 + g_{fs} R_s}$$

$$= \frac{150 \text{ k}\Omega + 10 \text{ k}\Omega}{1 + 5 \times 10} \doteqdot 3 \text{ k}\Omega$$

Note that as R_s increases R_{of} falls. The limits are

(a) when $R_s = 0$; $R_{of} = R_f$

(b) when $R_s = \infty$; $R_{of} = \dfrac{1}{g_{fs}}$

$$R_{of}(Y-Y) = R_{of} \parallel 10 \text{ k}\Omega = 2.3 \text{ k}\Omega$$

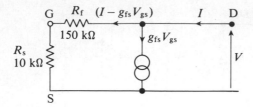

Fig. 5.33   Method for Finding R_{of}

Shunt Current NFB

The amplifying block of Fig. 5.34 has input current I_i and output current I_o so its current gain without NFB is A_i where

$$A_i = \frac{I_o}{I_i}$$

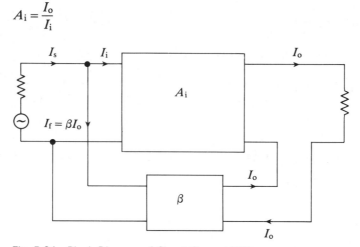

Fig. 5.34   Block Diagram of Shunt Current NFB

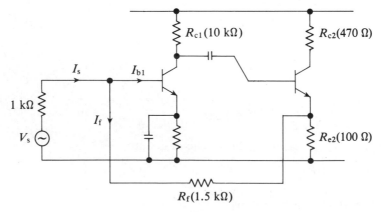

Fig. 5.35   Shunt Current NFB Amplifier

The β feedback network operates in such a way that the feedback current $I_f = \beta I_o$ is subtracted from the signal input current I_s in determining I_i, i.e.

$$I_i = I_s - I_f$$

Checking the feedback in Fig. 5.35, note that the current in R_f is dependent on the current in R_{e2} rather than the output voltage V_o which is developed across R_{c2}. As I_s increases, V_{ce1} and V_{e2} fall, so I_f increases and $I_i = I_s - I_f$. Therefore the feedback is shunt applied and negative as well as being current derived. Note also that as R_{e2} is not decoupled, the output stage also has series current NFB.

☐ **Example 5.11**

Using the component values given in Fig. 5.35, calculate A_{vf}, A_{if}, R_{if} and R_{of}, and also, V_o/V_s. BJT data $h_{fe} = 80$; $h_{ie} = 2$ kΩ.

To simplify the problem bias components have been omitted but these can easily be combined with h_{ie} in the usual manner. To calculate β note the equivalent circuit of Fig. 5.36 and the output circuit detail in Fig. 5.37. This shows actual current directions with I_{b1} positive. Then using the approximation that $V_{e2} \gg V_{be1}$,

$$\frac{I_o - I_f}{I_f} = \frac{1/R_{e2}}{1/R_f} = \frac{R_f}{R_{e2}}$$

$$\beta = \frac{I_f}{I_o} = \frac{R_{e2}}{R_{e2} + R_f} = \frac{1}{16}$$

$$I_o = -h_{fe} I_{b2} = -h_{fe} \frac{R_{c1}}{R_{c1} + R_{i2}} \times -h_{fe} I_{b1}$$

$$R_{i2} = h_{ie} + (1 + h_{fe})(R_e \| R_f) \doteq 9.6 \text{ k}\Omega$$

Therefore $I_o = 80 \times \dfrac{10}{10 + 9.6} \times 80\, I_{b1} \doteq 3270\, I_{b1}$

Therefore $I_f = \beta I_o = 204\, I_{b1}$

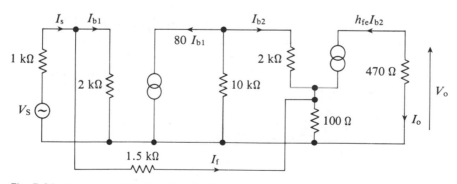

Fig. 5.36 Equivalent Circuit of Fig. 5.35

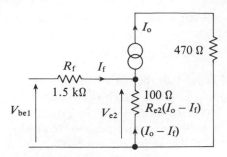

Fig. 5.37   Detail of Fig. 5.36 Showing Actual Current Directions with I_{b1} Positive

Therefore $I_s = 205\ I_{b1}$

$$A_{if} = \frac{3270}{205} \doteq 16 \qquad \left(\text{i.e. approx. } \frac{1}{\beta}\right)$$

$$R_{if} = \frac{V_{be1}}{205\ I_{b1}} = \frac{h_{ie}}{205} = 9.7\ \Omega$$

$$A_{vf} = \frac{V_o}{V_{be1}} = \frac{I_o R_{c2}}{I_s R_{if}} = \frac{16 \times 470}{9.7} = 775$$

$$\frac{V_s}{V_{be1}} = \frac{R_s + R_{if}}{R_{if}} = 104$$

Therefore $\dfrac{V_o}{V_s} = \dfrac{775}{104} = 7.5$

Since $h_{oe} = 0$, both R_o and R_{of} are infinite. Looking back into the 470 Ω load then R_o and R_{of} will both be 470 Ω.   □

5.4   USE OF NFB TO EXTEND BANDWIDTH

In Chapter 3 general expressions (equations (3.14) and (3.18)) were developed to characterize LF response and HF response. Generally, as amplification A falls with frequency, then the amount of NFB $(1 + \beta A)$ also falls and therefore the loss in amplification due to NFB will be less pronounced. As an example, consider the characteristic equation developed for HF loss (equation (3.18)):

$$A_{vh} = \frac{A_{vo}}{1 + jf/f_2} \tag{5.25}$$

where $A_{vh} = $ HF gain

$A_{vo} = $ MF gain

$f_2 = $ upper cutoff (3 dB) frequency.

Then applying the general expression for voltage gain with NFB,

$$A_{vf} = \frac{A_v}{1 + \beta A_v}$$

to the HF case,

$$A_{vhf} = \frac{A_{vh}}{1 + \beta A_{vh}} = \frac{A_v/(1 + jf/f_2)}{1 + \beta\{A_v/(1 + jf/f_2)\}}$$

$$= \frac{A_v}{(1 + \beta A_v) + jf/f_2} \tag{5.26}$$

At MF when the j term is negligible,

$$A_{vof} = \frac{A_{vo}}{1 + \beta A_{vo}}$$

which is the standard expression for gain of a feedback amplifier using A_{vo} to indicate MF gain. At HF the gain without NFB is given by equation (5.25) and will be 3 dB down when

$$A_{vh} = \frac{A_v}{1 + j} \qquad \text{i.e. when } f = f_2$$

The gain with NFB at HF will be 3 dB down when the real and imaginary parts of equation (5.26) are equal, i.e. when,

$$\frac{f}{f_2} = 1 + \beta A_v$$

and therefore

$$f = (1 + \beta A_v)f_2$$

which shows that the 3 dB cutoff frequency has been extended by a factor $(1 + \beta A_v)$. For an untuned amplifier the lower cutoff frequency will only be a few Hz or tens of Hz and is insignificant when specifying bandwidth, which is therefore approximately equal to the upper cutoff frequency. Bandwidth then, is increased by a factor $(1 + \beta A_v)$ as MF gain is reduced by $(1 + \beta A_v)$, i.e. the gain and bandwidth product is a constant. In this context the phrase 'constant gain × bandwidth product' means that, for a given shunt capacitance – representing transistor HF performance and/or stray capacitance – use of NFB allows us to trade gain for bandwidth.

A similar effect can be deduced for phase shift by referring to equation (5.26). This shows that while phase shift between input and output changes by $45°$ at frequency f_2 without NFB, then with NFB the $45°$ phase shift frequency will be $(1 + \beta A_v)f_2$.

Applying NFB to the LF equation (equation (3.14)),

$$A_{vl} = \frac{A_{vo}}{1 - jf_1/f}$$

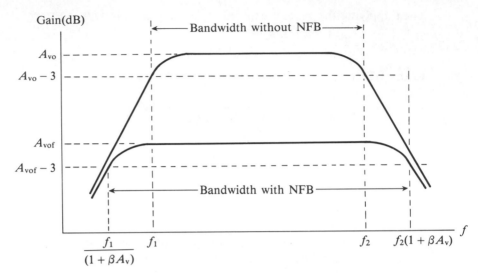

Fig. 5.38   Increasing Bandwidth at the Expense of Gain by Using NFB

will show the lower cutoff frequency being extended down by a factor $(1 + \beta A_v)$. Bode diagrams for an amplifier with and without NFB are shown in Fig. 5.38.

5.5   USE OF NFB TO REDUCE HARMONIC DISTORTION

Harmonic distortion occurs when a transistor is operated into nonlinear sections of its characteristics. This is most likely to happen in output stages where signal levels − peak-to-peak voltage and current values − are large. Figures 5.39a and b indicate the principle by which NFB reduces harmonic distortion. Figure 5.39a is for an amplifier without NFB. The input V_s is sinusoidal but one half-cycle of the output V_o has a flattened peak which is a characteristic shape of even harmonic distortion. In Fig. 5.39b − the waveforms after application of NFB − V_s has been increased to maintain V_o at the original level. The signal fed back, V_f, will, being subtracted from V_s, be in antiphase with it. Its negative peak will be flattened as it is a proportion (β) of V_o. Therefore V_i, which can be drawn by combining V_s and V_f, will have a sharply peaked positive half-cycle because there is less feedback. At this point then, the transistor will be driven harder to counteract the flattening.

To quantify the reduction in harmonic distortion consider an amplifier which, without NFB, produces, in addition to the signal required, a harmonic signal D. With NFB applied let the harmonic signal be D_f. V_s must be increased to maintain V_o at the required level and so D is still generated within the amplifier. However, since D_f is the distortion with NFB then $-\beta D_f$ is fed back to the input and appears at the output as $-\beta A D_f$.

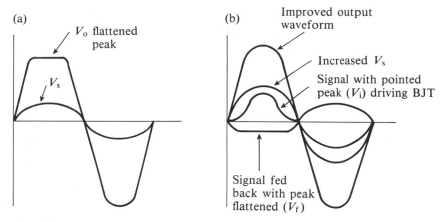

Fig. 5.39 Illustrating the Use of NFB to Reduce Harmonic Distortion (a) Without NFB (b) With NFB

Then $D_f = D - \beta A D_f$

Therefore $D_f = \dfrac{D}{1 + \beta A}$

Note that the analysis only holds if V_s is increased to compensate for the loss of amplification incurred by using NFB. But since the object is to reduce harmonic distortion for a given output signal level then the analysis represents the required condition. It is not difficult to increase V_s by extra amplification where signal levels are lower and therefore unlikely to become harmonically distorted.

5.6   USE OF NFB TO REDUCE NOISE

As regards noise, signal-to-noise ratio S/N is the significant quantity in specifying amplifier performance rather than absolute noise levels: provided the ratio is high then noise will be masked. When noise is generated within a section of an amplifier then application of NFB will reduce this noise provided the input signal level can be increased to counteract the loss of amplification caused by NFB. The analysis is similar to that used for harmonic distortion.

Without NFB let output noise $= N$

With NFB let output noise $= N_f$

Then $N_f = N - \beta A N_f$

$$N_f = \frac{N}{1 + \beta A}$$

However, the exercise is pointless unless the circuit which is used to increase the input signal does so without causing any deterioration of S/N. This can be achieved by an input stage especially designed to give high S/N by, for example, using FETs, screened leads, noise-free power supplies, etc.

5.7   STABILITY

Feedback is negative if it causes a reduction in gain. This will be the case if, in the general expression $A_f = A/(1 + \beta A)$, the value of $|1 + \beta A| > 1$. In most of the examples given in this chapter, βA, the loop gain, has been a positive real number and therefore $|1 + \beta A| > 1$. However, phase shift around the loop can lead to $|1 + \beta A|$ being less than 1, in which case $|A_f| > |A|$ and the benefits of NFB – for example, gain stability – will be lost. Further, in some circumstances, the value of βA can be such that the circuit will become unstable and burst into unwanted oscillation quite independently of the input signal. This is deliberately contrived in oscillator design (see Chapter 10), where $\beta A = -1$, therefore $A/(1 + \beta A) = \infty$ and the circuit gives an output with no input. Unwanted oscillation in an amplifier renders it unusable.

Nyquist Diagrams

The possibility of an amplifier becoming unstable can be estimated by drawing a Nyquist diagram which, in this context, is a plot of βA against frequency on a complex plane. In Fig. 5.40 the horizontal axis (R) represents the real part of βA and the vertical axis (I) the imaginary. For the single-stage amplifier with one RC coupling, whose Bode diagram is given in Fig. 3.32, then at MF, A is real and if β is derived from, say, a simple potential divider then OM represents βA. Being real, it lies along the R axis. Its length represents $|\beta A|$. As f falls, A, and therefore βA, fall and the phase angle leads so OL would represent βA at some frequency f_L. (If this were the 3 dB cutoff frequency, then $|\mathrm{OL}| = |\mathrm{OA}|/\sqrt{2}$ and arg OL $= 45°$.) As f continues to fall arg βA approaches $90°$ and $|\beta A|$ approaches 0. For increasing

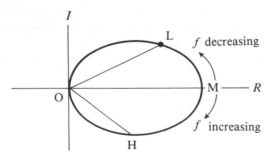

Fig. 5.40   Nyquist Diagram for a Single-stage Amplifier with One RC Coupling

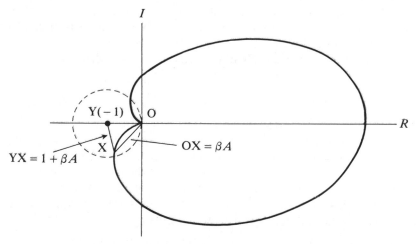

Fig. 5.41 Nyquist Diagram Showing that if βA is Inside Circle of Unity Radius and Center $(-1, 0)$ then $|1 + \beta A| < 1$

f, arg βA lags and the Nyquist diagram moves from M through H, and eventually to $\beta A = 0 \angle -90°$. Note that for this amplifier the whole of the Nyquist diagram is on the right of the I axis which, as we shall see, denotes good stability.

For a multistage amplifier with several RC couplings and shunt capacitances, arg βA can be greater than $90°$ so part of the Nyquist diagram will be to the left of the I axis as shown in Fig. 5.41. Consider any point X on the diagram so that

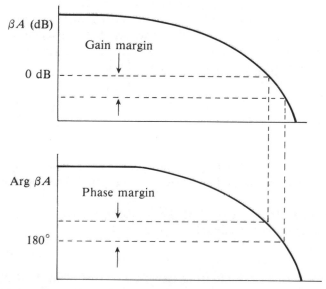

Fig. 5.42 Illustrating Gain and Phase Margins

OX represents βA. Also note the point Y $(-1, 0)$ where $OY = -1$. Then $YX = 1 + \beta A$ which is the feedback factor. If X lies inside a circle of unity radius with center Y then $|1 + \beta A| < 1$ and $|A_f| > |A|$ which is generally undesirable. Instability occurs when the Nyquist diagram encloses the point $(-1, 0)$.

Gain and Phase Margins

Rather than draw Nyquist diagrams it is more common to use Bode diagrams to specify two quantities which indicate the extent to which a feedback amplifier is safe from being oscillatory. First, there is gain margin, defined as the number of dBs that the loop gain $|\beta A|$ has fallen below 0 dB when the loop phase shift arg βA is $180°$. Second, there is phase margin which is defined at $|\beta A| = 0$ dB and is the angle by which arg βA is less than $180°$. The two quantities are illustrated in Fig. 5.42. Gain and phase margins should be at least -10 dB and $45°$ respectively.

SUMMARY

Instead of quantities such as gain and input and output resistance being dominated by transistor parameters, they can be determined, by using NFB, to meet specific requirements. NFB can also be used to stabilize gain, to extend frequency response and to improve, under certain conditions, signal-to-noise ratio and harmonic distortion content. These benefits are acquired by sacrificing gain but the modest cost of extra transistor stages is usually justified and, as we shall see in Chapter 8 (on operational amplifiers) gain is often available in abundance. The circuits which have been analyzed are no more than a representative selection of circuits which will be found in practice, but the techniques used are generally applicable. Usually, the designer with a specification to meet will employ these techniques on a trial-and-error basis.

REVISION QUESTIONS

5.1   An amplifier with an open loop gain $A_v = 5000 \pm 500$ is available. It is necessary to have an amplifier whose voltage gain varies by no more than $\pm 1\%$.
(a) Find the feedback factor β of the feedback network used.
(b) Find the gain with feedback.

(0.0018; 500)

5.2   An amplifier with an open loop voltage gain of 1000 delivers 10 W of output power at 10% harmonic distortion when the input signal is 10 mV. If 40 dB negative voltage series feedback is applied and the output power is to remain at 10 W, determine (a) the required input signal, (b) the percentage harmonic distortion.

(1 V; 0.1%)

5.3 An amplifier has an amplification of $1000 + j0$ at 1000 Hz and $300 - j500$ at 20 kHz. Calculate the value of the amplification at these two frequencies if 2.5% of the output voltage is fed back to the input as negative feedback.

$(38.5 + j0; 38.5 - j2.2)$

5.4 An amplifier without feedback is liable to changes of $\pm 10\%$ in gain as a result of fluctuations in supply voltage. If a feedback amplifier is required to operate from this supply to give a gain of $150 \pm 1\%$ calculate the inherent gain of the amplifier and the feedback factor.

(1500; 0.006)

5.5 A negative feedback amplifier is required to have an overall gain of 100. The amplifier is to be designed so that the internal gain of each of its three stages may fall by 10% without causing more than a 1% reduction in the overall gain. Calculate how large an internal gain (i.e. the gain without feedback) must be used.

(2700)

5.6 When an amplifier's load resistance is increased, its output voltage will rise. Use this as a starting point to argue qualitatively that voltage NFB decreases and current NFB increases the amplifier's output resistance.

5.7 In Fig. Q.5.7 the 2 kΩ source is connected to the 200 Ω load via an emitter follower. Calculate V_o. What would be the value of V_o if the source were connected direct to the load?
BJT data: $h_{fe} = 100$; $h_{ie} = 2.5$ kΩ.

(0.77 V; 0.09 V)

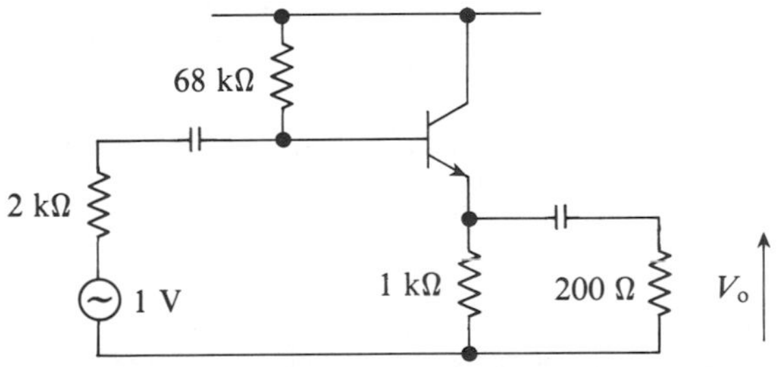

Fig. Q.5.7

5.8 In Fig. Q.5.8 calculate the source follower output voltage V_o and also the value of V_o if the signal source is connected direct to the 150 Ω load.

FET data: $g_{fs} = 4$ mS

(0.29 V; 0.015 V)

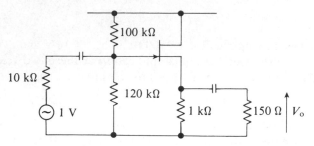

Fig. Q.5.8

5.9   For Fig. Q.5.9 calculate A_{vf}, R_{if} and R_{of}.
      BJT data: $h_{fe} = 80$; $h_{ie} = 1.5$ kΩ.

(−9.7; 7.5 kΩ; 10 kΩ)

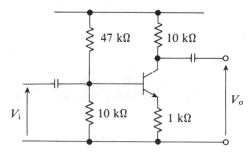

Fig. Q.5.9

5.10 For Fig. 5.26 calculate A_{vf}, R_{if} and R_{of} if the 20 kΩ source resistance is
      replaced by 50 kΩ, the 47 Ω feedback resistor by 100 Ω, and $h_{ie} = 1.5$ kΩ.

(27; 17.3 kΩ; 45 Ω)

5.11 For Fig. 5.35 calculate A_{vf}, A_{if}, R_{if} and R_{of} if $h_{fe} = 150$, $h_{ie} = 3$ kΩ, $R_{e2} = 50$ Ω
      and $R_f = 1$ kΩ.

(254; 20.7; 38.3 Ω; 470 Ω)

5.12 State and justify whether the negative feedback introduced to the circuit of
      Fig. Q.5.12 by the 100 kΩ resistor is voltage or current derived and whether
      it is series or shunt applied. Calculate the mid-frequency gains V_o/V_1 and
      V_o/V_s. Explain the effect of negative feedback in this case on the LF response
      and calculate the lower cutoff frequency.

FET data: $g_m = 5$ mS; $r_d = 40$ kΩ

$(-21; \ -66; \ 54 \text{ Hz})$

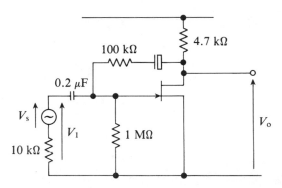

Fig. Q.5.12

6

Power Amplifiers

Chapter Objectives
6.1   Class A Amplifiers
6.2   Class B Complementary Symmetry Push–Pull Amplifier
6.3   Class AB
6.4   Thermal Resistance and Heat Sinks
6.5   Integrated Circuit Power Amplifiers
        Summary
        Revision Questions

For many applications an amplifier may be required to deliver a substantial amount of power, for example, to a loudspeaker, or to produce a large peak-to-peak voltage swing, for example, the drive to a CRT. In such cases small-signal equivalent circuits cannot be used for design and analysis and it is customary to establish the operating principles of power amplifiers – the usual term for large-signal amplifiers – by using transistor characteristics. An important aspect of power amplifier design is electrical efficiency, not because we are concerned at the waste of electrical power as such, but because of the problem of avoiding the overheating associated with the dissipation of wasted power and because of the cost involved in using transistors and associated components of unduly large power specification.

Integrated circuit power amplifiers are available for a range of power handling capabilities and bandwidths. Their designs are based on the general principles described in this chapter.

6.1   CLASS A AMPLIFIERS

Figure 6.1 is a common emitter amplifier with bias components omitted – but assume they have been chosen to make $V_{CN(Q)} = V_{CC}/2$, in this case 10 V. This quiescent condition permits the maximum possible undistorted output voltage swing. If the ac output power P_o is to be delivered to the collector load R_L, then

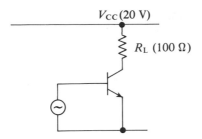

Fig. 6.1   Class A Amplifier

Fig. 6.2 is the load line for this circuit and it is drawn on collector characteristics which are ideal in that the minimum value of v_{CE} ($V_{CE(SAT)}$) is virtually 0 V. Assuming that the circuit is linear until v_{CE} falls to 0 V or rises to 20 V, then with a sine wave input signal, the maximum peak-to-peak output voltage swing is 20 V, i.e.

$$V_{o\ pk-pk} = V_{CC} = 20\text{ V}$$

and the maximum peak-to-peak output current is

$$I_{o\ pk-pk} = \frac{V_{CC}}{R_L} = 0.2\text{ A}$$

Then the maximum ac output power $P_{o(max)}$ delivered to R_L is the product of the

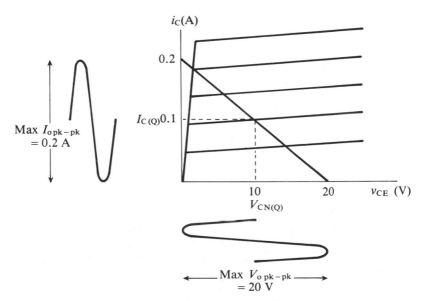

Fig. 6.2   Illustrating Maximum Peak-to-peak Output Voltage and Current Swings in a Class A Amplifier with an Ideal Transistor

rms output voltage and rms output current when maximum, i.e.

$$P_{o(max)} = \frac{V_{CC}}{2\sqrt{2}} \frac{V_{CC}}{R_L} \frac{1}{2\sqrt{2}} = \frac{V_{CC}^2}{8R_L} \tag{6.1}$$

All power supplied P_s to the circuit comes from the dc source V_{CC}.

Therefore $P_s = V_{CC}I_{c(av)} = V_{CC}I_{C(Q)}$

$$= V_{CC} \frac{1}{2} \frac{V_{CC}}{R_L} = \frac{V_{CC}^2}{2R_L} \tag{6.2}$$

Note that this power will be supplied even if there is no output signal, i.e. $P_o = 0$.

Amplifier efficiency η is defined as $\dfrac{P_o(max)}{P_s} \times 100\%$

Then the maximum efficiency $\eta_{(max)}$ occurs when $P_o = V_{CC}^2/8R_L$ and even this is unrealistically favorable in that it assumes an ideal transistor. Then

$$\eta_{(max)} = \frac{V_{CC}^2/8R_L}{V_{CC}^2/2R_L} \times 100\% = 25\% \tag{6.3}$$

In Fig. 6.1, $P_{o(max)} = 0.5$ W, $P_s = 2$ W.

One of the disadvantages of this circuit is that the quiescent current passing through R_L does not produce useful ac signal power. Therefore to find the power dissipated by the transistor, P_D, which is needed to specify the power rating of the transistor, we note that

$$P_D = P_s - P_o - P_{dc}$$

where P_{dc} is the dc power dissipated in R_L. This will be constant at

$$I_{C(Q)}^2 R_L = \left(\frac{V_{CC}}{2R_L}\right)^2 R_L = \frac{V_{CC}^2}{4R_L}$$

Therefore $P_{D(max)}$ occurs when $P_o = 0$, i.e.

$$P_{D(max)} = P_s - P_{dc} = \frac{V_{CC}^2}{2R_L} - \frac{V_{CC}^2}{4R_L} = \frac{V_{CC}^2}{4R_L} \tag{6.4}$$

in this example, 1 W.

It is possible to couple the transistor to the load by using a transformer with its primary in series with the collector. This improves the maximum efficiency to 50% and avoids having dc through the load, but the technique is not favored nowadays because of the availability of complementary (i.e. matched npn and pnp pairs) transistors, a concept which cannot be realized with valves.

6.2 CLASS B COMPLEMENTARY SYMMETRY PUSH–PULL AMPLIFIER

In Fig. 6.3, Q_1 conducts for only positive half-cycles of the input signal v_s and Q_2 for only the negative half-cycles. This mode of operation is called class B. In this circuit, note that the transistors are complementary, i.e. Q_1 is npn and Q_2 is pnp. There is zero dc bias between base and emitter of each transistor so they only conduct when driven in the appropriate polarity by the input signal. When v_s goes positive, Q_1 conducts and Q_2 is driven further into cutoff. Current i_{C1} flows from $V_{CC1} \rightarrow Q_1 \rightarrow R_L \rightarrow V_{CC1}$, so v_o goes negative. When v_s goes negative, Q_2 only conducts and i_{C2} flows from $V_{CC2} \rightarrow R_L \rightarrow Q_2 \rightarrow V_{CC2}$, so v_o goes positive. The waveforms in Fig. 6.4 show each transistor operating in class B. The load current ($i_L = i_{C2} - i_{C1}$) consists of a half-cycle in one direction from Q_1 and a half-cycle in the opposite direction from Q_2. The term 'push–pull' is derived from this action. Transistors are supplied with matched characteristics for this type of circuit.

Power Output P_o

The load line diagram for Q_1 (Fig. 6.5) shows the operating point P at V_{CC1}. When Q_1 is driven, v_{CE1} falls as i_{C1} rises. If the peak value of the ac component of v_{CE1} is V_p, then the peak value of i_{C1} will be

$$I_p = \frac{V_p}{R_L}$$

Given that the two transistors are matched then similar conditions obtain for Q_2 and for sine wave operation we have that

$$P_o = \frac{V_p}{\sqrt{2}} \frac{I_p}{\sqrt{2}} = \frac{V_p^2}{2R_L} \tag{6.5}$$

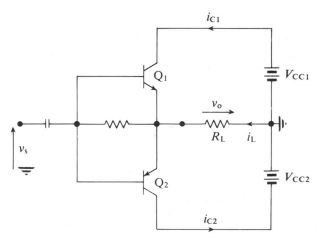

Fig. 6.3   Basic Form of a Class B Complementary Push–Pull Amplifier

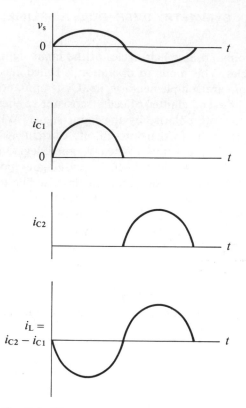

Fig. 6.4   Waveforms for Fig. 6.3

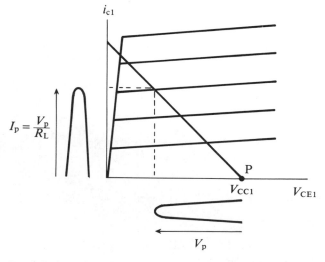

Fig. 6.5   Load Line Diagram for npn Transistor Operating in Class B Mode

If the two dc supply voltages are equal to V_{CC}, then assuming ideal transistors, the maximum value of $V_p = V_{CC}$ and therefore,

$$P_{o(max)} = \frac{V_{CC}^2}{2R_L} \tag{6.6}$$

Each of the transistors of Fig. 6.3 are operating as CE amplifiers and therefore the circuit can have a substantial voltage gain as well as a substantial current gain. Since neither power supply is earthed neither can be used for other parts of a total system.

Efficiency

Clearly class B operation is more efficient than class A because the amplifier only draws current from the dc supplies when there is an input signal and when signal power is to be delivered to the load. However, when this occurs there will be power loss in the transistors. Referring to Fig. 6.5, the power drain from each dc source is a series of half sine waves, therefore the total average current supplied is

$$I_{av} = \frac{2}{\pi} \frac{V_p}{R_L}$$

Therefore the total power supplied from the two dc sources is

$$P_s = V_{CC} I_{av} = V_{CC} \frac{2}{\pi} \frac{V_p}{R_L} \tag{6.7}$$

and is maximum when $V_p = V_{CC}$.

$$\text{Efficiency } \eta = \frac{P_o}{P_s} = \frac{(V_p^2/2R_L)}{V_{CC}(2/\pi)(V_p/R_L)} \times 100\% = \frac{\pi}{4} \frac{V_p}{V_{CC}} \times 100\%$$

and is clearly maximum in the maximum power output condition when $V_p = V_{CC}$.

$$\text{Therefore } \eta_{(max)} = \frac{\pi}{4} \times 100\% = 78\% \tag{6.8}$$

Transistor Power Dissipation

In the class A amplifier the transistor has to dissipate maximum power when the output power P_o is zero. In the class B amplifier no power is supplied and therefore none dissipated in the transistor when P_o is zero. However, P_D is not maximum when P_o is maximum. To find $P_{D(max)}$ note that at all times, for the amplifier,

$$P_D = P_s - P_o$$

and therefore for sine wave operation, using equations (6.7) and (6.5),

$$P_D = V_{CC} \frac{2}{\pi} \frac{V_p}{R_L} - \frac{V_p^2}{2R_L} \tag{6.9}$$

$$\frac{dP_D}{dV_p} = \frac{2}{\pi} \frac{V_{CC}}{R_L} - \frac{V_p}{R_L} = 0 \quad \text{for } P_{D(max)}$$

and

$$V_p = \frac{2}{\pi} V_{CC}$$

Substituting in equation (6.9),

$$P_{D(max)} = \frac{2}{\pi} \frac{2}{\pi} \frac{V_{CC}^2}{R_L} - \frac{4}{\pi^2} \frac{V_{CC}^2}{2R_L}$$

$$= \frac{2}{\pi^2} \frac{V_{CC}^2}{R_L} \doteq 0.2 \frac{V_{CC}^2}{R_L}$$

Therefore, using equation (6.6),

$$P_{D(max)} = 0.4 \, P_{o(max)}$$

and this is the dissipation for the amplifier, i.e. the total for the two transistors. Therefore, for each transistor,

$$P_{D(max)} = 0.2 \, P_{o(max)}$$

In other words, the power rating for each transistor must be at least one-fifth of the maximum sinusoidal output power.

☐ **Example 6.1**

In Fig. 6.3, $V_{CC1} = V_{CC2} = 20$ V, $R_L = 50 \, \Omega$.

Calculate the transistor power ratings.

$$P_{o(max)} = \frac{V_{CC}^2}{2R_L} = 4 \text{ W}$$

$$P_{D(max)} = 0.2 \, P_{o(max)} = 0.8 \text{ W} \qquad\qquad\qquad ☐$$

6.3 CLASS AB

The waveforms of Fig. 6.4 are idealized. In practice if the input signal is a voltage sine wave then the base currents at the low voltage parts of the input waveform are disproportionately small because of the nonlinearity of the transistor characteristics.

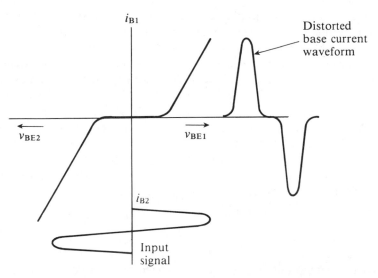

Fig. 6.6 Production of Crossover Distortion in Class B Amplifier

This results in base current (and therefore collector current and output voltage) waveforms shaped as in Fig. 6.6. Odd harmonic distortion is signified by the symmetry of the two half-cycles. The effect is called 'crossover distortion' and can be eliminated by supplying a small dc base bias to each transistor so that the input signal only operates on the linear parts of the input characteristics. This mode of operation is class **AB**.

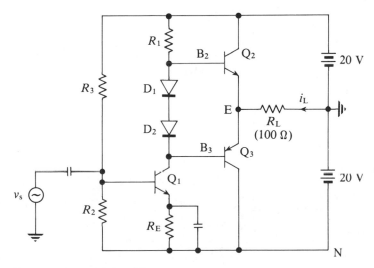

Fig. 6.7 Class AB Amplifier

A common method of implementing class AB operation is illustrated in Fig. 6.7. Q_1 will be biased for class A operation with R_1 as its collector load. Then in the quiescent condition ($v_s = 0$), $I_{L(Q)} = 0$ and $V_{EN(Q)} = 20$ V. The current through D_1 and D_2 produces a voltage $V_{D1} + V_{D2} = V_{B2E} + V_{EB3}$. In the quiescent condition this current will be $I_{C1(Q)}$ and can be chosen so that Q_2 and Q_3 are biassed for class AB operation. Taking $V_{B2E} = 0.8$ V then $I_{C1(Q)}$ and R_1 are chosen to make $V_{B2N} = 20.8$ V, giving $V_{EN(Q)} = 20$ V as required. It is common practice to mount the diodes on the same heat sink as Q_2 and Q_3 so that as they heat up, $V_{D1} + V_{D2}$ falls and reduces the quiescent current: in essence this is thermal NFB.

In ac operation, as v_s goes positive, v_{CN1} falls. This increases the current supplied to R_L by Q_3, while Q_2 is cut off. As v_s goes negative v_{CN1} rises which cuts off Q_3 and increases the current from Q_2 through R_L.

☐ **Example 6.2**

Design the Q_1 driver stage for Fig. 6.7.

When v_s is maximum negative, Q_1 approaches cutoff and v_{EN} (Fig. 6.8, which is a detail of Fig. 6.7) and v_{B2N} will approach the upper supply rail (+ 40 V). In this condition i_{B2} is maximum and will flow through R_1. In choosing a value for R_1 it must be remembered that any volt drop across it when i_{B2} is maximum will have to be deducted from 40 V in determining the output voltage swing available. Allowing 2 V, then when i_{B2} is maximum,

$$v_{B2N} = 38 \text{ V}$$

and if $v_{B2E} = 1$ V in this condition (we should expect it to be higher than the usual 0.8 V) then,

$$v_{EN} = 37 \text{ V}$$

Then $i_{C2(max)} = \dfrac{17 \text{ V}}{100 \text{ }\Omega} = 170 \text{ mA}$

Taking $h_{FE} = 100$, $i_{B2(max)} = 1.7$ mA and,

$$R_1 = \frac{2 \text{ V}}{1.7 \text{ mA}} = 1.2 \text{ k}\Omega.$$

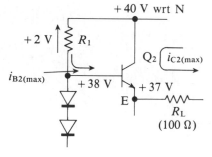

Fig. 6.8   Detail of Fig. 6.7 with v_s Maximum Negative

When $v_s = 0$, $v_{B2N} = 20.8$ V.

Then $i_{C1} = \dfrac{19.2 \text{ V}}{1.2 \text{ k}\Omega} = 16$ mA.

This is the quiescent current which will flow through D_1 and D_2 and they should be chosen by inspecting the B–E characteristics of Q_2 and Q_3 and finding the diodes which give class AB bias conditions.

Allowing, say, 1.5 V across R_E,

$$R_E = \frac{1.5 \text{ V}}{16 \text{ mA}} = 94 \ \Omega$$

Now R_2 and R_3 can be chosen by the methods set out in Chapter 2 (1.5 kΩ and 22 kΩ would be satisfactory).

The output transistors Q_2 and Q_3 are operating as emitter followers, so this push pull stage has no voltage gain. There must therefore be enough gain in the whole system up to and including Q_1 to produce the output voltage swing required.

$\square$

Improved Complementary Symmetry Amplifier

One limitation of Fig. 6.7 concerns the minimum value of R_L which can be driven. For example, if R_L were 8 Ω (a standard loudspeaker resistance) then the peak-to-peak current in each of the output transistors would be 2.5 A and the driver transistor Q_1 would have to handle a proportionally larger current. The current gain of

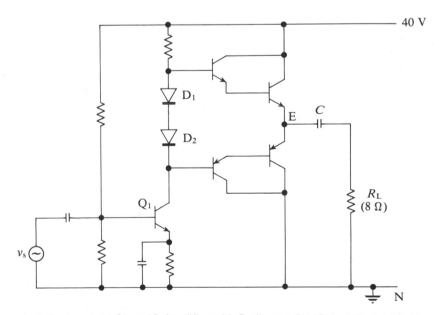

Fig. 6.9 Improved Class AB Amplifier with Darlington Pair Output Transistors for Driving Low Load Resistance and a Single Power Supply with Negative Side to Ground

the output stage can be increased by using Darlington pairs (see Chapter 7) as in Fig. 6.9. Now the current gain of each pair is approximately $h_{FE}{}^2$. Another disadvantage of Fig. 6.7 is the need for two dc power supplies. The technique of Fig. 6.9 can be used regardless of whether the push–pull stage uses Darlington pairs or single transistors. With Q_1 biased to make $V_{EN(Q)} = 20$ V then C will be charged up to 20 V with $v_s = 0$. If $X_c \ll R_L$ at the lowest required frequency, then C will retain the 20 V and the operating voltage for the push–pull transistors will be the same for Q_2 and Q_3 in Fig. 6.7.

6.4   THERMAL RESISTANCE AND HEAT SINKS

Manufacturers specify maximum junction temperatures T_j for transistors – usually $150\,^\circ$C for silicon – above which irreparable damage occurs. They will also quote a total device dissipation P_D at what is usually taken as ambient temperature, $T_A = 25\,^\circ$C. For example, the Motorola BD 135 is quoted as

$$P_D = 1.25 \text{ W at } T_A = 25\,^\circ\text{C}$$

Quoting P_D at ambient temperature implies that the device is operating in free air without a heat sink. Also quoted will be the thermal resistance between junction and free air, θ_{JA}. Thermal resistance is a measure of rise in temperature ($^\circ$C) per watt of power dissipated, giving the relationship

$$T_j - T_A = \theta_{JA} P_D \tag{6.10}$$

For the BD 135, $\theta_{JA} = 100\,^\circ$C/W and at maximum device dissipation $P_D = 1.25$ W, and $T_A = 25\,^\circ$C,

$$T_j - 25 = 100 \times 1.25 = 125\,^\circ\text{C}$$

and $T_j = 150\,^\circ$C, which is the maximum junction temperature for silicon. If T_A is higher than $25\,^\circ$C then P_D must be derated. For example, if $T_A = 60\,^\circ$C and T_j must not exceed $150\,^\circ$C then,

$$P_D = 0.9 \text{ W.}$$

θ_{JA} is not always quoted for a BJT but another quantity, θ_{JC}, the thermal conductivity between a transistor junction and its case, is invariably quoted for a power transistor and for the BD 135 is $10\,^\circ$C/W. If the case could be held at a constant temperature of $25\,^\circ$C then

$$T_j - 25 = \theta_{JC} P_D$$

and

$$P_D = 12.5 \text{ W.}$$

It is not usually practicable to hold the transistor case at ambient temperature but use of a heat sink does allow the device to be operated at a higher power than is permissible in free air. Thermal resistance now becomes a complex quantity and using the general equation

$$T_j - T_A = \theta P_D,$$

θ is the sum of several thermal resistances. Simply using a heat sink we have

$$\theta = \theta_{JC} + \theta_{CA}$$

where θ_{CA} is the thermal resistance between the transistor case and free air, i.e. it is the thermal resistance of the heat sink. Using the BD 135 with a heat sink of $3\,^{\circ}\text{C/W}$ with $T_A = 25\,^{\circ}\text{C}$,

$$T_j - 25 = (10 + 3)P_D$$

and

$$P_D \doteq 9.5 \text{ W}$$

Again, if T_A is higher than $25\,^{\circ}\text{C}$ the calculation has to be modified. Sometimes an electrically insulating washer has to be placed between heat sink and case, which increases the total thermal conductivity. The effect can be minimized using silicon grease and the total increase is typically $1\,^{\circ}\text{C/W}$.

6.5   INTEGRATED CIRCUIT POWER AMPLIFIERS

Power amplifiers are available in IC form with output capabilities ranging from around 2 W to over 100 W. In addition to having class AB characteristics – high power gain, high efficiency, low total harmonic distortion (THD) – IC power amplifiers can have built-in features such as high voltage gain, current limiting and automatic shutdown to prevent overheating.

A circuit using the LM 380, which will deliver 2 W into an 8 Ω load, is shown in Fig. 6.10 with all of the external components required. The IC itself is in the form of a 14 pin DIL package with $\theta_{JA} = 100\,^{\circ}\text{C/W}$. Clearly 2 W dissipation would damage the device but 7 of the 14 pins have to be grounded and by connecting some of these to the copper strip of a PCB then θ can be reduced – 4 in^2 of PCB reduces θ to $50\,^{\circ}\text{C/W}$.

A more powerful device is the HY 60 which is, as its code letters suggest, a hybrid IC. A hybrid IC is one which contains passive components made from thick or thin film circuit techniques as well as active silicon devices; a monolithic IC contains only active silicon devices even though many of them may be serving as resistors. The HY 60 has an integral heat sink and can deliver 30 W into an 8 Ω load. It does require a ± 25 V dc supply but, as Fig. 6.11 shows, the circuitry for an audio amplifier is very simple.

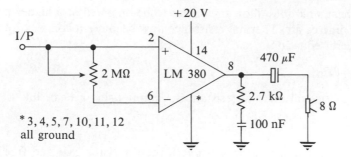

Fig. 6.10   2 W AF Amplifier (Courtesy RS Components Ltd)

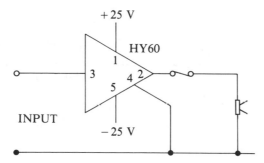

Fig. 6.11   30 W AF Amplifier (Courtesy RS Components Ltd)

SUMMARY

In this chapter we have looked at some of the techniques for amplifying large signals. Class A operation suffers from the disadvantage that the transistors are required to dissipate a large amount of power even when there is no input signal. In class B push pull operation the transistors do not have to dissipate as much power as in class A and when there is no input signal the class B operated transistors dissipate hardly any power at all. Crossover distortion is inherent in class B operation but can be eliminated by providing a small quiescent current, which puts the transistors into class AB operation. In spite of high power efficiency, class B and AB operated transistors do have to dissipate power and manufacturers provide data on thermal conductivity and heat sinks which enable designers to provide circuits in which transistors are protected from overheating.

REVISION QUESTIONS

6.1 A power amplifier stage has an efficiency of 30%. The average signal power in the load is 5 W. What is the average current taken from the 24 V supply?

(0.69 A)

6.2 A power amplifier stage has a single transistor which is dissipating 2 W. The 30 V power supply is giving an average current of 80 mA. What is the ac output power?

(0.4 W)

6.3 A single-transistor power-amplifier stage has an efficiency of 20%. The total power dissipated in the output transistor is 10 W. What is the ac output power?

(2.5 W)

6.4 A power amplifier stage takes 150 mA from a 20 V supply. The power dissipated in the output transistors is 1.8 W. What is the efficiency?

(40%)

6.5 For the complementary symmetry class B amplifier shown in Fig. Q.6.5, calculate, from first principles:
(a) the total collector power dissipation when the output power is maximum;
(b) the output power when the collector power dissipation is maximum.
Assume that the input signal is sinusoidal and that the transistor characteristics are ideal. Explain why class B operation produces crossover distortion and briefly describe how this distortion can be avoided.

(23 W; 33 W)

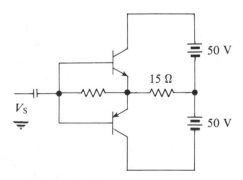

Fig Q.6.5

6.6 Explain the action of the power amplifier circuit of Fig. Q.6.6. What are the functions of the network DR_2?

The circuit is to be designed so that the maximum possible amount of power is to be delivered to the 40 Ω load under the sinusoidal conditions. Choose suitable values for R_1, R_2 and R_3, making the quiescent voltage between the bases of the output transistors 1.5 V. Calculate the maximum output power.

Transistor data: $V_{BE(ON)} = 0.8$ V $= V_{DIODE}$

$$h_{FE} = 75$$

(470 Ω; 22 Ω; 68 Ω; 1.8 W calculated allowing 2 V across R_1 when output positive and maximum and $V_{BE(max)} = 1$ V)

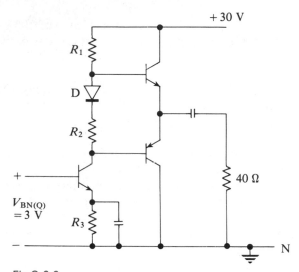

Fig Q.6.6

6.7 A BJT has a thermal resistance $\theta_{JA} = 50\,°C/W$. Assuming a maximum permissible junction temperature of 150 °C, calculate the maximum power dissipation when the ambient temperature is:

(a) 25 °C;
(b) 70 °C.

(2.5 W; 1.6 W)

6.8 A BJT is quoted as having $\theta_{JC} = 5\,°C/W$. What must be the thermal resistance of the heat sink on which the BJT must be mounted to run the BJT at 10 W in an ambient temperature of 70 °C? Assume $T_{J(max)} = 150\,°C$.

(3 °C/W)

7

Analog Circuits

Chapter Objectives
7.1   High Input Resistance Bootstrap Circuit
7.2   Darlington Pair
7.3   Current Sources
7.4   Differential Amplifier
7.5   Cascode Amplifier
7.6   Voltage Level Shifting
        Summary
        Revision Questions

Digital electronics continue to be used for many applications traditionally effected by analog techniques. Nevertheless, for many requirements analog electronics still provide the better, and indeed sometimes the only, solution. Moreover, analog circuit development has not languished: remarkable ingenuity by analog designers has resulted in the development of the IC op amp and a range of more specialist analog ICs. It is necessary that some of the circuits used within these devices and in discrete component design should be described in a book of this nature.

7.1 HIGH INPUT RESISTANCE BOOTSTRAP CIRCUIT

One useful property of the emitter follower (Section 5.3) is its high input resistance R_i. In its basic form (Fig. 5.10) $R_i \doteqdot h_{fe}R_e$, but in its practical form R_i will be shunted by bias resistors: for some applications this will result in an unacceptably low value of input resistance. The term 'bootstrapping' is applied in widely different circuit applications (e.g. see the bootstrap sweep generator in Chapter 11). In Fig. 7.1 it is a method of biasing the emitter follower so that the reduction of input resistance is insignificant. As V_i increases, V_o follows it and this signal is fed back to the junction of R_1, R_2 and R_3. Without R_3, R_1 and R_2 would be in parallel across the input. With R_3 and C_2, then apart from I_b the only other component of input

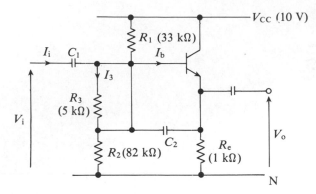

Fig. 7.1   High Input Resistance Bootstrap Circuit

signal current I_i is I_3 which must pass through R_3. If the gain $A_v = V_o/V_i$ were unity then there would be no signal voltage across R_3, and I_3 would be zero. This would mean that the bias chain would have no effect on the input resistance which could then be taken as $h_{fe}R_e$. Allowing for $A_v < 1$, then

$$I_3 = \frac{V_i - A_v V_i}{R_3} = \frac{V_i(1 - A_v)}{R_3}$$

so the shunting effect of the bias network would be

$$R_b = \frac{V_i}{I_3} = \frac{R_3}{1 - A_v} \tag{7.1}$$

$$R_i = h_{fe}R_e \parallel R_b \tag{7.2}$$

As A_v approaches unity the effect of R_b becomes negligible. Although R_1 and R_2 do not appear in this analysis it should be noted that, for signal conditions, they are in parallel with one another and with R_e, and therefore they affect A_v. For dc conditions the Thevenin equivalent of the bias resistors R_B (see Section 2.9) will be

$$R_B = R_3 + R_1 \parallel R_2$$

and the bias calculations should be modified accordingly.

☐   **Example 7.1**

Using the component values given in Fig. 7.1 and taking $h_{ie} = 2$ kΩ, $h_{FE} = h_{fe} = 50$, calculate R_i, A_v and $V_{EN(Q)}$.

$R_1 \parallel R_2 \doteqdot 24$ kΩ so their shunting effect on R_e is negligible.

Then $A_v = \dfrac{(1 + h_{fe})R_e}{h_{ie} + (1 + h_{fe})R_e}$ (equation (5.13))

$= 0.96$

Using equation (7.1),

$$R_b = \frac{R_3}{1 - A_v} = \frac{5 \text{ k}\Omega}{0.04} = 125 \text{ k}\Omega$$

$$R_i = h_{fe}R_e \| R_b = 50 \text{ k}\Omega \| 125 \text{ k}\Omega = 36 \text{ k}\Omega$$

Using the same values of R_1 and R_2 without bootstrapping, R_i would be 16 kΩ. Using equation (2.7),

$$V_T \quad = 7.1 \text{ V}, \; R_T = R_1 \| R_2 + R_3 = 29 \text{ k}\Omega, \text{ and taking } V_{BE} = 0.8 \text{ V},$$

$$I_E \quad = 4 \text{ mA; therefore}$$

$$V_{EN(Q)} = 4 \text{ V} \qquad\qquad\qquad\qquad\qquad\qquad \square$$

7.2 DARLINGTON PAIR

A pair of transistors connected as in Fig. 7.2 gives, in effect, a single transistor with a very high value of h_{fe}. The arrangement is called a Darlington pair and is often manufactured as an encapsulated package with just three terminals. Darlington pairs are often used in linear ICs such as op amps and in power amplifier output stages. The idea can be extended to three transistors but this is usually the limit.

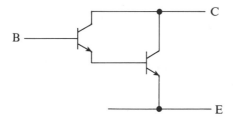

Fig. 7.2   Darlington Pair

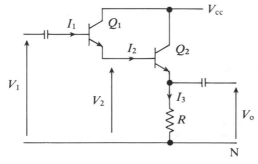

Fig. 7.3   Darlington Pair Emitter Follower

Figure 7.3 is an emitter follower and is the most common application of the Darlington pair. For a given load resistance R the arrangement gives a higher value of input resistance than a single transistor so it can either be seen as a very high input resistance circuit or as a means of driving small load resistances.

Analysis

Using the emitter follower relationship established in Section 5.3, then for Q_2,

$$A_{i2} = \frac{I_3}{I_2} = (1 + h_{fe2}) \doteq h_{fe2}$$

in which h_{oe2} is neglected.

$$R_{i2} = \frac{V_2}{I_2} \doteq h_{fe2}R$$

For Q_1 the emitter load is R_{i2}. The emitter current of Q_1 is I_2 and flows into R_{i2}: it is therefore the base current of Q_2. R_{i2} will be high, so it is necessary to include h_{oe1} in the analysis. In Fig. 7.4 we can see that $h_{fe1}I_1 (I_1 = I_{b1})$ divides between R_{i2} and h_{oe1}.

Then $I_2 = \dfrac{h_{fe1}I_1 \times (1/R_{i2})}{(1/R_{i2}) + h_{oe1}} = \dfrac{h_{fe}I_1}{1 + h_{oe1}R_{i2}}$

Then $A_i = \dfrac{I_3}{I_1} = \dfrac{I_3}{I_2}\dfrac{I_2}{I_1} = \dfrac{h_{fe}{}^2}{1 + h_{oe}h_{fe}R}$ 　　　　　　　　　　　　　(7.3)

where the transistors are assumed to have identical parameters. To find the input resistance,

$$V_1 = h_{ie}I_1 + A_{i1}I_1R_{i2}$$

$$R_i = \frac{V_1}{I_1} = h_{ie} + \frac{h_{fe}}{1 + h_{oe}h_{fe}R}\, h_{fe}R \qquad\qquad (7.4)$$

in which h_{ie} will be negligible.

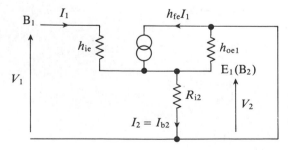

Fig. 7.4   Illustrating Division of $h_{fe}I_1$

☐ **Example 7.2**

Assuming identical transistors in Fig. 7.3 with $h_{fe} = 80$, $h_{oe} = 25\ \mu S$ and $R = 1\ k\Omega$, find A_i and R_i.

$$h_{oe}h_{fe}R = 2$$

Then using equations (7.3) and (7.4), $A_i = \dfrac{80^2}{3} = 2140$

and $R_i = \dfrac{80^2 \times 1\ k\Omega}{3} = 2.1\ M\Omega$ ☐

Biasing the Darlington Emitter Follower

Using standard biasing techniques described in Chapter 2 clearly reduces R_i. If this is unacceptable then a method such as bootstrapping must be used.

7.3 CURRENT SOURCES

The essential characteristic of a current source – more fully, a constant current source – is that it must have a high source (i.e. output) resistance. To take the case of a simple dc current source, consider Fig. 7.5. If $R_L = 0$ then $I_0 = 10\ \mu A$. If R_L is increased then as long as it does not exceed 10 kΩ, I_0 will not fall below 10 μA by more than 1%. To increase the current to, say 1 mA, would require a voltage source of 1000 V and such voltages are not usually available in electronic circuits. By using transistors to generate the required currents, high source resistance can be achieved with a modest supply voltage.

Common emitter and common drain amplifiers have output resistances of $1/h_{oe}$ and r_{ds} respectively if the collector and drain resistances are taken to be the loads. For a higher output resistance then in Fig. 7.6, I_0 is set by R_s and since it is not decoupled it is providing current NFB which increases R_o, as demonstrated in Chapter 5.

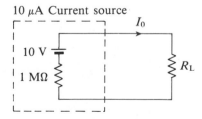

Fig. 7.5   The 10 V, 1 MΩ can be Regarded as a 10 μA Constant Current Source

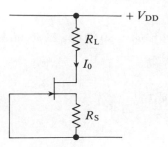

Fig. 7.6   FET Current Source

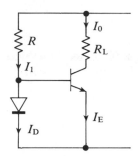

Fig. 7.7   BJT Current Source

For a BJT a constant current I_0 can be supplied to the load R_L in the circuit of Fig. 7.7. If the diode has the same I/V characteristic as B–E of the transistor then $I_D = I_E$ and neglecting base current, $I_0 = I_1$.

Good matching is achieved in an IC by using two transistors and connecting one as a diode, as in Fig. 7.8. This circuit is called a 'current mirror' and the current setting transistor (Q_1) can be used to control current sources for several loads as in Fig. 7.9. For each load $I_0 = I_1$ and

$$I_1 = \frac{V_{cc} - V_{BE}}{R}$$

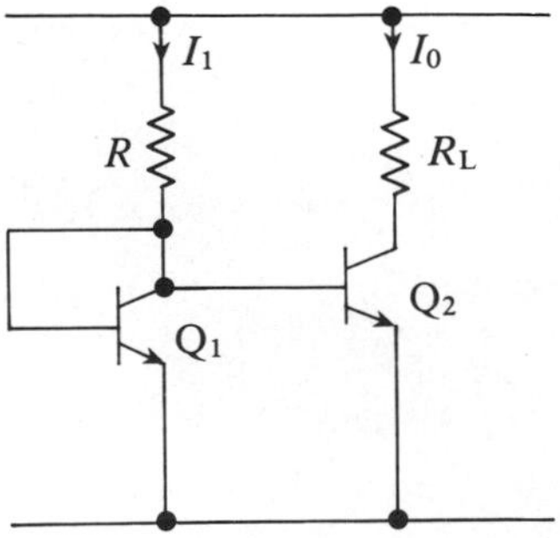

Fig. 7.8   BJT Current Mirror

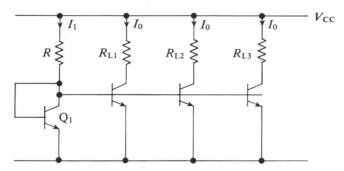

Fig. 7.9   Current Mirror Setting the Current in Several Different Loads

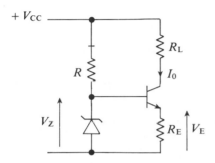

Fig. 7.10   Current Source Using Zener Diode and Emitter Resistor to Set Output Current

As the collector current for Q_1 has to provide base current for all BJTs the approximation that base current can be neglected becomes less valid.

Figure 7.10 is a current source in which a zener diode and R_E are used to set I_0. R is chosen to limit the diode current to a safe value and also to ensure that it is operating at its constant voltage value V_Z. Then $V_E = V_Z - V_{BE}$ and neglecting base current, $I_0 = V_E/R_E$. An example of the design of one of these circuits will be included in Section 7.4 when designing a differential amplifier circuit.

7.4   DIFFERENTIAL AMPLIFIER

The differential amplifier, also known as the long-tailed pair or source coupled amplifier (emitter coupled in BJT version), is an important circuit because it is invariably used as the means of producing high voltage gain in an IC op amp. In this context one of its virtues is that it is dc coupled – a necessity because large capacitors are not practicable in ICs and also, for many applications, dc coupling is a requirement. Having opted for dc coupling then resistance to drift, i.e. the resistance to changes in quiescent conditions as temperature changes, is an import-

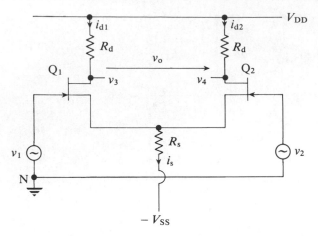

Fig. 7.11 FET Differential Amplifier

ant property and in this respect the differential amplifier is superior to dc coupled CS and CE amplifiers.

Figure 7.11 is an FET version of a differential amplifier. The term 'differential' signifies that the output signal is proportional to $v_d = v_2 - v_1$, the difference between the two input signals. To explain the action of the circuit, suppose that $v_o = v_4 - v_3$, the voltage difference between the two drains. Now let us consider three separate input cases, assuming in each one that the FETs are biased for linear operation and that the signal voltages v_1, v_2, etc., are the instantaneous values of signal components.

Case 1: $v_1 = v_2$

Suppose that v_1 and v_2 both increase by the same amount, i.e. the differential input voltage $v_d = 0$. As we shall see, it is desirable that R_s should have a high resistance. Then both drain currents tend to increase with v_1 and v_2 but because $i_s = i_{d1} + i_{d2}$, then the increase will be small because of the current NFB produced by R_s. However, any increase in the drain currents will cause v_3 and v_4 to fall, and with matched FETs these two voltages will be equal, so $v_o = v_4 - v_3 = 0$.

Case 2: $v_1 = -v_2$, i.e. $v_d = -2v_2$

If v_1 increases by, say, 5 mV and v_2 falls by 5 mV, then the differential input v_d will be -10 mV. Now i_{d1} rises and i_{d2} falls so there will be no change in i_s. Therefore R_s does not affect the operation of the circuit, so v_3 falls and v_4 rises, and there will be a substantial rise in the output voltage $v_o = v_4 - v_3$.

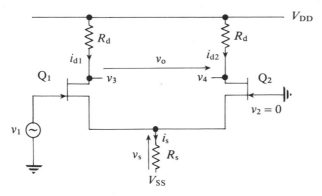

Fig. 7.12   Differential Amplifier with One Input Zero

Case 3: $v_2 = 0$, therefore $v_d = v_1$ (Fig. 7.12)

This is the case which is the most difficult to understand but which reveals most about the action of the circuit. If v_1 is a sine wave (Fig. 7.13) then i_{d1} will also be a sine wave in phase with v_1, and v_3 will be a sine wave in antiphase with v_1. The source current i_s and therefore v_s will tend to follow v_1 and i_{d1}. Consider the first half-cycle with v_1 rising. Then i_s and v_s will also tend to rise but with $v_2 = 0$, the rise in v_s will make v_{gs2} fall. Thus any change in Q_1 is coupled to Q_2 via the signal across the source resistor R_s and causes an opposite effect. The fall in v_{gs2} causes i_{d2} to fall and therefore offsets the increase in i_s. If it were not for Q_2 then with a large value of R_s, v_s would follow v_1 and v_{gs1} would be virtually unchanged. However as v_s tries to rise, it produces an equal but opposite effect in Q_2 with the result that v_{gs2} falls by the same amount that v_{gs1} rises, and i_s remains virtually constant. Referring to Fig. 7.14, $v_1 = v_{gs1} - v_{gs2}$ but because $v_{gs2} = -v_{gs1}$, $v_1 = 2v_{gs2}$. Then with only half of v_1 appearing between G and S,

$$\frac{v_3}{v_1} = \frac{-g_{fs}R_d}{2} \tag{7.5}$$

$$\frac{v_4}{v_1} = \frac{g_{fs}R_d}{2}$$

$$\frac{v_o}{v_1} = \frac{v_4 - v_3}{v_1} = g_{fs}R_d$$

The same result could be obtained more readily using case 2.

Let $v_1 = \dfrac{v_i}{2}; \ v_2 = \dfrac{-v_i}{2}$ (Fig. 7.15)

As the two inputs are in antiphase, then i_{d1} and i_{d2} are in antiphase and

$$i_s = i_{d1} + i_{d2} = \text{constant.}$$

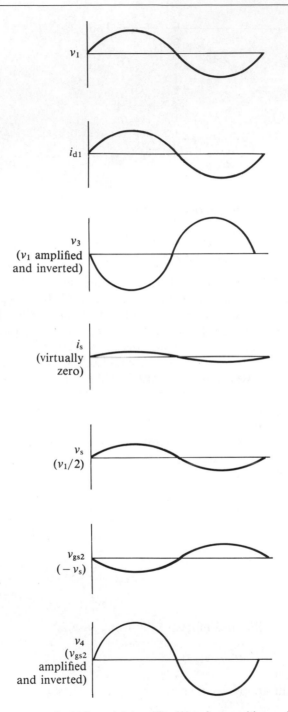

Fig. 7.13   Differential Amplifier Waveforms with $v_2 = 0$

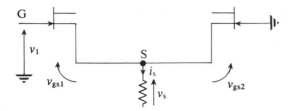

Fig. 7.14   Illustrating that in a Differential Amplifier with $v_2 = 0$, $v_1 = v_{gs1} - v_{gs2}$

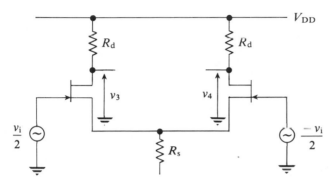

Fig. 7.15   Differential Amplifier with Equal but Antiphase Inputs

In these circumstances the two halves of the circuit could be said to operate independently, and taking just one side of the circuit (Fig. 7.16),

$$\frac{v_3}{v_1} = \frac{-g_{fs}R_d}{2}$$

From the last two cases we can see that an output can be obtained with balanced

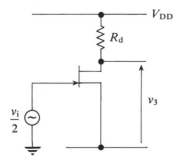

Fig. 7.16   To Evaluate v_3/v_1 in Fig. 7.15, Only One Side of the Circuit Needs to be Considered

inputs, i.e. v_1 and v_2 in antiphase, or unbalanced, i.e. one input grounded. Similarly, the output can be taken as the difference between the drain voltages or taken as one of the drain voltages wrt ground.

Common Mode Rejection Ratio (CMRR)

The case when both inputs are equal needs elaboration. While it is true that the differential output $v_0 = v_4 - v_3 = 0$ when the two inputs are equal, single-ended outputs such as v_4 wrt ground will be finite. However, the larger R_s, the smaller the change in i_{d2} and the less v_4 will be. Figure 7.11 can be redrawn as in Fig. 7.17 and when the two inputs are equal then the output at either drain wrt N can be calculated by simply looking at one side of the circuit as in Fig. 7.18, and using the type of analysis carried out in Section 5.3 for Fig. 5.21, which resulted in equation (5.19),

$$\frac{v_3}{v_1} = \frac{-R_d}{2R_s} \tag{7.6}$$

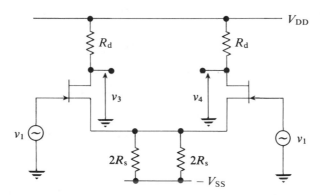

Fig. 7.17   Differential Amplifier with Common Mode Input and Source Resistor Shown as Two Resistors in Parallel

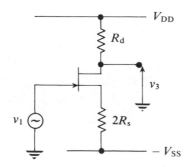

Fig. 7.18   Illustrating that the Common Mode Gain can be Evaluated by Considering One Side of Fig. 7.17

This is called the common mode gain and when an equal signal is applied to the input it is called a common mode signal. We therefore have two different gains:

(a) the difference gain A_d which should be high, and
(b) the common mode gain A_c which ideally should be zero.

The common mode rejection ratio (CMRR) is defined as the ratio of the difference and common mode gain expressed in dB, i.e.

$$\text{CMRR} = 20 \log_{10} \frac{A_d}{A_c} .$$

Using single-ended outputs to define each gain, then from equations (7.5) and (7.6),

$$\text{CMRR} = 20 \log_{10} g_{fs}R_s \tag{7.7}$$

This equation emphasizes the importance of R_s being a high value. Using similar analysis for a BJT version of a differential amplifier, then from Fig. 7.19,

$$A_d = \frac{h_{fe}R_c}{2h_{ie}} \tag{7.8}$$

and from Fig. 7.20,

$$A_c = \frac{R_c}{2R_e} \tag{7.9}$$

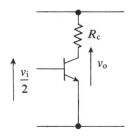

Fig. 7.19   Circuit for Evaluating Difference Gain of a BJT Differential Amplifier

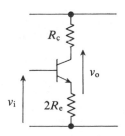

Fig. 7.20   Circuit for Evaluating Common Mode Gain of a BJT Differential Amplifier

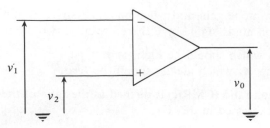

Fig. 7.21   Representation of a Differential Amplifier with Inverting $(-)$ and Noninverting $(+)$ Inputs

Therefore

$$CMRR = 20 \log_{10} \frac{h_{fe}R_e}{h_{ie}} \tag{7.10}$$

If source resistance is significant it must be added in with h_{ie}.

CMRR is important in that it is a measure of how successfully a differential amplifier rejects unwanted signals. For example, any unwanted signals from power supplies, e.g. mains hum or switching spikes, are likely to appear as common mode signals. Drift can be regarded as an unwanted common mode signal: an increase in temperature will tend to increase the drain currents in both FETs equally and therefore has the same effect as a common mode signal.

Representing a differential amplifier as a linear amplifying block (Fig. 7.21), assume that v_0 is a single-ended output, say v_4 in Fig. 7.11. Then a signal on v_1 will be inverted at v_0 and a signal on v_2 will be in phase at v_0. Hence v_1 and v_2 are known as the inverting and noninverting inputs respectively, and the $(-)$ and $(+)$ signs are meant to indicate this. Now ideally any signal v_0 will only be the result of amplifying the difference signal $v_d = v_2 - v_1$. However, if there is a common mode component in v_1 and v_2 it will have some effect at v_0. We can therefore write that

$$v_0 = A_d v_d + A_c v_c \tag{7.11}$$

where v_c is the common mode input. It is defined as

$$v_c = \frac{v_2 + v_1}{2} \tag{7.12}$$

as it would be illogical to count a signal appearing on both inputs twice. The following example demonstrates the meanings of some of the terms which have been introduced in describing the differential amplifier.

☐   **Example 7.3**

In Fig. 7.21 consider the following two sets of inputs and outputs:

(a) $v_1 = -1$ mV, $v_2 = +1$ mV, $v_0 = 1$ V
(b) $v_1 = 9$ mV, $v_2 = 11$ mV, $v_0 = 1.05$ V

Calculate A_d, A_c and CMRR

For (a),

$$v_d = v_2 - v_1 = 2 \text{ mV},$$

$$v_c = \frac{v_2 + v_1}{2} = 0$$

Using equation (7.11),

$$1 \text{ V} = A_d \times 2 \text{ mV} + 0$$

Therefore $A_d = 500$.

For (b),

$$v_d = 11 - 9 = 2 \text{ mV}$$

$$v_c = \frac{11 + 9}{2} = 10 \text{ mV}$$

Therefore $1.05 = 500 \times 2 \times 10^{-3} + A_c \times 10 \times 10^{-3}$

from which $A_c = 5$.

Then CMRR $= 20 \log_{10} \dfrac{500}{5} - 40 \text{ dB}$

This would be regarded as inadequate for many applications with 60 dB generally being a minimum: the next example illustrates the difficulty in achieving a high CMRR. □

□ **Example 7.4**

For Fig. 7.22 the BJT data is,

$$h_{fe} = 100; \; h_{ie} = 1 \text{ k}\Omega; \; V_{BE(ON)} = 0.7 \text{ V}$$

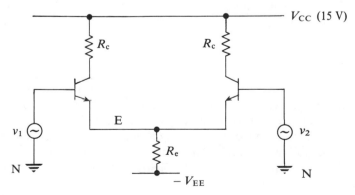

Fig. 7.22   Circuit for Example 7.4

It is required that $A_d = 200$ and CMRR $= 60$ dB.
Choose values for R_d, R_e and $-V_{EE}$.

Using equation (7.8),

$$A_d = \frac{100 \times R_c}{2 \times 1 \text{ k}\Omega} = 200$$

Therefore $R_c = 4$ kΩ.

With CMRR $= 60$ dB, then using equation (7.10),

$$1000 = \frac{100 \, R_e}{1 \text{ k}\Omega}$$

Therefore $R_e = 10$ kΩ.

With v_1 and v_2 referenced to ground, then

$$V_{BN(Q)} = 0 \text{ V}$$

Therefore $V_{EN(Q)} = -0.7$ V

Note that with a negative supply rail, base bias resistors are not required. To allow a substantial output voltage swing make $V_{Rc} = 8$ V.

Then $V_{CE(Q)} = 7.7$ V

and $\quad I_{C(Q)} = \dfrac{8 \text{ V}}{4 \text{ k}\Omega} = 2$ mA

and $\quad I_{E(Q)} \doteq 2 I_{C(Q)} = 4$ mA

The voltage across R_e is

$$-0.7 - V_{EE} = I_{E(Q)} R_e = 40 \text{ V}$$

Therefore $V_{EE} = -40.7$ V.

This is an inconveniently large voltage for the negative supply rail. To increase CMRR by a further 20 dB would require V_{EE} to be increased to about -400 V, which is obviously unacceptable. □

Use of Constant Current Source to Increase CMRR

The last example illustrates that to increase CMRR by using a large value of R_e creates practical difficulties. This brings us back to the current source circuits of Section 7.3. By replacing R_e with a current source such as the zener diode version of Fig. 7.10 we have, in effect, a very high emitter resistance whose value is independent of the voltage across it. Therefore the penalty of having to use a large negative supply rail voltage is avoided. Figure 7.23 is a complete differential amplifier with high CMRR.

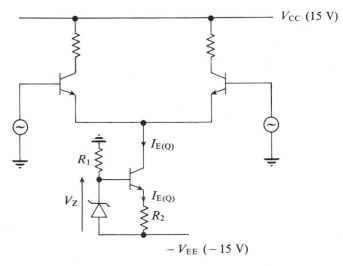

Fig. 7.23 Differential Amplifier with Constant Current Source Emitter Circuit

☐ **Example 7.5**

In the last example considerations unconnected with CMRR dictated that $I_{E(Q)} = 4$ mA. Given that $V_{EE} = -15$ V, design the emitter circuit for Fig. 7.23.

A Motorola zener diode M-ZPD 6.8 has $V_Z = 6.8$ V at 1 mA.
Then $V_{R1} = 8.2$ V, so $R_1 = 8.2$ kΩ is a satisfactory value.
Allowing $V_{BE} = 0.7$ V, $V_{R2} = 6.1$ V

Then $\qquad R_2 = \dfrac{6.1 \text{ V}}{I_{E(Q)}} \doteqdot 1.5 \text{ k}\Omega$

The effective value of the emitter resistance is the output resistance of the current source circuit which is a series current NFB amplifier. Using typical parameter values,

$$h_{fe} = 100; \; h_{oe} = 25 \; \mu S; \; h_{ie} = 1.5 \text{ k}\Omega = R_2,$$

then using equation (5.20) and ignoring the first term which is insignificant,

$$R_{of} = \frac{R_2}{h_{ic} + R_2} \frac{h_{fe}}{h_{oe}} = 2 \text{ M}\Omega.$$

If a single resistor R_e of 2 MΩ were used instead of the current source then $-V_{EE} = I_{E(Q)}R_e = 8000$ V.

Using equation (7.10),

$\qquad$ CMRR = 96 dB. $\qquad\qquad\qquad\qquad\qquad\qquad\qquad$ ☐

7.5   CASCODE AMPLIFIER

HF response is often limited by Miller effect (Section 3.9). This is especially true for a CS amplifier in which C_{gd} is significant. In the FET version of the cascode amplifier (Fig. 7.24), C_{gd} will only have a serious effect if V_2/V_1 is high. In this circuit we shall show that V_2/V_1 is low but by Q_2 acting as the drain load for Q_1 the overall amplification V_0/V_1 is comparable with that of a single-stage CS amplifier. Both FETs are biased into their active states, Q_1 by R_g and R_s, and Q_2 by R_1 and R_2. The signal input to G_2 is held at 0 V by C_1. Therefore any input to Q_2 is applied to its source.

Analysis

Referring to Fig. 7.25:

For Q_1, $V_{gs} = V_1$

For Q_2, $V_{gs} = -V_2$

$$I_0 = \frac{-V_0}{R_L} = \frac{V_0 - V_2}{r_{ds}} - g_{fs}V_2 \qquad\qquad \text{(A)}$$

$$I_0 = \frac{-V_0}{R_L} = \frac{V_2}{r_{ds}} + g_{fs}V_1 \qquad\qquad \text{(B)}$$

From (A),

$$-V_0\left[\frac{1}{R_L} + \frac{1}{r_{ds}}\right] = -V_2\left[\frac{1}{r_{ds}} + g_{fs}\right]$$

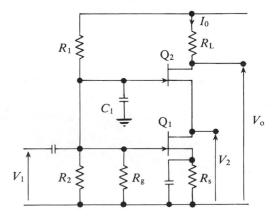

Fig. 7.24   FET Cascode Amplifier

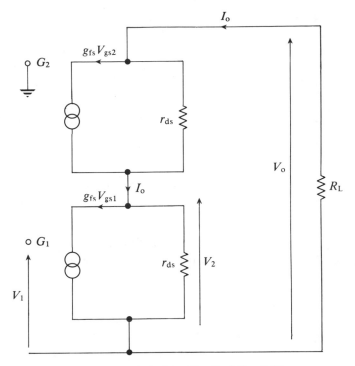

Fig. 7.25   Small-signal Equivalent Circuit of Fig. 7.24

If $r_{ds} \gg R_L$ and $g_{fs} \gg \dfrac{1}{r_{ds}}$, then

$$\frac{V_o}{V_2} \doteq g_{fs} R_L$$

which is the same as the approximate gain of a CS amplifier but without the phase reversal.

Substituting for V_o in (B),

$$-g_{fs} V_2 = \frac{V_2}{r_{ds}} + g_{fs} V_1$$

$$-V_2 \left[g_{fs} + \frac{1}{r_{ds}} \right] \doteq -V_2 g_{fs} = g_{fs} V_1$$

Therefore $\dfrac{V_2}{V_1} \doteq -1$

Therefore the ratio of V_2/V_1 is small, which eliminates the Miller effect and

$$\frac{V_o}{V_1} = \frac{V_o}{V_2}\frac{V_2}{V_1} = -g_{fs} R_L$$

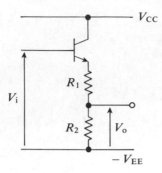

Fig. 7.26   Use of Emitter Follower for Shifting Quiescent Voltage Level – Note Attenuation of Signal Voltage

An alternative approach to the analysis of the BJT differential amplifier is to use the medium frequency hybrid Π equivalent circuit established in section 3.10. For this circuit,

$$g_m = \frac{h_{fe}}{h_{ie}}$$

Then, from equations 7.8 and 7.10,

$$A_d = \frac{g_m R_c}{2} \quad \text{and}$$

$$\text{CMRR} = 20 \log_{10} g_m R_e$$

7.6   VOLTAGE LEVEL SHIFTING

Biasing requirements for linear amplification inevitably lead to signals being superimposed on some quiescent voltage. In an ac coupled amplifier this quiescent voltage will be removed by the coupling capacitors. In a dc coupled amplifier it is usually desirable to set the output at 0 V when there is no input signal. Voltage level shifting or translation can be achieved in small steps using diodes and transistor B–E junctions, but a shift of several volts is more difficult.

The emitter follower of Fig. 7.26 can be used for the purpose but clearly, the signal voltage developed across R_1 will be lost. A better circuit using the current mirror (Section 7.3) will be explained by example.

☐   **Example 7.6**

The input to the emitter follower stage Q_1 in Fig. 7.27 is superimposed on 5 V dc. Choose a value for R to make $V_{o(Q)} = 0$ V.

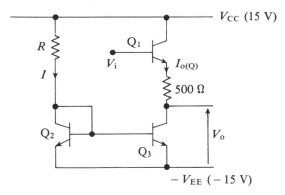

Fig. 7.27   Use of Current Source to Facilitate Voltage Level Shifting

Assuming that for all transistors $V_{BE} = 0.6$ V, the pd across the 500 Ω resistor must be 4.4 V. Then $I_{o(Q)} = 8.8$ mA.

With the current mirror,

$$I_{o(Q)} = I = \frac{30 - V_{BE}}{R}$$

Therefore $R = 3.3$ kΩ. □

SUMMARY

For analog circuit design it is first necessary to understand the basic techniques of amplification and negative feedback described in the first five chapters. It is then important to be able to deal with specific problems such as the need for very high input resistance, dc coupling, differential amplification, and so forth. Much of the stimulus for modern analog circuit techniques has been provided by the needs of circuits for IC op amps in particular, and linear ICs generally, but many of the circuits are useful for discrete component design.

REVISION QUESTIONS

7.1   Calculate R_i for Fig. Q.7.1. BJT data: $h_{ie} = 2$ kΩ; $h_{fe} = 80$

(74 kΩ)

7.2   Calculate A_i and R_i for the Darlington emitter follower in Fig. Q.7.2. BJT data $h_{ie} = 2$ kΩ; $h_{fe} = 50$; $h_{oe} = 100$ μS.

(1670; 167 kΩ)

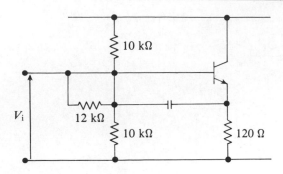

Fig. Q.7.1

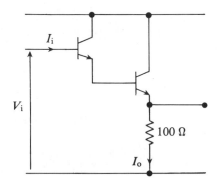

Fig. Q.7.2

7.3 Why is CMRR an important property of a differential amplifier? Given that v_s in Fig. Q.7.3 is a sine wave, sketch the two collector and the emitter waveforms on the same time scale as v_s.
Show that:

$$A_d = \frac{h_{fe}R_c}{2h_{ie}} \quad \text{and} \quad A_c = \frac{R_c}{2R_e}$$

and hence calculate CMRR. BJT data: $h_{fe} = 120$; $h_{ie} = 3$ kΩ.

(58 dB)

7.4 For Fig. Q.7.4 it is required that,

$$A_d = 250; \text{ CMRR} = 66 \text{ dB}; V_{CN(Q)} = 5 \text{ V}.$$

Choose values for R_c, R_e and $-V_{EE}$. BJT data: $h_{fe} = 50$; $h_{ie} = 1$ kΩ.

(10 kΩ; 40 kΩ; -40.7 V)

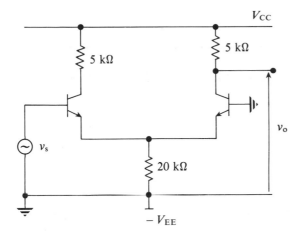

Fig. Q.7.3

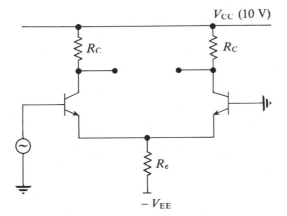

Fig. Q.7.4

7.5 For Fig. Q.7.4 design a circuit to replace R_e which will work at $-V_{EE} = -10$ V and maintain the required specification.

7.6 Show that the voltage gain of the cascode circuit of Fig. 7.24 is approximately $-g_{fs}R_L$. Why does the circuit eliminate Miller effect?

8

Operational Amplifier Properties

Chapter Objectives
8.1 General Features of Op Amps
8.2 Practical Properties
8.3 Frequency Response
8.4 Slew Rate
Summary
Revision Questions

In previous chapters we have described discrete component circuits, i.e. the use of transistors and passive components interconnected as required for particular applications. These techniques will continue to be used for the foreseeable future but the development of a wide range of linear ICs, in particular the IC op amp, has had a major impact on analog circuit design. The term 'op amp' was originally the name of a circuit used for carrying out mathematical operations such as summation and integration on input signals. An essential component for this purpose was an amplifier with high voltage gain. More recently the term op amp has been given to the linear IC which provides, among other things, the high voltage gain and which can therefore be used, together with externally connected passive components, for mathematical operations. Although its range of applications is wide, to be used effectively, it is necessary to understand its properties. This chapter is confined to a description of these properties, while the following chapter covers some op amp applications.

8.1 GENERAL FEATURES OF OP AMPS

An ideal op amp has the following properties:

(a) infinite open-loop voltage gain A_{vol};
(b) infinite input resistance R_i;

(c) zero output resistance R_o;

(d) infinite bandwidth;

(e) infinite CMRR.

It is also dc coupled.

In practice op amps do not, of course, have ideal properties, but common types nevertheless have impressive specifications. For the 741, typical values are quoted as

$$A_{vol} = 200\ 000$$
$$R_i = 2\ M\Omega$$
$$R_o = 75\ \Omega$$

Where typical values are given in this chapter they are quoted from 741 specifications unless a different type is given for comparison. More recent types such as the 3140 have FET input stages which give the circuits a higher input resistance than the 741 but with which they are pin compatible.

Type Numbers

Manufacturers add prefixes to type numbers such as 741 to indicate their own codings even though the specifications are similar, e.g.

μA 741 Fairchild
LM 741 National Semiconductors

An extra letter is sometimes added to indicate a temperature specification, e.g.

μA 741 A – military specification for guaranteed operation between $-55\,^{\circ}$C and $+125\,^{\circ}$C.

μA 741 C – commercial specification for temperature range $0\,^{\circ}$C to $70\,^{\circ}$C.

DC Supplies

The input stage of an op amp is usually a differential amplifier of the type described in Section 7.4 and drawn, in its basic form, in Fig. 8.1. For a dc coupled amplifier requiring an output which can swing positively or negatively wrt ground it is evident that the circuit requires positive and negative dc supplies, V_+ and V_-, referenced to some intermediate voltage level which will usually be ground. These voltages will have specified maxima and for the μA 741C they are ± 18 V. Figure 8.2 shows the pin layout for the dual in line version of a 741 op amp with pin numbers corresponding to the terminals in Fig. 8.1. Pins 1 and 5 are used for 'offset' purposes to be explained. Op amps are available to operate with a single supply referenced to ground, e.g. μA 124, but clearly the output voltage swing cannot move beyond the supply rail voltages, e.g. if the supply rail is positive then the output cannot swing below ground level.

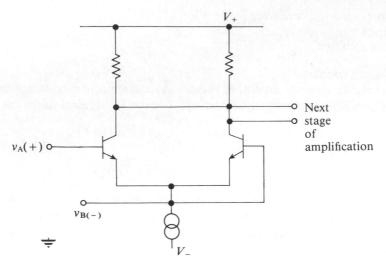

Fig. 8.1   Basic Differential Amplifier Input Stage for Op Amp

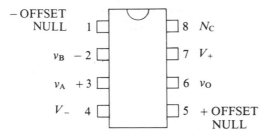

Fig. 8.2   Pin Out for Dual In Line Version of 741 Op Amp

8.2   PRACTICAL PROPERTIES

Open-loop Voltage Gain A_{vol}

As with the differential amplifier of Fig. 8.1, the op amp amplifies the difference v_d between the voltage on the noninverting (+) and inverting (−) terminals (Fig. 8.3). The term 'open loop' signifies that there is no external feedback connection between the output and either of the inputs. A_{vol} is defined as the ratio of the change in output voltage to the change in differential input voltage, usually for a load resistance of no less than 2 kΩ. As noted, 200 000 is typical for A_{vol}.

Input Resistance R_i

Input resistance is the open-loop incremental resistance looking into the two input

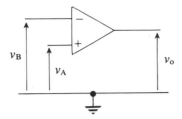

Fig. 8.3   Op Amp Symbol with Signal Voltages

terminals, and is typically 2 MΩ. Sometimes manufacturers quote the resistance between inputs and ground.

Output Resistance R_o

The open-loop output resistance is usually between 50 Ω and 500 Ω with the typical value for the 741 being 75 Ω.

At this stage we can see that Fig. 8.4 is the representation of an op amp as a circuit element.

Common Mode Rejection Ratio (CMRR)

When the input signals are equal and greater than zero, i.e. there is a common mode input voltage, there should not be any output voltage because $v_d = 0$. In general the common mode input v_c is defined as $(v_A + v_B)/2$ and the difference signal to be amplified, v_d, is $v_A - v_B$. Common mode gain A_c is defined as the ratio v_o/v_c when

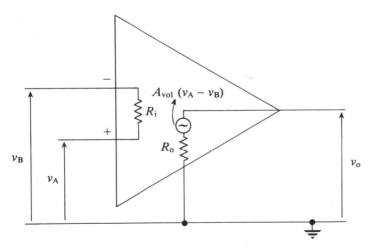

Fig. 8.4   Op Amp as a Circuit Component

$v_d = 0$. Common mode rejection ratio, CMRR, is defined as A_{vol}/A_c but is usually expressed in dB, i.e.

$$\text{CMRR} = 20 \log_{10} \frac{A_{vol}}{A_c}$$

It is typically 90 dB, i.e. $\dfrac{A_{vol}}{A_c} \doteq 32\ 000$.

Input Offset Voltage V_{OS}

When both inputs are tied to ground, i.e. both differential and common mode inputs are zero, the output should be zero. In practice there will be mismatches in amplifier components and if there is a mismatch in an input stage the effect will be amplified and lead to a significant output voltage. Input offset voltage V_{OS} is the differential input voltage required to make the output zero, and is typically 1 mV. With some op amps, e.g. the 741, two voltage offset terminals are provided and by connecting a potentiometer between them and taking the slider to the specified dc supply rail, as in Fig. 8.5, then the potentiometer can be adjusted to zero the offset voltage.

Input Bias Current I_B

Op amps with bipolar input stages of the basic form of Fig. 8.1 are biassed for linear operation by having the quiescent base voltage at ground voltage and the common emitter point negative. By operating at extremely low quiescent current values the base bias currents will be low but they do have to be taken into consideration. Input bias current I_B for an op amp is defined as the average of the two input currents with the inputs grounded (Fig. 8.6), i.e.

$$I_B = \frac{I_{B1} + I_{B2}}{2}$$

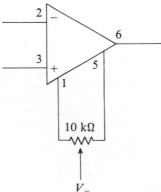

Fig. 8.5 'Offset Null' Terminals on 741 Connected to 10 kΩ Potentiometer and Negative Supply Rail to Zero the Offset Voltage

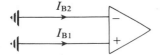

Fig. 8.6   Bias Currents in Op Amp

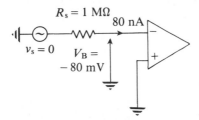

Fig. 8.7   Illustrating the Bias Current Generates an Input Voltage by Developing a pd across the Source Resistance

The typical value for a μA 741 is 80 nA. For an op amp with an FET input stage, I_B is obviously much less: 50 pA is quoted for a μA 771. Input bias current has an adverse effect when the resistance of the source feeding the op amp is large. In Fig. 8.7 the noninverting input is grounded and the inverting input is connected to a source of 1 MΩ resistance and of voltage v_s momentarily at 0 V. Then an 80 nA bias current generates a voltage of -80 mV on the noninverting input.

Input Offset Current I_{OS}

The problem referred to in the last paragraph would not arise if both inputs were connected to equal resistances (one of which could be a passive resistor) as shown in Fig. 8.8, provided that $I_{B1} = I_{B2}$. Then, using the same values as before, each input would be at -80 mV and the differential input would be zero. However, if they are unequal then there will be a finite differential input voltage. It is therefore necessary to specify the difference between the bias currents, and this is the input offset current I_{OS},

$$I_{OS} = I_{B1} - I_{B2}.$$

A typical value for I_{OS} is 20 nA.

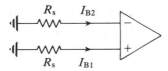

Fig. 8.8   With Equal Source Resistances There Will Be a Differential Input Voltage Due to the Bias Currents if They are Unequal

Both I_B and I_{OS} are usually measured with the output at 0 V but in practice the output voltage has little effect.

Power Supply Rejection Ratio (PSRR)

PSRR is a measures of an op amp's ability to disregard changes in power supply voltage. It is specified by the change in offset voltage V_{OS} for a 1 V change in dc power supply and is usually expressed in μV/V. A typical figure is 15 μV/V.

Maximum Differential Input Voltage

This is the maximum value of differential input voltage $v_A - v_B$ that can be applied without damaging the op amp.

Maximum Common Mode Input Voltage

This is the maximum voltage that the two inputs can be raised above ground potential before the op amp becomes nonlinear.

Output Voltage Swing

Ideally this will be equal to the difference between the two supply rail voltages but in practice it will be a few volts less.

8.3   FREQUENCY RESPONSE

Having an op amp with high open-loop voltage gain gives us flexibility in that the amount of NFB can then be chosen to bring down the gain, i.e. the closed-loop gain, to any value required. The danger of using NFB in a high gain system is, as noted in Section 5.7, instability. This is not a problem at LF in op amps because they are dc coupled and therefore there are no CR couplings to produce the phase shift which leads to instability. At HF, transistor performance at each amplifying stage not only causes reduced gain but also, phase lag (Section 3.9): at frequencies where the gain per stage is falling at 6 dB/octave (20 dB/decade) the phase lag approaches $90°$. With two stages, then at frequencies where the phase lag approaches $180°$, the gain will be too low to cause instability but with three stages there will be substantial gain where the phase lag is $180°$ and the probability of oscillation increases with the amount of NFB applied.

Internal Frequency Compensation

Some op amps such as the 741 have internal CR networks which are deliberately designed to reduce gain at HF. The result, as shown in Fig. 8.9, is that the open-loop

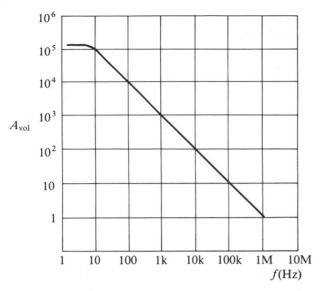

Fig. 8.9   Open-loop Gain (A_{vol}) Falling at 6dB/Octave (i.e. $A_{vol} \propto 1/f$) in an Internally Compensated Op Amp

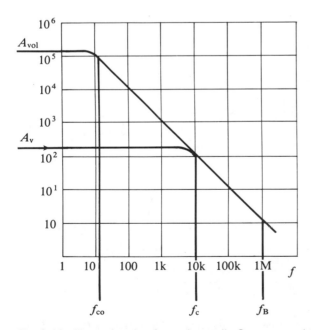

Fig. 8.10   Illustrating that for an Internally Compensated Op Amp, $A_{vol}f_{co} = A_v f_c = f_B$

gain begins to fall at a few Hz and then has a characteristic falling at 6 dB/octave (i.e. gain $\propto 1/f$) until eventually the gain is unity (0 dB) at about 1 MHz. Although the frequency response has been limited, the op amp will be stable for any NFB connection because it has been made to have gain and phase responses characteristic of a single-stage amplifier and therefore the maximum phase shift around the NFB loop will only be $90°$. When NFB is used then gain can be traded against bandwidth, as explained in Section 5.4. The gain $\times$ bandwidth product will be constant for any particular op amp and equal to the unity gain frequency f_B, i.e. about 1 MHz· for the 741 (Fig. 8.10). In general,

$$A_{vol} f_{co} = A_v f_c = f_B$$

where A_v is a particular value of closed-loop gain with corresponding cutoff frequency f_c, and f_{co} is the open-loop cutoff frequency.

External Frequency Compensation

For applications requiring a more extended HF response there are op amps with no internal compensation and whose HF response is only limited by transistor performance. It is therefore necessary to consider the likelihood of oscillation occurring when a certain amount of NFB is applied. External frequency compensation terminals are provided on these op amps (the 709 is a well known example) so that frequency response can be 'tailored' to avoid instability without the heavy degrading of frequency response incurred with internal compensation.

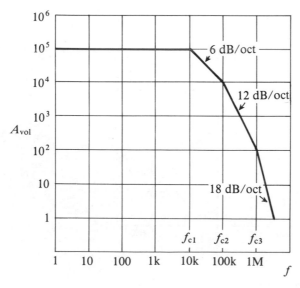

Fig. 8.11   Open-loop Frequency Response of a Three-stage dc Coupled Amplifier

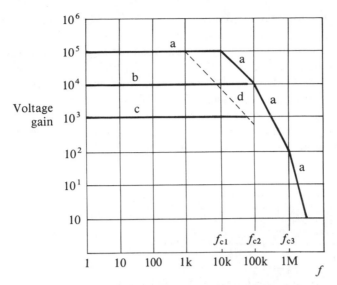

Curve (a) Open-loop Gain. Curve (b) Closed-loop Gain — stable if it cuts (a) at a frequency equal to or lower than f_{c2}. Curve (c) Closed-loop Gain — stable if Open-loop gain modified to (d)

Fig. 8.12 Conditions for Closed-loop Gain Stability

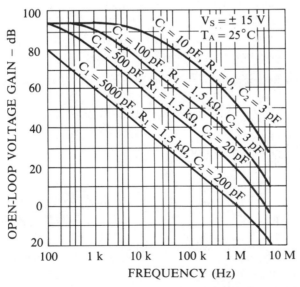

Fig. 8.13 Open-loop Frequency Response for Various Values of Compensation (709) (Courtesy Fairchild)

A brief insight into what is involved can be obtained by inspection of the open-loop frequency response of a three-stage dc coupled amplifier (Fig. 8.11). Each stage will have a different cutoff frequency so that when the highest, f_{c3}, is reached, each is contributing a 6 dB/octave roll-off, i.e. a total of 18 dB/octave. What is more significant is that beyond its cutoff frequency, each stage produces a phase lag approaching $90°$ and once the accumulated phase shift reaches $180°$ then the application of NFB gives rise to the probability of oscillation. Now it can be shown that if NFB is applied so that the closed-loop response (Fig. 8.12, curve b) intersects the open-loop response (curve a) at a frequency no higher than the second break point (f_{c2}) then the amplifier will be stable. With increased NFB (curve c) instability

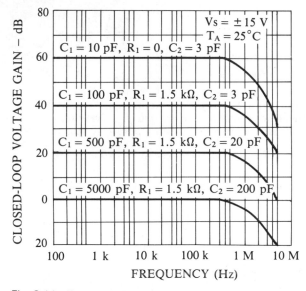

Fig. 8.14   Frequency Response for Various Closed-loop Gains (709) (Courtesy Fairchild)

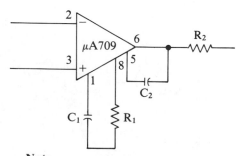

Note
Use $R_2 = 50\ \Omega$ when the amplifier is operated with capacitive loading.

Fig. 8.15   Frequency Compensation Circuit (Courtesy Fairchild)

can be avoided either by tailoring the open-loop response as in curve d or by connecting phase compensating components in the NFB network. Op amp manufacturers give open and closed responses plotted as in the examples for the μA 709 in Figs 8.13 and 8.14, for components connected to the frequency compensation terminals as shown in Fig. 8.15.

8.4 SLEW RATE

Slew (sometimes called slewing) rate is a measure of how fast the output voltage can change, i.e.

$$\text{Slew rate} = \frac{dv_0}{dt} \text{ (max)}$$

It is normally measured in response to a large input voltage step and is therefore usually associated with low closed-loop voltage gain. For a 741 the slew rate is 0.5 V/μs at $A_v = 1$, but for more recently developed op amps 5 to 10 V/μs is commonplace. The effect of slew rate in response to an input voltage step is illustrated in Fig. 8.16. With a sine wave input, slew rate limits a combination of maximum operating frequency and output voltage amplitude.

□ **Example 8.1**

What is the maximum frequency of output sine wave which can be produced at an amplitude of 1.5 V if the op amp slew rate is 0.5 V/μs?

$$v_o = V_m \sin \omega t$$

$$\frac{dv_o}{dt} = \omega V_m \cos \omega t$$

Fig. 8.16 Effect of Op Amp Slew Rate with Step Input Waveform

$$\text{Slew rate} = \frac{d_{vo(max)}}{dt} = \omega V_m = \frac{0.5}{10^{-6}}$$

$$\text{Therefore } \omega = \frac{0.5}{1.5 \times 10^{-6}}$$

and $f = 53$ kHz.   ☐

☐   **Example 8.2**

What is the maximum amplitude of output voltage sine wave that an op amp with slew rate of 0.5 V/μs can deliver at $f = 100$ kHz?

From the previous example,

$$\text{Slew rate} = \omega V_m = \frac{0.5}{10^{-6}}$$

$$\text{Therefore } V_m = \frac{0.5}{2\pi \times 10^5 \times 10^{-6}} = 0.8 \text{ V}. \qquad ☐$$

The effect of slew rate then, is to introduce nonlinearity if we attempt to make the output voltage change faster than the slew rate. Taking the figures from the last example, Fig. 8.17 shows the effect of overdriving the op amp at 100 kHz. At 0.8 V

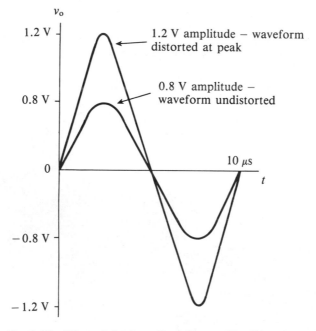

Fig. 8.17   Effect of Op Amp Slew Rate on Sine Wave when Overdriven

output amplitude the waveform is sinusoidal. At 1.2 V output the signal can only change at 0.5 V/μs from 0 V and clearly becomes distorted even though the op amp could obviously deliver sine waves of this amplitude at lower frequencies.

Slew rate occurs because at some stage in the amplifier a frequency-compensating capacitor will have to be charged and the limited charging current available restricts the maximum rate of change of capacitor voltage. With externally compensated op amps such as the 709, slew rate will depend on the value of the compensating capacitors and these in turn will be chosen on the basis of the closed-loop gain required. The lower the gain, the higher the compensating capacitors and hence the lower the slew rate. For a μA 709, slew rate is 0.3 V/μs at $A_v = 1$ and 1.5 V/μs at $A_v = 10$.

SUMMARY

An IC op amp provides high voltage gain at low cost and lends itself to a wide range of applications. However, its properties and limitations must be understood before it can be used effectively. Its main attributes – high open-loop voltage gain, high input resistance and low output resistance – approach the ideal but the effect of offsets and bias currents will produce unwanted outputs which may result in severe impairment unless steps are taken to minimize their effect. It is normal practice to use NFB so that its closed-loop performance meets a particular requirement but when using uncompensated op amps with NFB the possibility of oscillation must be taken into account and compensation provided to meet the circumstances. When specifying such output voltage requirements as rise time, output voltage and frequency, it is necessary to choose an op amp with a slew rate which meets the specifications.

REVISION QUESTIONS

8.1 State the properties of an ideal op amp and give typical values.

8.2 What does the term 'open loop' mean in the context of op amps?

8.3 Describe, with the aid of diagrams, the terms

>   input bias current
>   input offset voltage
>   input offset current
>   power supply rejection ratio

8.4 Under what circumstance does input bias current have the most adverse effect and how can this effect be minimized?

8.5   What does the term 'internal frequency compensation' mean when applied to an op amp? What are the benefits and penalties of internal frequency compensation?

8.6   Using the frequency response graphs of Fig. 8.14 for a 709 op amp, choose compensation components for the circuit to have a gain of 100 and frequency response up to 100 kHz.

(100 pF; 1.5 kΩ; 3 pF)

8.7   An op amp has a slew rate of 0.7 V/μs. What is the maximum amplitude of undistorted output sine wave that the op amp can produce at a frequency of 50 kHz?

What is the maximum frequency of undistorted output that the op amp will produce at an amplitude of 3 V?

(2.2 V; 37 kHz)

9

Applications of Operational Amplifiers

Chapter Objectives
9.1   Inverting Amplifier
9.2   Noninverting Amplifier
9.3   Voltage Follower
9.4   Inverting Adder
9.5   Noninverting Adder
9.6   Integrators
9.7   Differentiator
9.8   Analog Computers
9.9   High Input Resistance Differential Amplifier
9.10   Logarithmic Amplifier
        Summary
        Revision Questions

Op amps have many applications, and having covered the main properties of op amps in the last chapter some of these applications can now be described. This chapter concentrates on a limited range of op amp applications such as summing and integrating, but later chapters on specific topics such as sine wave oscillators and nonsinusoidal waveform generation will include the use of op amps in these contexts.

9.1 INVERTING AMPLIFIER

One common op amp circuit is shown in Fig. 9.1. In this particular circuit the noninverting input is connected to ground, i.e. $v_A = 0$ (all voltages in this chapter will be referenced to ground). Then the input current i will flow through R_1 in the direction shown and, because the op amp can be regarded as having infinite input resistance, i will also flow through R_2. Now imagine that the input signal v_s makes

229

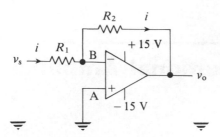

Fig. 9.1   Inverting Amplifier

v_B positive. Then v_o will go negative and increase i. This is the voltage shunt feedback effect in that, in going negative, v_o draws more current through the R_1R_2 resistor chain. There are now two possibilities. One is that if v_B remains finite, the infinite gain of the op amp will force v_o down to the negative supply rail voltage, i.e. $v_o = -15$ V. The other possibility is that i increases to the point where practically the whole of v_s is dropped across R_1 and v_B is 'virtually' zero. Then with 'virtually' infinite gain v_o will be finite and this is the condition for using an op amp as a linear amplifier. To find v_o, then noting that v_s is dropped across R_1,

$$i = \frac{v_s}{R_1}$$

With $v_B = 0$, $v_o = -iR_2 = -\dfrac{v_s}{R_1} R_2$

Then $\dfrac{v_o}{v_s} = -\dfrac{R_2}{R_1}$ \hfill (9.1)

and the value of v_o/v_s is independent of the op amp gain, the assumption being that it is very high. This circuit then, is simply an inverting amplifier but it is sometimes called a scale changer or simply, a scaler.

The action of the circuit results in v_B becoming so small that it can be regarded as zero for purposes of calculation. Because of this the circuit is sometimes called a virtual ground amplifier. However, the term is somewhat misleading because if $v_A \neq 0$ V as in Fig. 9.2 then i takes up a value that makes $v_B = v_A = 1$ V (in this case) and it is the differential input that becomes zero. This point is worth restating because it is fundamental to the action of an op amp as a linear amplifier. For the circuit to be linear, then current i takes up a value which makes the inverting inut voltage equal to the noninverting input voltage, i.e. the value of i is that which makes the differential input zero. If this condition cannot be met and the differential input is finite then v_o will be at the positive or negative supply rail voltage depending on the polarity of the differential input voltage. Once this principle is established then for a set of values such as those given in Fig. 9.2 the value of v_o can be found quite readily.

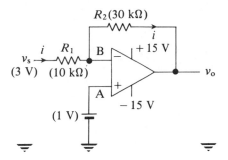

Fig. 9.2   Inverting Amplifier with Noninverting Input Positive

☐ **Example 9.1**

Using the values given in brackets in Fig. 9.2 calculate v_o.

$$V_B = V_A = +1 \text{ V}$$

$$i = \frac{v_s - V_A}{R_1} = \frac{3 - 1}{10 \text{ k}\Omega} = \frac{2 \text{ V}}{10 \text{ k}\Omega}$$

$$v_o = -iR_2 + V_B = \frac{-2}{10 \text{ k}\Omega} 30 \text{ k}\Omega + 1 \text{ V}$$

$$= -5 \text{ V}$$ ☐

Had the result of such a calculation given a value of v_o outside the rail voltages then v_o would be limited to the nearest rail voltage. The principle can be used to calculate v_o when v_s is an ac input.

☐ **Example 9.2**

If v_s in Fig. 9.2 is a sine wave of peak voltage 3 V, sketch v_o showing the significant voltage levels.

In Example 9.1 v_o has already been calculated for $v_s = +3$ V and is -5 V. When $v_s = 0$ V, i is reversed because V_B is positive wrt v_s, i.e.

$$i = \frac{-1 \text{ V}}{10 \text{ k}\Omega}$$

$$v_o = -iR_2 + V_B = -\left\{\frac{-1 \text{ V}}{10 \text{ k}\Omega}\right\} 30 \text{ k}\Omega + 1 = +4 \text{ V}$$

Note that a -3 V change in v_s produces a $+9$ V change in v_o, which corresponds to the inverting gain $-R_2/R_1 = -3$.

When $v_s = -3$ V, $i = \dfrac{-3 - 1}{10 \text{ k}\Omega} = \dfrac{-4}{10 \text{ k}\Omega}$

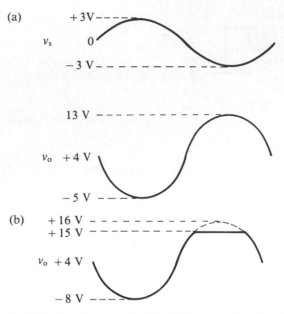

Fig. 9.3   Waveforms (a) for Fig. 9.2 when v_s is a Sine Wave of 3 V Peak; (b) if Fig. 9.2 is Modified by making $R_2 = 40$ kΩ

and $v_o = -iR_2 + V_B = 12 + 1 = 13$ V,

a gain of -3 times, as before. The waveforms v_s and v_o are sketched in Fig. 9.3a.

□

It is instructive to rework the problem with $R_2 = 40$ kΩ. Now the positive excursion of v_o attempts to go to $+16$ V but is limited by the positive supply rail (Fig. 9.3b). In a practical op amp the clipping level would be a volt or so below the rail voltage.

As noted in Section 8.2, a practical op amp does draw a small bias current into each input. For this reason it is normal practice to include a resistor R (Fig. 9.4) in

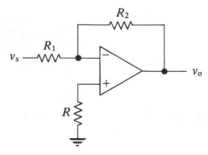

Fig. 9.4   Resistor R in Series with Noninverting Input to Minimize the Differential Input Voltage Generated by Bias Currents

series with the noninverting input. Then with both inputs at 0 V the output v_o should be 0 V. If the two bias currents are equal then if each is drawn through an equal resistance they will each generate equal pds and result in an equal voltage at each input, i.e. zero differential input voltage. To make the resistance 'seen' by each input equal then $R = R_1 \parallel R_2$. By using a variable resistance for R it can be adjusted to make $v_o = 0$ V when the bias currents are unequal if the application warrants such a refinement. Essentially, this is a method of generating an offset voltage to set v_o at 0 V.

9.2 NONINVERTING AMPLIFIER

In Fig. 9.5 the feedback chain is connected to the inverting input as with the inverting amplifier but the input signal v_s is connected to the noninverting input. Under linear op amp action $v_B = v_s$ and i will flow in the direction indicated in Fig. 9.5.

Then $i = \dfrac{v_B}{R_1} - \dfrac{v_s}{R_1}$

$$v_o = iR_2 + v_B = \frac{v_s}{R_1} \times R_2 + v_s$$

and $\dfrac{v_o}{v_s} = 1 + \dfrac{R_2}{R_1}$ (9.2)

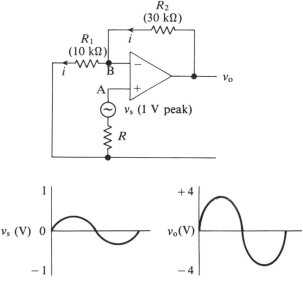

Fig. 9.5   Noninverting Amplifier

Then with $R_2 = 30$ kΩ and $R_1 = 10$ kΩ,

$$\frac{v_o}{v_s} = 4$$

Using the figures in brackets in Fig. 9.5, the waveforms of v_s and v_o will be as shown.

The expressions for inverting gain (equation (9.1)) and noninverting gain (9.2) hold whether the inputs are ac or dc.

☐ **Example 9.3**

Sketch the waveforms of v_s and v_o for Fig. 9.6.

In this example there is a constant voltage on the noninverting input and an alternating voltage on the inverting input. When

$$v_s = 0,$$

$$v_o = \left(1 + \frac{R_2}{R_1}\right)(-1.5) = -4.5 \text{ V}$$

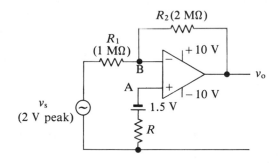

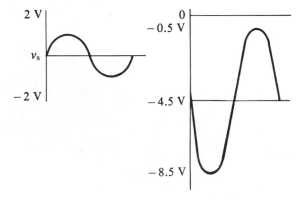

Fig. 9.6　Circuit for Example 9.3

and this will be the quiescent value of v_o. Note that the same result could have been found by making $V_B = -1.5$ V, calculating the current in R_1 and finding v_o as in Example 9.1. In this example v_s is connected to the inverting input and is therefore amplified -2 times. Therefore, at the output, the ac component will have a peak value of 4 V but be phase inverted wrt v_s and because of the dc component it will vary about -4.5 V as shown in Fig. 9.6. □

9.3   VOLTAGE FOLLOWER

The gain of the noninverting circuit of Fig. 9.7 is, using equation (9.2),

$$\frac{v_o}{v_s} = 1 + \frac{0}{R_1} = 1$$

and output and input are in phase. It can be deduced that the noninverting amplifiers employ series voltage NFB but, as v_s is in series with the noninverting input, the input resistance is high notwithstanding the fact that the series NFB also confers high input resistance. Thus with high input resistance, low output resistance and unity positive gain, the circuit has the same attributes as the transistor followers described in Section 5.3.

9.4   INVERTING ADDER

In Fig. 9.8 point B is at 0 V, so

$$i_1 = \frac{v_1}{R_1} \; ; \qquad i_2 = \frac{v_2}{R_2} \; ; \qquad i_3 = \frac{v_3}{R_3}$$

and $i = i_1 + i_2 + i_3$

and $v_o = -iR_4$

Therefore $\dfrac{v_o}{R_4} = -\left(\dfrac{v_1}{R_1} + \dfrac{v_2}{R_2} + \dfrac{v_3}{R_3}\right)$ (9.3)

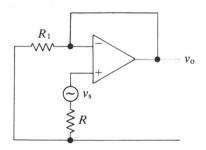

Fig. 9.7   Voltage Follower

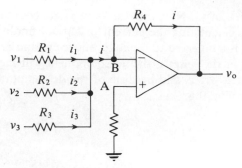

Fig. 9.8   Inverting Adder

If all resistors are equal then

$v_o = -(v_1 + v_2 + v_3)$

By resistor choice the inputs can have different weightings at the output, e.g. if $R_1 = R_2 = R_4 = R$ and $R_3 = 0.1\,R$,

then $v_o = -(v_1 + v_2 + 10v_3)$

The principle can be extended to make a simple digital-to-analog (D/A) converter. For example, if $R_4 = R_3 = R$; $R_2 = R/2$; $R_1 = R/4$, and the three inputs represent a three-digit binary number $ABC - A$ being the most significant bit (MSB) — in which inputs representing a 1 are at the same voltage, then from equation (9.3),

$$\frac{v_o}{R} = -\left(\frac{4A}{R} + \frac{2B}{R} + \frac{C}{R}\right)$$

$$v_o = -(4A + 2B + C)$$

Now v_o is a voltage which represents the value of ABC_2.

There can be more than three inputs to an inverting adder.

9.5   NONINVERTING ADDER

By feeding the output of an inverting adder into an inverter with a gain of -1 the overall result is noninverting addition. The circuit of Fig. 9.9 achieves the same result with a single op amp. In this circuit,

$$v_A = iR \qquad \text{where } i = i_1 + i_2.$$

$$i_1 = \frac{v_1 - v_A}{R} ; \qquad i_2 = \frac{v_2 - v_A}{R}$$

$$\text{Then } v_A = R\left(\frac{v_1 - v_A}{R} + \frac{v_2 - v_A}{R}\right)$$

$$v_A = v_1 + v_2 - 2v_A$$

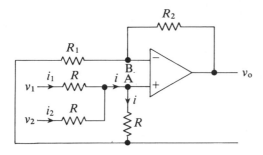

Fig. 9.9   Noninverting Adder

Therefore $v_A = \dfrac{v_1 + v_2}{3}$

Now $v_o = \left(1 + \dfrac{R_2}{R_1}\right) v_A$

So if $\dfrac{R_2}{R_1} = 2$,

$$v_o = 3v_A = v_1 + v_2$$

For each additional input to be added, R_2/R_1 must be increased by 1.

9.6   INTEGRATORS

Application of the principle of using integration for the generation of time base waveforms is described in Section 11.7. Integrators are used in a variety of measurement and signal processing applications: the basic op amp integrator is drawn in Fig. 9.10. Again, the important principle of the circuit action is that $v_B = v_A$, in this

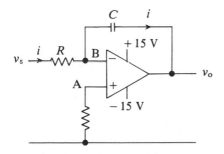

Fig. 9.10   Op Amp Integrator

case 0 V. Then

$$i = \frac{v_s}{R}$$

and

$$v_o = \frac{-1}{C} \int i \, dt \tag{9.4}$$

$$= \frac{-1}{CR} \int v_s \, dt \tag{9.5}$$

If $CR = 1$ s ($C = 1$ μF, $R = 1$ MΩ would be practical values), then

$$v_o = -\int v_s \, dt$$

If C were reduced to 0.1 μF,

$$v_o = -10 \int v_s \, dt$$

When integrating, an initial condition should be added which, in this circuit, is represented by the voltage across the capacitor when the operation begins.

☐ **Example 9.4**

Given that in Fig. 9.10, $C = 0.4$ μF, $R = 0.1$ MΩ, sketch v_o for a period of 0.5 s after the application of a constant input of 2 V at v_s. Assume that at the beginning of the operation C is discharged.

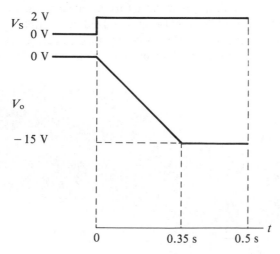

Fig. 9.11   Waveforms for Example 9.11

Using equation (9.5),

$$v_o = \frac{-1}{CR} \int_0^{0.5} v_s \, dt$$

$$= -50[t]_0^{0.5}$$

Since v_o is proportional to t it will be a linear ramp, negative going as indicated by the minus sign. However, evaluating v_o after 0.5 s gives $v_o = -25$ V. As the negative supply rail is only -15 V then v_o ramps down to this value, where it remains as shown in Fig. 9.11. □

It should be noted that equation (9.5) only holds if $v_A = 0$ V and in general it is safer to use equation (9.4) because one can usually calculate the current i quite readily.

An initial condition can be set by the circuit of Fig. 9.12. The two switches (which may be solid-state devices) are ganged and in the S (Set) position

$$V_o = \frac{-R_2}{R_3 + R_2} V_{REF}$$

because the 'virtual ground' will hold B at 0 V and so V_o will be below 0 V by the voltage across R_2. When the switch is put to I (Integrate) the integrated waveform will begin from the initial condition.

The circuit of Fig. 9.13 will begin its integration from V_A, the voltage on the noninverting input, but this will affect the charging current and therefore the slope of the output waveform.

□ **Example 9.5**

For Fig. 9.13 sketch v_o for 60 ms after S has been opened. Before S is opened

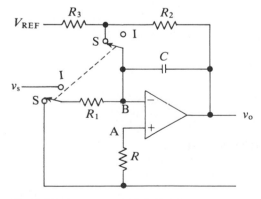

Fig. 9.12 Integrator with Initial Condition

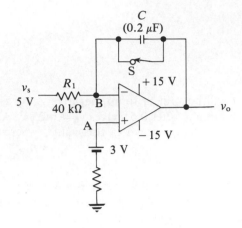

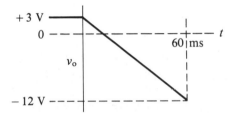

Fig. 9.13 Integrator with Positive Voltage on Noninverting Input, and Output Waveform for Example 9.5

$v_o = V_B = V_A = +3$ V. When S is opened $v_o = \dfrac{-1}{C} \int i \, dt + 3$ V.

$$i = \frac{5 \text{ V} - 3 \text{ V}}{40 \text{ k}\Omega} = 0.5 \times 10^{-4} \text{ A}$$

Using equation (9.4) and adding the initial condition,

$$v_o = \frac{-1}{0.2 \times 10^{-6}} [0.5 \times 10^{-4}]_0^{60 \text{ ms}} + 3$$

which is a negative-going linear ramp beginning at $+3$ V.

After 60 ms,

$$v_o = -5 \times 10^6 [0.5 \times 10^{-4} \times 60 \times 10^{-3}] + 3$$
$$= -12 \text{ V.}$$

The waveform is included in Fig. 9.13. □

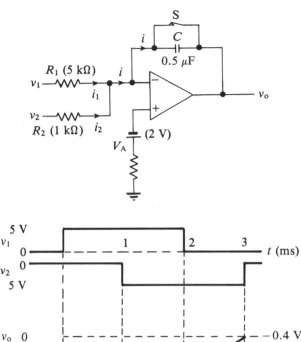

Fig. 9.14 Summing Integrator and Waveforms for Example 9.6

Summing Integrator

Addition and integration can be combined by the summing integrator of Fig. 9.14. In this circuit,

$$i = i_1 + i_2 \text{ and}$$

$$v_o = \frac{-1}{C} \int i \, dt + K$$

where K is the initial value of v_o.

☐ **Example 9.6**

Using the component values and input waveforms given in Fig. 9.14, sketch v_o given that S is opened at $t = 0$.

At $t = 0$

$$i = i_1 + i_2 = \frac{v_1 - V_A}{R_1} + \frac{v_2 - V_A}{R_2} = \frac{5 - (-2)}{5 \text{ k}\Omega} + \frac{0 - (-2)}{1 \text{ k}\Omega}$$

$$= 1.4 \text{ mA} + 2 \text{ mA} = 3.4 \text{ mA}$$

Then from $t = 0$ to 1 ms,

$$v_o = \frac{-1}{C} \int_0^{1 \text{ ms}} i \; dt - 2$$

$$= -2 \times 10^{-6} [3.4 \times 10^{-3} t]_0^{10^{-3}} - 2 \text{ V}$$

Which is a negative-going linear sweep beginning at -2 V as shown for the period 0 to 1 ms in Fig. 9.14. After 1 ms,

$$v_o = -6.8 - 2 = -8.8 \text{ V}.$$

From $t = 1$ ms to 2 ms

$$= i_1 + i_2 = 1.4 \text{ mA} + \frac{(-5 - [-2])}{1 \text{ k}\Omega} = 1.4 \text{ mA} - 3 \text{ mA}$$

$$= -1.6 \text{ mA}$$

$$v_o = \frac{-1}{C} \int_{1 \text{ ms}}^{2 \text{ ms}} i \; dt - 8.8 \text{ V}$$

as -8.8 V was the value of v_o at $t = 1$ ms.

$$\text{Then } v_o = -2 \times 10^{-6} \int_{1 \text{ ms}}^{2 \text{ ms}} -1.6 \times 10^{-3} \; dt - 8.8$$

$$= 2 \times 10^{-6} [1.6t]_{1 \text{ ms}}^{2 \text{ ms}} - 8.8 \text{ V}$$

which is a positive going linear sweep beginning at -8.8 V

At $t = 2$ ms,

$$v_o = 3.2 - 8.8 = -5.6 \text{ V}$$

For $t = 2$ ms to 3 ms

$$i = i_1 + i_2 = \frac{0 - (-2)}{5 \text{ k}} - 3 \text{ mA} = -2.6 \text{ mA}$$

and v_o is a positive going linear sweep beginning at -5.6 V and reaching $5.2 - 5.6 = -0.4$ V at $t = 3$ ms. The complete waveform for v_o is drawn in Fig.9.14.

$\square$

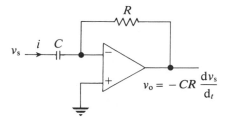

$$v_o = -CR\frac{dv_s}{d_t}$$

Fig. 9.15 Differentiator

9.7 DIFFERENTIATOR

For Fig. 9.15,

$$i = C\frac{dv_s}{dt}$$

Therefore $v_o = -iR = -CR\dfrac{dv_s}{dt}$ (9.6)

and if $CR = 1$ s then v_o is the differential of the input. A disadvantage of the circuit is that high-frequency noise and switching spikes are amplified. For example, if v_s includes a sinusoidal noise component $V_n \sin \omega t$ then at the output of the differentiator this will be $-\omega CR\ V_n \cos \omega t$, which is proportional to the frequency of the noise component. For this reason differentiators tend to be avoided.

9.8 ANALOG COMPUTERS

A physical system can be represented by a set of differential equations. Where it is difficult to set up such a system for experiment, for example, a large vibrating structure, then the equations can be modeled on an analog computer which uses continuously varying voltages to represent system variables. The equation can be solved by the computer and 'modeled' quantities readily varied by adjusting passive components on the computer. As noted, the mathematical functions required – integration, addition, scaling and inversion – can be provided by op amps.

For an equation such as

$$A\frac{d^2v_o}{dt^2} + B\frac{dv_o}{dt} + Cv_o = v_i$$ (9.7)

where A, B and C are constants, and v_o and v_i are time-dependent voltages which could be analogs of time-dependent variables such as displacement or rotation, then to solve the equation the following routine procedure can be adopted:

(a) Isolate the highest derivative on the LHS of the equation. Thus,

$$\frac{d^2v_o}{dt^2} = \frac{v_i}{A} - \frac{B}{A}\frac{dv_o}{dt} - \frac{C}{A}v_o$$ (9.8)

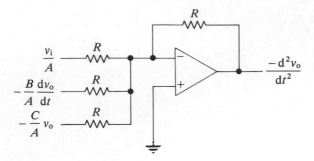

Fig. 9.16   Op Amp Adder for Solving Equation (9.8)

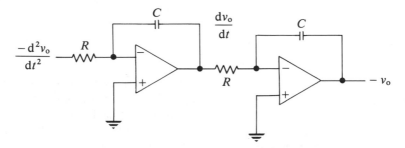

Fig. 9.17   Two Stages of Integration Yield the Required Variable v_o

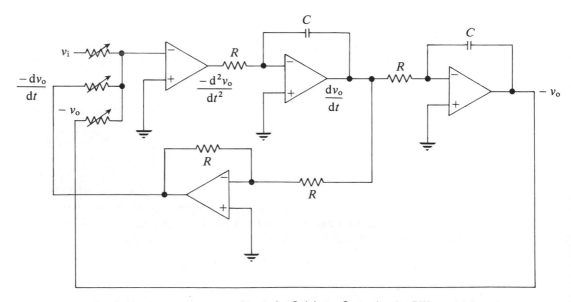

Fig. 9.18   Analog Computer Circuit for Solving a Second-order Differential Equation

(b) Sum the terms on the RHS of equation (9.8) in an op amp adder (Fig. 9.16) — the assumption that these terms are available is to be justified. Then the adder output will be $-d^2v_o/dt^2$.

(c) Successively integrate as many times as necessary to produce the problem variable v_o, in this case twice, as in Fig. 9.17. Here it is assumed that $CR = 1$ s although different values can be used for scaling purposes as mentioned in the next part of the procedure.

(d) Feed back to the adder inputs $-dv_o/dt$ and v_o and scale them with input potentiometers. Then apply and scale v_i and the various terms of the original differential equation (equation (9.7)) will be available for display on a CRO including the required term v_o, as shown in Fig. 9.18. The negative sign is of no consequence although it can easily be eliminated by inversion if necessary.

9.9 HIGH INPUT RESISTANCE DIFFERENTIAL AMPLIFIER

One disadvantage of some of the circuits described in this chapter is that their input resistances are too low for many applications. For example, because of the virtual ground effect, the circuit of Fig. 9.1 has an input resistance $R_i = R_1$. In Fig. 9.19 both inputs are connected to the noninverting inputs to which there is no NFB connection so the input resistance is as high as the input resistance of the op amp itself.

In Fig. 9.19,

$$v_{o1} = \left(1 + \frac{R_2}{R_1}\right) v_1$$

$$v_o = -\frac{R_1}{R_2} v_{o1} + \left(1 + \frac{R_1}{R_2}\right) v_2$$

$$= \left(-\frac{R_1}{R_2} - 1\right) v_1 + \left(1 + \frac{R_1}{R_2}\right) v_2$$

$$= \left(1 + \frac{R_1}{R_2}\right)(v_2 - v_1) \tag{9.9}$$

This result shows that the circuit can be used as a differential amplifier with both inputs connected to high input resistance. Further, if $v_2 = 0$ then

$$v_o = -\left(1 + \frac{R_1}{R_2}\right)$$

and the circuit can be used as a high input resistance inverting amplifier.

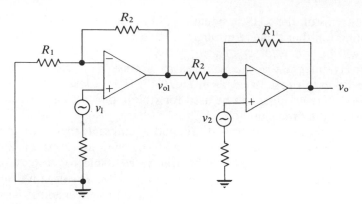

Fig. 9.19   High Input Resistance Differential Amplifier

9.10 LOGARITHMIC AMPLIFIER

The dynamic range of a signal is the ratio of the signal at its highest level to the signal at its lowest level: it is usually expressed in dB. Dynamic range can be compressed by increasing the lower-level signals. The benefit of compression is that if it is carried out before transmitting the signal over some communication channel then the ratio of signal to noise picked up on the channel at the lower signal levels is improved and it is at low signal levels that noise causes the greatest subjective disturbance. Compression is a form of nonlinear distortion and should be compensated for after reception by expanding the signal to its original dynamic range. The overall process is called companding.

One type of compressor is the logarithmic amplifier of Fig. 9.20. For a pn diode,

$$I = I_o(e^{KV} - 1) \tag{9.10}$$

(see Section 1.7). For all but small values of V the exponential term is dominant; therefore

$$I \doteqdot I_o e^{KV}$$

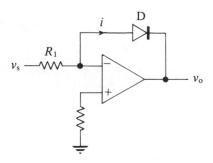

Fig. 9.20   Logarithmic Amplifier

and

$$\ln \frac{I}{I_o} = KV \qquad \text{or} \qquad V = \frac{1}{K} \ln \frac{I}{I_o}$$

For Fig. 9.20,

$$i = \frac{v_s}{R_1}$$

Therefore $v_o = \dfrac{-1}{K} \ln \dfrac{v_s}{R_1 I_o}$

The logarithmic relationship between v_o and v_s implies a compression as required. To produce the compensating expansion it is only necessary to interchange D and R_1.

A diode is not ideal for a logarithmic amplifier, first because of temperature instability and second because practical pn junction diodes do not exhibit the theoretical characteristics suggested by equation (9.10). Better results can be obtained using a BJT as the nonlinear element.

9.11 INSTRUMENTATION AMPLIFIER

Measurement transducers are usually two terminal devices, the voltage across which will be determined by a physical quantity such as pressure, temperature or displacement. In some applications neither terminal will be earthed. They are often used in electrically hostile environments which give rise to common mode voltages.

In Fig. 9.21 the input voltages v_1 and v_2 are from the transducer terminals. The circuit is designed to produce an output voltage v_3 which is the difference voltage $(v_1 - v_2)$ amplified.

Using superposition,

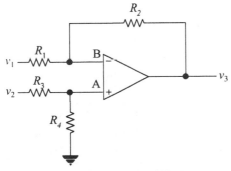

Fig. 9.21   Instrumentation Amplifier

with $\quad v_2 = 0, \; v_3' = -\dfrac{R_2}{R_1} \cdot v_1$

with $\quad v_1 = 0, \; v_A = \dfrac{R_4}{R_3 + R_4} \cdot v_2$

therefore $v_3'' = \left(1 + \dfrac{R_2}{R_1}\right) \left(\dfrac{R_4}{R_3 + R_4}\right) v_2.$

If $\dfrac{R_2}{R_1} = \dfrac{R_4}{R_3},$

$$v_3'' = \left(1 + \dfrac{R_2}{R_1}\right) \left(\dfrac{1}{\dfrac{R_1}{R_2} + 1}\right) v_2 = \dfrac{R_2}{R_1} v_2$$

Then $v_3 = v_3' + v_3'' = \dfrac{R_2}{R_1} (v_2 - v_1)$

Thus the difference signal will be amplified: any common mode signal will not produce an output.

9.12 ACTIVE FILTERS

Filters are used to pass parts of the signal frequency spectrum and attenuate others. Fig. 9.22 shows the frequency responses of ideal and practical low pass filters. The ideal filter passes all signals up to the cut off frequency f_1, and totally eliminates frequencies above f_1. The response of a practical filter changes more gradually and does not completely eliminate signals in the stop band. Practical filters can be designed to have a sharper transition than that suggested in Fig. 9.22.

There are four basic types of filter — low pass, high pass, band pass and band stop. Their responses are determined by passive resistive and reactive components.

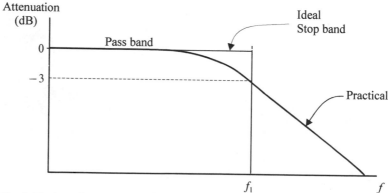

Fig. 9.22 Low Pass Filter Frequency response

For steep gradients of response, and for band pass and band stop filters, inductors and capacitors are required in combination if the filter is to be made up solely of passive components.

Active filters use op amps as well as passive components and some of their advantages are as follows:

(a) Inductors, which are difficult to accommodate in many electronic circuits, are not required.
(b) The passive components can be isolated from the load: with passive filters, loads can adversely affect frequency response.
(c) They can be designed to produce gain in the pass band.

Against these, active filters are limited by the high frequency capability of op amps which is currently a few megahertz.

First Order (Single Pole) Filters

The passive components of a simple low pass filter (LPF) are shown in Fig. 9.23. For this,

$$\frac{V_o}{V_s} = \frac{\dfrac{1}{j\omega C}}{\dfrac{1}{j\omega C} + R} = \frac{1}{1 + j\omega CR} \tag{9.10}$$

At LF where $\omega CR \ll 1$, $\left|\dfrac{V_o}{V_s}\right| \doteqdot 1$

and there is no attenuation.

When $\omega = \omega_1 = \dfrac{1}{CR}$, $\left|\dfrac{V_o}{V_s}\right| = \dfrac{1}{\sqrt{2}}$

i.e. $\left|\dfrac{V_o}{V_s}\right|$ is 3 dB down on its LF value.

At ω_1, the frequency $f_1 = \dfrac{1}{2\pi CR}$ and is called the cut off frequency.

At HF when $\omega CR \gg 1$, $\left|\dfrac{V_o}{V_s}\right| \alpha \dfrac{1}{f}$ and falls at 6 dB/oct.

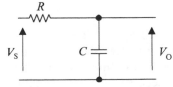

Fig. 9.23   Passive Low Pass Filter

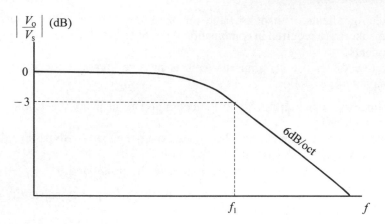

Fig. 9.24 Frequency Response of First Order Low Pass Filter

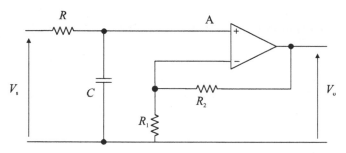

Fig. 9.25 Active Low Pass Filter

The frequency response is shown in Fig. 9.24.

This passive network can be connected into a noninverting op amp circuit to form the active filter of Fig. 9.25. At LF, V_s appears at the noninverting input A without attenuation. As frequency rises the input at A falls, as in equation (9.10). The circuit gain is determined by R_1 and R_2, as shown in section 9.2. Thus, the overall gain of the circuit will be,

$$\frac{V_o}{V_s} = \left(1 + \frac{R_2}{R_1}\right) \left(\frac{1}{1 + j\omega CR}\right)$$

Note that C and R, the components which determine the frequency response, are connected to the noninverting input whose input resistance is virtually infinite.

The circuit can be converted into a high pass filter by interchanging C and R. The analysis is the same as that given in section 3.9 for a coupling capacitor.

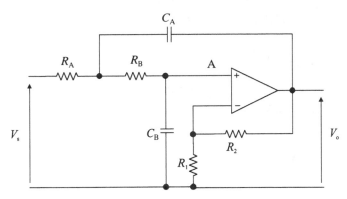

Fig. 9.26   Sallen and Key Low Pass Filter

Higher Order Active Filters

As noted, with a maximum rate of attenuation of 6 dB/oct, the application of first order filters is limited. The rate of attenuation can be doubled by cascading two of these active filters.

A more compact second order active filter is the Sallen and Key circuit of Fig. 9.26. Both pairs of passive components, $R_A C_A$ and $R_B C_B$, have an integrating action whose response falls with frequency. At HF, each pair of components contributes a 6 dB/oct fall, i.e. 12 dB/oct in total for the filter. Interchange of resistors and capacitors gives a high pass filter. Interchanging only R_B and C_B results in a band pass filter.

SUMMARY

Having described the basic properties of op amps in Chapter 8, this chapter has been used to introduce the op amp for linear circuit applications. Important in this context is the 'virtual ground' principle. This term must be used with caution and it is better to understand that in linear applications the circuit must operate in such a manner as to make the differential input zero. Once this principle is established then the operation of most circuits can be worked out from basic electrical circuit theory.

REVISION QUESTIONS

9.1   Explain why it is that in an op amp inverter the differential input must be zero.

9.2   In Fig. Q.9.2, for v_s equal to:

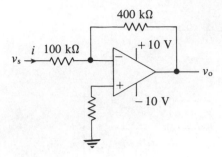

Fig. Q.9.2

9.3   In Fig. Q.9.3, for v_s equal to:

(a) $+2$ V
(b) 0 V
(c) -2 V
find i and hence v_0 in each case.
Sketch v_0 if v_s is a sine wave of 2 V peak.

$$(30 \ \mu A; \ 10 \ \mu A; \ -10 \ \mu A; \ -13 \ V; \ -5 \ V; \ +3 \ V)$$

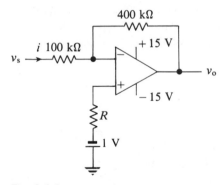

Fig. Q.9.3

9.4   Suggest a suitable value for R in Fig. Q.9.3.

$$(80 \ k\Omega)$$

9.5   For Fig. Q.9.5 sketch the waveforms of i_1, i_2, i_3 and i, and hence the waveform of v_0.

9.6   Design a three-input noninverting adder.

9.7   Repeat Example 9.4 but with $C = 0.5 \ \mu A$ and $R = 0.3 \ M\Omega$,

(a) with v_s as given;
(b) with v_s as given and v_0 initially at -2 V;
(c) with v_s inverted.

(a) $+2$ V

(b) 0 V

(c) -2 V

find i and hence v_o in each case.

If v_s is a sine wave of 2 V peak, sketch v_o.

Repeat for v_s as a sine wave of 3 V peak.

$$(20 \ \mu A; \ 0; \ -20 \ \mu A; \ -8 \ V; \ 0; \ +8 \ V)$$

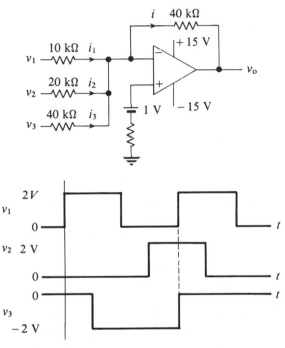

Fig. Q.9.5

9.8 Repeat Example 9.5 but with the noninverting input at -3 V. In this case, how long does it take to reach -15 V?

$$(12 \ \text{ms})$$

9.9 Repeat Example 9.6 but exchange the values of R_1 and R_2.

9.10 Design a single pole, high pass filter which has a cut off frequency of 2 kHz, and a gain of 12 dB in the pass band.

$$(0.017 \ \mu F; \ 4.7 \ k\Omega; \ 3.3 \ k\Omega; \ 10 \ k\Omega)$$

10

Sinusoidal Oscillators

Chapter Objectives
10.1   Conditions for Oscillation
10.2   RC Oscillators
10.3   LC (Tuned) Oscillators
        Summary
        Revision Questions

The process of using negative feedback to improve amplifier performance was described in Chapter 5, the benefits being secured by sacrificing gain: negative feedback is said to be degenerative. In this chapter we are dealing with the opposite situation: feedback is made positive, i.e. regenerative, and gain is increased to the extent that an amplifier will produce an output signal with no input. In other words, it is an oscillator. This chapter is confined to describing the generation of sine waves, Chapter 11 being reserved for nonsinusoidal waveform generation. Apart from the generation of sine waves at audio frequencies for test purposes, oscillators are used up into the gigahertz (10^9 Hz) frequency range for a wide variety of applications, notably in radio and in telecommunication transmission systems.

10.1 CONDITIONS FOR OSCILLATION

In an NFB amplifier (Fig. 10.1) the signal fed back, V_f, is connected in such a way as to oppose the externally applied input signal V_s, i.e. in determining the amplifier input V_i, the signal fed back has to be subtracted from the input signal. Consider the feedback circuit of Fig. 10.2 in which there is no externally applied input signal. There is, however, always electrical noise even in passive components and conductors, so assume that there is some signal V_i at the amplifier input. After amplification and feedback the signal appears back at the input as $\beta A_v V_i$. If this is in phase with (i.e added to) V_i and the loop gain $\beta A_v > 1$, then the signal builds up

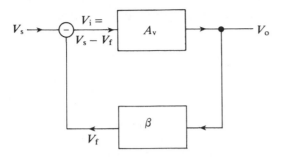

Fig. 10.1 NFB Amplifier in which $V_i = V_s - V_f$

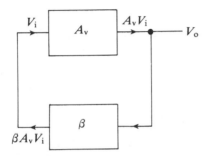

Fig. 10.2 Feedback Amplifier with no Externally Applied Input Signal

and the circuit bursts into oscillation (the Nyquist stability criterion discussed in Chapter 5 is contravened). Nonlinearity in the amplifier will limit the extent to which oscillations build up, i.e. the loop gain βA_v will fall to unity and the circuit will settle down so that it is producing its own input V_i and output $V_o = A_v V_i$.

The requirement that the input and fed-back signals must be in phase implies that the total phase shift round the loop must be zero or an exact multiple of $360°$. In a sine wave oscillator this condition will only be met at one frequency and this will define the operating frequency of the oscillator. The condition for oscillation then is that $\beta A_v = 1$. This can be expressed in polar coordinates as

$$|\beta A_v| = 1; \; \text{Arg } \beta A = 0 \quad \text{or} \quad 2n\pi \tag{10.1}$$

or in rectangular coordinates as

$$R_p(\beta A) = 1; \; I_p(\beta A) = 0 \tag{10.2}$$

The condition that, for oscillation, $\beta A_v = 1$ is known as the Barkhausen criterion.

In Chapter 5 the equation for the voltage gain of an NFB amplifier,

$$A_{vf} = \frac{A_v}{1 + \beta A_v}$$

was developed by subtracting the signal fed back at the input. If instead the signal fed back had been added, then the equation would have been,

$$A_{vf} = \frac{A_v}{1 - \beta A_v}$$

and if $\beta A_v = 1$,

$$A_{vf} = \infty$$

i.e. the amplifier would produce an output with no input. This is something of an abstraction and it is more meaningful to use the concept that in amplifying signals round the feedback loop there will be a frequency at which $\beta A_v = 1$ and this will be the oscillator frequency.†

There are two types of sine wave oscillator,

(a) RC or phase shift oscillator;
(b) LC or tuned oscillator, a type which includes crystal oscillators.

10.2 RC OSCILLATORS

Figure 10.3 is a block diagram of an amplifier with feedback. We must first try and dispose of a potential source of confusion – the use of the words 'input' and 'output'. In Fig. 10.3, V_i, which is the input to the amplifier, is also the output of the feedback network. Similarly, V_o, the amplifier output, is the feedback network input. In this chapter we shall be concentrating on the characteristics of feedback networks and then choosing amplifiers that, in combination with the network, fulfill the Barkhausen criterion for oscillation that $\beta A_v = 1$. We shall use the variable V_1 for the network input and V_2 for the network output. Having found $\beta = V_2/V_1$ for a network then the amplifier gain $A_v = V_o/V_i$ must be the inverse of V_2/V_1, so that the loop gain $\beta A_v = 1$.

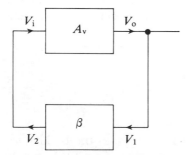

Fig. 10.3   Illustrating that the Amplifier Output is the Feedback Network Input and Vice Versa

† In some texts feedback theory is unified by using $A_{vf} = A_v/(1 + \beta A_v)$ for all types of feedback, and that for oscillation, $\beta A_v = -1$.

10.2.1 Ladder Network RC Oscillators

Each section of the network of Fig. 10.4 introduces a phase shift which makes the signal progressively lead V_1 as it progresses along the network. At a certain frequency f_o the phase shift will be $180°$. If all Rs and all Cs are equal then it can be shown that

$$\frac{V_2}{V_1} = \frac{1}{1 - 5/(\omega CR)^2 - j(6/\omega CR - 1/(\omega CR)^3)} \qquad (10.3)$$

One of the conditions for oscillation is that the imaginary term must be zero and this can be used to find the frequency of oscillation f_o. In this condition

$$\frac{6}{\omega CR} = \frac{1}{(\omega CR)^3}$$

Therefore $(\omega CR)^2 = \dfrac{1}{6}$

and $$f_o = \frac{1}{2\pi CR\sqrt{6}}$$

Now with the j term zero, substitute for $(\omega CR)^2$ in equation (10.3) and we have

$$\frac{V_2}{V_1} = \frac{1}{1 - 5 \times 6} = \frac{-1}{29}$$

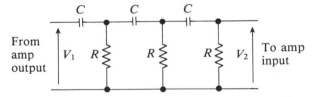

Fig. 10.4   An RC Ladder Network for Producing a $180°$ Phase Shift at One Frequency f_o

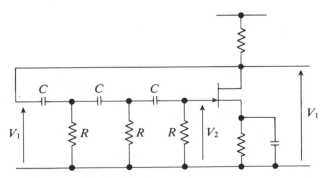

Fig. 10.5   An FET Ladder Network

the minus sign confirming the $180°$ phase shift in the network at f_o. To compensate for the $180°$ phase shift in the network the amplifier must also have a $180°$ phase shift, i.e. it must be an inverter. It must also have a gain of 29 to compensate for the loss in the network. Figure 10.5 is an FET version of a ladder RC oscillator. The ladder network can be constructed with the Rs and Cs interchanged and with unequal values of resistors and capacitors, but the operating frequency and modulus of gain required must be recalculated.

10.2.2 Wien Bridge RC Oscillator

A more common oscillator phase shift network is shown in Fig. 10.6. For this network,

$$\frac{V_2}{V_1} = \frac{\dfrac{R_2}{1 + j\omega C_2 R_2}}{R_1 + \dfrac{1}{j\omega C_1} + \dfrac{R_2}{1 + j\omega C_2 R_2}}$$

$$= \frac{R_2}{R_1 + \dfrac{C_2 R_2}{C_1} + j\omega C_2 R_2 R_1 + \dfrac{1}{j\omega C_1} + R_2}$$

For the network to have a phase shift of $0°$ or $180°$ then,

$$I_p\left(\frac{V_2}{V_1}\right) = 0$$

Therefore $\omega C_2 R_2 R_1 = \dfrac{1}{\omega C_1}$

and $$f_o = \frac{1}{2\pi\sqrt{R_1 R_2 C_1 C_2}} \tag{10.4}$$

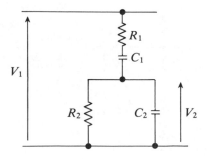

Fig. 10.6   Wien Bridge Network

At f_o,

$$\frac{V_2}{V_1} = \frac{R_2}{R_1 + C_2 R_2/C_1 + R_2}$$

$$= \frac{1}{R_1/R_2 + C_2/C_1 + 1}$$

Therefore, for the amplifier,

$$A_v = \frac{C_2}{C_1} + \frac{R_1}{R_2} + 1$$

and since V_2/V_1 is positive at f_o the phase shift of the network is zero so the amplifier must be noninverting.

If $C_1 = C_2 = C$ and $R_1 = R_2 = R$, then

$$f_o = \frac{1}{2\pi CR} \tag{10.5}$$

and $A_v = 3$

10.2.3 Op Amp Wien Bridge Oscillator

One way of obtaining the gain required for a Wien bridge oscillator is by using the op amp circuit of Fig. 10.7. In this case we have the simplified Wien network with

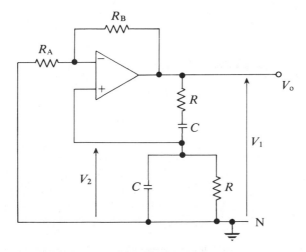

Fig. 10.7 Op Amp Wien Bridge Oscillator

equal values of resistance (R) and capacitance (C) so to compensate for the loss of the phase shift network the op amp must have a noninverting gain of 3. Now the amplifier input will be V_2 which is connected to the noninverting input. To provide a gain of 3,

$$1 + \frac{R_B}{R_A} = 3$$

Therefore $\quad R_B = 2R_A$

Note that in selecting components for the Wien network the output resistance of the amplifier must be included in the value of the series resistance and the input capacitance of the op amp must be included in the value of the shunt capacitance — the op amp input resistance is normally so large as to be of no significance. These quantities limit the upper operating frequency of the circuit. However, the $R_A R_B$ chain provides voltage-derived NFB so output resistance will be small.

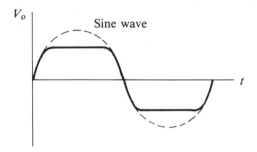

Fig. 10.8   Output Waveform Peaks Flattened by Oscillations Building Up

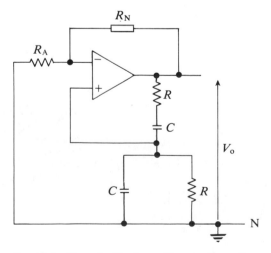

Fig. 10.9   Use of a Nonlinear Element R_N to Control Gain. In this Position R_N Must Fall with Output Voltage

A disadvantage of this circuit is that the gain cannot be made exactly 3 (nor can the gain of the FET amplifier of Fig. 10.5 be exactly -29). If the gain is less than 3 then the circuit will not oscillate. If it is greater than 3 then the amplitude of the oscillations builds up until nonlinearity reduces the gain, but this will result in the waveform peaks being flattened (Fig. 10.8). In other words, the waveform will not be sinusoidal, i.e. harmonic distortion is added. The solution is to introduce a nonlinear component into the negative feedback network which determines the gain.

Figure 10.9 is the op amp version of the Wien bridge oscillator but with R_B replaced by a nonlinear component whose resistance is R_N. With R_N replacing R_B in Fig. 10.7 it is necessary for R_N to fall as the rms voltage of the sine wave across it increases. Now if the component is chosen so that when the circuit is powered $R_N > 2R_A$, then $A_v > 3$ and the circuit will oscillate. However, as the oscillations build up R_N will fall and the circuit will stabilize when $R_N = 2R_A$, i.e. when the gain $A_v = 3$. A suitable nonlinear component for this purpose is a thermistor.

☐ **Example 10.1**

Design a Wien bridge oscillator with $f_o = 10$ kHz and $V_o = 2$ V rms using a 741 op amp and an RA 54 thermistor whose R_{TH} vs. V_{TH} characteristic is given in Fig. 10.10.

For the Wien bridge, making the two Rs and the two Cs equal, then,

$$f_o = 10^4 = \frac{1}{2\pi CR}$$

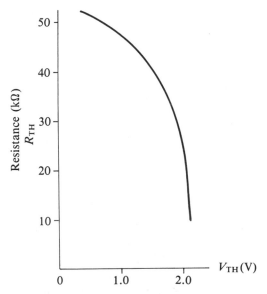

Fig. 10.10 R_{TH} vs. V_{TH} Characteristic of a Thermistor (*Courtesy STC Components Ltd.*)

Making $R = 10 \text{ k}\Omega$, then the 741 op amp output resistance at nominally $150 \, \Omega$ has little effect on f_o. (If required, this could be included in the series R.)

Then, $C = \dfrac{1}{2\pi f_o R} = 1.6 \text{ nF}$

Although the parallel C is across the op amp input, at $C_{in} = 1.4 \text{ pF}$ for the 741 its effect on f_o is insignificant.

Referring now to Fig. 10.9. note that V_o is distributed across R_N and R_A. For $A_v = 3$, $R_N = 2R_A$. Therefore two-thirds of V_o is dropped across R_N and one-third across R_A. In this case, the rms voltage across R_N − which is to be the thermistor − is $\frac{2}{3} \times 2 \text{ V} = \frac{4}{3} \text{ V}$. Referring to the RA 54 characteristic (Fig. 10.10), when $V_{TH} = \frac{4}{3} \text{ V}$, $R_{Th} = 42 \text{ k}\Omega$. Now if we choose $R_A = 21 \text{ k}\Omega$, then when the circuit is powered, the gain will automatically be set to 3 by R_{TH} becoming $42 \text{ k}\Omega$ which will be when $V_o = 2 \text{ V}$ as required. $\qquad\qquad\qquad\qquad\qquad\qquad\qquad\qquad\qquad\qquad\qquad$ □

An alternative to replacing R_B in Fig. 10.7 with a component whose resistance falls with voltage is to replace R_A with a component whose resistance rises with voltage. One such component is simply a lamp, because filament resistance always rises with temperature and therefore with applied voltage. Better than a lamp, which requires a high current and whose resistance will be influenced by ambient temperature, is the use of an FET as a voltage-controlled resistance (Section 4.4). In Fig. 10.11, if the FET gate voltage is zero then the drain to source resistance R_{DS} will be at its lowest value. When the circuit is powered it will oscillate if $R_B/R_{DS} > 2$. As oscillation builds up, the negative half-cycles of V_o pass through D and charge C_f up to the negative peak voltage (see Section 19.3 on rectifiers), R_f simply allows the pd

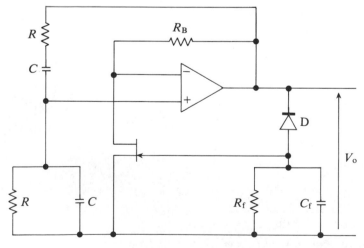

Fig. 10.11   Use of FET to Stabilize Wien Bridge Oscillator Amplitude

across C_f to fall if V_o falls. The resulting negative voltage on the FET gate will increase the value of R_{DS} until the amplitude stabilizes, which will be when $R_{DS} = R_B/2$. The arrangement can be modified to make a variable amplitude oscillator by making R_f a potential divider with the slider connected to the gate. As the slider is moved down, R_{DS} will fall so oscillations will increase in amplitude to restore R_{DS} to $R_B/2$.

10.3 LC (TUNED) OSCILLATORS

As their name suggests, these oscillators use, in conjunction with amplifiers, LC circuits working at resonant frequencies. They are usually used to generate higher frequencies than RC oscillators. At very high frequencies the inductances and capacitances may not be standard components but instead be derived from tuned cavities. The two most common LC oscillators are called, after their inventors, the Hartley oscillator and the Colpitts oscillator. It is normal practice to carry out general analyses, one for FET tuned oscillators and one for BJT. In each case the resulting formulas can be used to find the frequency of oscillation f_o and the transistor parameter requirement for both Hartley and Colpitts oscillators.

10.3.1 Generalized Approach to FET LC Oscillators

Figure 10.12 is an FET LC oscillator in which, to simplify the analysis, all bias components and dc supplies have been omitted. Also, the three components, Z_1, Z_2 and Z_3 are pure reactances. The object of the analysis is:

(a) to show what type of reactance – inductance or capacitance – the components must be:
(b) to relate the values of the components to the oscillation frequency f_o; and
(c) to deduce the FET parameters required to sustain oscillation.

In Fig. 10.12 Z_1 is connected between gate and source so by definition, the voltage across it is V_{gs}. This is also shown in the small-signal equivalent circuit of Fig. 10.13, which will be used for the analysis. In the context of the condition for oscillation specified in Section 10.1, V_{gs} is the amplifier input. This voltage, in conjunction with the FET, produces a current $g_{fs}V_{gs}$ which feeds the three arms of the network of Fig. 10.13, and which will give rise to a voltage across Z_1. Again, this voltage is, by definition, V_{gs} and if it is equal in amplitude and phase to the original value of V_{gs} then the loop gain is unity and the circuit will oscillate. The process of analysis then, is to find the voltage across Z_1 as a function of $g_{fs}V_{gs}$ and the circuit components, and equate it to V_{gs}. This will give the condition for oscillation. In Fig. 10.13,

$$I = \frac{-1/(Z_1 + Z_3)}{1/(Z_1 + Z_3) + 1/Z_2 + 1/r_{ds}}\, g_{fs}V_{gs}$$

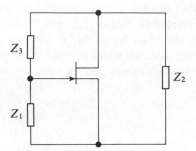

Fig. 10.12   Basic Form of FET LC Oscillator without Bias Components

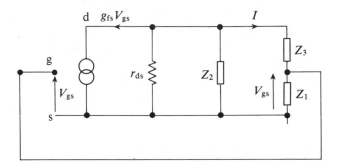

Fig. 10.13   Small-signal Equivalent Circuit of FET LC Oscillator

For oscillation,

$$IZ_1 = V_{gs}.$$

Canceling V_{gs} we have

$$\frac{-g_{fs}Z_1}{1 + (Z_1 + Z_3)/Z_2 + (Z_1 + Z_3)/r_{ds}} = 1 \tag{10.6}$$

With all Zs pure reactance, the numerator of this expression is imaginary and therefore the real terms in the denominator must add up to zero.

Therefore $\dfrac{Z_1 + Z_3}{Z_2} + 1 = 0$

and $Z_1 + Z_2 + Z_3 = 0$ \hfill (10.7)

With the real terms in equation (10.6) equal to zero,

$$-g_{fs}Z_1 = \frac{Z_1 + Z_3}{r_{ds}}$$

Therefore, substituting for equation (10.7),

$$g_{fs}r_{ds}Z_1 = Z_2 \qquad\qquad (10.8)$$

Since $g_{fs}r_{ds}$ is positive, Z_1 and Z_2 are the same type of reactance, i.e. both inductive or both capacitive. Substituting for Z_2 in equation (10.7),

$$Z_1 + g_{fs}r_{ds}Z_1 = -Z_3$$

i.e. $Z_3 = -(1 + g_{fs}r_{ds})Z_1$

Therefore Z_3 is the opposite type of reactance to Z_1 and therefore to Z_2.

For a Colpitts oscillator Z_1 and Z_2 are both capacitive and Z_3 inductive, i.e.

$$Z_1 = \frac{1}{j\omega C_1}; \; Z_2 = \frac{1}{j\omega C_2}; \; Z_3 = j\omega L_3$$

Substituting in equation (10.7),

$$\frac{1}{j\omega C_1} + \frac{1}{j\omega C_2} + j\omega L_3 = 0$$

from which

$$\omega^2 = \frac{C_1 + C_2}{L_3 C_1 C_2}$$

and $\displaystyle f_o = \frac{1}{2\pi\sqrt{\dfrac{L_3 C_1 C_2}{C_1 + C_2}}}$ \qquad\qquad (10.9)

The FET parameter requirements to sustain oscillation can be deduced from equation (10.8).

$$\frac{g_{fs}r_{ds}}{j\omega C_1} = \frac{1}{j\omega C_2}$$

Therefore $\displaystyle g_{fs}r_{ds} = \frac{C_1}{C_2}$ \qquad\qquad (10.10)

Given values of C_1 and C_2 this equation gives the minimum value of $g_{fs}r_{ds}$ required for oscillation. Normally the practical value will be greater than the design value so that when the circuit is powered oscillations will build up until limited by the FET. However, the tuned circuit will act as a filter which selects the fundamental — in this case f_o — and rejects the harmonics so that the oscillator output will be a pure sine wave.

Figure 10.14 is a practical form of Colpitts oscillator. For signal voltages V_{DD} can be regarded as being connected to ground and therefore to the gate through C_g. R_g holds the gate to ground for dc purposes. L_s is an inductor chosen to be, in effect, an open circuit at f_o so that the junction of C_1 and C_2 is connected to source as required but not to ground, which is connected to gate. For dc purposes R_s provides

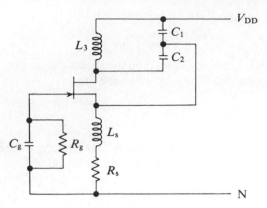

Fig. 10.14   FET Colpitts Oscillator

a bias voltage. With the junction of L_3 and C_2 connected to drain, all signal conditions are fulfilled in accordance with Fig. 10.12 and the types of reactance required for an FET Colpitts oscillator. For a Hartley FET oscillator,

$$f_o = \frac{1}{2\pi\sqrt{C_3(L_1 + L_2)}} \tag{10.11}$$

as we shall show in the section on BJT oscillators, and $g_{fs}r_{ds} = L_2/L_1$ to sustain oscillation, which can easily be deduced from equation (10.8).

☐   **Example 10.2**

The design of an LC oscillator such as a Colpitts often begins with the availability of suitable coils. For example, given that a 50 μH coil is available, choose values of C_1 and C_2 for the oscillator of Fig. 10.14 to run at 1 MHz.

Substituting the values given for L_3 and f_o in equation (10.9),

$$\frac{C_1 C_2}{C_1 + C_2} = 500 \text{ pF}$$

This is the expression for two capacitors in series in which the smaller capacitor will dominate the combined value. Referring to equation (10.10), it is to be expected that $g_{fs}r_{ds} > 10$ so if we make $C_1 = 10C_2$ then the FET will sustain oscillation. Then substituting for C_1 in the expression for two capacitors in series,

$$\frac{10\ C_2{}^2}{11\ C_2} = 500 \text{ pF}$$

$$C_2 = 550 \text{ pF}$$

$$C_1 = 5500 \text{ pF} \qquad\qquad ☐$$

10.3.2 BJT LC Oscillator

For a BJT tuned oscillator the procedure and results are similar to those for an FET. Figure 10.15 is the basic oscillator and Fig. 10.16 its equivalent circuit using the simplified h parameter model. For oscillation, the signal loop is as follows. The signal base current I_b flowing into h_{ie} gives rise to a current $h_{fe}I_b$ which feeds the network. For oscillation the resulting current in h_{ie} must be equal to the original value of I_b, both in magnitude and phase. So if we write equations for the network in which the current generator $h_{fe}\,I_b$ produces a current I_b in h_{ie}, then the current is self-sustaining and the circuit oscillates.

Again Z_1, Z_2 and Z_3 are pure reactances and the analysis must show whether they are inductances or capacitances, the oscillating frequency f_o and the relationship between the transistor parameters and the reactances to sustain oscillation. In Fig. 10.16,

$$I_1 Z_1 = I_b h_{ie}$$

Therefore $I_i = \dfrac{h_{ie}}{Z_1}\,I_b$ (10.12)

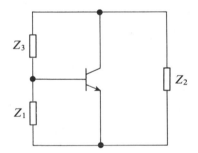

Fig. 10.15 Basic Form of a BJT Oscillator without Bias Components

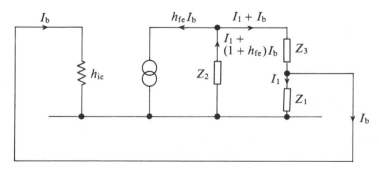

Fig. 10.16 Equivalent Circuit of BJT LC Oscillator

Applying KVL to the loop containing Z_2, Z_3 and h_{ie},

$$[I_1 + (1 + h_{fe})I_b]Z_2 + (I_1 + I_b)Z_3 + I_b h_{ie} = 0$$

Substituting for I_1 from equation (10.12) leads to

$$I_b \left[h_{ie} \frac{Z_2}{Z_1} + h_{ie} \frac{Z_3}{Z_1} + (1 + h_{fe})Z_2 + Z_3 + h_{ie} \right] = 0$$

in which $I_b \neq 0$; therefore the term inside the square brackets is zero. With all Zs pure reactances, the 1st, 2nd and 5th terms are real and sum to zero, and the 3rd and 4th terms are imaginary and also sum to zero.
Therefore,

$$h_{ie} \left[\frac{Z_2}{Z_1} + \frac{Z_3}{Z_1} + 1 \right] = 0$$

and since $h_{ie} \neq 0$ the term in brackets is zero, from which,

$$Z_1 + Z_2 + Z_3 = 0 \tag{10.13}$$

Summing the imaginary terms to zero,

$$(1 + h_{fe})Z_2 = -Z_3 \tag{10.14}$$

Therefore Z_2 and Z_3 are opposite types of reactance.
Combining equations (10.13) and (10.14),

$$Z_1 = -Z_2 - Z_3 = -Z_2 + (1 + h_{fe})Z_2 = h_{fe}Z_2 \tag{10.15}$$

Therefore Z_1 and Z_2 are the same type of reactance.
For a Hartley oscillator, Z_1 and Z_2 are inductive and Z_3 capacitive, and from equation (10.13),

$$j\omega L_1 + j\omega L_2 - \frac{j}{\omega C_3} = 0$$

$$\omega^2 = \frac{1}{C_3(L_1 + L_2)}$$

$$f_o = \frac{1}{2\pi\sqrt{C_3(L_1 + L_2)}}$$

From equation (10.15),

$$j\omega L_1 = h_{fe}j\omega L_2$$

$$h_{fe} = \frac{L_1}{L_2}$$

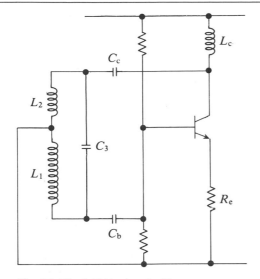

Fig. 10.17 BJT Hartley Oscillator

to sustain oscillation. Again this will be the minimum value of h_{fe}, so harmonics produced by the excess gain will be filtered out by the tuned circuit.

Figure 10.17 is a practical BJT Hartley oscillator. The choke L_c isolates the collector from the dc supply rail at f_o so the signal paths from collector to base and emitter are through L_2 and C_3 as required, C_c and C_b being signal frequency short-circuits. Note that, as required by the basic oscillator layout, the L_2C_3 junction is connected to collector, L_1C_3 to base, and L_1L_2 to emitter via the bias resistor R_e which will be a small value. If L_1 and L_2 are coupled by a mutual conductance M then effective inductances are $(L_1 + M)$ and $(L_2 + M)$.

10.3.3 Crystal Oscillators

There will always be some discrepancy between the design and actual operating frequencies of an oscillator because of factors such as component tolerance. Further, oscillator frequency may drift during operation because, for example, of the effect of temperature variation on component values. Where a high degree of frequency stability and predictability are required a piezoelectric crystal may be used as part of the frequency-determining circuit.

The piezoelectric effect is exhibited in certain crystals, notably quartz. Mechanical force across one pair of faces produces a voltage across another pair. This effect is used in crystal microphones and gramophone pick-ups where mechanical vibrations are transduced into an electrical signal. For oscillators the converse effect is used, i.e. when a voltage is applied across one pair of faces then mechanical deformation occurs between another pair. The mechanical effect as such is not used, but

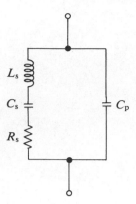

Fig. 10.18   The Effective Electrical Circuit of a Piezoelectric Crystal

a crystal very readily vibrates, i.e. resonates, when the applied voltage is at the resonant frequency. In consequence the crystal behaves as a very high Q (typically 10^4 to 10^6) series-tuned circuit load to a signal source connected across it. In Fig. 10.18 this tuned circuit is represented by L_s, C_s and R_s. To apply the voltage the crystal is held between two plates which form a capacitance C_p across the tuned circuit as shown.

Figure 10.19 shows how the reactance of a piezoelectric crystal varies with frequency. A series-tuned circuit is capacitive below its series resonant frequency f_s and inductive above it. Therefore, at some frequency f_p which will be higher than f_s, the crystal will act as a parallel-tuned circuit as the series circuit resonates with C_p. The series resonant frequency is given by

$$f_s = \frac{1}{2\pi\sqrt{L_s C_s}}$$

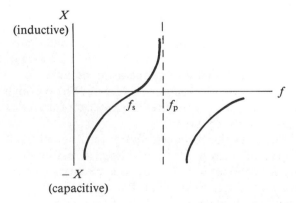

Fig. 10.19   Reactance vs. Frequency Characteristic of a Piezoelectric Crystal. In Practice f_s and f_p are Close Together

and the parallel resonance by

$$f_p = \frac{1}{2\pi\sqrt{L_s C}}$$

where $C = \dfrac{C_s C_p}{C_s + C_p}$ (10.16)

The effective values of the three series components of crystals are not typical of those found when using coils and capacitors for series-tuned circuits. For example, a Cathodeon HC-18 enclosure 10 MHz crystal has $L_s = 12.7$ mH, $C_s = 19.9$ fF ($F \times 10^{-15}$) and $R_s = 8\ \Omega$, which gives $Q = 99\,000$. A conventional series circuit for the same resonant frequency would have an inductance value a factor of about 1000 less than the crystal inductance and a capacitance correspondingly higher. With the same value of R_s, Q would be down to 99. As the value of C_s in a crystal is so small it dominates the value of C in equation (10.16). For the crystal quoted, $C_p = 5.7$ pF. Then substituting the values of C_s and C_p in equation (10.16),

$C = 19.8$ fF

which is 0.5% less than C_s. The value of C_p then is not critical and as a result the difference between f_s and f_p is negligible.

Resonant frequencies are accurately determined when crystals are ground. These frequencies are subject to small fluctuations with temperature change although the fluctuations are much less than for conventional tuned circuits. A quartz crystal watch with an oscillator stability of 1 in 10^6 will keep time to better than 1 minute a year. For very high stability of frequency, crystals can be located in temperature-controlled 'ovens' in which frequency stability of 1 in 10^8 is commonplace.

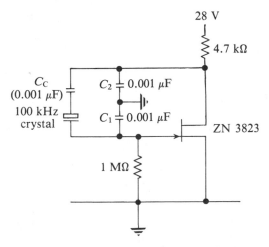

Fig. 10.20   Crystal Controlled Colpitts Oscillator (Courtesy National Semiconductors)

An example of a crystal oscillator is the FET Colpitts circuit of Fig. 10.20. The junction of capacitors C_1 and C_2 is connected to the FET source and they act as Z_1 and Z_2 in the generalized LC configuration of Fig. 10.12. C_c is just a blocking capacitor to prevent dc from passing through the crystal so the crystal is between drain and gate and forms Z_3. It must therefore be inductive for the circuit to oscillate, which will be the case for frequencies between f_s and f_p. As noted, the difference between these two frequencies will be small so f_o will be accurately fixed.

SUMMARY

Sine wave oscillators are usually classified into RC and LC types – some texts preferring the alternative titles phase shift and tuned respectively. These classifications refer to the make-up of the passive feedback network which is used to determine the frequency of oscillation. Broadly speaking, RC types are used for the LF including the audio end of the spectrum, and LC types for HF. Having chosen the frequency-determining network, the next problem is to find the amplifier requirement which makes up for the loss in the network and which results in an overall feedback loop gain of unity. Secondary problems of oscillators are concerned with amplitude stability which can be resolved by including a nonlinear element in the amplifier gain determining network, and frequency stability which can be made very high using a piezoelectric crystal.

REVISION QUESTIONS

10.1 For the Wien bridge oscillator of Fig. Q.10.1 find the relationship between the frequency of oscillation and CR. If $R = 10$ kΩ find C if the oscillating frequency is to be 2 kHz. Suggest values for R_1 and R_2.

(8 nF; 50 kΩ; 100 kΩ)

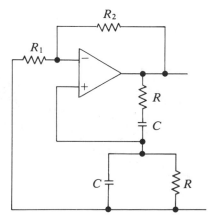

Fig. Q.10.1

10.2 Calculate the frequency of oscillation for Fig. Q.10.2 and the ratio R_2/R_1 necessary for oscillation. What would be the effect of interchanging the two capacitors?

(360 Hz; 13.6)

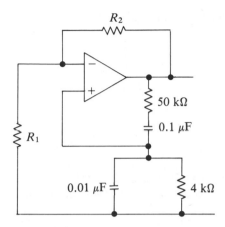

Fig. Q.10.2

10.3 Calculate the frequency of oscillation of Fig. Q.10.3a and the ratio of R_2/R_1 required to ensure that oscillation is sustained. If R_2 is a thermistor whose characteristic is given in Fig. Q.10.3b calculate the value of R_3 if V_o is to be stabilized at 4 V rms.

(1.12 kHz; 13.3 kΩ)

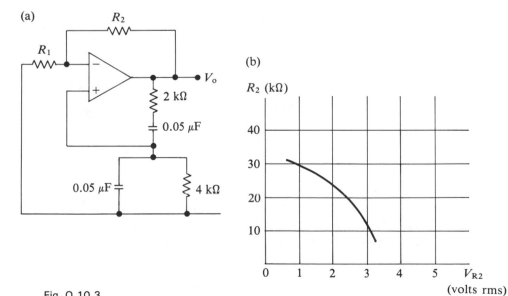

Fig. Q.10.3

10.4   The Colpitts oscillator of Fig. Q.10.4 is required to oscillate at 1 MHz. Calculate A and L from first principles.

(10; 0.55 mH)

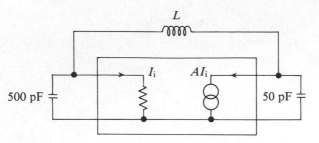

Fig. Q.10.4

10.5   For the Hartley oscillator of Fig. Q.10.5 find from first principles the frequency of oscillation and the value of h_{fe} required for oscillation.

(1.55 MHz; 20)

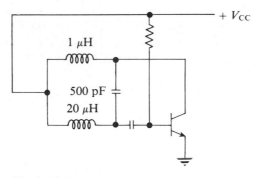

Fig. Q.10.5

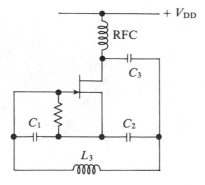

Fig. Q.10.6

10.6 Fig. Q.10.6 is a Colpitts oscillator in which the frequency-determining components are C_1, C_2 and L_3. Describe the function of the RF choke and C_3. Draw the equivalent circuit of the oscillator and from it find suitable values for C_1, C_2 and L_3 for the circuit to oscillate at 1.3 MHz. What should be the FET parameters?

(Suggested solution: $C_1 = 1630$ pF; $C_2 = 163$ pF; $L_3 = 100 \ \mu$H; $g_{fs}r_{ds} \geqslant 10$)

11

Nonsinusoidal Waveform Generators

Chapter Objectives
11.1 Rectangular Waveform Definitions
11.2 Multivibrators
11.3 Op Amp Multivibrators
11.4 BJT Multivibrators
11.5 Schmitt Trigger
11.6 Voltage Time Base Circuits
11.7 Miller and Bootstrap Time Bases
       Summary
       Appendix
       Revision Questions

Generation of rectangular waveforms involves the use of semiconductor devices in a manner totally different than we have considered up to this point – op amps will usually have significant differential input voltages so that their outputs will be at one of the supply rail voltages; transistors will either be operated at cutoff or in saturation. Timings will be determined by charge and discharge in CR circuits, basically from dc supplies. Triangular waveform generation involves the use of amplifiers to maintain constant charging currents through capacitors. In some circumstances discrete components have advantages over ICs.

11.1 RECTANGULAR WAVEFORM DEFINITIONS

A rectangular waveform is one which switches between two voltage levels V_1 and V_2 (Fig. 11.1). The voltage levels may be of the same or opposite polarity; in many cases one of them will be 0 V. For many purposes it is necessary for the transition time from V_1 to V_2 and the return from V_2 to V_1 to be as short as possible. The definition of rise and fall time will be given at the end of the chapter. Apart from

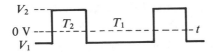

Fig. 11.1   Rectangular Waveform

the values of V_1 and V_2 the significant voltage definition is the difference between V_1 and V_2 which is called the peak-to-peak voltage (V_{pk-pk}).

Mark to Space ratio M/S: This term originates from telegraphy. If we consider that V_1 in Fig. 11.1 occurs by setting a switch OFF and V_2 by setting a switch ON, then T_2 is the mark duration and T_1 the space duration, so

$$M/S = \frac{T_2}{T_1}.$$

Duty Cycle: Duty cycle is defined as the ratio of ON time to TOTAL time, so for Fig. 11.1,

$$\text{Duty cycle} = \frac{T_2}{T_1 + T_2}$$

A waveform with M/S = 1, i.e. duty cycle = 0.5, is usually called a square wave.

11.2 MULTIVIBRATORS

A multivibrator (M/V) is a circuit which can, at any instant, have only one of two output voltage levels, except during a transition, and which, when switching between these levels, does so rapidly. Thus the circuit of Fig. 11.2a might produce a waveform such as Fig. 11.2b where v_o will, at any instant, be $+5$ V or 0 V wrt N. These two voltages are called the two states of the circuit which can be classified as follows:

(a) Stable — in which case the circuit will not change to the other state unless externally triggered.

(b) Quasistable — in which case the circuit will remain at that particular output

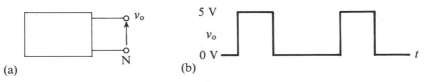

Fig. 11.2   (a) Multivibrator (b) M/V Output Waveform

voltage for a period (say T_1) determined by the circuit components before switching to the other state without any external trigger.

When the circuit is switching between states it is said to be unstable or in a transient condition.

These definitions of states lead us to the following definitions of circuits.

Bistable M/V or Flip Flop: Both states are stable and the circuit will only change state when triggered, as indicated by the waveforms of Fig. 11.3a. Flip flops are described in Chapter 14.

Monostable M/V or One-shot: One state (V_1 in Fig. 11.3b) is stable and the other (V_2) is quasistable. The circuit therefore rests in its stable state until triggered into the quasistable state where it remains for the predetermined period T_1 before returning to the stable state. Note that while T_1 is fixed, the time between pulses depends on the triggers.

Astable M/V: Both states are quasistable, so the circuit simply switches between the two states (Fig. 11.3c), each of which having its own fixed duration. The circuit is therefore free running and does not require triggering. It can, however, be triggered to bring it into synchronization with another signal.

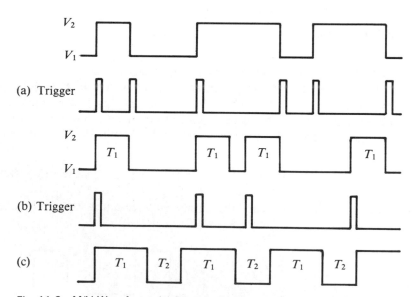

Fig. 11.3   M/V Waveforms (a) Bistable (b) Monostable (c) Astable

11.3 OP AMP MULTIVIBRATORS

11.3.1 Op Amp Astable M/V

To explain this circuit we begin with an ideal op amp (Fig. 11.4) having equal but opposite dc supplies $+V_D$ and $-V_D$. These and the three terminal voltage waveforms v_A, v_B, and v_F, are referenced to the earth rail N. Before considering the astable M/V the main point to keep in mind is that the op amp is a differential amplifier. Therefore, if v_{AB} is positive, $v_F = +V_D$ because the op amp is assumed to have infinite gain, so v_F will limit at the positive supply voltage V_D. Likewise, if v_{AB} is negative, $v_F = -V_D$. For example, even if both v_A and v_B are positive but $v_A = +3$ V and $v_B = +3.1$ V, $v_{AB} = -0.1$ V, so $v_F = -V_D$.

When the astable M/V of Fig. 11.5 is switched on then, because of imbalance, offsets, etc., the output v_F will go to one of the supply voltages, say $+V_D$. If we now look at the noninverting input terminal A we see that it is connected to the $R_1 R_2$ potential divider chain with v_A being equal to the voltage across R_2. Defining the potential division ratio as

$$\beta = \frac{R_2}{R_1 + R_2},$$

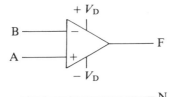

Fig. 11.4 Op Amp Terminals

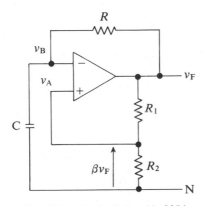

Fig. 11.5 Op Amp Astable M/V

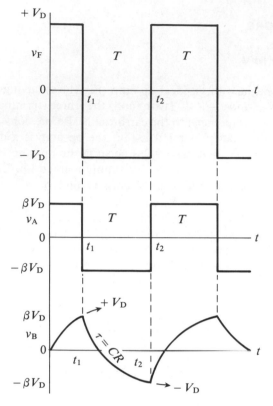

Fig. 11.6   Op Amp Astable M/V Waveforms

v_A is always equal to βv_F. At switch-on then, $v_A = +\beta V_D$. These switch-on voltages for v_F and v_A are marked on the respective waveforms in Fig. 11.6 at $t = 0$. At switch-on, C is uncharged, so $v_B = 0$ V as indicated.

The essential part of the M/V is shown in Fig. 11.7. There will be a voltage V_D across R at switch-on, so a current i flows through R and begins to charge C. Therefore, v_B begins to rise exponentially towards V_D with a time constant $\tau = CR$. However, when v_B passes through βV_D it becomes more positive than v_A. This means that v_{AB} is now negative, so the circuit switches, at t_1, to $v_F = -V_D$. It is from this instant that the circuit begins it normal switching sequence and from which we can calculate the state durations. The voltages on the CR network are now as shown in Fig. 11.8 with

$$v_F = -V_D$$
$$v_B = +\beta V_D$$

Also $v_A = \beta v_F = -\beta V_D$

Current i through C and R is now reversed and noting the voltage at each end of

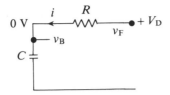

Fig. 11.7 Timing Network of M/V at Switch-on

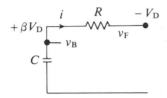

Fig. 11.8 Timing Network of M/V Immediately After Switching to $-V_D$

R, we can see that the current in R will be $(V_D + \beta V_D)/R$. v_B will now fall exponentially towards $-V_D$ (note v_B in Fig. 11.6 from t_1 to t_2) but as soon as it passes through $-\beta V_D$, v_{AB} becomes positive, so v_F switches back to $+V_D$ and v_A to $+\beta V_D$.

Clearly, both states are quasistable, so the circuit is an astable M/V. It is also symmetrical, in this case, so the positive and negative pulse durations are equal. To calculate the output pulse duration, refer to Fig. 11.8. At t_1,

$$i = \frac{V_D + \beta V_D}{R}.$$

Immediately before the transition at t_2, v_B has fallen to $-\beta V_D$. Therefore the voltage across R is $V_D - \beta V_D$ and i has fallen to

$$\frac{V_D - \beta V_D}{R}.$$

Using these two current values and the current exponential decay equation,

$$\frac{V_D}{R}(1 - \beta) = \frac{V_D}{R}(1 + \beta)e^{-T/CR}.$$

$$e^{T/CR} = \frac{1 + \beta}{1 - \beta}$$

$$T = CR \ln \frac{1 + \beta}{1 - \beta} \tag{11.1}$$

where T is the time of a half-cycle of the complete waveform. The waveform frequency

$$f = \frac{1}{2T}.$$

Asymmetrical Op Amp M/V – Direct Analysis

Given a particular circuit to analyze, it is not recommended that a general expression such as equation (11.1) is developed, followed by the substitution of component values. It is better to treat the op amp as a device which gives, at its output, one of two dc voltages depending on the polarity of v_{AB}, and to establish waveform timings from these voltages and the circuit components.

☐   **Example 11.1**

Calculate the frequency and M/S for Fig. 11.9 output.

In this circuit M/S $\neq 1$ because the supply voltages are not of equal magnitude. The part cycle drawn in Fig. 11.6 can now be ignored, i.e. we assume that the circuit is oscillating so we can begin with the assumption that the M/V has just switched to, say, $v_F = -9$ V.

Now $\beta = \dfrac{20\ k\Omega}{20\ k\Omega + 10\ k\Omega} = \dfrac{2}{3}$

so $v_A = -6$ V.

As the circuit has just switched, v_B must have just reached $\beta \times 15$ V. This is so because immediately before the transition $v_F = 15$ V and $v_A = \beta \times 15$ V. So for Fig. 11.10 at $t = 0$, $v_B = \frac{2}{3} \times 15$ V $= +10$ V.

The timing network with voltages at $t = 0$ is shown in Fig. 11.11 and with current i flowing in the direction shown, v_B will fall exponentially from $+10$ V

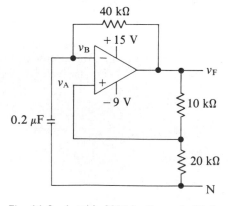

Fig. 11.9   Astable M/V for Example 11.1

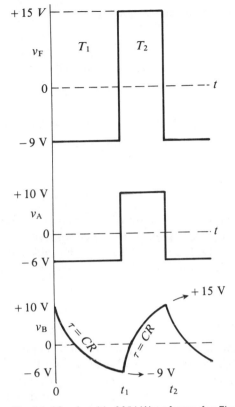

Fig. 11.10   Astable M/V Waveforms for Fig. 11.9

towards -9 V with $\tau = 0.2 \times 10^{-6} \times 40 \times 10^{3} = 8$ ms. However, with $v_A = -6$ V, the M/V will switch v_F back to $+15$ V as v_B passes through -6 V.

With the voltages as in Fig. 11.11,

$$i = \frac{19 \text{ V}}{40 \text{ k}\Omega}.$$

When v_B reaches -6 V at $t = t_1$,

$$i = \frac{3 \text{ V}}{40 \text{ k}\Omega}.$$

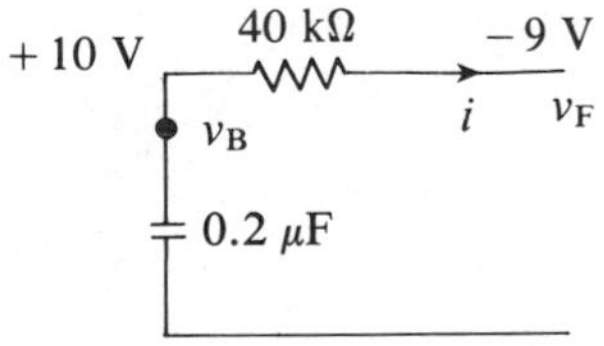

Fig. 11.11   Timing Network for Fig. 11.10 at $t = 0$

Using the exponential decay relationship and canceling the 40 kΩ,

$$3 = 19e^{-T_1/CR}$$

$$T_1 = CR \ln \frac{19}{3} = 8 \times 10^{-3} \times 1.8 \text{ s}$$

$$= 15 \text{ ms}$$

The reader should now complete the problem by replacing voltages given in Fig. 11.11 by $v_F = +15$ V, $v_B = -6$ V and establishing that $T_2 = 11$ ms and that M/S = 0.73. □

Alternative Asymmetrical M/V

Instead of using unequal supply voltages, the use of two diodes as shown in Fig. 11.12 will make an op amp M/V asymmetrical. When v_F is positive, D_1 is OFF and $\tau = CR_2$. When v_F is negative, D_2 is OFF and $\tau = CR_1$. Another alternative is to use only one diode, say D_1, and then the timing resistance is R_2 when the diode is OFF, and $R_1 \parallel R_2$ when it is ON.

11.3.2 Monostable Op Amp M/V

Making the previous circuits monostable is simply a matter of connecting a diode D across C as in Fig. 11.13. The direction of D will determine which of the two output states is stable. To determine the stable state, assume $v_F = +15$ V. Then current i flows through R and charges C, but as soon as v_B goes positive (assume that for D, $V_{ON} = 0$ V) D conducts and the charging process ceases. With $\beta = 0.5$, v_A is held at $+7.5$ V, so the M/V is stable. If $v_F = -15$ V, current i is reversed, so D goes OFF and v_B falls exponentially towards -15 V, but when it reaches -7.5 V (the voltage on v_A) the circuit switches back to $v_F = +15$ V. Therefore, $v_F = -15$ V is a quasistable state.

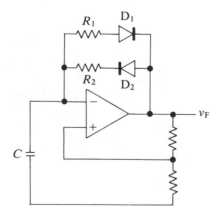

Fig. 11.12   Asymmetrical M/V

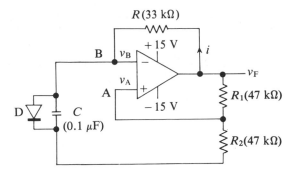

Fig. 11.13 Monostable M/V

Pulse Duration of Op Amp Monostable M/V

To calculate the monostable pulse duration T, v_B will begin at 0 V (a more accurate result will be obtained allowing for the voltage across D, in which case v_B begins at, say, $+0.8$ V). Assume that the circuit is triggered into its quasistable state by a negative pulse on A which is large enough to make v_{AB} momentarily negative. Then v_F goes to -15 V and v_A to -7.5 V and, as with the astable M/V, v_B begins an exponential run-down towards -15 V (Fig. 11.14). As noted, when v_B reaches

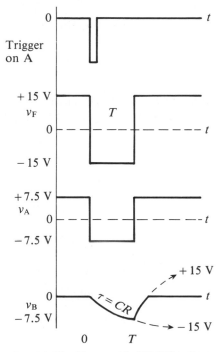

Fig. 11.14 Monostable M/V Waveforms

-7.5 V, the circuit switches back to its stable state. v_B now begins to rise from -7.5 V up towards $+15$ V but D comes ON as it tries to go positive and v_B is clamped at 0 V until the next trigger occurs.

At $t = 0$, the instant of the trigger in Fig. 11.14, the current i in R is

$$i = \frac{-15 \text{ V}}{R}$$

At $t = T, i = \dfrac{-7.5 \text{ V}}{R}$

Therefore,

$$7.5 = 15e^{-T/CR}$$

$$T = CR \ln \frac{15}{7.5}$$

$$\beta = 0.1 \times 10^{-6} \times 33 \times 10^3 \times 0.69 \text{ s}$$

$$= 2.3 \text{ ms.}$$

A general expression can be derived, giving

$$T = CR \ln \frac{1}{1 - \beta}.$$

However, implicit in this is the approximation which neglects the diode voltage. It is better to use the first approach where we can allow for the diode voltage by subtracting 0.8 V from the voltage across R when calculating the current at $t = 0$ and $t = T$.

Trigger Circuit for Monostable M/V

In Fig. 11.15 only the components concerned with triggering the M/V have been drawn. An incoming pulse train at TR will be differentiated by $C_T R_T$ if the time

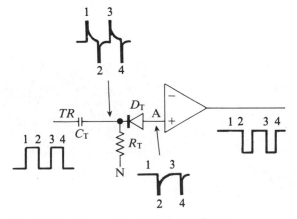

Fig. 11.15   Monostable M/V Trigger Circuit

constant is considerably less than the pulse train period. D_T allows only the negative going spikes to pass to A as required for triggering the M/V. By preventing the positive spikes from reaching A we avoid the possibility of triggering the M/V back to its stable state before the completion of the output pulse. The input pulse amplitude must be large enough to make v_{AB} go negative.

11.4 555 TIMER

This is essentially an IC multivibrator. External components can be chosen to operate the device in either monostable or astable mode. It will produce pulses down to about 1 μs in duration. Fig. 11.16 shows the circuit connected as a monostable. The section inside the dotted box is a simplified version of the 555 IC.

The labels on the external terminals DIS, TH and TR are abbreviations for Discharge, Threshold and Trigger. V_{CC} can be as high as 15 V which is the value we shall use in describing the circuit operation. Note that the chain of three equal resistors holds the inverting input to COMP 1 and the noninverting input to COMP 2

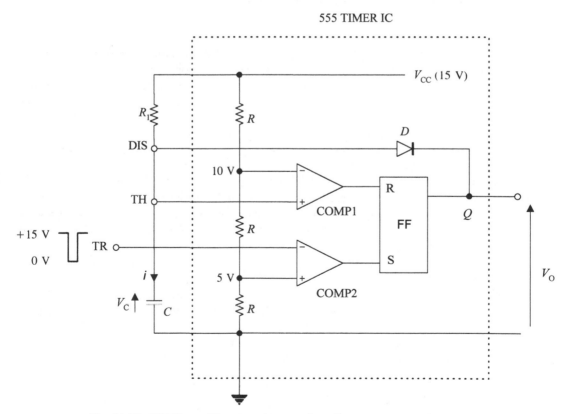

Fig. 11.16   555 Timer with external connections for monostable operation

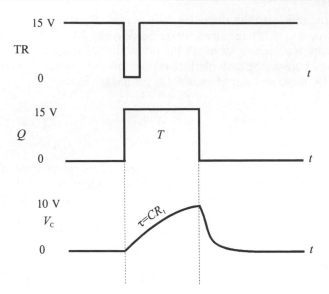

Fig. 11.17   555 Timer — Monostable Waveforms

at $+10$ V and $+5$ V respectively. COMPS 1 and 2 are simply op amp comparators whose outputs switch to their positive or negative supply voltages as the polarity of the voltage between their noninverting and inverting inputs changes. The RSFF (flip flop) is a digital device which will be described in section 14.2. All we need to know for this purpose is that if R goes positive, $Q = 0$ V, and if S goes positive, $Q = 15$ V. With both R and S negative, the flip flop does not change state.

Assume, then, that R and S are both negative and that $Q = +15$ V. Then D is OFF and C charges through R_1 so that v_C rises exponentially towards $+15$ V. When it (i.e. TH) reaches $+10$ V, COMP 1 will reset Q to 0 V. Therefore, this state is not stable. If $Q = 0$ V, D is ON and holds v_C at 0 V (approx.). Therefore, $R = 0$ V. In the absence of a trigger pulse, TR is positive so $S = 0$ V. Therefore, the state is stable. Figure 11.17 shows the important voltages in the timer before it is triggered from the stable state to the quasistable state at $t = 0$.

The timer is triggered by the input pulse taking TR down to 0 V. When this happens, COMP 2 sets Q to $+15$ V. D will now be OFF so C begins to charge through R_1. Therefore, v_C rises from 0 V towards $+15$ V but, when it reaches $+10$ V, COMP 1 resets the flip flop and Q falls to 0 V. D will now be ON so C discharges.

To calculate T, note that at $t = 0(+)$,

$$i = I_0 = \frac{15\text{ V}}{R_1}$$

At $t = T(-)$, i will have fallen to $\dfrac{5\text{ V}}{R_1}$.

Therefore, $\dfrac{5}{R_1} = \dfrac{15}{R_1}\, e^{-T/CR_1}$

so $T = CR_1.\ln 3 = 1.1\,CR_1$.

This analysis involves approximations, e.g. that there is no voltage across D when it conducts. However, for rapid discharge of C, a transistor circuit is used instead of a diode which improves the approximation.

555 Timer as an Astable Multivibrator

To change from monostable to astable operation (Fig. 11.18), note that only external components and connections have been altered. Terminals TR and TH are joined together, and C now has to charge through resistors R_1 and R_2, but discharge through R_2 only.

First check that neither $Q = 0$ V nor $Q = 15$ V is a stable state. If $Q = 15$ V, D is OFF and C will charge through R_1 and R_2. When v_C reaches 10 V, COMP 1 resets Q to 0 V. If $Q = 0$ V, D is ON and C discharges through R_2 and D. When v_C has fallen to 5 V, COMP 2 sets Q to $+15$ V. Neither state is stable so the timer is free running.

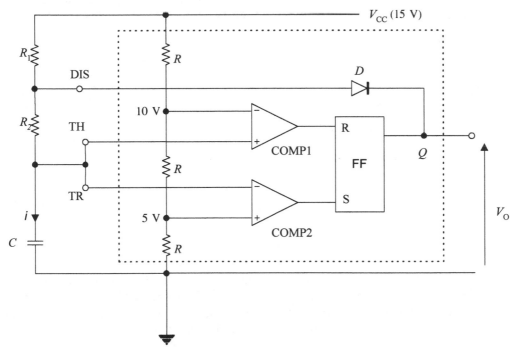

Fig. 11.18   555 Timer — Astable Operation

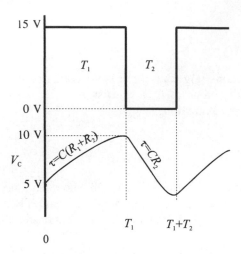

Fig. 11.19   555 Timer — Astable Waveforms

To calculate the pulse times, note the waveforms of Fig. 11.19. From the foregoing it can be seen that when Q goes to $+15$ V at $t = 0(+)$, $v_C = 5$ V. Then C begins to charge towards 15 V through R_1 and R_2. When v_C reaches 10 V, COMP 1 resets Q to 0 V.

At $t = 0(+)$, $v_C = 5$ V so the voltage across $R_1 + R_2$ is 10 V.

Then $i = I_0 = \dfrac{10 \text{ V}}{R_1 + R_2}$.

At $t = T_1(-)$, v_C has risen to 10 V, so

$$i = \dfrac{5 \text{ V}}{R_1 + R_2}.$$

Therefore, $\dfrac{5}{R_1 + R_2} = \dfrac{10}{R_1 + R_2}\, e^{T_1/C(R_1+R_2)}$

and $T_1 = C(R_1 + R_2) \ln 2 = 0.7C(R_1 + R_2)$.

With $Q = 0$ V, D is ON and C discharges from 10 V to 5 V through R_2. Note that with $v_C = 10$ V and $Q = 0$ V, the voltage across R_2 at $t = T_1(+) = 10$ V. Therefore, the initial and final currents for T_2 are the same as for T_1 but, in the opposite direction, only the time constant has changed. This gives

$$T_2 = 0.7CR_2.$$

The free running frequency $f = \dfrac{1}{T_1 + T_2}$

$$= \frac{1}{0.7C(R_1 + 2R_2)}$$

The analysis shows that one limitation of the 555 timer astable is that the mark : space ratio $T_1 : T_2$ is always greater than 1.

11.5 BJT MULTIVIBRATORS

11.5.1 BJT as a Switch

Op amps have limited operating frequencies and relatively slow pulse edges because of their slew rates. For high-frequency pulse trains and fast edges, BJT M/Vs are superior. Unlike their use in analog circuits, BJTs in switching circuits are operated in only two states. To explore the idea of a BJT as a switch, first note the performance of the conventional switch S in Fig. 11.20.

When S is open (OFF),

$V_R = 0 \qquad I = 0 \qquad V_S = E$

When S is closed (ON),

$V_R = E \qquad I = \dfrac{E}{R} \qquad V_S = 0$

These relationships are established on the assumption that the switch S is ideal, that is to say that when open it is an open circuit (no leakage current) and when closed it is a short circuit (zero contact resistance).

In Fig. 11.21a the BJT replaces switch S and it is opened or closed by the absence or presence of base current. The two states or the circuit can be established in conjunction with the load line diagram of Fig. 11.21b and they are:

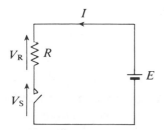

Fig. 11.20   DC Circuit with a Conventional Switch

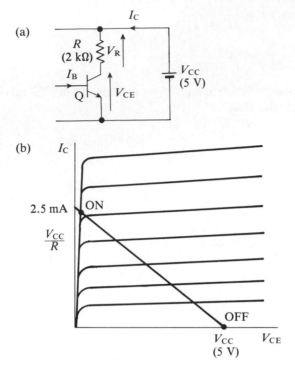

Fig. 11.21   BJT as a Switch with Associated Load Line

(a) OFF determined by $I_B = 0$ and therefore $I_C = 0$. Then $V_{CE} = V_{CC} = 5$ V using the values marked on the circuit diagram.

(b) ON, in which case the transistor is said to be saturated (or bottomed) so that $V_{CE(ON)}$ approaches 0 V and therefore

$$I_{C(ON)} \doteq \frac{V_{CC}}{R} = 2.5 \text{ mA}$$

In practice $V_{CE(ON)}$ must be nonzero but for a switching transistor it will be as low as 0.1 V to 0.2 V. In this chapter it will be assumed that

$$V_{CE(ON)} = 0 \text{ V and therefore,}$$

$$I_{C(ON)} = \frac{V_{CC}}{R}$$

The conditions for ensuring that a BJT is saturated may be deduced with reference to the circuit of Fig. 11.22. If an external input is holding V_{BE} negative (or even 0 V) then Q is OFF. Alternatively, if the input has become an open circuit

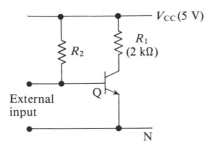

Fig. 11.22 R_2 Must Be Chosen to Saturate Q when External Input is an Open Circuit

– for example, by another BJT having been turned OFF – then Q will be saturated if R_2 is correctly chosen. In this circuit it is required that

$$I_{C(ON)} = 2.5 \text{ mA}$$

Suppose for the particular model of BJT in use the minimum dc current gain is quoted as $h_{FE(MIN)} = 50$.

Then if $I_B \geqslant \dfrac{I_{C(ON)}}{h_{FE(MIN)}} = \dfrac{2.5 \text{ mA}}{50} = 50 \ \mu\text{A}$,

then Q will be saturated.
Assuming $V_{BE(ON)} = 0.7$ V,

$$\text{then } R \leqslant \frac{4.3 \text{ V}}{50 \ \mu\text{A}} = 86 \text{ k}\Omega.$$

The nearest 10% preferred value is 82 kΩ, but as this could be as high 90 kΩ, good design practice would lead to the choice of 68 kΩ for R_2.

In calculating the value of R_2 a significant error would be introduced by making the approximation that in the ON state $V_{BE} \doteq 0$ V. However, using this approximation does simplify the explanation of BJT operation. The approximation will therefore be used for this purpose and for the calculation of quasistable state duration, although a more precise approach is not particularly difficult. The approximation will not be used in calculating resistor values because it may lead to the design of an M/V that will not operate at all.

11.5.2   BJT Astable M/V

The circuit of Fig. 11.23 is essentially two cross-coupled common emitter amplifiers with the output of Q_1, v_A, connected, via C_1, to the input of Q_2 and vice versa. Suppose at switch-on that both transistors go into their active states. Then, if v_A is rising, v_C will follow it. Since each amplifier is an inverter, v_D will fall, causing v_B to fall, which reinforces the original rise in v_A. We have, therefore, positive feed-

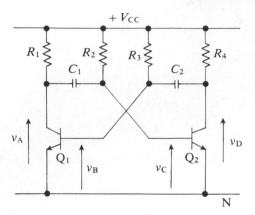

Fig. 11.23   BJT Astable M/V

back, and if the components are correctly chosen, one transistor, say Q_1, is driven into saturation (ON) and the other, Q_2, is driven OFF. The action is called a regenerative switch. Using the approximate voltages suggested in the last paragraph: for the ON transistor,

$$v_A = V_{CE(ON)} = 0 \text{ V}$$

$$v_B = V_{BE(ON)} = 0 \text{ V}$$

for the OFF transistor,

$$v_C < 0 \text{ V}$$

$$v_D = V_{CC}$$

These voltages are marked on the waveform diagrams of Fig. 11.24 at $t = 0$. Now note that with the top end of R_2 at V_{CC} and the bottom (v_C) negative, current will flow through it. With Q_2 OFF this current must flow through C_1 and Q_1, which is ON, to earth.

The essential parts of the M/V at this time are shown in Fig. 11.25. v_A is fixed at about 0 V because Q_1 is saturated, so v_C rises exponentially from its negative value towards V_{CC} with $\tau = C_1 R_2$ – this assumes that the collector–emitter resistance of Q_1 is negligible when it is saturated.

When v_C reaches 0 V, Q_2 comes ON, so v_D falls and the reverse transition occurs to make Q_1 OFF, Q_2 ON. We must now look at the four waveforms in Fig. 11.24 in detail. Since Q_2 has been switched ON, its collector voltage v_D falls from V_{CC} to 0 V. This fall is conducted to v_B by C_2, so v_B also falls by V_{CC}, but since it begins at 0 V it goes to $-V_{CC}$. Although v_A rises from 0 V to $+V_{CC}$, v_C cannot follow it because, having reached 0 V and brought Q_2 ON, v_C is now clamped since the base emitter junction of Q_2 is conducting. We must, therefore, contrast the way in which v_A rises and v_D falls. When the right-hand side of C_2 (Fig. 11.23) falls, the left-hand side can follow it because Q_1 is OFF, i.e. an open circuit. But when the left-hand

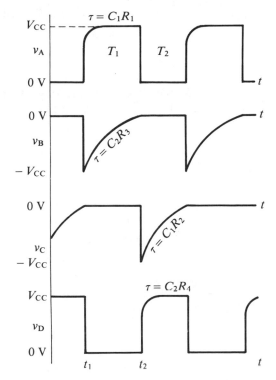

Fig. 11.24   BJT Astable M/V Waveforms

side of C_1 rises with v_A, the right-hand side cannot follow it because v_C is clamped by base current in Q_2. It follows that C_1 must be charged through R_1 by V_{CC} for v_A to rise, hence the $\tau = C_1 R_1$ on the v_A waveform. Now that the M/V has switched to Q_1 OFF, Q_2 ON, note that v_B is at $-V_{CC}$. Current now flows through R_3 and C_2 through Q_2, so v_B rises exponentially from $-V_{CC}$ towards $+V_{CC}$ with $\tau = C_2 R_3$.

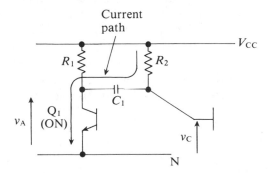

Fig. 11.25   Current Path Through C_1 when Q_1 is ON

When v_B reaches 0 V, Q_1 comes back into conduction and the M/V switches back to Q_1 ON, Q_2 OFF.

Calculation of Pulse Duration

Referring to waveform v_B in Fig. 11.24, the time T_1 is the time it takes v_B to complete half of the rise from $-V_{CC}$ to $+V_{CC}$, i.e. the time it takes to reach 0 V.

Current in R_3 at $t_1 = \dfrac{2\,V_{CC}}{R_3}$

Current in R_3 at t_2 immediately before the transition $= \dfrac{V_{CC}}{R_3}$

Therefore, $\dfrac{V_{CC}}{R_3} = \dfrac{2\,V_{CC}}{R_3}\,e^{-T_1/R_3 C_2}$

$$T_1 = R_3 C_2 \ln 2 \doteqdot 0.7\,R_3 C_2$$

Similarly, $T_2 = 0.7\,R_2 C_1$

If the circuit is symmetrical, i.e.

$$R_2 = R_3 = R;\ C_1 = C_2 = C$$

Total duration $\doteqdot 1.4\,CR$

Frequency $\doteqdot \dfrac{1}{1.4\,CR}$

Design of a BJT Astable M/V

An astable M/V is required which runs at 200 kHz with M/S = 4/1. The dc supply rail is to be $+5$ V. A suitable transistor is 2N3809 which has $h_{FE}(\text{min}) = 100$.

First we must choose the four resistors (Fig. 11.23) which will ensure that when a transistor is ON, it is saturated. To ensure rapid transition between states necessitates each stage having adequate gain, so the collector resistors must be large enough for this purpose. However, the larger these resistors, the slower the rise times of v_A and v_D, but if they are low in value then the saturation currents $I_{C(ON)}$ and $I_{B(ON)}$ will be high. Making an arbitrary but typical choice of 1.8 kΩ for R_1 and R_4, we can proceed with the design and see if it results in an acceptable circuit. R_2 must be small enough to ensure that Q_1 is saturated when ON. Allowing $V_{BE(ON)} = 0.7$ V (Fig. 11.26), the voltage across R_2 will be $V_{CC} - 0.7 = 4.3$ V.

In saturation $I_{C(ON)} = \dfrac{5\,\text{V}}{1.8\,\text{k}\Omega} = 2.8$ mA.

Therefore, $I_{B(ON)} \geqslant \dfrac{I_{C(ON)}}{h_{FE}(\text{min})} = \dfrac{2.8\,\text{mA}}{100} = 0.028$ mA.

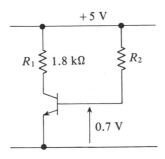

Fig. 11.26   Circuit for Choice of R_2

Therefore, $\qquad R_2 \leqslant \dfrac{4.3 \text{ V}}{0.028 \text{ mA}} \doteq 150 \text{ k}\Omega.$

If we use a 10% preferred value of 150 kΩ the actual value used might be 10% greater than this, so for a safe design it would be better to use 120 kΩ. The same consideration governs the choice of R_3. Since the choice of R_2 and R_3 has been constrained by the requirement for transistor saturation, C_1 and C_2 must fit in with these values to meet the pulse duration specification.

With frequency = 200 kHz, overall duration is 5 μS. With M/S = 4/1, one of the outputs, say v_D must be at $+5$ V for 4 μS and at 0 V for 1 μS.

Therefore, $0.7 \, C_1 R_2 = 4$ μs, from which

$$C_1 = 48 \text{ pF}$$

and $\qquad 0.7 \, C_2 R_3 = 1$ μs, from which

$$C_2 = 12 \text{ pF}$$

11.5.3   BJT Monostable Multivibrator

In the astable M/V of Fig. 11.23, both couplings between the transistors are RC, i.e. they are ac couplings. In the monostable M/V of Fig. 11.27 the coupling from Q_2 output (v_D) to Q_1 input (v_B) is via resistors R_4 and R_5, i.e. it is a dc coupling, so v_B is not a function of time for either of the two states of v_D. To determine the stable state we note that if Q_1 is ON, C can charge via R_2 and Q_1 as with the astable M/V. v_C will therefore rise until Q_2 comes ON. Then v_D falls from V_{CC} to 0 V and Q_1 is cut off. With v_D at 0 V,

$$v_B = \frac{R_4}{R_4 + R_5} \times (- V_{BB})$$

and is therefore negative. Using the bracketed component values in Fig. 11.27, $v_B = -1.2$ V and Q_1 is therefore securely held OFF.

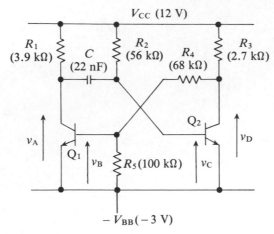

Fig. 11.27 BJT Monostable M/V

There is nothing in the circuit to make v_B change, so the two states of the M/V are

Q_1 OFF, Q_2 ON – stable

Q_1 ON, Q_2 OFF – quasistable.

Monostable M/V Waveforms and Pulse Duration

Just before the M/V in Fig. 11.23 is triggered at $t = 0$ the dc levels are as shown in Fig. 11.28. When it is triggered to Q_1 ON, Q_2 OFF the transition mechanism is similar to that of the astable M/V. However, v_D, in rising, is not delayed by having to charge a timing capacitor. v_C begins an exponential rise from $-V_{CC}$ towards $+V_{CC}$ but initiates the transition when it reaches 0 V. The calculation for pulse duration is the same as for the astable M/V, so $T = 0.7\,CR_2$, which gives 0.86 ms for the M/V of Fig. 11.27.

It is a worthwhile exercise to find the value of h_{FE}(min) required for Q_1 and Q_2. For Q_2 the method is similar to that used to check that the transistors of the astable M/V were saturated when ON. Thus,

$I_{C2(ON)} = 4.4$ mA   (neglecting current in R_4R_5 chain)

Allowing $V_{BE2(ON)} = 0.7$ V, $I_{B2} = 0.2$ mA

Therefore, for Q_2, h_{FE}(min) = 22

For Q_1, $I_{C1(ON)} = \dfrac{12\text{ V}}{3.9\text{ k}\Omega} = 3.1$ mA

To calculate I_{B1} refer to Fig. 11.29 which assumes that Q_1 is ON and Q_2 OFF and which, therefore, is excluded from the diagram. With Q_1 ON, $V_{BE1} = 0.7$ V, so the voltage across the R_3R_4 chain is 11.3 V and the voltage acrosss R_5 is 3.7 V.

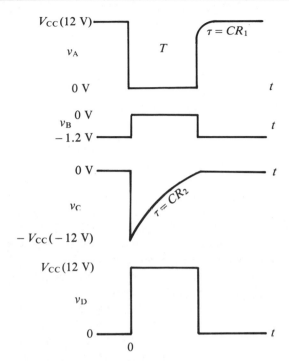

Fig. 11.28   BJT Monostable M/V Waveforms

Therefore,

$$I_1 = \frac{11.3 \text{ V}}{70.7 \text{ k}\Omega} \doteq 0.16 \text{ mA}$$

$$I_2 = \frac{3.7 \text{ V}}{100 \text{ k}\Omega} \doteq 0.04 \text{ mA}$$

$$I_{B1} = I_1 - I_2 = 0.12 \text{ mA}$$

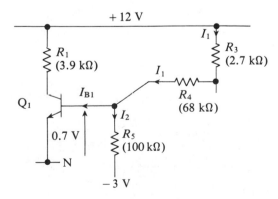

Fig. 11.29   Circuit for Calculating I_{B1}

Therefore,

$$h_{FE}(\text{min}) = \frac{3.1}{0.12} = 26.$$

11.5.4 Bistables

The previous circuits can be modified to form a bistable M/V by using dc couplings between both transistors. Bistable M/Vs (sometimes just bistables) are more commonly called flip flops and are often formed by cross-coupling logic gates, a subject to which some detail is devoted in Chapter 14. However, cross-coupling of transistors to form bistables is used in certain memory ICs (see Section 17.7.1 – 'Static RAM Cells'). In Fig. 11.30 we have an n channel MOSFET bistable in which the drain loads themselves are MOSFETs. These load MOSFETs (Q_3 and Q_4) have their gates held at $+V_{DD}$ and are therefore conducting. If Q_1 is ON, Q_2 gate will be low and therefore Q_2 will be OFF. Therefore Q_2 drain can be regarded as being almost at $+V_{DD}$. With both Q_1 and Q_3 having their gates at $+V_{DD}$, then if they were identical devices, Q_1 drain would be at $V_{DD}/2$. However, with, for example, different channel dimensions, it can be arranged that in the ON state Q_1 has a lower resistance between drain and source than Q_3 and therefore the voltage on its drain will be low enough to hold Q_2 OFF. The circuit is symmetrical and can be triggered into the state Q_1 OFF, Q_2 ON. The subject of different MOSFET configurations in logic ICs is discussed in Chapter 16.

11.6 SCHMITT TRIGGER

One method of producing a square wave is to use a suitable circuit called a Schmitt trigger. It does not run freely like an astable M/V but converts a periodic waveform

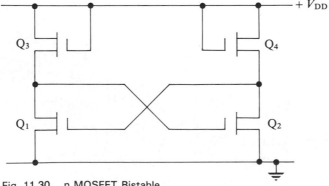

Fig. 11.30   n MOSFET Bistable

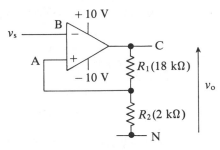

Fig. 11.31   Schmitt Trigger

such as a sine wave into a square wave. Schmitt triggers may be discrete transistor or IC types. An example of an op amp version is shown in Fig. 11.31. Potential divider chain R_1R_2 feeds back

$$\beta = \frac{R_2}{R_1 + R_2}$$

(0.1 in this circuit) of the output voltage v_o to the noninverting terminal A. The input waveform v_s may be a sine wave of say, 5 V peak as shown in Fig. 11.32a.

Assume that initially, when $v_s - 0$ V, $v_o - +10$ V. Then $v_A - +1$ V. When v_s reaches $+1$ V the op amp will switch to -10 V because, as with the M/V circuits, v_{AB} has become negative. Now note that, since $v_{AN} = -1$ V, v_o will not switch when v_o falls back to $+1$ V but waits until it reaches -1 V before going back to -10 V.

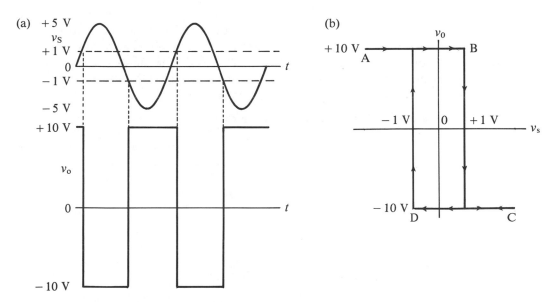

Fig. 11.32   Schmitt Trigger (a) Waveforms (b) Hysteresis

It will return to $+10$ V when v_s reaches $+1$ V. The output will be a square wave as shown in Fig. 11.32a.

A Schmitt trigger is said to have hysteresis. In this context this means that when -1 V $< v_s < +1$ V, the value of v_o depends on whether the last switch was to $+10$ V or to -10 V. The circuit will work equally well with v_s connected to A and the feedback to B.

The hysteresis effect is illustrated in the v_o/v_s characteristic of Fig. 11.32b. When $v_s < -1$ V, $v_o = +10$ V, but as v_s increases from A it will not cause v_0 to switch to -10 V until it reaches $+1$ V (B). Reducing v_s from C will not cause the circuit to switch to $+10$ V until $v_s = -1$ V. One advantage of the hysteresis effect is that it reduces the probability of an unwanted input signal causing the current to switch. For example, if $v_0 = +10$ V with $v_s = 0$ V then the unwanted signal must increase v_s to at least $+1$ V before v_o switches to -10 V.

Schmitt triggers are often used in, for example, time base generators, where there is a need to convert a slowly varying input signal into a square wave for accurate triggering.

11.7   VOLTAGE TIME BASE CIRCUITS

The most important feature of the triangular waveform of Fig. 11.33 is the forward sweep section T_1. This should rise (or fall) linearly with time, so that when it is used, for example, in a CRO time base generator, it will cause the spot to move across the screen at a constant speed and thus facilitate accurate time measurement. In period T_2 we have the flyback or restoration part of the waveform. For many purposes it is not necessary for this to be linear or of shorter duration than T_1, although television time bases are one exception because the system only permits a short time for flyback. Our task, then, is to describe methods of generating a repetitive waveform, one section of which varies linearly with time.

11.7.1   Charging a Capacitor with a Constant Current

Figure 11.34 shows a capacitor C being charged through a resistor R from a square voltage source v_I. It is clear that the output v_0 is not linear but rises and falls

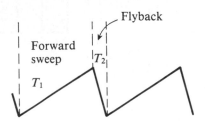

Fig. 11.33   Triangular (Sweep) Waveforms

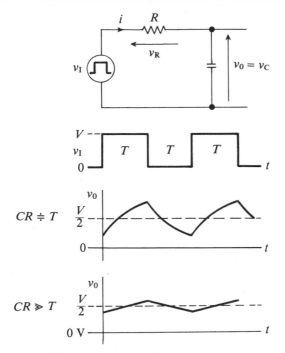

Fig. 11.34   *CR* Circuit and Waveforms

exponentially. As the time constant CR is increased relative to T, the duration of each half-cycle, then, although the amplitude of v_0 falls, it does become more linear. However, as a means of generating a linear sweep it is unsatisfactory except for the crudest of time bases.

The waveform is nonlinear because, as v_0 rises, so v_R falls and, therefore, $i = v_R/R$ falls. As i falls, the *rate* of rise of v_0 falls, i.e. it is nonlinear. If i could

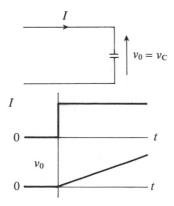

Fig. 11.35   Constant Current Charging a Capacitor

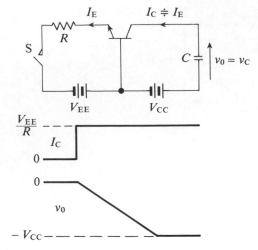

Fig. 11.36   Common Base Circuit Time Base

be kept constant at I as in Fig. 11.35 then

$$v_0 = \frac{1}{C} \int i \, dt = \frac{It}{C} = kt$$

and v_0 rises linearly with respect to time as shown. Generally, the basis of voltage time base circuits is that of devising methods of charging a capacitor with a constant current.

11.7.2   Common Base Amplifier as a Constant Current Source

A common emitter amplifier can be made to produce a reasonably constant collector current but the common base amplifier shown in Fig. 11.36 is superior in this respect. When forward biased the B–E junction has a low resistance ($\simeq 10 \, \Omega$), so that if R is several kilohms,

$$I_E \simeq \frac{V_{EE}}{R} \text{ when S is closed.}$$

Now $I_C \simeq I_E$ and is virtually constant if the C–B junction is at anything above 0 V. Therefore, C charges linearly and v_0 will be a negative-going linear sweep terminating at $-V_{CC}$, as shown in Fig. 11.36 when v_{CB} has fallen to 0 V.

11.8   MILLER AND BOOTSTRAP TIME BASES

In Section 11.7.1 it was noted that, as v_0 rises, so v_R falls because at all times $v_R = v_I - v_C$ (Fig. 11.37) and since $i = v_R/R$, then this charging current falls. If we

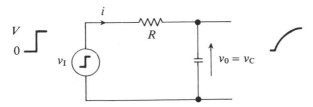

Fig. 11.37   Series *CR* Circuit

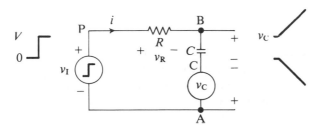

Fig. 11.38   Generating a Constant Current in a *CR* Circuit

could devise a method of adding an extra voltage generator of voltage v_C in series with the circuit and with the polarity shown in Fig. 11.38 then, at all times $v_R = v_I$ and $i = v_R/R = v_I/R$ and is constant. To check that $v_R = v_I$, go around the loop PACB adding the voltages. The waveform v_c then will be a positive-going linear sweep and the extra generator its inverse.

Figure 11.39 will be recognized as an op amp integrator. It follows from the 'virtual earth' principle that the output is $-v_C$, and if we compare the terminals A, B and C in Figs 11.38 and 11.39 and go around the loop PACB in each case to find v_R, then we shall see that the op amp output is the extra generator required. Therefore $v_R = v_I$, i is constant at v_I/R, and the sweep is linear. Taking the op amp output as the sweep then, in this case, it will be negative-going.

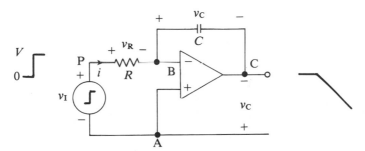

Fig. 11.39   Op Amp Integrator

Transistor Miller Sweep

Figure 11.39 is accurate as an integrator in so far as the op amp voltage gain approaches infinity and point B is a 'virtual earth'. For all practical purposes it may be so regarded. In Fig. 11.40 a common emitter amplifier with a capacitor providing shunt voltage negative feedback is also a Miller integrator. Point B is a 'virtual earth' if the voltage gain is large. This will necessitate a high current gain transistor and a large value of R_1. If the gain is large then the output v_0 will be approximately $-v_C$, and comparing the voltages around the loop PACB then Fig. 11.40 does produce the configuration required by Fig. 11.38 for a linear sweep.

Alternative Description of Miller Integrator

For a more direct explanation of the Miller integrator refer again to Fig. 11.40. When v_I rises from 0 to V, terminal B will tend to rise but because of the inverting voltage gain, point C falls. This draws current i through C, charging it, and the current thus drawn causes a volt drop across R which holds B virtually at 0 V. This input current will be constant at V/R and since it (with the exception of a small base current) flows through C then v_0 will be a negative-going linear sweep. If the initial collector current, when $v_I = 0$ V, is zero, then v_o falls from V_{CC} to almost 0 V.

An FET Miller Sweep

Figure 11.41 is a simplified diagram of a CRO time base using a common source amplifier to obtain the required voltage gain. Instead of driving the input with a rectangular pulse, the blanking pulse at B holds D_1 OFF when it is positive and ON when negative. When D_1 is OFF the pnp transistor is ON, so the Q_2 gate is clamped just above 0 V. D_2 will be ON, so Q_2 drain is at about $+2$ V. When the blanking pulse goes negative, D_1 is turned ON, and Q_1 and D_2 are turned OFF, so Q_2 gate and drain are no longer clamped. Current now flows from the drain through C and R down to the -13 V supply, tending to make the gate negative, so a positive-going

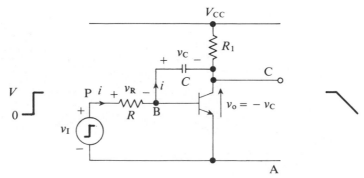

Fig. 11.40   Transistor Miller Sweep

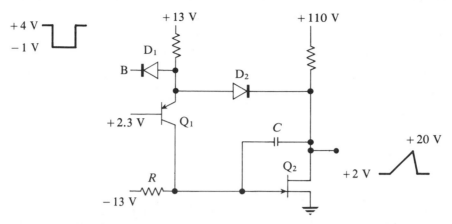

Fig. 11.41   CRO Time Base. Based on Telequipment D61a Oscilloscope (Reproduced by permission of Tektronix UK Ltd.)

Miller sweep is initiated which ends when the blanking pulse returns to $+4$ V. C and R can be switched to several values to give a range of sweep speeds.

Bootstrap Sweep

An alternative way of achieving the configuration of Fig. 11.38 is shown in Fig. 11.42. The amplifier is an emitter follower with voltage gain approximately $+1$. Therefore, the voltages v_C and v_0 are equal. Again, checking the loop PACB shows that $v_R = v_I$, so when v_I goes positive the sweep will also be positive. The idea behind the name of the circuit is that the output, in rising, pulls up the input, as it were, 'by its own bootstraps'.

A practical problem arises because the input v_I and the output v_0 cannot share a common earth. One solution is to make v_I a dc source V_I as shown in Fig. 11.43 and to discharge C with a switch S. When S opens, V_I charges C through R and v_0 follows v_C. v_R will be constant at V_I as required.

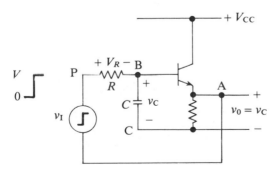

Fig. 11.42   Basic Bootstrap Sweep

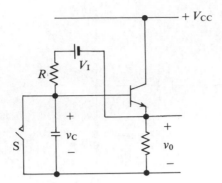

Fig. 11.43   Bootstrap Sweep with dc Source for Input

Practical Bootstrap Sweep

To avoid using the dc source V_I, consider Fig. 11.44 in which C_S is used to store a constant voltage. A further but separate feature is to use Q_2 for discharging C. When the gate waveform is positive Q_2 is ON and holds C discharged, so v_o will be 0 V. At the same time C_S will be charged via D and R_e to almost V_{CC}. When the gate waveform goes to 0 V, Q_2 goes OFF and C begins to charge through D and R, but as soon as the bootstrapping action makes v_o follow v_C, both sides of C_S rise in voltage, so D cuts off and C_S acts as the constant voltage source. C_S is now providing charge for C, so to maintain its voltage reasonably constant – a condition for a linear sweep – C_s must be much greater than C. Whatever charge it loses in the forward sweep is replenished during the flyback, as indicated.

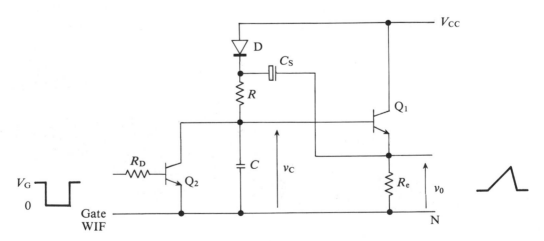

Fig. 11.44   Practical Bootstrap Sweep

Design Problem

A bootstrap sweep as illustrated in Fig. 11.44 is required to meet the following specification:

Forward sweep   0 V to 8 V in 1 ms
Flyback   8 V to 0 V within 0.2 ms
Supply rail $V_{CC} = 15$ V
Gate pulse $+5$ V when ON.
Transistor data:
$h_{FE}(min) = 100$
$I_C(max) = 30$ mA
$V_{BE}(ON) = 0.7$ V $= V_{DIODE}(ON)$.

Many of the component choices are interdependent and if we begin with an arbitrary choice of one component and on completion we find, for example, that transistor ratings are exceeded, then we may have to alter the original choice. The forward sweep will depend on C, R, and V_{CC} – the section of the time base shown in Fig. 11.45. At the beginning of the sweep $v_C = 0$ V, and $v_R = 14.3$ V.

$$I = \frac{14.3}{R}$$

$$v_C = \frac{I \times t}{C} = \frac{14.3 \times t}{CR} \quad \text{and must reach 8 V in 1 ms.}$$

Therefore,

$$CR = 1.8 \text{ ms.}$$

If R is small, C and I will be large and on flyback Q_2 will have to take a large discharge current. A value of R in kilohms gives I in mA which is the right order of magnitude, so begin with a somewhat arbitrary choice of $R = 3.3$ kΩ.

Then $C = \dfrac{1.8 \times 10^{-3}}{3.3 \times 10^3} = 0.55 \ \mu$F.

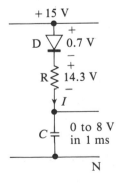

Fig. 11.45   *CR* Part of Time Base

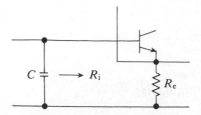

Fig. 11.46 Input Resistance to Emitter Follower Must be High Enough to Prevent any Discharge of C During Forward Sweep

Now C must not suffer any significant discharge during the forward sweep by the input resistance R_i of the emitter follower (Fig. 11.46). Therefore, if $CR_i = 25\ T_f$, where T_f is the forward sweep time, then the discharge will be no more than 0.1%.

$$R_i \geqslant \frac{25 \times 10^3}{0.55 \times 10^{-6}} = 45\ \text{k}\Omega$$

Now $R_i \simeq h_{FE} R_e$ and $h_{FE}(\min) = 100$

Therefore,

$R_e = 450\ \Omega$, so $470\ \Omega$ will be satisfactory.

Making $C_S = 100 \times C = 55\ \mu\text{F}$ will be sufficient to ensure that C_S does not significantly discharge during the sweep.

Now consider the flyback requirements.

C must discharge from 8 V to 0 V in 0.2 ms
Therefore, $I_{C2} \times T_D = C \times V$

Where $I_{C2} = Q_2$ collector discharge current

T_D = discharge (flyback) time

$$I_{C2} = \frac{0.55 \times 10^{-6} \times 8}{0.2 \times 10^{-3}} = 22\ \text{mA}.$$

If this had been greater than the permissible limit for Q_2 it would be necessary to begin again by increasing R and reducing C. To find R_D, then

$$V_{R_D} = V_{GATE} - V_{BE} = 5 - 0.7 = 4.3\ \text{V}$$

$$I_B = \frac{I_C}{h_{FE}} = \frac{22\ \text{mA}}{100} = 0.22\ \text{mA}.$$

Therefore, $R_D = \dfrac{4.3\ \text{V}}{0.22\ \text{mA}} \simeq 18\ \text{k}\Omega.$

One modification which may be needed is the inclusion of a diode D_S shown in Fig. 11.47.

If R_e is large, C_S may not be able to acquire enough charge during flyback to

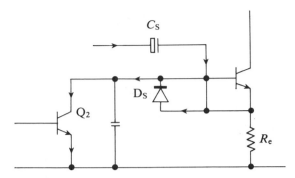

Fig. 11.47   During Flyback C_s is Recharged via D_s and Q_2

replace that lost in the forward sweep. During flyback D_s and Q_2 come ON and provide a lower resistance path than R_e for C_s to recharge.

SUMMARY

Op amps are the basis of simple circuits for generating rectangular waveforms but where high operating frequencies or fast edges are required then more traditional circuits using discrete components are superior.

Linearity of voltage time bases depends on the extent to which the current charging a capacitor can be held constant. This can be successfully achieved with circuits such as the Miller and bootstrap sweeps which have been described in this chapter in discrete component form, although they will often be found encapsulated into integrated circuits.

APPENDIX

Rise Time and Bandwidth

The rise time of a pulse is defined as the time it takes the leading edge to rise from 0.1 to 0.9 of its amplitude, as illustrated in Fig. 11.48.

When a voltage step of amplitude V is impressed on the input to a CR circuit (Fig. 11.49) the rise time of the output v_o can be evaluated in terms of CR, the circuit time constant.

$$0.1\ V = V(1 - e^{-T_1/CR})$$

from which

$$T_1 \doteq 0.1\ CR$$

$$0.9\ V = V(1 - e^{-T_2/CR})$$

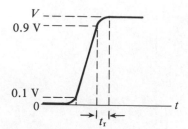

Fig. 11.48   Illustrating Rise Time t_r

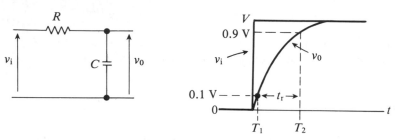

Fig. 11.49   Evaluating Rise Time of a *CR* Circuit

from which

$$T_2 \doteqdot 2.3\, CR$$

$$t_r = T_2 - T_1 = 2.2\, CR.$$

To evaluate the bandwidth, consider a sine wave input of rms voltage V_S replacing the step input (Fig. 11.50).

Then $V_o = \dfrac{V_S \times 1/j\omega C}{R + 1/j\omega C}$

$$= \frac{V_S}{j\omega CR + 1}$$

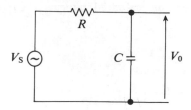

Fig. 11.50   Evaluating Bandwidth of a *CR* Circuit

V_o will be 3 dB down when

$$|V_o| = \frac{V_S}{\sqrt{2}},$$

i.e. when $\omega CR = 1$,

$$f_B = \frac{1}{2\pi CR}$$

where f_B = bandwidth.

The $f_B = \dfrac{2.2}{2\pi t_r} = \dfrac{0.35}{t_r}$

REVISION QUESTIONS

11.1   For each circuit of Fig. Q.11.1 draw the waveforms v_A, v_B and v_o, marking the value of the voltages of the waveform extremities. Use the waveforms to calculate the output waveform frequency and duty cycle.

(15.6 Hz; 0.5; 890 Hz; 0.42; 5.4 kHz; 0.75)

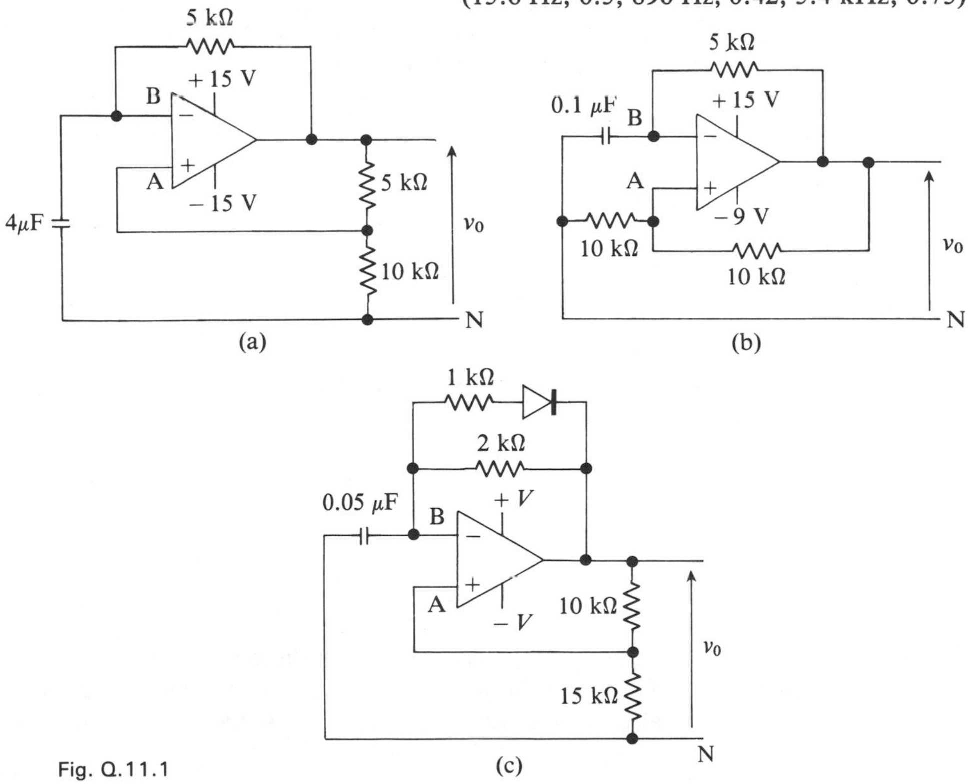

Fig. Q.11.1

11.2   For the multivibrator circuit shown in Fig. Q.11.2, determine the pulse-repetition frequency of the output waveform. Assume the amplifier is ideal. (*Hint:* apply Thevenin to v_o, 2 kΩ, 8 kΩ part of circuit.)

(2.13 kHz)

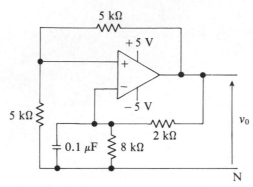

Fig. Q.11.2

11.3   For the multivibrator circuit of Fig. Q.11.3, sketch the resultant waveforms at A, B, and the output. Determine the pulse-repetition frequency and mark–space ratio of the output waveform.

(732 Hz; 0.405 : 1)

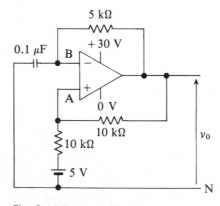

Fig. Q.11.3

11.4   For the circuit of Fig. Q.11.4, determine the output stable state assuming that the amplifier and diode are ideal. Draw a suitable triggering arrangement, and calculate the output pulse width of the monostable.

(329 μs)

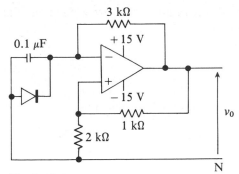

Fig. Q.11.4

11.5 For the circuit shown in Fig. Q.11.5, derive an expression for the pulse width at the output, assuming that the amplifier is ideal and $C_1 R_3 \ll$ output pulse width.

$$\left(\tau = C(R_1 2 R_2) \log_e \frac{30\beta}{V_1} \text{ s} \right)$$

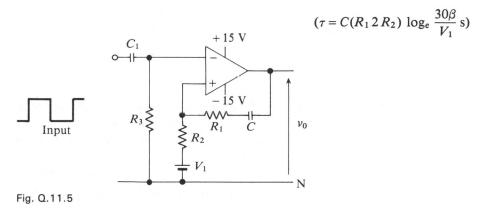

Fig. Q.11.5

11.6 Calculate the largest 10% tolerance, preferred value of R that can be used to ensure that the transistor in Fig. Q.11.6 is saturated.

Assume $V_{BE(sat)} = 0.8$ V

$$V_{CE(sat)} = 0 \text{ V}$$

$$h_{FE(min)} = 40$$

(56 kΩ)

Fig. Q.11.6

11.7 Sketch, for Fig. Q.11.7, v_1, v_2, v_3 and v_4 on a common time axis.
For v_1 calculate:
(a) high duration,
(b) low duration,
(c) mark–space ratio,
(d) duty cycle,
(e) $h_{FE(min)}$ for transistors assuming $V_{BE(on)} = 0.8$ V.

$$(0.94 \text{ ms}, 4.7 \text{ ms}, 0.2, 0.17, 21)$$

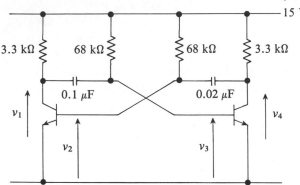

Fig. Q.11.7

11.8 Design a collector coupled astable multivibrator of PRF 0.4 MHz, M/S = 3/1, peak-to-peak output voltage 10 V, using transistor with $h_{FE(min)} = 30$, collector load 2.7 kΩ ± 10%. The resistors available are preferred values of 10% tolerance.

$$(12 \text{ V}; 56 \text{ k}\Omega; 16 \text{ pF}; 48 \text{ pF})$$

11.9 For Fig. Q.11.9, which transistor is ON in the stable state?
Calculate, for each transistor, $h_{FE(min)}$ to guarantee saturation in its ON state.
Assume $V_{BE(sat)} = 0.8$ V.

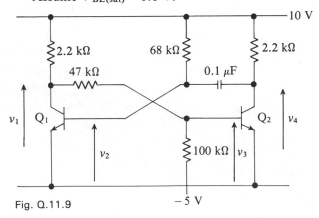

Fig. Q.11.9

Draw the waveforms v_1, v_2, v_3 and v_4 from the instant the circuit is triggered into its quasistable state until it returns to the stable state. Hence calculate the duration of the output pulse.

$$(Q_1, 33, 35, 4.7 \text{ ms})$$

11.10 For the circuit of Fig. Q.11.10, given that $V_{BE}(ON) = 0.8$ V, sketch v_o for a period of 30 ms after S is opened. Assume current gain is unity.

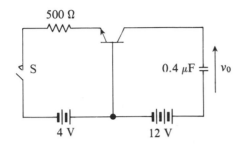

Fig. Q.11.10

11.11 For the Miller sweep of Fig. Q.11.11, sketech v_o for the period 0 to 5 ms.

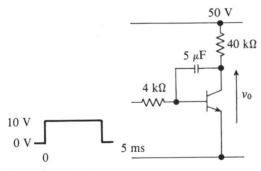

Fig. Q.11.11

11.12 For the bootstrap time base circuit of Fig. 11.39 choose components to give a 0 to 5 V sweep in 2 ms with flyback in 0.4 ms given the following data:
Gate pulse amplitude = 5 V
$$V_{CC} = 10 \text{ V}$$
$$V_{BE}(ON) = 0.8 \text{ V}$$
$$h_{FE}(min) = 30$$

11.13 Use a 555 timer to design an astable multivibrator which produces an output waveform of frequency 20 kHz, and mark/space ratio of 2 : 1.

$$(0.017 \ \mu\text{F}; \ 2.2 \text{ k}\Omega; \ 2.2 \text{ k}\Omega)$$

12

An Introduction to Digital Electronics – Combinational Logic

Chapter Objectives

12.1   Analog and Digital Signals
12.2   Digital Circuits
12.3   Logic Functions and Logic Gates
12.4   Binary Numbers
12.5   Combinational Logic – Canonical Form
12.6   Minimization of Logic Functions
12.7   NAND/NOR Logic – Implementation of Logic Functions in One Type of Gate
12.8   Exclusive OR Logic
12.9   Pulse Trains as Input Variable to Combinational Logic Circuits
12.10   Static Hazards
           Summary
           Revision Questions

The remarkable growth of the digital computer industry has led to the availability of relatively cheap digital integrated circuits for many applications traditionally realized by analog techniques. For example, the telecommunication network of the UK will be almost completely digital by the end of the century. Information technology, a technology made possible by the availability of digital computers, is considered to be capable of quite enormous industrial development. Digital electronics is the technology used in the design of digital computers but it is also used in a wide range of control and signal processing applications.

There is a danger with digital electronics of regarding it as an enclosed world to be described in abstract and symbolic terms: this is unwise for the electronic engineer who ultimately has to relate digital design to the constraints of analog devices, transducers, instruments, and so forth. In covering the basic theory of digital electronics we shall continually refer to the practical reality which makes it possible.

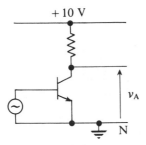

Fig. 12.1 Analog Circuit

12.1 ANALOG AND DIGITAL SIGNALS

An analog signal can have any value within certain operating limits. For example, in the common emitter amplifier of Fig. 12.1, the output v_A can have any value between about 0 V and 10 V. A digital signal can only have a number of fixed values within certain tolerances. In Fig. 12.2, V_D can have only two values, 0 V when S is open and 5 V when S is closed. It is known as a two-state or binary logic circuit.

12.2 DIGITAL CIRCUITS

In digital electronics the nominal values of the logic levels are usually standardized for certain types of circuit within specified tolerances. Although it is not an electronic circuit, the principle can be understood from Fig. 12.2. When S is closed V_D could be less than 5 V if the switch contact resistance is significant in relation to R. Then with S closed V_D might be between 4 V and 5 V. Further, S, when open, might have a finite leakage resistance, so that V_D would be between 0 V and 1 V. Provided that V_D can never, except during a transition, be such that $1 V < V_D < 4 V$, then the circuit is digital, with one state between 0 V and 1 V and the other between 4 V and 5 V.

The two nominal voltage levels are labeled 0 and 1. When the more positive voltage is 1 then the logic is said to be positive: otherwise it is negative. Table 12.1 gives examples of nominal positive and negative logic voltages.

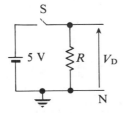

Fig. 12.2 Digital Circuit

Table 12.1

	Positive Logic				Negative Logic		
5 V	1	0 V	1	5 V	0	0 V	0
0 V	0	− 10 V	0	0 V	1	− 10 V	1

The foregoing is somewhat abstract because the opening and closing of switches and so forth usually determines or represents the state of a two-state system. Thus the two logic voltage levels we now call 0 and 1 may represent pairs of states, e.g. an alarm may be ON or OFF, a switch or door OPEN or CLOSED, a statement TRUE or FALSE. It will be a matter for the designer to determine the logic voltages, which of a pair is to be represented by the more positive voltage, and whether the logic labels 0 and 1 are allocated to make the logic positive or negative. In this chapter positive logic is implied unless stated otherwise.

12.3 LOGIC FUNCTIONS AND LOGIC GATES

OR Function

In the 19th century George Boole invented a system of algebra for analyzing problems in logic. This Boolean algebra is used for the analysis of binary logic circuits. It contains certain basic statements, one of which is the OR function which may be stated as follows. A function F of, say, two binary variables A and B will be 1 if either A OR B OR both is 1. Otherwise $F = 0$. Two circuits in which the OR function can be physically realized are shown in Fig. 12.3.

With the switch circuit (Fig. 12.3a) if either A OR B (OR both) is closed, then $V_0 = 10$ V, i.e. $F = 1$. Otherwise $V_0 = 0$ V, i.e. $F = 0$.

In the diode circuit (Fig. 12.3b) A and B represent binary logic voltages. Thus

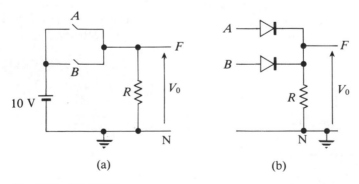

(a)   (b)

Fig. 12.3   OR Gates

Table 12.2 Truth Table for OR Function

A	B	F
0	0	0
0	1	1
1	0	1
1	1	1

A can either be 0 V (0) or $+10$ V (1) wrt N. The same applies to B. Then neglecting the pd across a conducting diode, if either A OR B (OR both) is 1 ($+10$ V) then $V_0 = +10$ V, i.e. $F = 1$. Otherwise $F = 0$.

The two circuits are called OR gates. In them we have defined 1 as the more positive voltage so they are called positive OR gates. Symbolically, the OR function is written

$$F = A + B$$

and can be extended to any number of variables.

The specification of the OR function and the operation of the OR gates can be summarized in a truth table (Table 12.2) which gives F for every possible combination of A and B.

AND Function

Another Boolean function is the AND function which can be defined by the statement that for a function F of two variables A and B then $F = 1$ if both A AND B are 1, otherwise $F = 0$. This is summarized in Table 12.3 for two variables. For any higher number of variables then F will only be 1 when *all* inputs are 1.

Switch and diode AND gates are shown in Fig. 12.4a and b respectively. The operation of the switch circuit is obvious. In the diode circuit, when A AND B are both 1 (10 V) the diodes will not conduct and with no current in R, $V_0 = 10$ V, i.e. $F = 1$.

Table 12.3 Truth Table for AND function

A	B	F
0	0	0
0	1	0
1	0	0
1	1	1

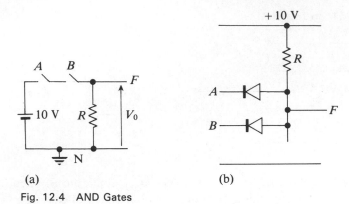

(a)   (b)

Fig. 12.4   AND Gates

If either (or both) input(s) is 0, the corresponding diode(s) is conducting and neglecting the diode volt drop, F will be at the same voltage as the input to the conducting diode, i.e. $F = 0$.

Symbolically the AND function is written $F = A \cdot B$ although the dot is usually omitted, i.e. $F = AB$.

REVERSING THE LOGIC

It could be said that these circuits are 'unaware' that positive logic has been designated. If, in both truth tables (Tables 12.2 and 12.3) all 0s are replaced by 1s and vice versa, then we have negative logic, and inspection will reveal that the positive OR gate (Fig. 12.3) is a negative AND gate and vice versa for Fig. 12.4.

NOT Function

The third fundamental Boolean function is the NOT function which is simply that if A is 1 NOT A is 0 and vice versa. NOT A is written $\bar{A}$ (spoken 'A bar') or A' (spoken 'A dash') and is called the complement of A.

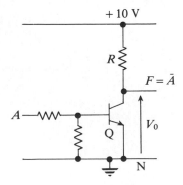

Fig. 12.5   NOT Circuit

Table 12.4 Truth Tables for AND, NAND, OR, NOR

A	B	AB	$\overline{AB}$	$A+B$	$\overline{A+B}$
0	0	0	1	0	1
0	1	0	1	1	0
1	0	0	1	1	0
1	1	1	0	1	0

The common emitter circuit of Fig. 12.5 is a NOT circuit. When $A = 0$ V (0), Q is OFF and $V_0 = 10$ V (1). When $A = 10$ V (1), then the circuit components can be chosen to ensure that 10 V is dropped across R, i.e. Q is saturated, and that $V_0 = 0$ V (0). Therefore $F = \bar{A}$.

NAND and NOR Functions

Two important derived Boolean functions are NOT AND, called NAND, and NOT OR, called NOR. They are the complements of AND and OR and are written:

$F = \overline{AB}$ (NAND)

$F = \overline{A+B}$ (NOR)

Table 12.4 summarizes their specification.

When analyzing NAND logic it is useful to remember that any 0 input gives a 1 output and with NOR logic any 1 input gives a 0 output. Again note that if the logic is reversed, a positive NAND is a negative NOR and vice versa.

Exclusive OR Function

Another useful Boolean function is the Exclusive OR function, abbreviated XOR. It can be summarized by saying that if $F = 1$ when either A OR B, but not both, is 1, then $F = A$ XOR B, symbolically $F = A \oplus B$. The function is specified in Table 12.5.

Table 12.5 XOR Truth Table

A	B	$A \oplus B$
0	0	0
0	1	1
1	0	1
1	1	0

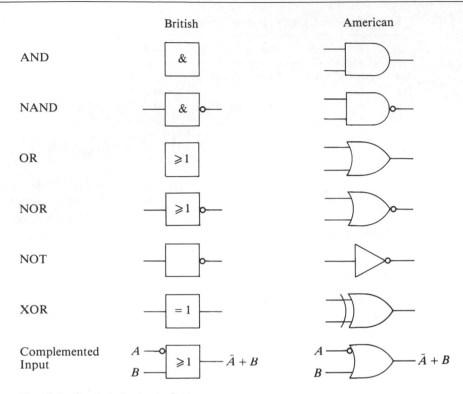

Fig. 12.6   Symbols for Logic Gates

Logic Symbols

Figure 12.6 shows the British and the American symbols for logic circuits. The latter are now dominant and will be used throughout this book.

☐   **Example 12.1 – Applications of Boolean Functions**

1. A machine tool requires that to be operated a power switch must be on ($P = 1$). As a safety feature a guard barrier must also be closed ($G = 1$). Therefore to operate the machine $M = PG$ (Fig. 12.7a).
2. If any one or more of three people A, B and C support a certain candidate, then he will be elected (E) if they set their particular switch to 1, the result being signified by lighting a lamp ($E = 1$). Then

$$E = A + B + C \quad \text{(Fig. 12.7b)}$$

However, in a variation of this procedure, support by the president P is additionally essential. In this case election E_M is given by:

$$E_M = P(A + B + C) \quad \text{(Fig. 12.7c)}.$$

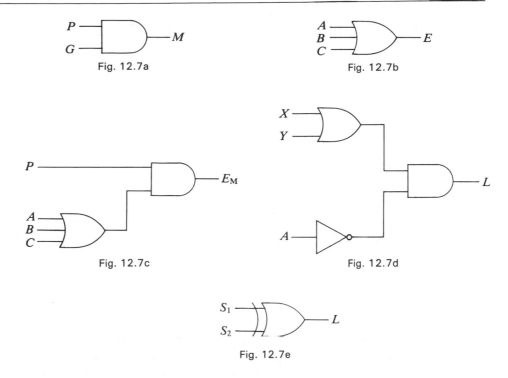

Fig. 12.7a

Fig. 12.7b

Fig. 12.7c

Fig. 12.7d

Fig. 12.7e

3. Either or both of two switches X and Y will start a lift unless an alarm key A is activated. Therefore to make the lift start ($L = 1$) either X or Y or both must be 1 and A must be 0. Then

$$L = (X + Y)\bar{A} \qquad \text{(Fig. 12.7d)}$$

4. A two-way light system has switches S_1 and S_2, one and only one of which is to be 1 for the light L to be on. Then

$$L = S_1 \oplus S_2 \qquad \text{(Fig. 12.7e)} \qquad \Box$$

12.4 BINARY NUMBERS

The first part of this chapter has covered basic binary logic operations and the way they can be used to express the simple two-state logic systems. A particularly important use of binary logic systems is in the manipulation of numbers for arithmetic purposes in digital computers and for processing numbers for a variety of other purposes, e.g. display, transmission, etc. Electronic logic circuits are two-state devices; therefore it is more convenient if the digits of the number system that they represent do not have 10 possible values as with the decimal system, but just 2 as in the binary system.

Table 12.6 Three-bit binary code and its decimal equivalent.

Decimal	A	B	C
0	0	0	0
1	0	0	1
2	0	1	0
3	0	1	1
4	1	0	0
5	1	0	1
6	1	1	0
7	1	1	1

The decimal number 275_{10} is shorthand for $2 \times 10^2 + 7 \times 10^1 + 5 \times 10^0$. In the binary number system the radix is 2 rather than 10, so a number such as 173_{10} is represented in the binary system as:

$1\ 0\ 1\ 0\ 1\ 1\ 0\ 1_2$, i.e.

$$1 \times 2^7 + 0 \times 2^6 + 1 \times 2^5 + 0 \times 2^4 + 1 \times 2^3 + 1 \times 2^2 + 0 \times 2^1 + 1 \times 2^0$$

Arithmetic processes in binary numbers such as addition will be dealt with as required. For present purposes it is sufficient to note that when we state that a binary digit – a 'bit' – is 1 or 0 it can either represent the state of a device (e.g. OPEN or CLOSED) or it can actually be the binary numbers 1 or 0. Pure binary code can be built up in a table such as Table 12.6, where all possible combinations of a three-bit binary number ABC_2 and its decimal equivalent are displayed.†

In general, an N-bit binary number can have 2^N combinations and the highest number it can represent is $2^N - 1$. For example, a binary number $ABCD$ can have 16 combinations and the highest number it can represent is $1111_2 = 15_{10}$.

Serial and Parallel

Binary numbers of several bits – generally called 'words' – can occur in either serial or parallel form. A group of 8 bits is called a byte. In serial form bits occur one at a time on a single line. In Fig. 12.8 bits occur at each of the times t_1 to t_4 with 5 V representing 1, and 0 V representing 0. Assuming that the most significant bit (MSB) occurs first then the number $ABCD$ is 1101.

To represent the same number in parallel form then separate lines will carry the appropriate logic voltages wrt a reference – usually earth – as shown in Fig. 12.9.

Clearly serial form is cheaper but slower than parallel form.

---

† There is some merit in making A represent the least significant bit (LSB) and adding letters as the digit weighting increases but the pros and cons are balanced. A more informative but cumbersome method is to use suffixes, e.g. A_0, A_1, A_2 etc where A_0 represents the presence or absence of 2^0 and so forth.

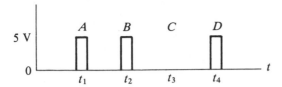

Fig. 12.8   Bits Occurring in Serial Form

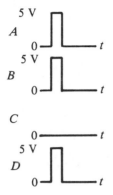

Fig. 12.9   Bits Occurring in Parallel Form

12.5   COMBINATIONAL LOGIC – CANONICAL FORM

The type of circuit deduced for the problems of Example 12.1 are known as combinational logic circuits. That is to say, at any instant the output depends on the values of the binary input variables at that instant. In the case of the machine tool it would operate provided that both Power (P) and Guard (G) variables were 1. There are also sequential circuits in which the sequence of operations is important, e.g. it could be a requirement that the guard had to be closed before the mains was switched on, otherwise the machine tool would not operate. This chapter is only concerned with the first type of problem, i.e. combinational logic.

The previous examples were solved intuitively. Our next task is to describe a method of formally writing Boolean expressions for the solution of design problems and then to consider how they may be implemented in hardware.

☐   **Example 12.2**

A system having three binary inputs A, B and C is required to give an output $F = 1$ when any two and only two of the inputs are 1. This could be used for a process of packaging any two different items from 3 sources.

The most straightforward method is to draw up a truth table (Table 12.7) in

Table 12.7

A	B	C	F
0	0	0	0
0	0	1	0
0	1	0	0
0	1	1	1
1	0	0	0
1	0	1	1
1	1	0	1
1	1	1	0

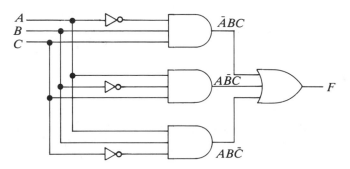

Fig. 12.10   Logic Diagram

which $F = 1$ when any two inputs are 1. Then taking the first row which gives $F = 1$ we note that if we had an AND gate with inputs $\bar{A}$, B, C then it would give an output 1 when $A = 0$, $B = 1$, $C = 1$ as required. In this way we can build up, from the three rows in which F is 1, that

$$F = \bar{A}BC + A\bar{B}C + AB\bar{C} \tag{12.1}$$

and that this function will be realized by the circuit of Fig. 12.10.   □

A Boolean expression such as equation (12.1) which consists of a series of terms each containing all of the variables (complemented or uncomplemented) is said to be in *canonical* form. In this case it is written as a *sum* (OR) of products (ANDs) and each product is called a *minterm*. An alternative and shorter way of writing an expression in canonical form is to equate the function with the decimal values of the binary combinations giving $F = 1$. Thus for the previous function,

$$F = \sum (3, 5, 6)$$

This notation links the ideas of using combinational logic circuits for logical operations such as the last example, or for arithmetic operations based on the binary numbering system previously described. For example, if the three binary variables A, B, C represent binary code as set down in Table 12.6 and we require the system to produce a logical 1 output when the decimal equivalent of the input is 3, 5, or 6, then the same combinational logic can be used equally well.

A Boolean function can also be written as a product of sums. From the same truth table, the terms for which $F = 0$ yield the complement of F, i.e.

$$\bar{F} = \bar{A}\bar{B}\bar{C} + \bar{A}\bar{B}C + \bar{A}B\bar{C} + A\bar{B}\bar{C} + ABC$$

from which it can be shown (see de Morgan's theorem in the next section) that

$$F = (A + B + C)(A + B + \bar{C})(A + \bar{B} + C)(\bar{A} + B + C)(\bar{A} + \bar{B} + \bar{C})$$

Each bracketed sum is called a *maxterm*.

Functions written in canonical form are often unnecessarily complex and an important feature of binary logic circuit design is that of minimization to economize on gates and interconnections.

12.6 MINIMIZATION OF LOGIC FUNCTIONS

Where a Boolean function can be simplified from canonical form there is usually a saving in the hardware required to implement the function. We shall look at minimization:

(a) by using theorems of Boolean algebra; and
(b) by using Karnaugh maps.

12.6.1 Theorems of Boolean Algebra

These consist of a number of statements, some of which are self-evident and some requiring proof. They are, with the exception of the first, presented in pairs, one being basically an OR statement and the other, its dual, in AND form

1. If $A = 1$, $\bar{A} = 0$ and vice versa; $\quad\quad \bar{\bar{A}} = A$
2. $A + 0 = A;$ $\quad\quad\quad\quad\quad\quad\quad\quad A \cdot 0 = 0$
3. $A + 1 = 1;$ $\quad\quad\quad\quad\quad\quad\quad\quad A \cdot 1 = A$
4. $A + A = A;$ $\quad\quad\quad\quad\quad\quad\quad A \cdot A = A$
5. $A + \bar{A} = 1;$ $\quad\quad\quad\quad\quad\quad\quad A \cdot \bar{A} = 0$

These five theorems are regarded as self-evident but can, if required, be proved by what is known as perfect induction. A feature of binary logic is that each variable can have only two values, so the validity of a theorem can be tested by substituting all possible combinations of 1s and 0s. Thus, in single-variable theorems, for example the five above, there are two combinations; there will be four for two variables, eight for three variables, and so on. Then with theorem 2, for the function $A + 0$,

if $A = 1$ $A + 0 = 1$

if $A = 0$ $A + 0 = 0$

then $A + 0$ is always the same as A and the theorem is proved by perfect induction.

6. The distribution laws for Boolean algebra are

$$A(B + C) = AB + AC \quad ; \quad (A + B)(A + C) = A + BC$$

The second of this pair can be proved by perfect induction (try it) or by manipulation as follows:

$$(A + B)(A + C) = AA + AB + AC + BC = F$$
$$F = A + AB + AC + BC \quad \text{(since } AA = A\text{)}$$
$$F = A(1 + B + C) + BC$$

but $1 + X = 1$

therefore $F = A \cdot 1 + BC = A + BC$

7. $A + AB = A \quad ; \quad A(A + B) = A$

8. $A + \bar{A}B = A + B \quad ; \quad A(\bar{A} + B) = AB$

Only the first of theorem 8 in this group is not evident, so we shall prove it by perfect induction.

A	B	$\bar{A}B$	$A + \bar{A}B$	$A + B$
0	0	0	0	0
0	1	1	1	1
1	0	0	1	1
1	1	0	1	1

The two right-hand columns are equal so the theorem is proved.

9. De Morgan's theorem – this theorem is important in NAND and NOR logic.

$$\overline{A + B} = \bar{A}\bar{B} \qquad \overline{AB} = \bar{A} + \bar{B}$$

These can also be stated:

$$A + B = \overline{\bar{A}\bar{B}} \qquad AB = \overline{\bar{A} + \bar{B}}$$

The statements can be proved by perfect induction or by Venn diagrams.

Using the Venn diagram method for the first theorem, in Fig. 12.11 the left-hand circle represents A. Everything within the rectangle outside of A represents $\bar{A}$. Similarly for the right-hand circle which represents B. The overlapping region is AB. The complete area enclosed by the two circles is $A + B$. Therefore, in Fig. 12.11a the region outside the two circles is $\overline{A + B}$. In Fig. 12.11b the diagonal shading in one direction represents $\bar{A}$ and in the other direction $\bar{B}$. Therefore where

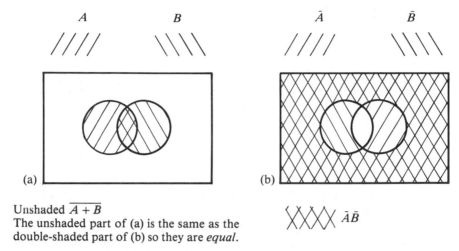

Unshaded $\overline{A + B}$
The unshaded part of (a) is the same as the
double-shaded part of (b) so they are *equal*.

$\bar{A}\bar{B}$

Fig. 12.11   Venn Diagram Proof of First of De Morgan's Theorems

the shadings overlap we have $\bar{A}\bar{B}$. This is identical with the unshaded area $\overline{A + B}$ in
Fig. 12.11a. Therefore, $\overline{A + B} = \bar{A}\bar{B}$.

Proof of the second theorem should be attempted following the same method.

De Morgan's theorem can be demonstrated by everyday example. Using the
second statement, to smoke, S, you need cigarettes, C, AND matches, M.

Therefore, $S = CM$
and $\bar{S} = \overline{CM}$

Alternatively, if (a) you have no cigarettes, $\bar{C}$, OR (b) you have no matches, $\bar{M}$, then
you cannot smoke $\bar{S}$.

Therefore, $\bar{S} = \bar{C} + \bar{M} = \overline{CM}$.

12.6.2   Karnaugh Maps

Experienced designers often use theorems for minimization of Boolean functions,
but a more straightforward technique is the use of Karnaugh (K) maps. They are
most useful for functions of up to four variables — the limit of this chapter — but
they can be used for five and six variables.

The K map is based on the proposition that when, in a sum of two products,
the products differ only in one variable being complemented, then that variable is
superfluous. For example,

$$F = AB + A\bar{B} = A(B + \bar{B}) = A \cdot 1 = A$$

For such an expression it is scarcely worth drawing a K map but it does illustrate
the idea: in Fig. 12.12 the 1s have been entered into the squares corresponding to

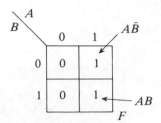

Fig. 12.12   K Map for $A\bar{B} + AB$

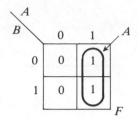

Fig. 12.13   Looping 1's to Give $F = A$

$A\bar{B}$ (top right) and AB (bottom right). The key point to notice is that the terms in every adjacent pair of squares (but not diagonally opposite) differ by only one variable being complemented. Looping adjacent squares containing 1s implies minimization and it should be evident from Fig. 12.13 that the loop represents A. Since the loop contains all the 1s we can conclude that the *minimal* form of F is A. These loops are often called prime implicant loops.

Three-variable K Maps

For a three-variable function there are eight possible minterms so the K map has eight squares. In Fig. 12.14 the minterm corresponding to each square is labeled. Note that each square has a corresponding binary number which can be read off the axes, e.g. $AB\bar{C}$ is also 110.

Also note that not only do adjacent squares (vertically or horizontally) differ by only one variable being complemented, but that the same applies to opposite ends. Thus the top end squares are $\bar{A}\bar{B}\bar{C}$ (000) and $A\bar{B}\bar{C}$ (100). This does not apply to diagonally opposite squares. Other arrangements of variables around the map are acceptable provided that the adjacency relationship for eliminating variables is maintained.

Given a function such as $F = AB\bar{C} + \bar{A}\bar{B}C + ABC + A\bar{B}C$ then these four minterms can be represented on a K map as shown in Fig. 12.15. The two loops reveal that C is redundant for the two terms in the AB column and A is redundant for the loop taking in the bottom ends. For the former the loop clearly represents AB, while for the latter the remaining variables, after eliminating A, are $\bar{B}$ and C,

C \\ AB	00	01	11	10
0	$\bar{A}\bar{B}\bar{C}$	$\bar{A}B\bar{C}$	$AB\bar{C}$	$A\bar{B}\bar{C}$
1	$\bar{A}\bar{B}C$	$\bar{A}BC$	ABC	$A\bar{B}C$

Fig. 12.14   Three-variable K Map

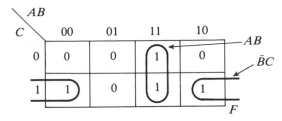

Fig. 12.15   K Map Looping

so the loop represents $\bar{B}C$ and the function is reduced to

$$F = AB + \bar{B}C$$

This requires two 2-input AND gates, one 2-input OR gate, and one inverter. The original function requires four 3-input AND gates, one 4-input OR gate, and three inverters. In Fig. 12.15 it would have been possible to loop the bottom right-hand pair of 1s but the other two 1s could not be looped to one another so this would not produce a minimal solution.

Figure 12.16 illustrates a case where one minterm can be used twice to advantage. The original function would be

$$F = \bar{P}Q\bar{R} + PQ\bar{R} + PQR$$

Since $X + X = X$

Then $X = X + X$

and we can repeat a term of a function.

Therefore, $F = \bar{P}Q\bar{R} + PQ\bar{R} + PQ\bar{R} + PQR$
$$= Q\bar{R}(\bar{P} + P) + PQ(\bar{R} + R)$$
$$= Q\bar{R} + PQ$$

The result is confirmed by the loops in Fig. 12.16 but the preceding algebra is unnecessary as the minimal result could have been read directly from the K Map.

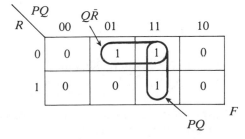

Fig. 12.16   Including a Minterm in Two Loops

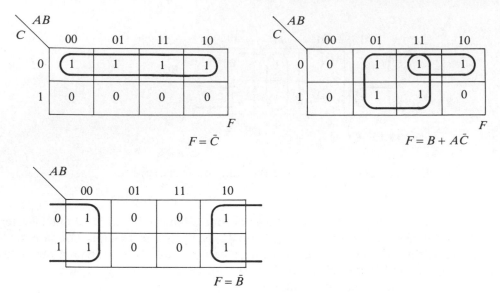

$$F = \bar{C}$$

$$F = B + A\bar{C}$$

$$F = \bar{B}$$

Fig. 12.17   Three-variable K Maps

Where four terms can be looped either in a square or a line, the resulting term is reduced from a sum of four terms each having three variables to a single term of one variable. Examples are shown in Fig. 12.17.

If all squares contain 1 then $F = 1$.

Four-variable K Maps

A function of four variables has 16 possible minterms and a corresponding map (Fig. 12.18). Squares containing 1s can be looped in groups of 2^N where N is the number of variables by which the resulting term is reduced from the minterms. Again any two adjacent terms can be looped including opposite ends but not two diagonally opposite corners. Loops of four must be lines or squares including all four corners, and eights will be 2×4 rectangles. Examples are shown in Fig. 12.19.

Summary of Looping Rules

1. Only adjacent squares can be looped, but this includes corresponding end squares.
2. Terms can only be looped in groups of 2^N, where N is an integer.
3. The term resulting from a 2^N grouping has N less variables than the terms it groups.
4. When four squares are looped they must be a line of four or a square of four – in a four-variable map this includes the four corners.
5. The resulting function must include all 1s whether they can be looped or not.

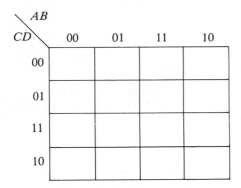

Fig. 12.18   Four-variable K Maps

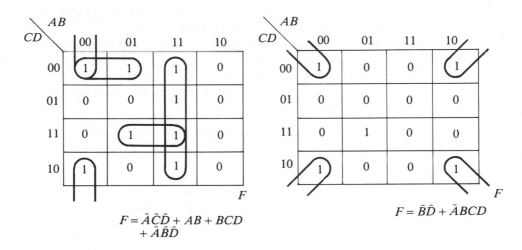

$$F = \bar{A}\bar{C}\bar{D} + AB + BCD + \bar{A}\bar{B}\bar{D}$$

$$F = \bar{B}\bar{D} + \bar{A}BCD$$

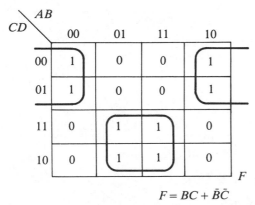

$$F = BC + \bar{B}\bar{C}$$

Fig. 12.19   Four-variable K Maps

'Can't Happen' States

In practical digital systems there are cases where particular combinations of 0s and 1s cannot occur. Consider, for example, the case of three push buttons A, B and C, only two of which will lock down in the ON position simultaneously. In this case, the combination $ABC = 111$ cannot occur. Such a state is called a 'can't happen' or 'don't care' state. With this particular example suppose we require logic to detect the condition

$$ABC = 011.$$

On Table 12.8 the state 011 is duly marked and gating $F = \bar{A}BC$ gives the circuit required. The 'can't happen' state $ABC = 111$ has been marked X. By inspection it can be seen that there are only two cases where B and C are both 1: these are the required state and the 'can't happen' state. Therefore, if the simpler logic $F = BC$ is used, the only state it will detect is $ABC = 011$ because $ABC = 111$ cannot occur.

Entering these two states on a K map, we note that they are in adjoining squares and by looping to include both, we quickly deduce the required logic $F = BC$. Note that it is not necessary to include all Xs in the resulting logic function. In the push button example, if you could not have the condition where no buttons were pressed then $ABC = 000$ would be a 'can't happen' state and the top left-hand square in the K map would have X in it. This X could not be looped with the required 1 and would therefore be ignored.

Table 12.8

A	B	C	F
0	0	0	0
0	0	1	0
0	1	0	0
0	1	1	1
1	0	0	0
1	0	1	0
1	1	0	0
1	1	1	X

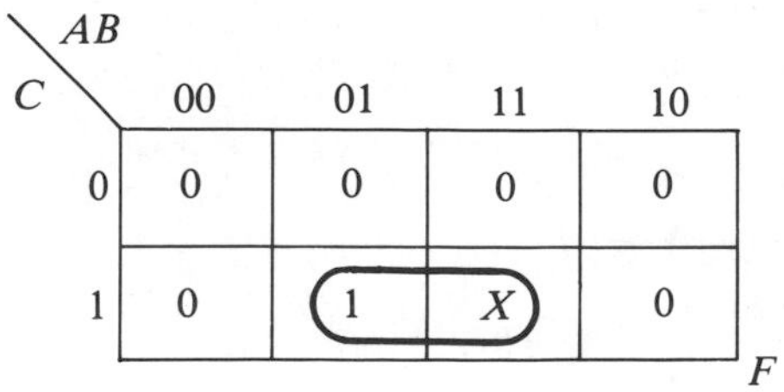

C \ AB	00	01	11	10
0	0	0	0	0
1	0	1	X	0

F

Fig. 12.20   Including a 'Can't Happen' State in a Loop

☐ **Example 12.3   Using 'Can't Happen' States in a Four-variable K Map**

In a binary coded decimal (BCD) sequence the four bits $PQRS$ represent the decimal numbers 0–9. Numbers 10–15 cannot occur. It is required to design a minimal circuit which gives a 1 output when any number in the input sequence is divisible by 3.

For 3, 6, 9, $F = 1$, so enter 1s on the K map at 0011, 0110 and 1001 (Fig. 12.21). For 0, 1, 2, 4, 5, 7, 8, $F = 0$, so enter corresponding 0s.

For 10–15 the states cannot occur so enter Xs on the K map.

When looping, the Xs can be included if they lead to minimization. The resulting minimal function then is:

$$F = PS + \bar{Q}RS + QR\bar{S}$$ ☐

GROUPING 0s

If, in determining the output from a K map, we take all of the 0s instead of the 1s, then we have the complement $\bar{F}$. Redrawing the first Fig. 12.19 map in Fig. 12.22 and taking the 0s, we have

$$\bar{F} = A\bar{B} + \bar{B}D + \bar{A}\bar{C}D + \bar{A}BC\bar{D}$$

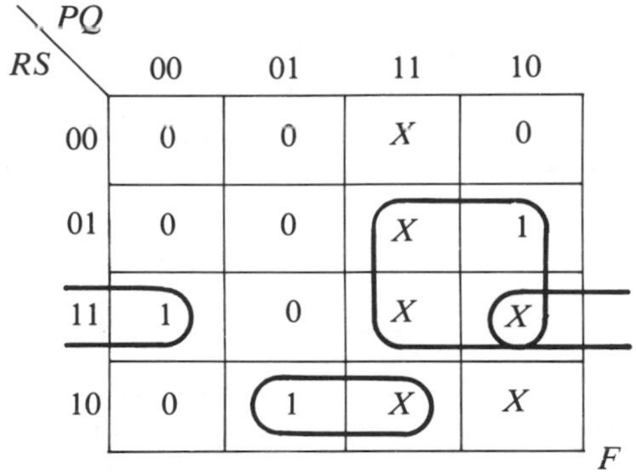

Fig. 12.21   K Map for Numbers Divisible by 3 in BCD Sequence

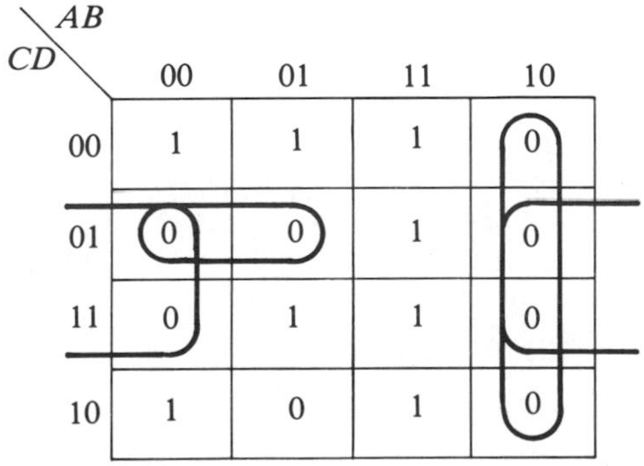

Fig. 12.22   Grouping 0s for $\bar{F}$

12.7 NAND/NOR LOGIC – IMPLEMENTATION OF LOGIC FUNCTIONS IN ONE TYPE OF GATE

It can be shown that any Boolean function, for example, a sum of products including complemented variables, can be realized by using nothing but NAND, or nothing but NOR gates. This can be economical if you only want to stock one type of gate, or if you have available spare gates on multiple-gate ICs. For example, one of the cheapest and most widely available ICs is the SN 7400 which contains four 2-input positive NAND gates. Further, the IC manufacturer can produce arrays of identical gates which can be interconnected for the functions required by the user.

12.7.1 NAND Logic

The three basic logic functions NOT, AND and OR can be realized with NAND gates as follows:

NOT Function

If the two inputs to a NAND gate are tied together as in Fig. 12.23 then in effect only the first and last lines of the NAND truth table (Table 12.9) are pertinent because $A = B$. Then if $A = 0$, $F = 1$ and if $A = 1$, $F = 0$; therefore $F = \bar{A}$. A similar result could be achieved by tying B to logical 1. Now only the second and fourth lines of the truth table are pertinent, and again $F = \bar{A}$.

AND Function

Having produced NOT from NAND, then Fig. 12.24 shows the realization of AND. Gate X gives $\overline{AB}$. Gate Y complements the output of gate X so $F = \overline{\overline{AB}} = AB$.

Table 12.9   NAND Truth Table

A	B	$F = \overline{AB}$
0	0	1
0	1	1
1	0	1
1	1	0

$$A \quad\quad F = \bar{A}$$

Fig. 12.23   NAND Gate Connected as a NOT Circuit

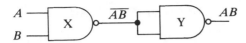

Fig. 12.24 AND from NAND

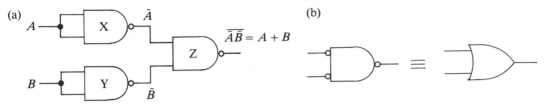

Fig. 12.25 OR from NAND (a) NAND Gates (b) Logical Equivalence

OR Function

Using de Morgan to convert OR into the NAND form, we have $A + B = \overline{\bar{A}\bar{B}}$. In Fig. 12.25a gate X gives $\bar{A}$, gate Y gives $\bar{B}$, and gate Z gives $\overline{\bar{A}\bar{B}} = A + B$. The logical equivalence of OR from NAND is expressed symbolically in Fig. 12.25b.

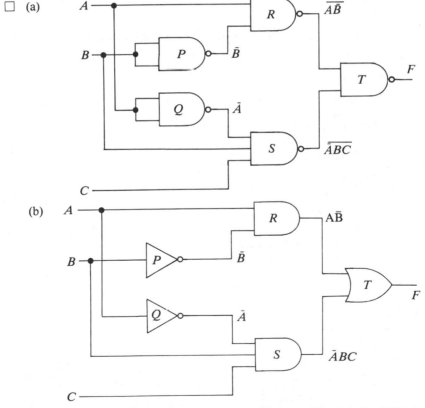

Fig. 12.26 (a) Sum of Products in NAND Gates (b) Sum of Products in NOT, AND, OR Gates

Example 12.4   Realization of a Sum of Products in NAND form

Suppose the function $F = A\bar{B} + \bar{A}BC$ is required using only NAND gates. Then let $A\bar{B} = P$, and $\bar{A}BC = Q$. Then

$$F = P + Q = \overline{\bar{P} \cdot \bar{Q}} = \overline{\overline{A\bar{B}} \cdot \overline{\bar{A}BC}}$$

The NAND gate configuration is shown in Fig. 12.26a. Comparing it with the straightforward NOT, AND, OR realization (Fig. 12.26b), it will be observed that the various gates have all simply been replaced by NAND gates.   □

12.7.2   NOR Logic

NOT Function

Again NOT can be produced by commoning the inputs to a NOR gate (Fig. 12.27) – a fact which can be confirmed by reference to the first and last lines of the NOR truth table (Table 12.10). NOT can also be produced by tying the unused input to 0.

OR Function

OR is NOT NOR, i.e. $A + B = \overline{\overline{A + B}}$ as shown in Fig. 12.28.

AND Function

Using de Morgan to obtain the AND function in NOR form we have:

$AB = \overline{\bar{A} + \bar{B}}$, which is realized in Fig. 12.29.

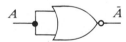

Fig. 12.27   NOR Gate Connected as a NOT Circuit

Table 12.10   NOR Truth Table

A	B	$F = \overline{A + B}$
0	0	1
0	1	0
1	0	0
1	1	0

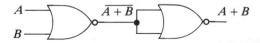

Fig. 12.28   OR from NOR

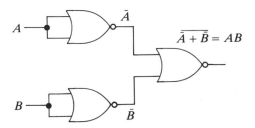

Fig. 12.29　AND from NOR

Realization of a Sum of Productions in NOR form

Using the same function that we used for the NAND form,

$$F = A\bar{B} + \bar{A}BC$$
$$= \overline{\bar{A} + B} + \overline{A + \bar{B} + \bar{C}}$$

Since $\bar{\bar{F}} = F$ then

$$F = \overline{\overline{\bar{A} + B} + \overline{A + \bar{B} + \bar{C}}}$$

The required circuit is shown in Fig. 12.30.

NOR Realization by finding $\bar{F}$ from K Map

An alternative NOR realization can be found by obtaining $\bar{F}$ from a K map. In Fig. 12.31 the previous function F has been entered on a K map but the 0s have been

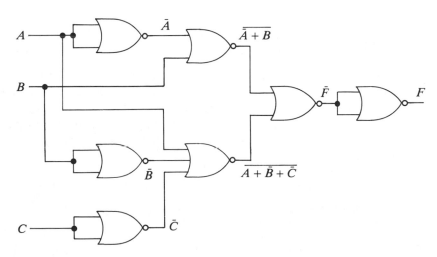

Fig. 12.30

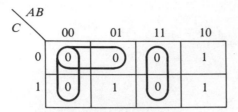

Fig. 12.31   Grouping 0s to Find $\bar{F}$

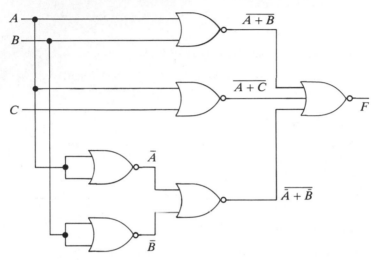

Fig. 12.32   Alternative NOR Circuit

used to give

$$\bar{F} = \bar{A}\bar{B} + \bar{A}\bar{C} + AB$$

$$\bar{F} = \overline{A+B} + \overline{A+C} + \overline{\bar{A}+\bar{B}}$$

$$F = \overline{\overline{A+B} + \overline{A+C} + \overline{\bar{A}+\bar{B}}}$$

The resulting circuit is shown in Fig. 12.32. This method certainly saves two inverters at the output but whether it is cheaper than the alternative solution depends on the original function. Another possibility would have been to loop the middle two squares of the top row of the K map instead of the left pair, but this would have required an extra inverter, a fact that the reader might care to demonstrate.

12.8   EXCLUSIVE OR LOGIC

XOR gates are used in the process of adding binary numbers electronically, a process described in Chapter 13. However, in the general context of minimization

Table 12.11   XOR Truth Table

A	B	F
0	0	0
0	1	1
1	0	1
1	1	0

of Boolean functions they can, for certain problems, provide a more economic implementation than by using other logic gates. Before illustrating the point by example, first consider Table 12.11, the truth table for the XOR function. Expressing the function as a sum of products we have

$$F = \bar{A}B + A\bar{B}$$

By observing the 0s in Table 12.11 we can also write down the complement of F as

$$\bar{F} = \bar{A}\bar{B} + AB$$

Consider now the following problem.

☐   **Example 12.5**

Design a circuit that will enable any one of three switches to control a light independently of the other two.

Table 12.12 is the truth table for this problem. It is drawn up in the order shown to emphasize the feature that whenever a switch changes from 0 to 1 or vice versa it changes the light. Drawing a K map (Fig. 12.33) from the truth table, it will be noted that no adjacent pairs of squares contain just 0s or 1s and therefore there can be no minimization from canonical form. However, writing down the function in

Table 12.12   Truth Table for Example 12.5

A	B	C	F
0	0	0	0
0	0	1	1
0	1	1	0
0	1	0	1
1	1	0	0
1	1	1	1
1	0	1	0
1	0	0	1

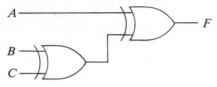

C \ AB	00	01	11	10
0	0	1	0	1
1	1	0	1	0

Fig. 12.33   K Map for Example 12.5

Fig. 12.34   Circuit for Example 12.5

canonical form, we have

$$F = \bar{A}\bar{B}C + \bar{A}B\bar{C} + ABC + A\bar{B}\bar{C}$$
$$= \bar{A}(\bar{B}C + B\bar{C}) + A(BC + \bar{B}\bar{C})$$
$$= \bar{A}(B \oplus C) + A(\overline{B \oplus C})$$
$$= A \oplus (B \oplus C)$$

which only requires two XOR gates, as shown in Fig. 12.34.

Reduction of a function to XOR form is characterized by a K map where 1s are diagonally opposite to each other as in Fig. 12.33.   □

The foregoing problem can be extended to four switches, the solution to which is

$$F = (A \oplus B) \oplus (C \oplus D)$$

Use of XOR Gate for NOT Function

An XOR gate can be used as a NOT circuit by holding one input at 1. Referring to the XOR truth table (Table 12.11), suppose that B is always 1. Then if $A = 0$, $F = 1$ and if $A = 1$, $F = 0$. Therefore, $F = \bar{A}$.

12.9 PULSE TRAINS AS INPUT VARIABLES TO COMBINATIONAL LOGIC CIRCUITS

Combinational logic circuits often have pulse trains as input variables. Figure 12.35 is an example of a NAND logic circuit with three input waveforms. For purposes

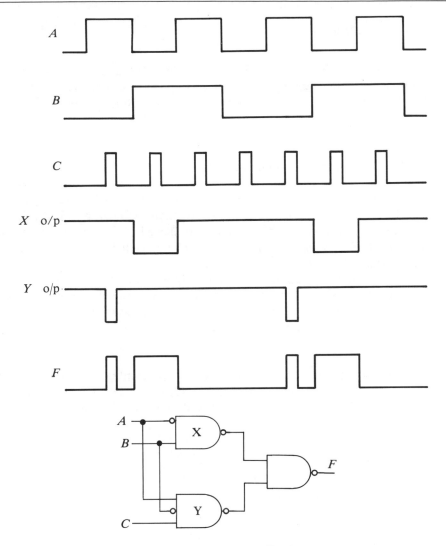

Fig. 12.35   Pulse Trains in Combinational Logic Circuits

such as fault finding it is often necessary to be able to predict the pulse trains at various points in the circuit. In this case, remembering the rule that for a NAND gate any one or more 0 inputs results in a 1 output, then the waveforms at gate X and gate Y outputs can be predicted. The output waveform F can be found in the same way and should confirm that

$$F = \overline{\overline{AB} \cdot \overline{A\bar{B}C}} = \bar{A}B + A\bar{B}C$$

The exercise should be repeated replacing all NAND gates with NOR gates.

12.10 STATIC HAZARDS

Section 12.6.2 described how K maps could be used to minimize functions. For example, suppose that the K map of Fig. 12.36 had been derived from a truth table, then by looping as shown the minimal sum of products would be

$$F = A\bar{B} + BC$$

Figure 12.37 is an implementation of this function.

Consider now the conditions when the inputs change from $ABC = 111$ to $ABC = 101$. When B changes to 0 the output of gate Y falls to 0 and that from gate X rises to 1 to maintain $F = 1$. However, because of delays through the gates, it takes time for an input change to propagate itself to the output. If the delay through X is longer than that through Y then, as indicated by the waveforms of Fig. 12.38, the output F falls, momentarily, to 0. This is called a static hazard. For some applications this might be unacceptable. For example, if the condition $F = 0$ triggers an alarm then it may do so unintentionally where there is a static hazard.

The characteristic feature of a static hazard is the existence of adjacent, non-overlapping loops as in Fig. 12.36. A hazard can be eliminated by adding a gate which creates an overlap on the K map, as in Fig. 12.39. Now,

$$F = A\bar{B} + BC + AC$$

and although the term AC is redundant in strict combinational logic, it does mean that, in the redesigned circuit of Fig. 12.40, when ABC changes from 111 to 101 the output of gate Z will remain at 1 and maintain F without a hazard.

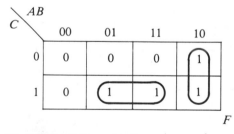

Fig. 12.36   K Map with Two Adjacent Nonoverlapping Loops

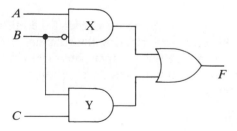

Fig. 12.37   Circuit for $F = A\bar{B} + BC$

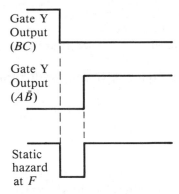

Fig. 12.38 Showing That When ABC (Fig. 12.37) Goes From 111 to 101, There May Be an Unwanted Transition at F

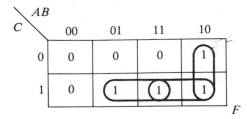

Fig. 12.39 Adding an Overlapping Loop to Eliminate Static Hazard

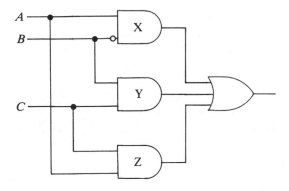

Fig. 12.40 Fig. 12.37 Redesigned to Eliminate Static Hazard

SUMMARY

At the end of this chapter readers should:

1. Know the meaning of AND, OR, NAND, NOR, NOT and XOR functions.

2. Be able to convert decimal numbers to direct binary code.
3. Be able to represent a combinational logic proposition in canonical form.
4. Be able to minimize Boolean sums of products of up to four variables using K maps.
5. Be able to implement logic functions in both NAND and NOR forms.

REVISION QUESTIONS

12.1   An exclusive OR circuit has two inputs and gives an output $F = 1$ when one and only one input is 1. Write down a truth table for F, express F as a sum of products, and draw the circuit diagram.

12.2   Write down an expression for F (Fig. Q.12.2) and draw up a truth table which specifies its performance.

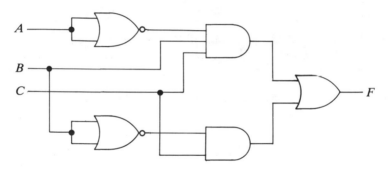

Fig. Q.12.2

12.3   Prove both parts of de Morgan's theorem:
(a) by perfect induction;
(b) by Venn diagram.

12.4   Simplify algebraically the expression:

$$F = (A + \bar{B})(A + \bar{C})$$

$$(A + \bar{B}\bar{C})$$

and use de Morgan's theorem to show that

$$\bar{F} = \bar{A}(B + C)$$

12.5   Devise a circuit which gives an output $F = 1$ when any two of its three inputs is 1.

$$(AB\bar{C} + A\bar{B}C + \bar{A}BC)$$

12.6 A 3-bit binary counter counts from 0 ($ABC = 000$) to 7 ($ABC = 111$). Deduce a sum of products expression which gives $F = 1$ for any prime number excluding 0 and 1 and hence draw a suitable circuit diagram.

$$(\bar{A}B + AC)$$

12.7 Design a logic circuit for a car starter S which will only operate if the key K is turned, the driver seat belt D is fastened and either there is no front seat passenger P or there is a front seat passenger with his seat belt F fastened.

Use any obvious 'can't happen' condition which will simplify the design.

$$(KD\bar{P} + KDF).$$

12.8 Write down the Boolean expressions which are represented by these Karnaugh maps:

(a)

C \ AB	00	01	11	10
0		1		1
1	1		1	

(b)

C \ AB	00	01	11	10
0	1			
1				1

(c)

C \ AB	00	01	11	10
0		1	1	
1		1	1	

12.9 Represent the following Boolean expressions on Karnaugh maps and deduce minimal sums of products.

(a) $F = \bar{A} \cdot B \cdot C + A \cdot \bar{B} \cdot C$
(b) $F = A \cdot \bar{B} \cdot \bar{C} + \bar{A} \cdot \bar{B} \cdot \bar{C} + A \cdot B \cdot C$
(c) $F = \bar{P} \cdot Q \cdot R + P \cdot Q \cdot \bar{R} + \bar{P} \cdot Q \cdot \bar{R} + P \cdot Q \cdot R$
(d) $F = \bar{A} \cdot B + A \cdot \bar{B} \cdot \bar{C}$

(no reduction; $\bar{B}\bar{C} + ABC$; Q; no reduction)

12.10 Draw Karnaugh maps to represent the functions given by these truth tables and deduce minimal sums of products.

(a)

A	B	C	O/P
0	0	0	0
0	0	1	0
0	1	0	0
0	1	1	0
1	0	0	1
1	0	1	0
1	1	0	1
1	1	1	0

$$(A\bar{C})$$

(b)

A	B	C	O/P
0	0	0	0
0	0	1	1
0	1	0	1
0	1	1	0
1	0	0	1
1	0	1	0
1	1	0	1
1	1	1	0

$$(A\bar{C} + B\bar{C} + \bar{A}\bar{B}C)$$

12.11 Minimize the following switching function using K-map:

$$F = \sum (0, 1, 5, 7, 8, 9, 12, 14, 15)$$

Don't care $= \sum (3, 11, 13)$

$$(D + AB + \bar{B}\bar{C})$$

12.12 Using the K map or otherwise, minimize the expression:

$$F = \sum (0, 2, 3, 7, 8, 10, 11, 15)$$

$$(\bar{B}\bar{D} + CD)$$

12.13 Simplify the following output equation:

$$Z = \bar{A}\bar{B}\bar{C}D + \bar{A}\bar{B}CD + \bar{A}B\bar{C}D + AB\bar{C}D + ABCD + A\bar{B}\bar{C}D + A\bar{B}CD$$
$$(\bar{C}D + \bar{B}D + AD)$$

12.14 Derive the minimal expression from the given Karnaugh map for the logic function F.

CD \ AB	00	01	11	10
00	1	1		1
01		1		
11		1	1	
10	1	1	1	1

F

Hence or otherwise obtain a minimal logic expression for $\bar{F}$.
Implement the function F using
(a) NAND gates only;
(b) NOR gates only.

$$\overline{\overline{A}B \cdot \overline{B}C \cdot \overline{B}\overline{D}}$$

$$\overline{\overline{B + \bar{D}} + \overline{\bar{A} + \bar{B} + C}}$$

12.15   Determine the gate output for the input waveforms in Fig. Q.12.15.

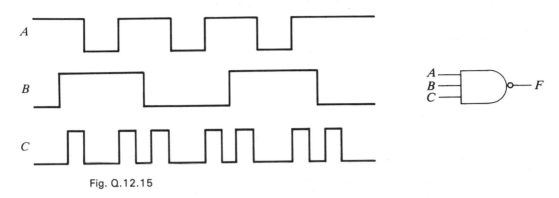

Fig. Q.12.15

12.16   Determine the output waveform in Fig. Q.12.16.

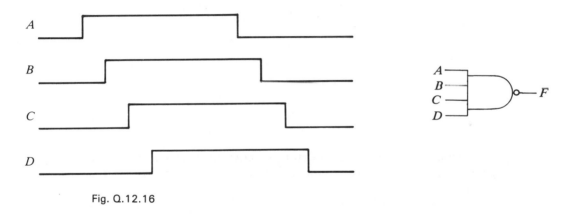

Fig. Q.12.16

12.17   Determine the output waveform in Fig. Q.12.17.

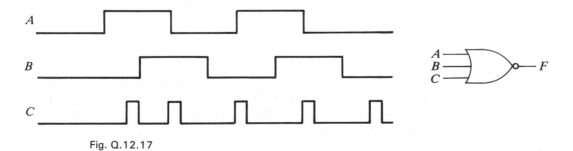

Fig. Q.12.17

12.18 A circuit is required which has four inputs A, B, C and D, and one output F which must obey the following truth table:

A	B	C	D	F
0	0	0	0	1
0	0	0	1	1
0	0	1	0	1
0	0	1	1	0
0	1	0	0	1
0	1	0	1	0
0	1	1	0	0
0	1	1	1	0
1	0	0	0	X
1	0	0	1	X
1	0	1	0	1
1	0	1	1	0
1	1	0	0	0
1	1	0	1	0
1	1	1	0	0
1	1	1	1	0

The inputs $ABCD = 1000$ and 1001 can never occur.

Assuming that inputs are available in both complemented and uncomplemented forms, design minimal circuits to obtain the required output using:
(a) AND and OR gates
(b) only NAND gates
(c) only NOR gates.

$$(\bar{B}\bar{C} + \bar{B}\bar{D} + \bar{A}C\bar{D};\ \overline{\overline{\bar{B}\bar{C}} \cdot \overline{\bar{B}\bar{D}} \cdot \overline{\bar{A}C\bar{D}}}$$

$$\overline{\bar{A} + \bar{B} + \bar{C} + \bar{D}} + \overline{\bar{B} + \bar{D}} + \overline{\bar{B} + C}\ \text{or}\ \overline{\overline{B + C} + \overline{B + D} + \overline{A + C + D}})$$

12.19 Circuits are required which will obey the truth table set out below. Inputs A, B, C, D are available in both uncomplemented and complemented forms. Draw minimal circuits which will obey the truth table using (a) all NAND gates; (b) all NOR gates

	Inputs			Output
A	B	C	D	F
0	0	0	0	0
0	0	0	1	0
0	0	1	0	X
0	0	1	1	0
0	1	0	0	0
0	1	0	1	1
0	1	1	0	1
0	1	1	1	1
1	0	0	0	X
1	0	0	1	1
1	0	1	0	1
1	0	1	1	1
1	1	0	0	0
1	1	0	1	1
1	1	1	0	1
1	1	1	1	1

Inputs for which F is marked X cannot occur.

$$(\overline{\overline{A\bar{B}} \cdot \overline{BD} \cdot \overline{C\bar{D}}}; \; \overline{\overline{A + B} + \overline{C + D}})$$

12.20 A network has four inputs, a, b, c, d and two outputs, f_1 and f_2. The network transmission function is expressed in tabular form:

a	b	c	d	f_1	f_2
0	0	0	0	1	0
0	0	0	1	1	0
0	0	1	0	0	0
0	0	1	1	0	0
0	1	0	0	1	0
0	1	0	1	0	0
0	1	1	0	0	1
0	1	1	1	x	x

a	b	c	d	f_1	f_2
1	0	0	0	1	1
1	0	0	1	1	0
1	0	1	0	0	0
1	0	1	1	0	0
1	1	0	0	1	1
1	1	0	1	0	0
1	1	1	0	0	1
1	1	1	1	x	x

The input combinations 0111 and 1111 cannot occur. Design a minimal

network having the defined transmission function using
(a) all NAND gates
(b) all NOR gates

$$\overline{\overline{cd} \cdot \overline{bc}}\ \overline{bc \cdot a\overline{c}d};\ \overline{c + \overline{b} + \overline{d}}\ \overline{\overline{d + a} + c + b + \overline{c}})$$

12.21   For the input waveforms in Fig. Q.12.21 what NAND logic circuit would generate the output waveform shown? (*Suggestion*: enter states directly on a K map.)

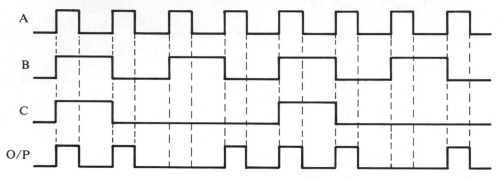

Fig. Q.12.21

12.22   For Fig. Q.12.22 sketch the waveforms at each numbered point in the correct relationship to one another. Deduce a logic function for the output and hence verify the output waveform.

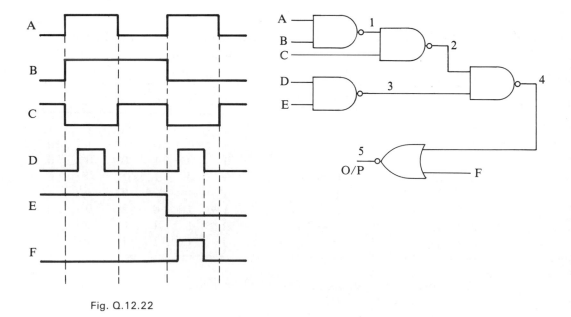

Fig. Q.12.22

12.23 A, B, C, D are four data bits to be transmitted.
Logic is required for a parity bit which will give an output $P = 1$ when there are an even number of 1s among the data bits including the all zeros condition. Design the logic for P using only XOR gates.

$$\overline{(A \oplus B) \oplus (C \oplus D)}$$

12.24   Explain the meaning of the term 'static hazard' in the context of Fig. 12.15. What is the equivalent logic function in which the hazard is eliminated?

$$(AB + \bar{B}C + AC)$$

13

Combinational Logic Applications

Chapter Objectives
13.1   Binary Addition and Subtraction
13.2   Binary Codes
13.3   Encoders
13.4   Decoders
13.5   Code Conversion
13.6   Multiplexer – Data Selector
13.7   Demultiplexer
        Summary
        Revision Questions

Having defined, in Chapter 12, the functions of binary logic and formulated gating arrangements for their efficient implementation, we can now describe some of the more common applications of combinational logic. A combinational logic circuit is one in which the values (0 or 1) of binary inputs such as A, B and C (Fig. 13.1) at any instant completely determine the value(s) of the binary output(s), F_1, etc., at that instant. This is in contrast to sequential logic circuits in which the output is not only determined by the present combination of input values, but also by their previous values.

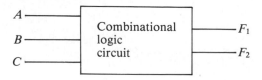

Fig. 13.1   In a Combinational Logic Circuit the Output at any Time is Solely Determined by the Inputs at that Time

13.1 BINARY ADDITION AND SUBTRACTION

Half Adder

A digital adder will add just two binary numbers. Consider the truth table for two binary digits (bits) A and B (Table 13.1). When two digits are added (binary or otherwise) two results are required: one is the sum S at that place of digit and the other is C_o, the carry out to the next place. So the result of adding the decimal digits 7 and 6 is $S = 3$, $C_o = 1$.

In Table 13.1 the sum S is 1 only when A or B (but not both) is 1, and C_o is only 1 when both A and B are 1. Therefore, in binary logic,

$$S = A \oplus B$$
$$C_o = AB$$

and the logic implementation is shown in Fig. 13.2. It can be realized in other gate forms, so for this text we shall simply use the block diagram form also shown in Fig. 13.2.

This circuit is called a half adder, 'half' because we have not taken into account the fact that, with the exception of the first place of digit, there may be a 1 to carry in (C_i) from the previous place. Addition therefore has to be carried out in two

Table 13.1 Truth Table for Adding Two Bits

A	B	S	C_o
0	0	0	0
0	1	1	0
1	0	1	0
1	1	0	1

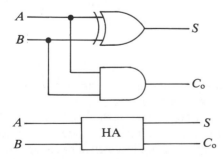

Fig. 13.2 Logical Implementation of a Half Adder and its Block Diagram Representation

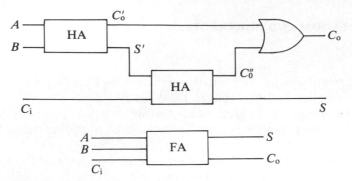

Fig. 13.3   Full Adder

Table 13.2   Full Adder Truth Table

A	B	C_i	S'	$C_o{}'$	$C_o{}''$	C_o	S
0	0	0	0	0	0	0	0
0	0	1	0	0	0	0	1
0	1	0	1	0	0	0	1
0	1	1	1	0	1	1	0
1	0	0	1	0	0	0	1
1	0	1	1	0	1	1	0
1	1	0	0	1	0	1	0
1	1	1	0	1	0	1	1

stages:

(a) add bits A and B, thereby producing a partial sum S';

(b) add to the result of (a) C_i from the previous place.

The idea of adding C_i to the partial sum S', as in the full adder of Fig. 13.3, is the same logical process as addition with pencil and paper. To justify C_o as the logical OR of the two half adder carry-outs, we can see from Table 13.2 that $C_o{}'$ and $C_o{}''$ can never both be 1 but that if either is 1 then $C_o = 1$.

☐   **Example 13.1**

Add the decimal numbers $A = 7$ and $B = 3$ in binary form, showing the values of $A, B, S', C_o{}', C_i, C_o{}'', C_o$ and S for each pair of bits.

The solution is given in Table 13.3. Note that for the LSBs, $S', C_o{}'$ C_i and $C_o{}''$ are all blank because only half addition is required at this stage with no carry in. At the MSBs, $C_o = 1$ becomes $S = 1$ for a fourth bit which is necessary for the result 1010 (decimal 10).

Table 13.3   Solution for Example 13.1

A		1	1	1
B		0	1	1
S'		1	0	–
C_o'		0	1	–
C_i		1	1	–
C_o''		1	0	–
C_o		1	1	1
S	1	0	1	0

Figure 13.4 is a four-stage parallel adder – parallel means that the four bits of each number are fed in simultaneously. Before the complete sum appears the addition (HA) of stage 0 has to produce and propagate its carry to stage 1, which then takes time to generate its carry to propagate to stage 2, and so on. Clearly the carrys are generated and move through the stages sequentially, hence the usual name of the circuit – ripple carry parallel adder. Figure 13.4 shows the steady state of all points in the circuit resulting from adding 1101 (A) to 1011 (B), the MSB of the result being the carry out of the final stage.

Two's Complement Arithmetic

Subtraction in digital systems is carried out by converting the number to be subtracted into a code which will be recognized as the negative of the number – often what is known as 2's complement form – and adding. If negative numbers are to be coded into binary form then the range represented by a certain number of bits will not be the same as for a code which only represents positive numbers. For

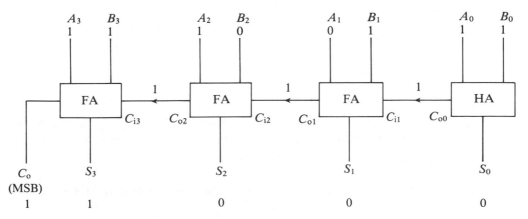

Fig. 13.4   Four-stage Ripple Carry Parallel Adder

example, if 8 bits are to be used only for positive numbers then the range is from 00000000 to 11111111, i.e. 0 to 255. In 2's complement arithmetic the MSB is used as a sign bit, 0 signifying positive and 1 negative. For positive numbers the remaining 7 bits are used as direct binary code, i.e. 00000000 to 01111111 represent $+0$ to $+127$. For negative numbers the 2's complement is taken as follows:

(a) Write the number in binary, including the sign bit as a positive number, e.g. $+25 = 0\ 0\ 0\ 1\ 1\ 0\ 0\ 1$.

(b) Take the 1's complement of this binary number, i.e. complement all bits, giving $1\ 1\ 1\ 0\ 0\ 1\ 1\ 0$.

(c) Convert this into the 2's complement by adding 1, i.e.

$$
\begin{array}{r}
1\ 1\ 1\ 0\ 0\ 1\ 1\ 0 \\
+ \quad\quad\quad\quad\quad 1 \\
\hline
1\ 1\ 1\ 0\ 0\ 1\ 1\ 1 = -25
\end{array}
$$

The range of numbers which are now represented by 8 bits using 2's complement arithmetic is from -128 to $+127$ with the extremes of the negative range being

$$1\ 0\ 0\ 0\ 0\ 0\ 0\ 0 = -128$$

and $1\ 1\ 1\ 1\ 1\ 1\ 1\ 1 = -1$

Now let us consider some examples of addition and subtraction using 2's complement arithmetic.

(a) $38 - 25$

$$
\begin{array}{rcl}
38 & = & 0\ 0\ 1\ 0\ 0\ 1\ 1\ 0 \\
-25 & = & 1\ 1\ 1\ 0\ 0\ 1\ 1\ 1 \\
\hline
\text{Add} & & 1\ 0\ 0\ 0\ 0\ 1\ 1\ 0\ 1
\end{array}
$$

The 9th bit will be ignored and the 8th, being a 0, indicates a positive result. The remaining 7 bits are equal to 13 in decimal which is the correct answer. Now reverse the sum, i.e.

(b) $25 - 38$

$$
\begin{array}{rcl}
25 & = & 0\ 0\ 0\ 1\ 1\ 0\ 0\ 1 \\
-38 & = & 1\ 1\ 0\ 1\ 1\ 0\ 1\ 0 \\
\hline
\text{Add} & & 1\ 1\ 1\ 1\ 0\ 0\ 1\ 1
\end{array}
$$

A negative result is indicated with bit 8 at 1. Therefore, to find what this is equivalent to as a decimal number we take the 2's complement of the result, i.e.

$$0\ 0\ 0\ 0\ 1\ 1\ 0\ 0 + 1 = 0\ 0\ 0\ 0\ 1\ 1\ 0\ 1$$

which is 13, again the correct answer – remembering that the result is, in fact, negative.

Figure 13.5 is an 8 bit adder/subtractor from which S_7 gives the sign of the

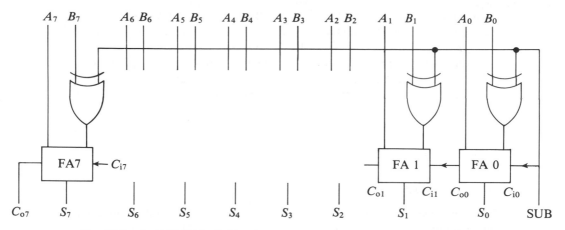

Fig. 13.5   An 8-bit Adder/Subtractor

result. The circuit adds when $SUB = 0$. In this condition the output of each XOR gate will be the same as its B input bit. The carry-in to FA0, C_{i0}, will be 0 so FA0 acts as a half adder and the circuit performs in the same manner as the ripple carry parallel adder of Fig. 13.4. To subtract B from A, $SUB = 1$. Then each XOR output will be the complement of its B input. (Check this by drawing the truth table for an XOR gate with one input held at 1.) Therefore the eight XOR output bits will be the 1's complement of B. But with $C_{i0} = SUB = 1$ the effect is to add 1 to the result, which has the same result as adding the 2's complement of B to A. When subtracting, C_{o7} will be ignored and if $S_7 = 1$ the result is negative and in 2's complement form. When adding, $C_{o7} = 1$ indicates an overflow which can be used as the carry-in to another adder for processing larger numbers.

Signed Binary Numbering

In this notation the MSB is used to indicate sign — 0 for positive and 1 for negative, as with 2's complement notation. The difference is that the other 7 bits represent the number in direct binary form. For example,

$$0\ 0\ 0\ 0\ \ 0\ 0\ 1\ 1_2 = +3_{10}$$
$$1\ 0\ 0\ 0\ \ 0\ 0\ 1\ 1_2 = -3_{10}$$

The range of numbers for signed binary with 8 bits will be -127 to $+127$.

Hexadecimal Numbering

It is standard practice to represent groups of 4 binary digits in hexadecimal form. Sixteen symbols are required to represent the decimal numbers 0 to 15, with 0 to 9 being used as in the decimal system and A to F representing 10 to 15. For an 8 bit number two hex digits are used and it is only necessary to convert each group of

4 bits into a hex digit irrespective of whether the 8 bits represent a signed or an unsigned number. For example, the decimal number 232 in unsigned binary is 1110 1000 which is E8 in hex. The same number E8 as a 2's complement number still represents the same 8 bits and in decimal would be -24. Symbols may be attached to a hex number to avoid ambiguity. For example, the hex number 36 (decimal 54) might be written 36_{16} or 36H or \$36.

13.2 BINARY CODES

8421 BCD Code

In digital systems, numbers and instructions are translated into binary code. The most obvious way of doing this is to translate numbers directly into binary form as we saw in Section 12.4, where N bits can represent decimal numbers up to $2^N - 1$. One common variant of direct binary code is the 8421 BCD – binary coded decimal – code. It is sometimes called naturally binary coded decimal (NBCD). When the term BCD is used on its own, 8421 code can be inferred. Each decimal digit of a multi-digit decimal number is represented by four bits counting in direct binary form as shown in Table 13.4. This BCD code is said to be 'weighted' in that the MSB represents 8, the next bit 4, and so on. As there are only ten decimal digits, six binary combinations are not used. To represent a number such as 387 the BCD code would be 0011 1000 0111.

Gray Codes

With devices such as shaft encoders, angular displacement is indicated by sensors reading off the bit values which change between 0 and 1 as the shaft rotates. In Fig. 13.6a the three circular sectors represent the bits with the LSB on the outer sector.

Table 13.4   8421 BCD Code

Decimal	A	B	C	D
0	0	0	0	0
1	0	0	0	1
2	0	0	1	0
3	0	0	1	1
4	0	1	0	0
5	0	1	0	1
6	0	1	1	0
7	0	1	1	1
8	1	0	0	0
9	1	0	0	1

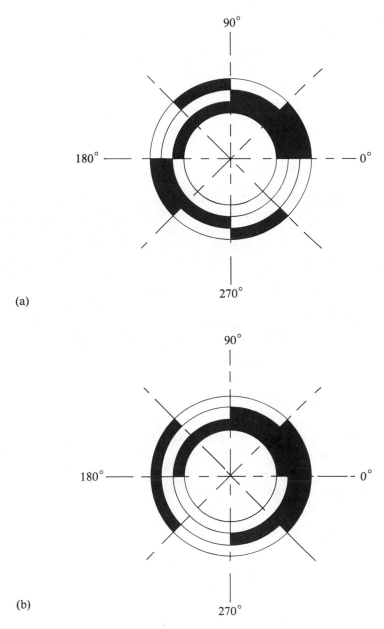

Fig. 13.6   Shaft Encoding (a) Direct Binary (b) Gray Code. Black = 0, White = 1, MSB on Inner Ring

Table 13.5   A 3 bit Gray Code

Decimal	A	B	C
0	0	0	0
1	0	0	1
2	0	1	1
3	0	1	0
4	1	1	0
5	1	1	1
6	1	0	1
7	1	0	0

This shaft is encoded in direct 3 bit binary form. At certain points in its rotation, more than one bit will change value. For example, as the shaft passes through the $90°$ point the code should change from 001 to 010, i.e. two bits change value. Now if the sensors are not in precise alignment and, for example, the middle bit changes value first, the sequence generated will be 001 011 010 – the second group of digits being spurious and a possible source of error. The distinguishing feature of Gray codes – for which there is no limit on the number of bits nor is there a unique code for a given number of bits – is that only one bit changes between adjacent values. Table 13.5 is a 3 bit Gray code representing the decimal numbers 0 to 7. A shaft encoder using this Gray code would be as in Fig. 13.6b in which it may be seen that the sensors would never encounter more than a single bit changing in any shaft position.

Excess 3 Code

This is a 4 bit binary code for the 10 decimal digits. Each combination is found by adding 3 to the decimal number being coded and translating the result into direct

Table 13.6   Excess 3 Code

Decimal	A	B	C	D
0	0	0	1	1
1	0	1	0	0
2	0	1	0	1
3	0	1	1	0
4	0	1	1	1
5	1	0	0	0
6	1	0	0	1
7	1	0	1	0
8	1	0	1	1
9	1	1	0	0

binary form. The excess 3 code is set down in Table 13.6. Its advantage is in digital computer arithmetic. It will be noted that the 1's complement of any excess 3 code is the 9's complement of the corresponding decimal number. For example, the 1's complement of 0101 (2) = 1010 (7) and $9 - 2 = 7$. This feature simplifies the subtraction of numbers.

13.3 ENCODERS

An encoder (or just coder) is a device which has a number of input lines equal to the number of code combinations to be generated: this will be ten for a BCD encoder as indicated in Fig. 13.7. It has a number of output lines equal to the number of bits in the code — four for BCD. The switches in Fig. 13.7 might be actuated by the decimal keys on a VDU keyboard. The logic required in the encoder can be ascertained from the BCD code in Table 13.4 which shows, for example, that A must be 1 if either K_8 or K_9 is pressed, which makes the K_8 and K_9 inputs 1, i.e.

$$A = K_8 + K_9$$

Similarly, the logic for the other outputs will be

$$B = K_4 + K_5 + K_6 + K_7$$
$$C = K_2 + K_3 + K_6 + K_7$$
$$D = K_1 + K_3 + K_5 + K_7 + K_9$$

resulting in Fig. 13.8 as a BCD encoder. An alternative is to use a read only memory (ROM) which will be described in Chapter 17.

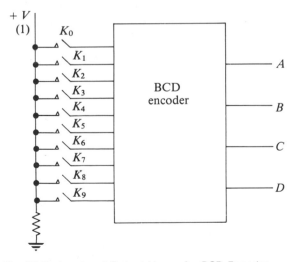

Fig. 13.7   Input and Output Lines of a BCD Encoder

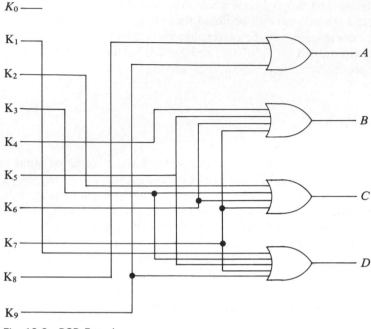

Fig. 13.8   BCD Encoder

13.4   DECODERS

A decoder performs the opposite function of an encoder. With a number of input lines equal to the number of bits in the code and as many output lines as there are combinations in the code, then an output line will go to, say, 1 when the appropriate input combination occurs. Table 13.7 is a direct 3 bit binary code and Fig. 13.9

Table 13.7   Direct 3 bit Binary Decoding

A	B	C	$Output = 1$
0	0	0	F_0
0	0	1	F_1
0	1	0	F_2
0	1	1	F_3
1	0	0	F_4
1	0	1	F_5
1	1	0	F_6
1	1	1	F_7

A
B
C
3
to
8
line
decoder
F_0
F_1
F_2
F_3
F_4
F_5
F_6
F_7

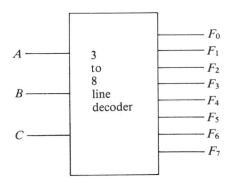

Fig. 13.9 Input and Output Lines of a 3 to 8 Decoder

shows the input and output lines of the required 3 to 8 line decoder. The decoder logic must be such that, for example, $F_3 = 1$ when $ABC = 011$ and therefore the decoder logic for output F_3 will be $F_3 = \bar{A}BC$. The complete decoder could be made up of eight 3-input AND gates and three inverters.

8421 BCD Decoder

For a 4 to 10 line BCD decoder the logic would be

$$F_0 = \bar{A}\bar{B}\bar{C}\bar{D};\ F_1 = \bar{A}\bar{B}\bar{C}D$$

and so forth, although the logic could be simplified by recognizing that the six highest numbers cannot occur, and K mapping for minimum logic accordingly.

Excess 3 Decoder

Applying this for an excess 3 decoder we should note the excess 3 decoder requirements in Table 13.8. Entering, as an example, the inputs for F_2 on a K map (Fig.

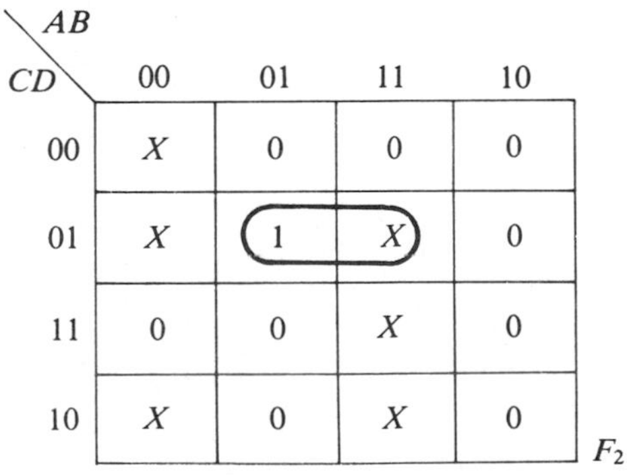

AB
CD
00 01 11 10
00 X 0 0 0
01 X 1 X 0
11 0 0 X 0
10 X 0 X 0
F_2

Fig. 13.10 K Map for F_2 Output of an Excess 3 Decoder

Table 13.8   Excess 3 Decoding

A	B	C	D	$Output = 1$
0	0	1	1	F_0
0	1	0	0	F_1
0	1	0	1	F_2
0	1	1	0	F_3
0	1	1	1	F_4
1	0	0	0	F_5
1	0	0	1	F_6
1	0	1	0	F_7
1	0	1	1	F_8
1	1	0	0	F_9

13.10) in which six combinations of input variables do not occur then,

$$F_2 = B\bar{C}D$$

($\bar{A}CD$ is an alternative but requires an extra inverter.)

BCD to 7 Segment Decoder

This favorite exercise in binary logic requires that the 7 segments in the display of Fig. 13.11 have to be activated to display decimal digits in accordance with a BCD input. For example, segment a must be activated for decimal digits $0, 2, 3, 5, 6, 7, 8$ and 9. Entering these on a K map (Fig. 13.12) and noting that the binary combinations for 10 to 15 do not occur, then the logic for segment a would be

$$a = A + C + BD + \bar{B}\bar{D}$$
$$= A + C + \overline{B \oplus D}$$

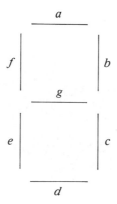

Fig. 13.11   Seven-segment Display

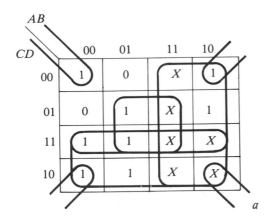

Fig. 13.12   K Map for Segment *a* of a Seven-segment Display Decoder

ECONOMICS OF MINIMIZATION

With the availability of MSI circuits in general and ROMs in particular, the benefits of minimization are questionable and designers often implement a logic function in minterm form.

The DM 74154 4 to 16 line Decoder

In this MSI circuit (Fig. 13.13) a line is selected when the circuit makes an output go to 0. The input variables are in the order $DCBA$, i.e. D is the MSB. To select, for example, output F8 (pin 9), $DCBA = 1000$. Tracing the gate lines through, it can be seen that only the gate feeding pin 9 has all inputs at 1 which is the requirement for a NAND gate output to be 0. It may also be noted that both complemented and uncomplemented routes to gate inputs have inverters. From a logic point of view it is obviously unnecessary to have the two inverters in the uncomplemented route but they are provided as buffers for the inputs. The circuit also has two enable inputs G_1 and G_2: these must both be at 0 to enable the decoder. If they are not 0 then all outputs will be 1 regardless of the value of $DCBA$.

13.5   CODE CONVERSION

A common requirement is conversion from one digital code to another. For example, BCD code is often readily available, but excess 3 code may be required. In Table 13.9 the available BCD code and the excess 3 code to be derived from it are shown side by side. The code converter (Fig. 13.14) will contain four sets of logic gates whose inputs will be A, B, C and D. Each set of gates will have a single output, one for each of the excess 3 code variables P, Q, R and S.

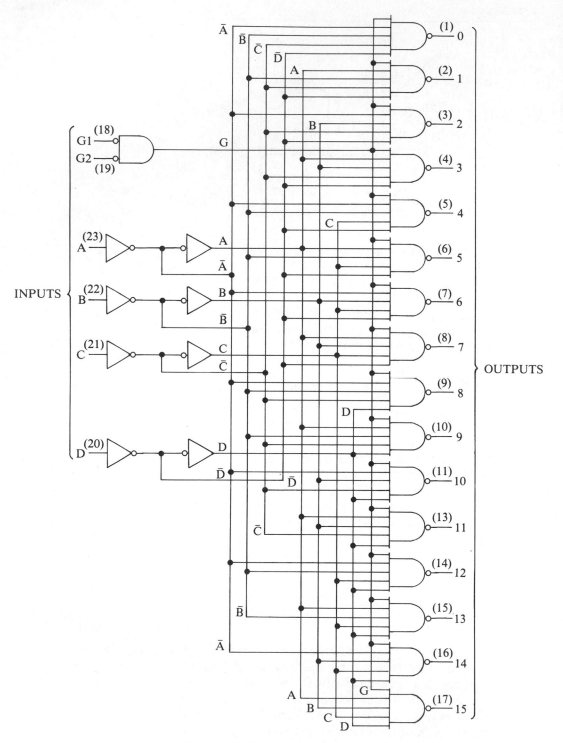

Fig. 13.13   DM 74154 4 to 16 Line Decoder (Courtesy National Semiconductors Ltd)

Table 13.9   BCD Code and Excess 3 Code to be Derived From it

	BCD				Excess 3		
A	*B*	*C*	*D*	*P*	*Q*	*R*	*S*
0	0	0	0	0	0	1	1
0	0	0	1	0	1	0	0
0	0	1	0	0	1	0	1
0	0	1	1	0	1	1	0
0	1	0	0	0	1	1	1
0	1	0	1	1	0	0	0
0	1	1	0	1	0	0	1
0	1	1	1	1	0	1	0
1	0	0	0	1	0	1	1
1	0	0	1	1	1	0	0

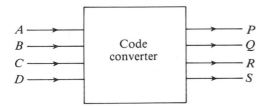

Fig. 13.14   Input and Output Lines for a BCD to Excess 3 Code Converter

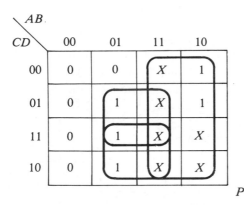

Fig. 13.15   K Map for *P* of BCD to Excess 3 Code Converter

Drawing a K map for the *P* variable (Fig. 13.15) and noting that the input combinations for decimal numbers 10 to 15 cannot occur, then

$$P = A + BC + BD.$$

Q, R and *S* can be derived in a similar manner.

An alternative is to connect the four BCD digits to the inputs of a 4 bit adder and to add 3, i.e. 0011, so that the adder output gives BCD + 3.

13.6   MULTIPLEXER – DATA SELECTOR

A multiplexer (MUX) is a device which accepts data from a number of sources and selects which one of these sources is switched to a single output line. A simple 4 to 1 multiplexer is illustrated in Fig. 13.16. There are four sources of data, D_0 to D_3. There are two select variables, A and B, which are made available in the multiplexer in both uncomplemented and complemented forms. Each data source is connected to a gate and a gate is only enabled for data to pass to the output when the select logic inputs are both 1. Thus for D_2 to be selected $AB = 10$, which enables G_2, and D_2 will appear at the output F. The equation for F is

$$F = \bar{A}\bar{B}D_0 + \bar{A}BD_1 + A\bar{B}D_2 + ABD_3$$

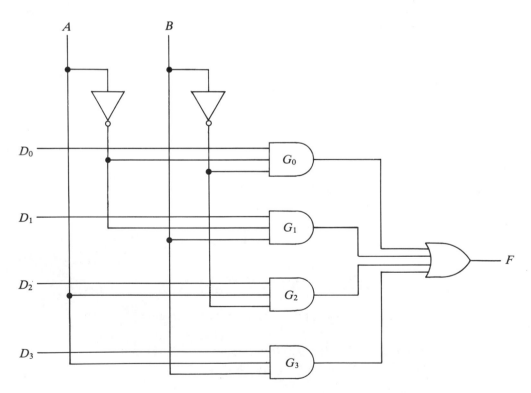

Fig. 13.16   A 4 to 1 Multiplexer

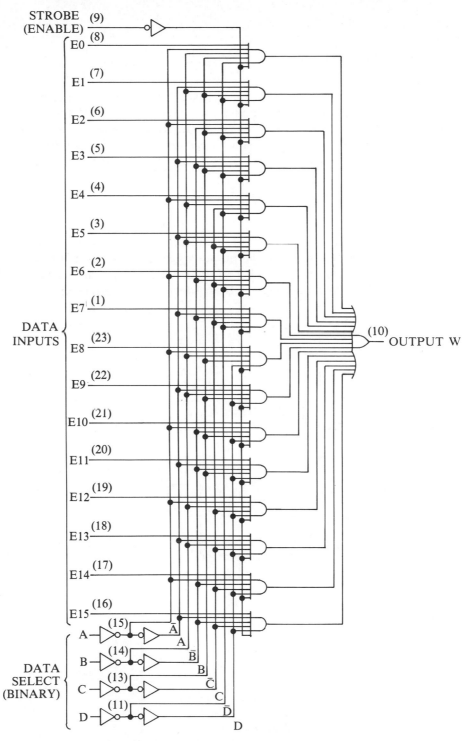

Fig. 13.17 16 to 1 Multiplexer DM 74150 (Courtesy National Semiconductors Ltd)

Parallel to Serial Conversion

A multiplexer such as the 4 to 1 device described can be used to sample its four data inputs sequentially and send data out at F in batches of four samples. For example, suppose that the data on lines D_0 to D_3 remain constant for T seconds. Then, if within that period of T seconds, the address at A and B is stepped through the sequence 00 01 10 11, the data on each of the four input lines would be sampled and fed out serially at F.

DM 74150

The DM 74150 (Fig. 13.17) is a 16 to 1 multiplexer with a strobe (enable) line which must be at 0 for the device to be enabled: this facility is provided to ensure that the multiplexer only produces an output when the input data is stable.

Expansion of the number of data sources to be multiplexed can be achieved with a layout such as Fig. 13.18 in which 64 lines are accepted into four 16 to 1

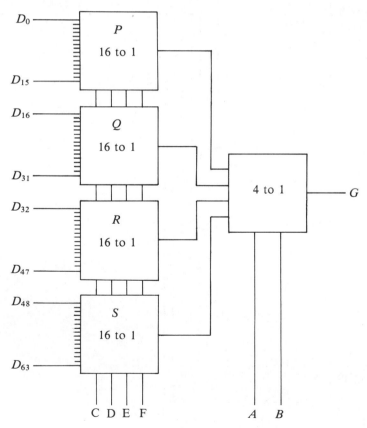

Fig. 13.18   An Arrangement for Multiplexing 64 to 1

multiplexers, and whose outputs are fed into a 4 to 1 multiplexer. If, for example, $CDEF = 0011$ then inputs D_3, D_{19}, D_{35} and D_{51} are all fed to the output multiplexer. If D_{19} is the data source required at G, then making $AB = 01$ will select only the output from multiplexer Q as required.

13.7 DEMULTIPLEXER

A demultiplexer (DEMUX) is a circuit which takes as inputs (a) a single source of data, and (b) select lines to which a binary code will be connected. The code is then used to select on which one of a number of output lines the data is to be sent. One of the simplest of demultiplexers is the 1 to 4 line arrangement of Fig. 13.19. The values of A and B will determine on which of the output lines L_0 to L_3 the data will be sent. If direct binary code is used for line selection then, for example, when $AB = 00$ data will be sent on L_0.

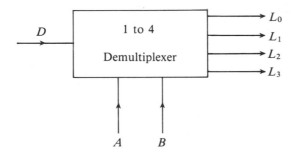

Fig. 13.19   Input and Output Lines of a 1 to 4 Demultiplexer

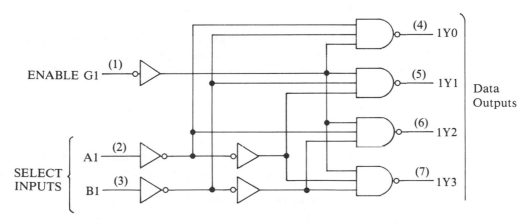

Fig. 13.20   2 to 4 Line Decoder/1 to 4 Line Demultiplexer (Half of DM 74139) (Courtesy National Semiconductors Ltd)

Conceptually a demultiplexer is similar to a decoder and many ICs are designed to perform both of these functions. A simple 2 to 4 line decoder/1 to 4 line demultiplexer is the circuit of Fig. 13.20 which is half of the dual IC package DM 74139. When used as a 2 to 4 line decoder the inputs are A_1 and B_1 and the enable line G_1 must be at 0 for the output gates to be enabled. When used as a 1 to 4 line demultiplexer, data is fed at the enable input G_1 and fed out on the line selected by the values of the logic combination A_1B_1. In a similar manner, the DM 74154 (Fig. 13.13) can be used as a 1 to 16 line demultiplexer as well as a 4 to 16 line decoder (Section 13.3) by using one of the enable inputs G_1 or G_2 as the data input.

SUMMARY

The circuits described in this chapter are usually subsystems which are likely to be encountered in many digital systems. As ICs they will be in the MSI category and will often be available in a range of complexity and facility, and in different logic families. The range of MSI devices given is not exhaustive, but is representative of those in common use.

REVISION QUESTIONS

13.1   Repeat Table 13.3 for adding the decimal numbers $A = 6, B = 7$.

13.2   Write down the following decimal numbers in 8 bit, 2's complement form:
1, $-1, 127, -127, 74, -74$.

(0000 0001; 1111 1111; 0111 1111; 1000 0001; 0100 1010; 1011 0110)

13.3   Carry out the following arithmetic using 8 bit, 2's complement numbers. In each case check your answer against the result obtained from working in decimals.

$74 - 29$
$29 - 74$
$-74 - 29$

13.4   Enter the binary numbers for $A = 15, B = 14$ to the inputs of the four-stage ripple carry adder of Fig. 13.4 and show the sums and carrys at each stage.

13.5   The 8 bit adder/subtractor of Fig. 13.5 has

(a) $A = \$3C$ $\quad$ $B = \$2D$

(b) $A = \$1E$ $\quad$ $B = \$AA$

In each case determine the sum as a hex number for SUB $= 0$ and SUB $= 1$.

($69; $0F; $C8; $74)

13.6 Derive the minimal logic functions for F_3 and F_9 for the excess 3 decoder described in Section 13.4.

$(BC\bar{D}; AB)$

13.7 Derive the minimal logic function for segment g of the 7 segment display described in Section 13.5.

$(A + C\bar{D} + B \oplus C)$

13.8 Derive Q for the BCD to excess 3 decoder of Section 13.5.

$(\bar{B}D + B\bar{C}\bar{D} + \bar{B}C)$

13.9 Derive the logic functions required to convert 3 bit binary code (ABC) into 3 bit Gray code (PQR) of Table 13.5.

$(P = A; Q = A \oplus B; R = B \oplus C)$

13.10 Draw a diagram showing the layout required to implement a 32 to 1 multiplexer using only 8 to 1 multiplexers.

13.11 Using only DM 74150 multiplexer ICs (Fig. 13.17), show how to implement a 256 to 1 multiplexer. State the input logic values to select inputs 28 and 228.

13.12 The input waveforms for the DM 74139 decoder/multiplexer (Fig. 13.20) are shown in Fig. Q.13.12. Sketch 1Y0, 1Y1, 1Y2 and 1Y3.

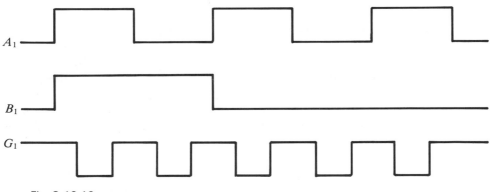

Fig. Q.13.12

14

Flip Flops, Counters and Registers

Chapter Objectives
14.1   Latches and Flip Flops
14.2   SR Flip Flop
14.3   JK Flip Flop
14.4   Master–Slave JK Flip Flop
14.5   Asynchronous Counters
14.6   Synchronour Counters
14.7   Applications of Counters
14.8   Shift Registers
        Summary
        Revision Questions

Astable and monostable multivibrators were described in Chapter 11. The bistable multivibrator – better known as the flip flop (FF) – is a basic component of digital systems and as such it is more appropriate to deal with it in this part of the book than as an aspect of waveform generation. A flip flop is the most common electronic circuit for storing a single binary digit (one bit) and is therefore an important aspect of memory. Storage allows us to progress from combinational circuits, in which the output is totally dependent on the values of the present inputs, to sequential circuits whose outputs depend, to some extent, on memorizing previous input combinations as well as those existing at present.

One of the more common logic families, which will be described in Chapter 16, is TTL. Its basic circuit, from which more complex configurations such as flip flops are derived, is a positive NAND gate, and flip flop circuits in this chapter will be described exclusively in terms of arrangements of NAND gates. It should be noted that flip flops can equally well be made up of NOR gates. However, the main emphasis of the chapter is on the operation of flip flops as digital elements, and two important applications – counters and registers – make up the second part of the chapter.

14.1 LATCHES AND FLIP FLOPS

Latches are bistable circuits which respond to changes of input logic levels as they occur, whereas flip flops will only respond to inputs when they are enabled by a clock pulse. However, in many respects, the circuits are similar, and frequently a latch is regarded as a type of flip flop. It helps to understand the operation of these circuits if we keep in mind two basic facts about a 2 input NAND gate (Fig. 14.1):

(a) If one of the inputs (say A) is 0 then the output $Q = 1$ regardless of the value of B: the gate is said to be disabled by the 0 on A.
(b) If one of the inputs (say A) is 1 then F is determined by B and in fact, $Q = \bar{B}$. The gate is said to be enabled by the 1 on A.

SR Latch

In the *SR* latch of Fig. 14.2 note that the output of gate A is fed back to the input of gate B and vice versa. Feedback is a characteristic feature of electronic memory circuits. In Fig. 14.2 let $SR = 00$ and assume $Q = 0$. Then $\bar{Q} = 1$ and therefore both inputs to gate A are 1 so $Q = 0$ is confirmed and the circuit is stable. There are two points to note here:

(a) Had it been assumed that $Q = 1$ then $\bar{Q} = 0$, i.e. the outputs are complementary and their labeling appropriate.
(b) That with $SR = 00$ the circuit is storing one bit of information, i.e. it is storing the fact that Q was either 0 or 1 before SR became 00.

Fig. 14.1   NAND Gate

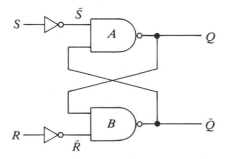

Fig. 14.2   *SR* Latch

If Q is 0 and only S changes to 1, the $\bar{S}$ input becomes 0 and Q goes to 1. Both inputs to gate B are now 1 so $\bar{Q}$ goes to 0. Equally, it is evident from the symmetry of the circuit that with $SR = 01$ then Q will go to 0 and $\bar{Q}$ to 1. It is said that,

making $SR = 10$ Sets the latch to $Q = 1$ and,
making $SR = 01$ Resets the latch to $Q = 0$.

Forbidden Condition $SR = 11$

When $SR = 11$, $\bar{S}\bar{R} = 00$ and $Q = \bar{Q} = 1$. If both S and R then go to 0 simultaneously both outputs start to fall, but in so doing counteract the fall in one another because of the cross-coupling. In this event there is said to be a *race* which ends by the individual gate characteristics determining which output becomes 0 and which becomes 1. This uncertainty is unsatisfactory in digital design because it is important to know, prior to the $SR = 00$ condition, whether the latch was set or reset, i.e. whether the latch is storing a 1 or a 0.

The 74279 of Fig. 14.3 is a quadruple SR latch which does not have inverters on the inputs which are therefore $\bar{S}$ and $\bar{R}$. However, the logic remains the same, e.g. if $SR = 10$ the latch will be set to 1, but to do this the inputs must be $\bar{S} = 0$ and $\bar{R} = 1$. Two of the latches have two $\bar{S}$ inputs, either of which can be 0 to set the circuit.

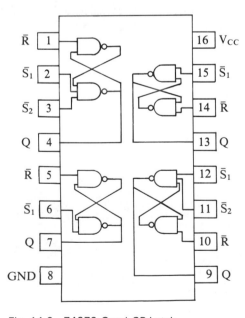

Fig. 14.3   74279 Quad *SR* Latch

14.2 *SR* FLIP FLOP

Logic inputs, such as S and R to the latch of Fig. 14.2, are vulnerable to unpredictable variations such as switching transients in between required changes of input. It is possible that during a change of input logic level there could be a spurious occurrence of the forbidden condition $SR = 11$. In a flip flop these changes are rendered ineffective until the S and R inputs have stabilized at their required levels. In Fig. 14.4 the input Ck (clock) only allows a change of state when Ck = 1. When Ck = 0 the input gates are disabled but when Ck = 1 they are effectively NOT gates and so S and R have the same effect as in the latch of Fig. 14.2. During operation the circuit should be organized so that only when S and R have stabilized will a positive-going clock pulse at Ck allow the circuit to be set or reset as required. Table 14.1 summarizes the specification for an SR flip flop; Q_n is the state of the circuit before a clock pulse and Q_{n+1} is the state of the circuit following a clock pulse. The first line simply shows that whatever the value of Q before the clock pulse, then when $SR = 00$, it will be the same after it.

Figure 14.5a is a symbol for an SR flip flop: the triangle on the Ck input indicates that the flip flop is triggered on the edge of a clock pulse, in this case on the positive-going edge. Negative-edge triggering is indicated by the ring on the Ck input in Fig. 14.5b.

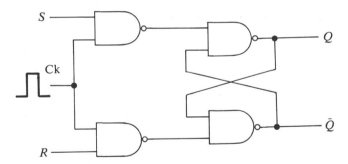

Fig. 14.4   *SR* Flip Flop

Table 14.1   *SR* Flip Flop Truth Table

S	R	Q_{n+1}
0	0	Q_n
0	1	0
1	0	1
1	1	?

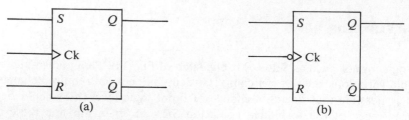

Fig. 14.5   Edge-triggered SR Flip Flop Symbols (a) Triggered on Positive Edge of Clock Pulse (b) Triggered on Negative Edge of Clock Pulse

14.3   *JK* FLIP FLOP

This should be regarded as being similar to an SR flip flop with J and K being equivalent to S and R respectively. The difference is that the input condition $JK = 11$ is allowed and in this condition, when the flip flop is clocked, the output always changes state – it is said to toggle. Table 14.2 is a summary of the specification of a JK flip flop.

Table 14.2   *JK* Flip Flop Truth Table

J	K	Q_{n+1}
0	0	Q_n
0	1	0
1	0	1
1	1	$\bar{Q}_n$

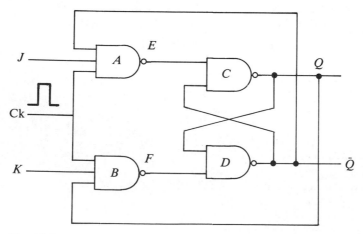

Fig. 14.6   *JK* Flip Flop

A *JK* flip flop is shown in Fig. 14.6 and it differs from an *SR* flip flop in that output Q is fed back to the K gate input and $\bar{Q}$ to the *J* gate input. To understand the operation of the circuit, assume that $Q = 0$, $\bar{Q} = 1$. Then gate B is disabled by $Q = 0$, i.e. $F = 1$. The only way to make the circuit change over is for gate A to be enabled by making $J = 1$ as well as the $\bar{Q}$ input which is already 1. This will be the case irrespective of the value of K. Then when Ck = 1 all inputs to gate A are 1 and E goes to 0, which makes $Q = 1$. With Q and F both 1, $\bar{Q} = 0$ so the flip flop has changed state.

14.4 MASTER–SLAVE *JK* FLIP FLOP

On the face of it there are no uncertainties with the *JK* flip flop; (a) because it is clocked, and (b) because apparently, its output state is predictable for all values of *J* and *K*. However, we have to examine what is known as a *race* associated with the condition $JK = 11$ and a clock pulse of finite duration. In the last paragraph it was stated that with gate B disabled by $Q = 0$, $F = 1$. Now, when the clock pulse occurs, E goes to 0, Q goes to 1, and $\bar{Q}$ to 0. If the clock pulse is maintained, then with A now disabled by $\bar{Q} = 0$ and B enabled by $Q = 1$, the circuit can change back again and indeed will continue to toggle between the two states for the duration of the clock pulse. In other words, for correct operation, it is assumed that the clock pulse is short enough to permit the flip flop to change only once. A race is more likely to occur with modern high-speed ICs although it can be eliminated by introducing delays in the feedback paths between outputs and the *J* and *K* inputs. A better solution to the problem is the master–slave *JK* flip flop of Fig. 14.7.

In this circuit gates *A*, *B*, *C* and *D* form the master flip flop, and *TUVW* form

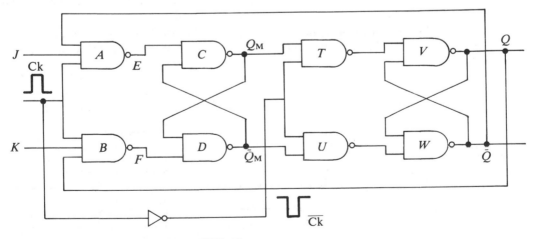

Fig. 14.7 Master–Slave *JK* Flip Flop

the slave. Note that when Ck = 1, enabling the master, $\overline{Ck} = 0$, which disables the slave. Suppose that before the occurrence of a clock pulse, the state of the flip flop is that $Q = 1$, $\bar{Q} = 0$ and $JK = 11$. B is enabled by $Q = 1$ so that when the clock pulse arrives, i.e. Ck = 1, F goes to 0 and $\bar{Q}_M$ to 1. Now, with $E = 1$ and $\bar{Q}_M = 1$, Q_M goes to 0 so the master has been reset. The slave however remains disabled until Ck goes to 0. At this stage we simply note that the slave is essentially an SR flip flop with inputs S and R equal to Q_M and $\bar{Q}_M$ respectively. Thus, when Ck goes to 0, $\overline{Ck}$ goes to 1 and the slave is now reset by its inputs $Q_M\bar{Q}_M$ being 01, i.e. Q goes to 0 and $\bar{Q}$ to 1. But the feedback to J and K cannot cause a race because with Ck = 0 the master is disabled.

Summarizing, the output of the master–slave JK flip flop can be predicted for all combinations of JK and for any duration of clock pulse, and because of this it is the most versatile and universal of flip flops. SR flip flops are also available in master–slave configuration.

☐   **Example 14.1**

Given the Ck, J and K waveforms of Fig. 14.8, draw waveform Q.

Since the initial state of Q is not given, it cannot be drawn in for clock pulse CP1 when $JK = 00$. During CP2, $JK = 10$ and therefore at the trailing edge $Q = 1$. At the trailing edge of CP4, Q goes to 0 with $JK = 01$. At CP10 and 11, $JK = 11$ so the output toggles.   ☐

Preset and Clear Inputs

Further sophistication of an SR or a JK flip flop can be achieved by providing the facility to preset (make $Q = 1$) or clear (make $Q = 0$) the circuit before the occurrence of a clock pulse. For the master–slave JK flip flop of Fig. 14.7 preset and clear facilities are provided with an extra input to each of the gates C and D as shown in Fig. 14.9. Note that the inputs are labeled in complementary form, i.e. $\bar{P}_r$ and $\bar{C}_r$, and this signifies that making $\bar{P}_r = \bar{C}_r = 1$ does not disable gates C and D and therefore has no effect on the circuit. With $\bar{P}_r = 0$ the output of gate C, i.e. Q_M, will

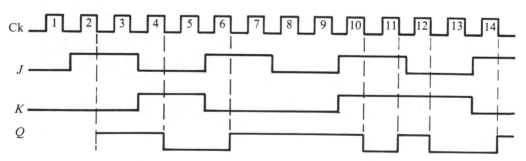

Fig. 14.8   Waveforms for Example 14.1

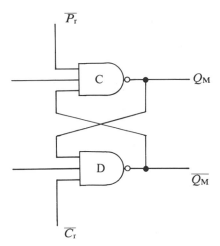

Fig. 14.9   Preset and Clear Inputs Added to Gates C and D of Fig. 14.7

be 1 and with $Ck = \bar{0}$ (i.e. *before* the occurrence of a clock pulse) Q, in Fig. 14.7, will be set. Likewise $\bar{C}_r = 0$ resets the flip flop. Terminologically, inputs with bars such as $\bar{P}_r$ and $\bar{C}_r$ are said to be *active low*.

Commercially Available Master–Slave *JK* Flip Flops

Figure 14.10 shows the pin connection of a DM 7476 dual master–slave *JK* flip flop and also the logic symbol for one of the flip flops. The Ck input is shown as negative edge triggering because any change in output Q will be coincident with the trailing edge of a clock pulse. These circuits can be clocked at about 25 MHz in the standard TTL form. CMOS versions are also available (74C76) with clock rates much lower at about 4 MHz if operated at 5 V which is necessary for compatibility with TTL. However, this can be increased to 11 MHz by operating at 10 V. CMOS gains against TTL by lower power consumption as a trade-off for its lower clock rate.

T Flip Flop

With a *JK* flip flop we have seen that with $JK = 11$ the output changes on every clock pulse. The change will be coincident with the clock pulse trailing edge and the flip flop is said to toggle, as shown in Fig. 14.11 when $T = 1$.

D Flip Flop

These flip flops store one bit which will be the logical value of D in Fig. 14.12. Clearly $K = \bar{J}$ so the *JK* inputs will always be 10 or 01. If $D = 1$ then $JK = 10$ and when the flip flop is clocked $Q = 1$. The opposite will be the case if $D = 0$. Therefore,

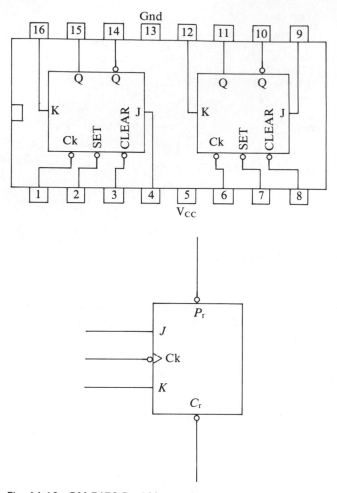

Fig. 14.10   DM 7476 Dual Master–Slave *JK* Flip Flop with Preset and Clear

whatever occurs at D will be stored at Q when the flip flop is clocked and will remain, even if D changes, until the next clock pulse.

14.5 ASYNCHRONOUS COUNTERS

A *JK* flip flop connected to toggle ($JK = 11$) makes one transition per clock pulse (CP) as shown in Fig. 14.13. In essence the output Q is counting negative edges of clock pulses but since the flip flop has only two states then Q is simply counting from 0 to 1 and back to 0. In its toggle mode the flip flop is sometimes called a scale of

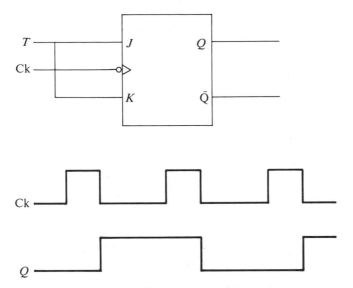

Fig. 14.11 *JK* Flip Flop Connected as a *T* Flip Flop

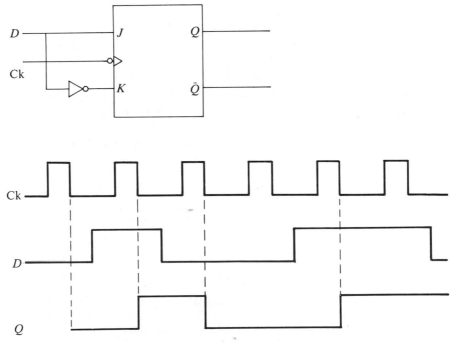

Fig. 14.12 *D* Flip Flop

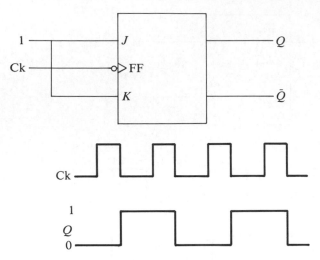

Fig. 14.13   A *JK* Flip Flop Connected to Toggle Counts 0-1-0-1 at Q

2 counter. It is also said to divide by 2 because it produces one negative edge at the output for every two negative edges of clock pulse input.

Ripple Counters

By connecting N toggling flip flops in cascade (three are shown in Fig. 14.14), then there are 2^N (8) combinations of output state and the circuit has the ability to count from 0 to $2^N - 1$ (7). Figure 14.14 is a scale of 8 counter. Note that the clock pulses only trigger FF_0, the following flip flops being triggered by the negative edges from the previous flip flop Q outputs. Beginning, arbitrarily, with all flip flops reset, Fig. 14.15 shows the complete sequence of waveforms for the three outputs Q_2, Q_1 and Q_0. Taking the initial state of the circuit as $Q_2Q_1Q_0 = 000$ it will be observed that the state of the circuit corresponds to the number of negative clock pulse edges

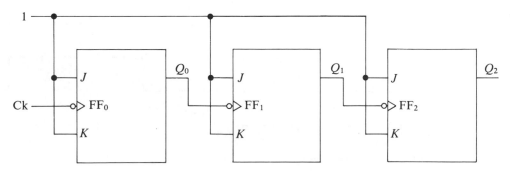

Fig. 14.14   Scale of 8 (3-bit) Ripple Counter

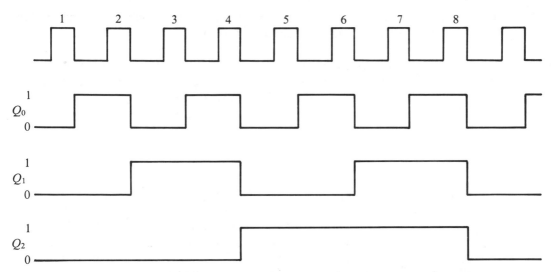

Fig. 14.15 3-bit Counter Waveforms Counting through the 8 States 000 (0) to 111 (7)

which have occurred at the input, up to $Q_2Q_1Q_0 = 111$, i.e. 7. Thus, after CP6, $Q_2Q_1Q_0 = 110$. Note that by numbering the flip flops so that the first one in the chain is FF$_0$, the next FF$_1$, and so forth, we achieve a systematic system of numbering the state of the circuit beginning with Q_0 as the LSB.

When CP8 occurs, Q_0 resets Q_1 and then Q_1 resets Q_2: the changes are said to 'ripple' through the counter. This cumulative delay limits the operating frequency.

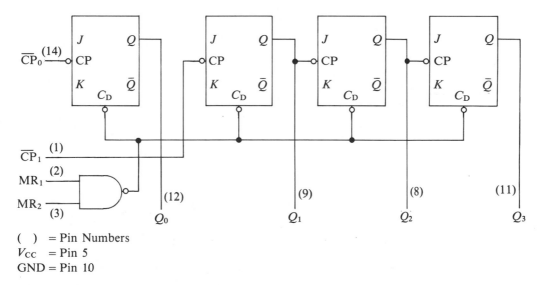

( ) = Pin Numbers
V_{CC} = Pin 5
GND = Pin 10

Fig. 14.16 7493 4-bit Binary Counter

Counters of this type are said to be asynchronous because only FF_0 is triggered by clock pulses; the others, being triggered from the previous flip flop in the chain, are subject to delay so the flip flops are not in synchronism.

Figure 14.16 shows the arrangement of a 7493 4-bit binary counter. Separated into two sections, a 1 bit and a 3 bit, the IC can be used as a scale of 2, a scale of 8, or if Q_0 (pin 12) is connected to $\overline{CP_1}$, a scale of 16 counter. With either of the MR (Clear) inputs at 0 the counter operates as described. With both MR inputs at 1 all four flip flops are reset.

Asynchronous Decade Counter

Counting to moduli other than 2^N is a frequent requirement, the most common being to count through the binary coded decimal BCD (8421) sequence. All that is required is a four-stage counter which, having counted from 0000 (0) to 1001 (9), i.e. ten states, resets to 0000 on the next clock pulse. The obvious solution is to allow the counter, on the tenth clock pulse, to go into the state 1010. If we now have a NAND gate with inputs $Q_3\bar{Q}_2Q_1\bar{Q}_0$, it will give an output 0 when the counter reaches 1010 and this can be used to clear the flip flops as shown in Fig. 14.17. The waveforms are shown in Fig. 14.18 in which the unwanted but brief occurrence of the state 1010 should be noted. A simpler solution is to recognize that the first time in the sequence that Q_3 and Q_1 both become 1 together is when the count reaches 1010, so the clearing logic can simply be $C_r = \overline{Q_3Q_1}$. However, the waveforms will remain as in Fig. 14.18. Using the 7493 referred to in the previous paragraph, a decade counter could be constructed by connecting MR_{01} and MR_{02} to Q_3 and Q_1.

For counting to any modulus other than 2^N, the principles are:

(a) Use N flip flops where 2^N is the next highest value of 2^N above the modulus

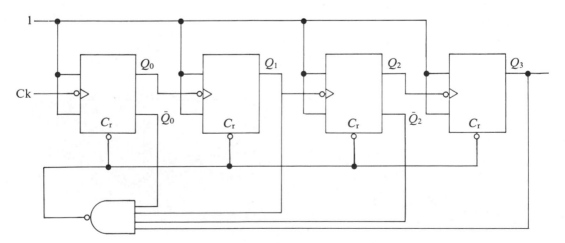

Fig. 14.17   Asynchronous Decade Counter

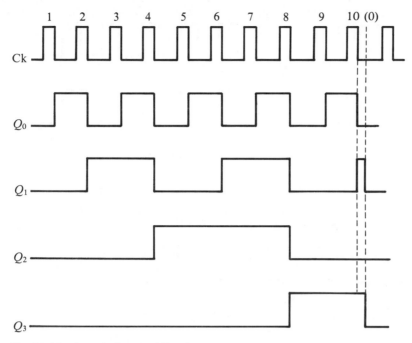

Fig. 14.18   Decade Counter Waveforms

required, e.g. for a scale of 27 counter use 5 flip flops because $2^5 = 32$ is the next 2^N value above 27.

(b) Use gating to detect when the counter has reached 27 to reset the counter. In this way the counter will count from 0 to 26 and reset to 0 on the 27th clock pulse.

Up-Down (Reversible) Counters

The waveforms of Fig. 14.19 represent the sequence required to count from 111 (7) down to 000 (0). Using three flip flops for this purpose note that:

(a) Q_o toggles and should therefore be driven by the clock as with the up counter.

(b) In contrast to the up counter, Q_1 and Q_2 change when the previous flip flop output is positive-going.

Using negative-edge triggered flip flops, the latter condition can be met by connecting FF_1 and FF_2 inputs to the complements of the previous stages, as shown in Fig. 14.20. Then, for example, at the negative edge of CP5, Q_o goes positive. But $\bar{Q}_o$ will be negative-going and trigger FF_1 as required. Figure 14.21 is a reversible counter, i.e. it can count up or down. When $P = 1$, AND gates A and B are enabled so FF_1 and FF_2 will be triggered by Q_0 and Q_1 and the count will be up. With $P = 0$, C and D are enabled and the count will be down.

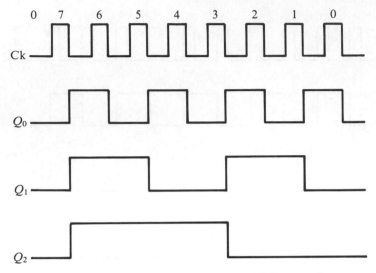

Fig. 14.19   Waveforms for Counting from 111 (7) Down to 0

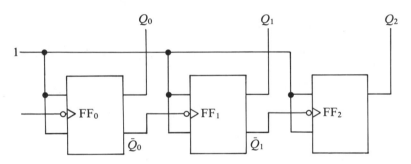

Fig. 14.20   Asynchronous 7 down to 0 Counter

14.6 SYNCHRONOUS COUNTERS

The 'ripple through' effect in asynchronous counters limits their operating frequencies. Further, because all Q output changes do not occur simultaneously (also as a consequence of the ripple-through effect) then a circuit decoding the counter output will be subject to momentary unwanted combinations which will result in decoding spikes. In a synchronous counter, logic at the JK inputs of flip flops determines whether or not a transition will occur, but when they do occur they will be triggered by the clock and will therefore be in synchronism with another.

Figure 14.22 is a 3 bit scale of 8 counter. Flip flops used in synchronous counter ICs are usually triggered by the positive edges of the clock pulse inputs, so Fig. 14.23 illustrates the waveforms required for the count to progress from 000 to 111. Q_0

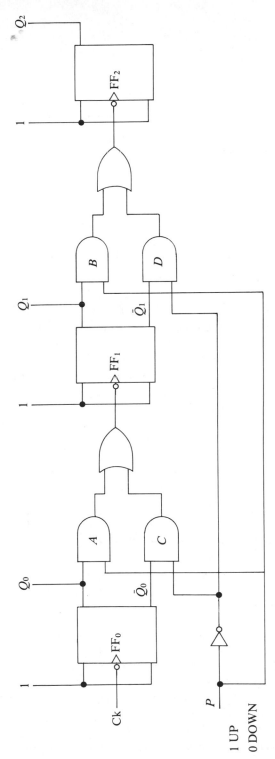

Fig. 14.21   Reversible Counter

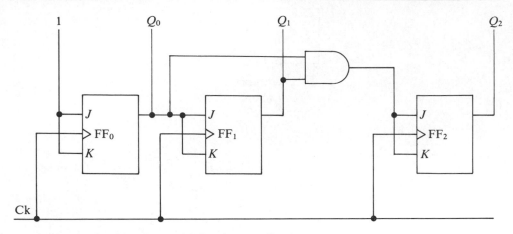

Fig. 14.22   3-bit, Scale of 8 Synchronous Counter

must change over at every clock pulse so FF_0 must be made to toggle: this it will do with $J_0 = K_0 = 1$. For the other two flip flops it is important to note that since the positive edge of a clock pulse initiates any transition, then if we are going to use flip flop outputs to provide the logic variables which will determine whether or not a clock pulse triggers a flip flop, we must look at the flip flop outputs *before* each clock pulse positive edge. For example, observing Fig. 14.23, Q_1 must always change if, before a clock pulse, $Q_0 = 1$, i.e. *before* CP2, CP4, CP6 and CP8. Therefore, by making $J_1 = K_1 = Q_0$, then if $Q_0 = 1$ before a clock pulse FF_1 will change over as required. Similarly, FF2 must change over at clock pulses following $Q_0Q_1 = 11$ (CP4 and CP8). Therefore $J_2 = K_2 = Q_0Q_1$.

Clearly, for a straightforward 0 to $2^N - 1$ binary counter the principle can be

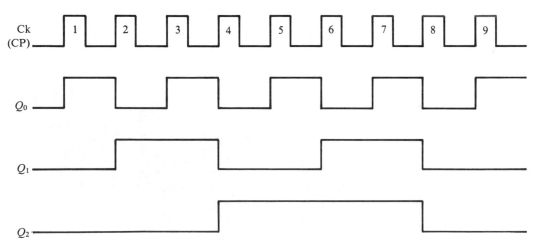

Fig. 14.23   Waveforms for Scale of 8 Synchronous Counter with Positive Edge Triggering

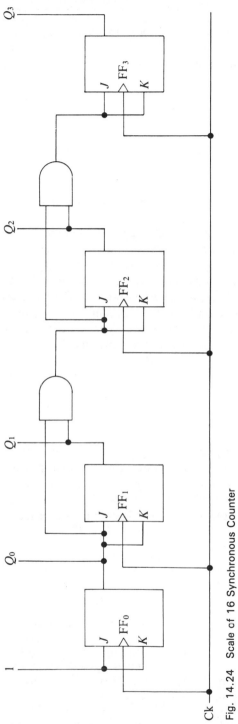

Fig. 14.24 Scale of 16 Synchronous Counter

extended and for the final stage of a 4 bit scale of 16 counter, $J_3 = K_3 = Q_0 Q_1 Q_2$. As the number of stages is extended, the gating becomes increasingly complex but it can be simplified if we note that

since $J_1 = K_1 = Q_0$

then $J_2 = K_2 = Q_0 Q_1 = J_1 Q_1$

and $J_3 = K_3 = Q_0 Q_1 Q_2 = J_2 Q_2$

A scale of 16 synchronous counter is shown in Fig. 14.24.

14.6.1 Systematic Design of Synchronous Counters

The method of determining the JK logic for a synchronous counter, just described, by observing the transitions in the required waveform chart is unsystematic and requires a great deal of intuition for the design of any counter other than a straight-forward 0 to $2^N - 1$ binary sequence. Two systematic methods for determining synchronous counter logic will now be described. The first involves writing down the value of J and K, i.e. 0 or 1, at each counter state in order to obtain the next state required. The second, which is a more elegant method of achieving the same end, requires the derivation of a Boolean expression called the *change function*. The first method will be explained by example.

☐   **Example 14.2**

A counter is required to count from 000 (0) through to 110 (6) before resetting to 000.

Clearly three flip flops are required. The seven states of the counter and the values of J and K during those states are set down in Table 14.3. For any row of the table the JK values must be those needed for the counter to move into the next required state when the flip flops are clocked. At the first row (decimal 0) J_2 and K_2 must be such that Q_2 remains at 0 when FF_2 is clocked into the next state (001). Therefore, J_2 must be 0 – otherwise FF_2 will be set – but the value of K_2 is

Table 14.3 J and K values for a scale of 7 counter

Decimal	Q_2	Q_1	Q_0	J_2	K_2	J_1	K_1	J_0	K_0
0	0	0	0	0	01	0	01	1	01
1	0	0	1	0	01	1	01	01	1
2	0	1	0	0	01	01	0	1	01
3	0	1	1	1	01	01	1	01	1
4	1	0	0	01	0	0	01	1	01
5	1	0	1	01	0	1	01	01	1
6	1	1	0	01	1	01	1	0	01

immaterial because whether it is 0 or 1, Q_2 will remain at 0. K_2 is therefore shown as 01. The same applies to FF_1 which must also remain reset, i.e. $J_1 = 0$, $K_1 = 01$. For FF_0, Q_0 must change from 0 to 1 when the flip flop is clocked, so J_0 must be 1. The value of K_0 is immaterial because if $K_0 = 0$, then with $J_0 = 1$, FF_0 will be set, and if $K_0 = 1$ as well as J_0 being 1, FF_0 will toggle and therefore Q_0 will change from 0 to 1 as required. Where a flip flop must remain set, i.e. $Q = 1$ (e.g. Q_1 at decimal 2), then K must be 0 to prevent the flip flop being reset but the value of J is immaterial. Where a flip flop must be reset so that Q goes from 1 to 0 (e.g. Q_0 at decimal 1), K must be 1 but the value of J is immaterial: with $JK = 01$ the flip flop is reset; with $JK = 11$ the flip flop will toggle taking Q from 1 to 0 as required. Summarizing the rules for entering 0, 1 or 01 in the JK columns for Table 14.3:

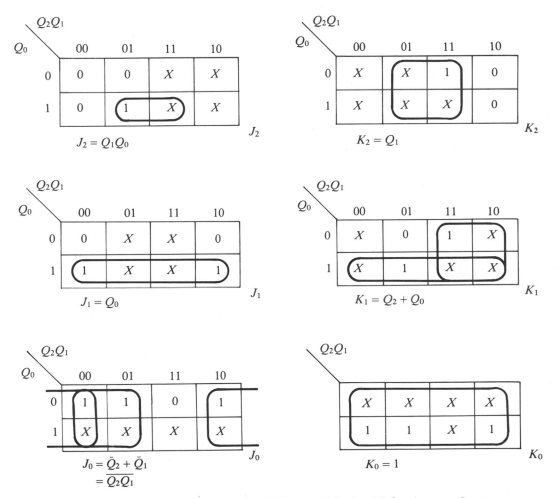

Fig. 14.25   Karnaugh Maps for J and K Inputs of Scale of 7 Synchronous Counter

(a) When $Q = 0$ and is to remain at 0 it is only essential for $J = 0$ to prevent the flip flop from being set. Therefore $J = 0$, $K = 01$.

(b) When $Q = 0$ and is to become 1 the only essential condition is for $J = 1$ to ensure that the flip flop will be set when clocked. Therefore $J = 1$, $K = 01$.

(c) When $Q = 1$ and is to remain at 1 it is only essential that $K = 0$ to prevent the flip flop from being reset. Therefore $J = 01$, $K = 0$.

(d) When $Q = 1$ and is to become 0, it is only essential that $K = 1$ to ensure that the flip flop is reset. Therefore $J = 01$, $K = 1$.

Having completed Table 14.3 we can now draw six Karnaugh maps (Fig. 14.25), one for each value of J and of K. For each map the input variables are Q_2, Q_1 and Q_0, the flip flop outputs. Whenever an output is 01, then X is entered on the K map because it is a 'doesn't matter' state. In this case the state $Q_2Q_1Q_0 = 111$ is also entered as an X on all six maps because it 'can't happen'. From the K maps, the following logic can be deduced for the flip flop inputs:

$$J_2 = Q_1Q_0; \; K_2 = Q_1$$
$$J_1 = Q_0; \; K_1 = Q_2 + Q_0$$
$$J_0 = \bar{Q}_2 + \bar{Q}_1 = \overline{Q_1Q_2}; \; K_0 = 1$$

The resulting scale of 7 counter is drawn in Fig. 14.26.

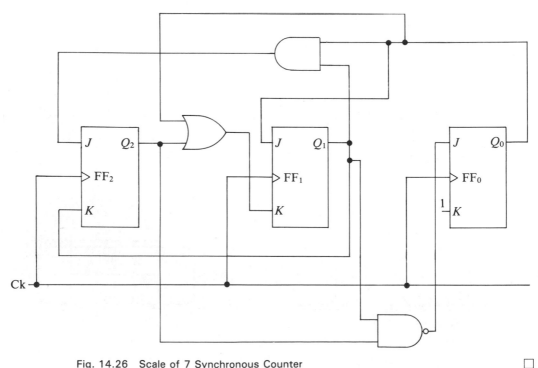

Fig. 14.26   Scale of 7 Synchronous Counter

Table 14.4 Truth Table for the Change Function of a *JK* Flip Flop

J	K	Q	CF
0	0	0	0
0	0	1	0
0	1	0	0
0	1	1	1
1	0	0	1
1	0	1	0
1	1	0	1
1	1	1	1

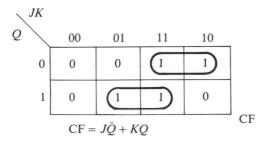

$$CF = J\bar{Q} + KQ$$

Fig. 14.27 K Map for the Change Function of a *JK* Flip Flop

14.6.2 Change Function Method of Designing Synchronous Counters

For a *JK* flip flop we can draw a truth table which shows, for every possible combination of *J*, *K* and *Q*, whether or not *Q* will change when the flip flop is clocked. For example, when $JKQ = 010$, *Q* will remain at 0 when the flip flop is clocked but when $JKQ = 011$ then with $K = 1$ the flip flop will reset so that *Q* changes from 1 to 0. Table 14.4 is the truth table in which the column CF is the change function: CF will be 1 for any combination of *JKQ* which will result in a change of *Q*.

Entering these conditions on a Karnaugh map (Fig. 14.27) for the change function gives us that

$$CF = J\bar{Q} + KQ \tag{14.1}$$

Note that the coefficient of $\bar{Q}$ is *J* and that of *Q* is *K*. Now let us see how the change function is used by once again designing a counter.

☐ **Example 14.3**

Repeat the problem of Example 14.2 which was to design a scale of 7 counter but this time using the change function.

First we draw up the truth table (Table 14.5) this time entering the change function for each flip flop. For example, for FF_2, $CF2 = 1$ at 011 because Q_2 must change from 0 to 1 at the next clock pulse. Similarly at 110, $CF2 = 1$ because Q_2 must change from 1 to 0.

The next step is to draw a K map for each change function in turn beginning with CF2 (Fig. 14.28a). However, instead of writing down the Boolean expression

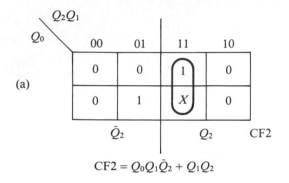

$$CF2 = Q_0 Q_1 \bar{Q}_2 + Q_1 Q_2$$

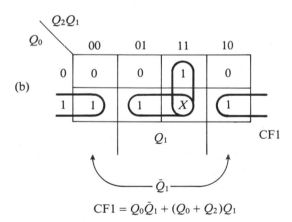

$$CF1 = Q_0 \bar{Q}_1 + (Q_0 + Q_2) Q_1$$

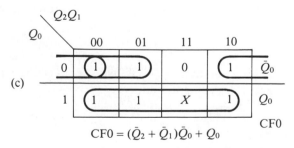

$$CF0 = (\bar{Q}_2 + \bar{Q}_1)\bar{Q}_0 + Q_0$$

Fig. 14.28 K Maps for the Three Change Functions of a Scale of 7 Counter. Note That in Each Case *CF* Must Be Read as Two Separate Expressions One From Each Half of the Map

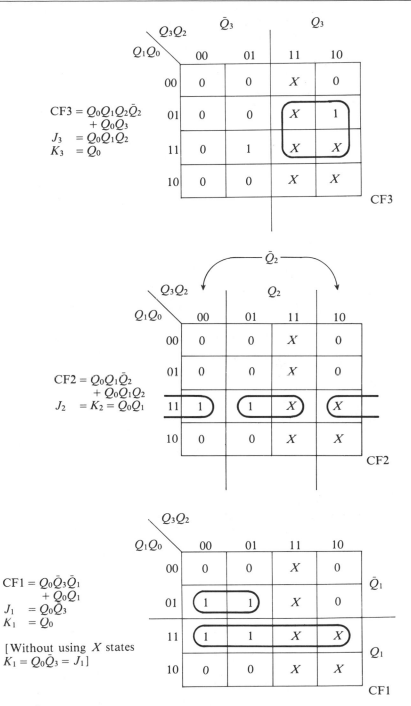

$CF3 = Q_0Q_1Q_2\bar{Q}_2$
$\quad\quad + Q_0Q_3$
$J_3 \quad = Q_0Q_1Q_2$
$K_3 \quad = Q_0$

$CF2 = Q_0Q_1\bar{Q}_2$
$\quad\quad + Q_0Q_1Q_2$
$J_2 \quad = K_2 = Q_0Q_1$

$CF1 = Q_0\bar{Q}_3\bar{Q}_1$
$\quad\quad + Q_0Q_1$
$J_1 \quad = Q_0\bar{Q}_3$
$K_1 \quad = Q_0$

[Without using X states
$K_1 = Q_0\bar{Q}_3 = J_1$]

Fig. 14.29   K Maps for Change Functions of a Decade Counter

Table 14.5 Truth Table for Scale of 7 Counter Showing Change Function for each Flip Flop

FF Outputs			FF Change Functions		
Q_2	Q_1	Q_0	$CF2$	$CF1$	$CF0$
0	0	0	0	0	1
0	0	1	0	1	1
0	1	0	0	0	1
0	1	1	1	1	1
1	0	0	0	0	1
1	0	1	0	1	1
1	1	0	1	1	0
0	0	0			

direct from the K map it is first divided into the two halves $\bar{Q}_2$ and Q_2 as shown. Reading CF2 for each half of the map, we have

$$CF2 = Q_0 Q_1 \bar{Q}_2 + Q_1 Q_2 \tag{14.2}$$

Now the change function for FF$_2$ will also be, using equation (14.1),

$$CF2 = J_2 \bar{Q}_2 + K_2 Q_2 \tag{14.3}$$

Comparing the coefficients of $\bar{Q}_2$ and Q_2 in equations (14.2) and (14.3) gives

$$J_2 = Q_0 Q_1 \quad \text{and} \quad K_2 = Q_1$$

which is the same result as derived using the previous method. Repeating the exercise for Q_1, the K map is shown in Fig. 14.28b, but is now divided into $\bar{Q}_1$ and Q_1. This

Table 14.6 Decade Counter Change Functions

Q_3	Q_2	Q_1	Q_0	$CF3$	$CF2$	$CF1$	$CF0$
0	0	0	0	0	0	0	1
0	0	0	1	0	0	1	1
0	0	1	0	0	0	0	1
0	0	1	1	0	1	1	1
0	1	0	0	0	0	0	1
0	1	0	1	0	0	1	1
0	1	1	0	0	0	0	1
0	1	1	1	1	1	1	1
1	0	0	0	0	0	0	1
1	0	0	1	1	0	0	1
0	0	0	0				

gives

$$CF1 = Q_0\bar{Q}_1 + Q_0Q_1 + Q_2Q_1$$
$$= Q_0\bar{Q}_1 + (Q_0 + Q_2)Q_1$$

from which

$$J_1 = Q_0 \quad \text{and} \quad K_1 = (Q_0 + Q_2)$$

Finally, drawing the K map for CF0 (Fig. 14.28c) and dividing it into $\bar{Q}_0$ and Q_0,

$$CF0 = \bar{Q}_2\bar{Q}_0 + \bar{Q}_1\bar{Q}_0 + Q_0$$
$$= (\bar{Q}_2 + \bar{Q}_1)\bar{Q}_0 + Q_0$$

from which,

$$J_0 = \overline{Q_2Q_1} \quad \text{and} \quad K_0 = 1 \qquad \qquad \square$$

The technique can readily be extended to a four-bit counter.

Table 14.6 is the truth table for a decade counter and Fig. 14.29 gives the K maps and logic required for the decade counter. Clearly since CF0 is 1 for every row in the table it is simply required to toggle so $J_0 = K_0 = 1$.

Clocking a Counter out of an Unwanted State

When a counter is powered it is possible that, if not all of the 2^N states of the N flip flops are required, then the circuit could be in an unwanted state when it is powered. We must therefore be certain that when the counter is clocked it will go to a required state and take up the sequence for which it was designed. Returning to the decade counter K maps of Fig. 14.29, note that when an unwanted state – marked X – is included in a prime implicant loop it means that the change function will be 1 if the circuit is in this state. For example, if the circuit were to go to state 1101 it can be seen on the K map for CF3, that CF3 = 1 in this state so FF_3 will change from 1 to 0. Also, because FF_0 toggles, it always changes. The counter therefore goes from 1101 to 0100 which is in the required sequence. For state 1110, only FF_0 changes so the counter goes to 1111 which is also not required. However, this state is in the prime implicant loops of all flip flops, so the next state is 0000, which is required. Where a design is such that an unwanted state does not lead to a required state logic must be added so that the change function includes the unwanted state. The fact that X states are included in change functions makes no difference to the counter once it is in its required sequence.

14.7 APPLICATIONS OF COUNTERS

Event Counters

Counters can be used to count physical events. For example, suppose items passing along a production line are to be counted into batches of ten. Then it can be arranged that each time an item passes a particular point it interrupts a beam of light passing into a photodetector and in so doing generates a pulse. The pulses can be used to clock a decade counter which gives one output pulse for every ten clock pulses.

Timers

The accuracy of a digital timer such as a watch depends on a crystal oscillator as its basic timing device. A small watch can only house a small oscillator and these will always be high-frequency devices. One common arrangement is to use a 6.5536 MHz crystal oscillator. Sixteen flip flops in cascade divide by $2^{16} = 65\,536$ so the output from a 16 stage binary counter clocked from the oscillator will be 100 Hz. Two decade counters bring this down to one pulse per second and two stages of divide by 60, each made up of a scale of 6 and a decade counter, give one pulse per minute and one pulse per hour. The arrangement is shown in Fig. 14.30.

Sequence Generators

Using the outputs of the stages of a binary counter as inputs to logic gates, sequences of signals of particular time intervals can be developed. The technique can be explained with the aid of an example.

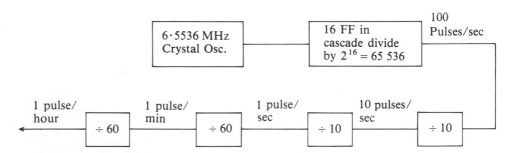

Fig. 14.30   Dividing Chain for Digital Watch

☐   **Example 14.4**

Consider the problem of generating three signals R, Y and G representing traffic lights, the lights being switched on when a signal is 1. Suppose a scale of 12 counter

Table 14.7   Truth Table for Example 14.4

Q_3	Q_2	Q_1	Q_0	R	Y	G
0	0	0	0	1	0	0
0	0	0	1	1	0	0
0	0	1	0	1	0	0
0	0	1	1	1	0	0
0	1	0	0	1	0	0
0	1	0	1	1	1	0
0	1	1	0	0	0	1
0	1	1	1	0	0	1
1	0	0	0	0	0	1
1	0	0	1	0	0	1
1	0	1	0	0	0	1
1	0	1	1	0	1	0

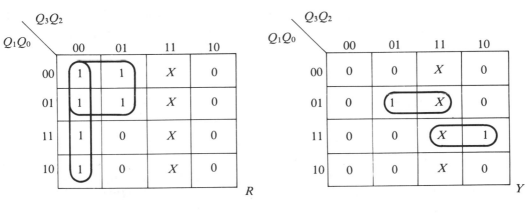

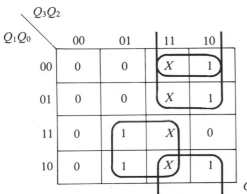

Fig. 14.31   K Maps for Example 14.4

is available, then the 12 states of the counter can be represented as in the first four columns of Table 14.7. If the counter is clocked once every 5 s then it could be used to generate the following sequence:

$R = 1$ 25 s
R and $Y = 1$ 5 s
$G = 1$ 25 s
$Y = 1$ 5 s

To achieve these timings the values of the three variables R, Y and G have been added to Table 14.7 at each stage of the count. For example, $R = 1$ for the first six states, i.e. 25 s for $R = 1$ and 5 s for R and $Y = 1$. To determine the logic required for the full sequence, then K maps have been drawn for each of the three variables in Fig. 14.31. The required logic is

$$R = \bar{Q}_3\bar{Q}_1 + \bar{Q}_3\bar{Q}_2$$
$$Y = Q_2\bar{Q}_1 Q_0 + \bar{Q}_3 Q_1 Q_0$$
$$G = Q_2 Q_1 + Q_3\bar{Q}_1 + Q_3\bar{Q}_0$$

Note that it was an arbitrary decision to begin the sequence at the 0000 state of the counter. Beginning at any other state would result in a different but equally effective design.   ☐

14.8   SHIFT REGISTERS

A register is a memory device consisting of a number of flip flops each of which will store one bit. Registers are often used in digital systems for the temporary storage of binary data. A single bit can be stored on a D flip flop (Fig. 14.12) or on an SR flip flop connected as a D flip flop (Fig. 14.32). When a positive edge occurs at Ck the logic level, 0 or 1, standing at A will appear at Q and its complement at $\bar{Q}$.

14.8.1   Parallel In, Parallel Out (PIPO) Registers

Figure 14.33 is a parallel in, parallel out (PIPO) register. The terminology signifies that the 4 bits of the binary word $ABCD$ are 'read in' to the register simultaneously

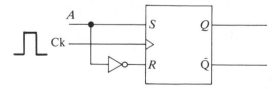

Fig. 14.32   SRFF Connected to Store One Bit

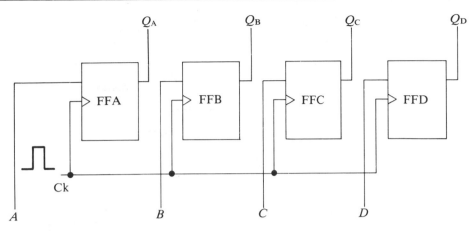

Fig. 14.33   PIPO Register

when Ck goes positive and that the 4 bits of the resulting stored output word Q_A Q_B Q_C Q_D are all available to be 'read out' after the clock pulse. Whatever the input logic levels become in between clock pulses does not affect the output.

14.8.2   Serial In, Serial Out (SISO) Shift Registers

There is a requirement to shift bits stored in a register from one flip flop to another, firstly so that data can be read from a single input line one bit at a time, i.e. serially, then held and 'clocked out' when required, and secondly for arithmetic and logic processing. For example, if the word 0110 is shifted left it becomes 1100 and has therefore been multiplied by 2. A right shift divides by 2. A serial in, serial out (SISO) register is shown in Fig. 14.34. Suppose that initially, all flip flops are reset and the word 1011 occurs serially at A, LSB first. The first 1 at A sets FFA when clocked. However, with master–slave flip flops the 1 will not appear at Q_A until the trailing edge of the clock pulse. Therefore flip flops B, C and D remain reset because Q_A, Q_B and Q_C were 0 when the clock went positive. At the next clock pulse FFA

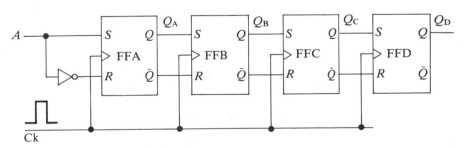

Fig. 14.34   SISO Shift Register

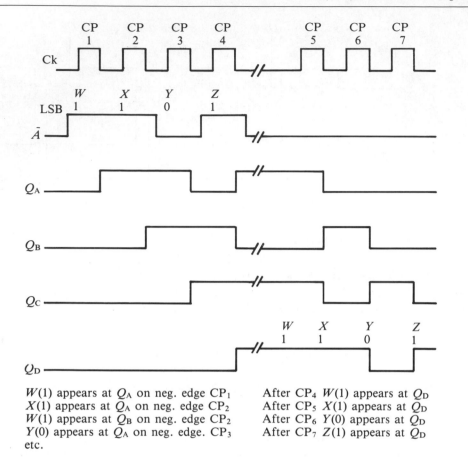

$W(1)$ appears at Q_A on neg. edge CP_1 After CP_4 $W(1)$ appears at Q_D
$X(1)$ appears at Q_A on neg. edge CP_2 After CP_5 $X(1)$ appears at Q_D
$W(1)$ appears at Q_B on neg. edge CP_2 After CP_6 $Y(0)$ appears at Q_D
$Y(0)$ appears at Q_A on neg. edge. CP_3 After CP_7 $Z(1)$ appears at Q_D
etc.

Fig. 14.35   Clocking 1011 Through a 4-bit Shift Register

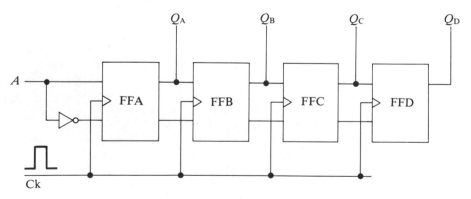

Fig. 14.36   SIPO Shift Register

is kept set by the second 1 of the data input word at A. FFB will be set by the 1 at Q_A while flip flops C and D remain reset. The third bit at A is a 0 which resets FFA but flip flops B and C are set by the 1s on Q_A and Q_B. The MSB, a 1, is clocked in by the fourth clock pulse which also results in the LSB appearing at Q_D. The complete word is now stored in the four flip flops with the LSB at Q_D. When the next three clock pulses occur the other three bits appear, one at a time, i.e. serially, at Q_D.

The process is presented pictorially in the waveforms of Fig. 14.35. In following the data through the register note that with master slave flip flops what exists at the input to a flip flop when the leading edge of a clock pulse occurs appears at the flip flop output when the trailing edge occurs.

14.8.3 Serial In, Parallel Out (SIPO) Shift Registers

An obvious application of an SIPO shift register is for taking serial data from a single link, e.g. the bits of an ASCII character from a keyboard, and making them available in parallel form to be read in to a computer. All that is required to convert the SISO device of Fig. 14.34 to SIPO is to make the output of each flip flop available as shown in Fig. 14.36. Although the bits are shifted through the register only as clock pulses occur, this does not matter if the device reading the output word is inhibited until the complete word is available.

14.8.4 Parallel In, Serial Out (PISO) Shift Registers

Taking parallel data from a computer to be fed out over a single transmission line requires a PISO device such as Fig. 14.37. When the input data are available, making S/L (shift/load) = 1 enables G_1 to G_4 so that the four flip flops are loaded on the next clock pulse. Also, S/L = 1 disables G_5 to G_7 so the shift mechanism does not operate. When S/L = 0, G_1 to G_4 are disabled and G_5 to G_7 enabled so that consecutive clock pulses shift the data out serially at Q_D.

14.8.5 Multimode Shift Registers

TTL 7496 5 Bit Shift Registers

Shift register ICs are commonly available which can be made to operate in more than one mode. Figure 14.38 shows the pin connection of a 7496 5 bit shift register which can be used in any combination of serial or parallel, input/output. The full specification is available in manufacturers' data sheets but briefly, to parallel load, the IC should first be cleared by making $C_r = 0$. Then with $C_r = 1$, $P_r = 1$ (preset enable), a 1 on any preset input will set the corresponding flip flop. None of these operations requires a clock pulse. To load serially, $C_r = 1$; $P_r = 0$ and then data at

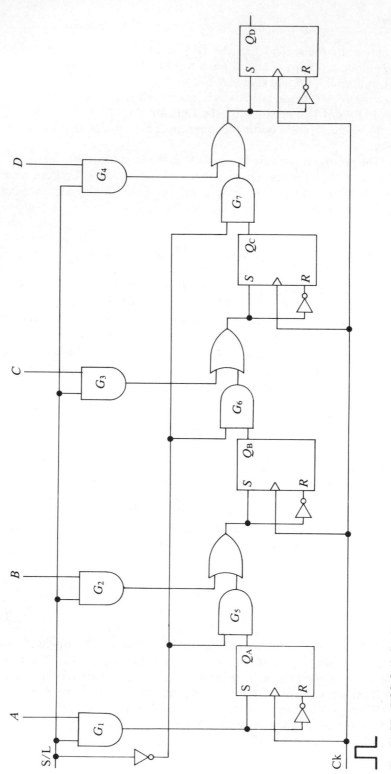

Fig. 14.37 PISO Shift Register

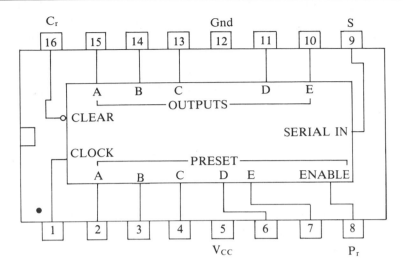

Fig. 14.38   7496 5-bit Shift Register

S will be clocked into FFA, all existing data being simultaneously shifted right and therefore out at E.

Bidirectional (Right Shift/Left Shift) Registers

ICs such as the TTL 74194 have virtually all of the features required of a multimode shift register. In addition to those mentioned in the last paragraph, Fig. 14.39 shows how the direction of shift can be controlled. With Right/Left $= 1$, G_1 to G_4 are enabled and G_5 to G_8 disabled so that data are clocked in at FFA and moved right with successive clock pulses. With Right/Left $= 0$, G_5 to G_8 will be the enabled gates so that data are clocked in at FFD and moved left.

SUMMARY

Memory is fundamental to sequential logic systems and this chapter has been used to introduce the flip flop as the most widely used element of electronic storage. A number of variants of the flip flops have also been considered, notably the *JK* master–slave version. Counters are an important class of sequential digital systems and methods of designing these in both asynchronous and synchronous forms have been described. The chapter concluded with an introduction to the shift register as a versatile word store which can be used to read and write digital information in both serial and parallel form.

412

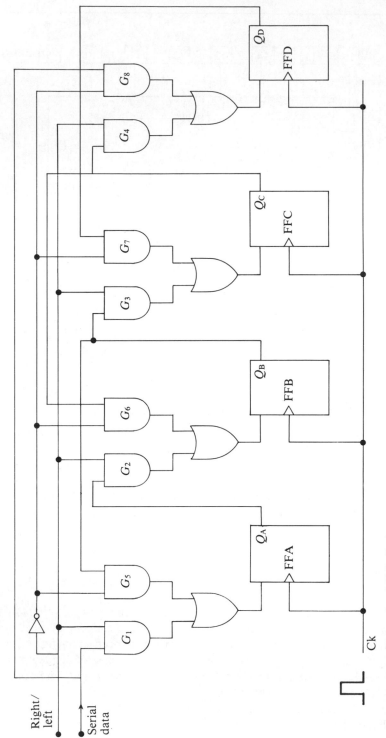

Fig. 14.39   4-bit Bidirectional Shift Register

REVISION QUESTIONS

14.1 Draw and describe the operation of an SR latch made up of logic elements. Why is the condition $SR = 11$ forbidden?

14.2 What is the difference between an SR latch and an SR flip flop?

14.3 State the properties of SR, JK, T and D flip flops.

14.4 What is the advantage of a master–slave JK flip flop over a standard JK flip flop?

14.5 The waveforms of Fig. Q.14.5 are the inputs to a master–slave JK flip flop. Draw the output waveform.

14.6 Sketch the flip flop output waveforms for a scale of 12 ripple counter and design the circuit logic. Explain how the counter can be made to count backwards.

$$(C_r = Q_3 Q_2 \bar{Q}_1 \bar{Q}_0)$$

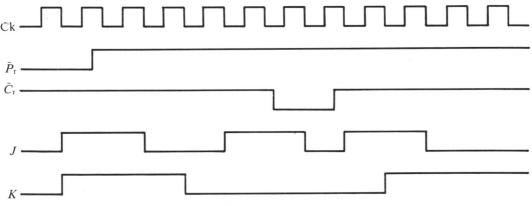

Fig. Q.14.5

14.7 What advantage does a synchronous counter have over an asynchronous counter?

14.8 Design a synchronous counter which counts backwards from 7 down to 0.

$$(J_2 = K_2 = \bar{Q}_1 \bar{Q}_0; \ J_1 = K_1 = \bar{Q}_0; \ J_0 = K_0 = 1)$$

14.9 Design a synchronous counter which counts from 0 to 4. Demonstrate that if the counter were to be in any of the unwanted states 5 to 7 when powered, it would be clocked into the required sequence.

$$(J_2 = Q_1 Q_0; \ K_2 = 1; \ J_1 = K_1 = Q_0; \ J_0 = Q_1; \ K_0 = 1)$$

14.10 Design a circuit which will count through the sequence 0, 1, 2, 5, 6, 7. If the circuit is in the state representing 3, which state will it be clocked through before entering the required sequence?

$$(J_2 = Q_1, K_2 = Q_1Q_0; J_1 = Q_0; K_1 = \bar{Q}_2 + Q_0; J_0 = K_0 = 1; \text{ 3 to 4 to 5})$$

14.11 Design a synchronous counter which counts through the sequence 0, 3, 6, 9, 12, 15, 0.

$$(J_3 = Q_2, K_3 = Q_1; J_2 = Q_0, K_2 = Q_1; J_1 = \bar{Q}_0, K_1 = Q_2; J_0 = K_0 = 1)$$

14.12 The following sequence of decimal numbers is to be generated in binary form:

0, 5, 7, 11, 25, 0, ...

A 3-stage scale of 5 counter is available in which the unused states do not occur. Design suitable logic.

$$(Q_2; Q_2 + Q_1Q_0; Q_1 \oplus Q_0; Q_1; Q_2 + Q_1 + Q_0)$$

14.13 A decade counter is pulsed every 10 s. The following traffic light sequence is required:

R	40 s
R and Y	10 s
G	40 s
Y	10 s

Design suitable logic.

$$(\bar{Q}_3\bar{Q}_2 + \bar{Q}_3\bar{Q}_1\bar{Q}_0; Q_2\bar{Q}_1\bar{Q}_0 + Q_3Q_0; Q_3\bar{Q}_0 + Q_2Q_0 + Q_2Q_1)$$

14.14 A 4 bit SISO shift register is initially cleared. Then the number 1101 is clocked in with LSB first. Draw the waveforms for all flip flop outputs for 7 clock pulses. Show that after clock pulse 4 the number is stored and that from clock pulses 4 to 7 the number occurs serially at the final flip flop output.

15

Asynchronous Sequential Circuits

Chapter Objectives
15.1   An Asynchronous Sequential Circuit
15.2   Analysis of an Asynchronous Sequential Circuit
15.3   Asynchronous Sequential Circuit with Output Logic
15.4   Design of an Asynchronous Sequential Circuit Using Gates
15.5   Design of an Asynchronous Sequential Circuit Using an SR Flip Flop
15.6   Design of an Asynchronous Sequential Circuit with Two Secondary Variables
15.7   Primary Plane Method for Asynchronous Sequential Circuit Design
        Summary
        Appendix
        Revision Questions

Chapters 12 and 13 were concerned with combinational logic – circuits whose outputs at any instant are wholly determined by the logical values of the inputs at that instant. The circuits of Chapter 14 were sequential but the function of circuits such as counters is simply to produce patterns of waveforms in response to an input train of clock pulse. This chapter is concerned with the output of circuits being determined, not only by the logical values of the inputs, but also by the sequence in which these values occur. For example, if a safe opens when two separate keys are turned in any order, then the system is combinational. But if the safe will open when both keys are turned but only if one particular key is turned before the other, then the system is sequential. It follows that a circuit which meets such a specification must have the capability of storing digital information related to the values of the inputs and outputs – the *state* of the circuit – before it changed. In circuits where the output will respond to input changes when they occur, as opposed to being clocked, then they are said to be asynchronous. They have a speed advantage over synchronous circuits in which the clock frequency must be low enough for the slowest circuit in the whole system to respond.

15.1 AN ASYNCHRONOUS SEQUENTIAL CIRCUIT

In the circuit of Fig. 15.1 the value of q can readily be determined for three combinations of A and B. If $B = 0$ then $q = 0$ whatever the value of A. If $AB = 11$ then $q = 1$. However, if $AB = 01$ then q can be either 0 or 1. For example, if $AB = 11$ then, as noted, $q = 1$. Now, if A changes to 0, q remains at 1 because the output of gate Y will remain at 1 and thus maintain q. The state of the circuit is $AB = 01$, $q = 1$. Alternatively, if $AB = 00$, $q = 0$, and if B now changes to 1, q remains at 0. Both AND gates give output 0 in both states. The state of the circuit is now $AB = 01$, $q = 0$. Therefore with the input condition $AB = 01$, the value of q depends on the previous state of the circuit. We can conclude that the value of q is always predictable with the proviso that in the case of $AB = 01$ the previous value of q (or AB) must be known. The only circumstance in which the value of q may not be predictable is when both inputs change simultaneously and therefore this occurrence is forbidden. It will be noted that, unlike a purely combinational circuit, this circuit has feedback from output to input, which is characteristic of sequential circuits.

Feedback Delay

It is inevitable that if q in Fig. 15.1 were to change because of a change in one of the inputs A or B then there would be a finite delay between the change of input and the consequent change of q. This could be depicted as in Fig. 15.2, where now q is shown as an output and Q is the same signal but which has been fed back having been delayed. (The delay is, however, distributed throughout the circuit and not just in the feedback path.) In an asynchronous sequential circuit this delay is not only inevitable but also necessary for the correct operation of the circuit, because any change of q is governed by changes of A or B and also, in the value of Q *before* the change in A or B took place.

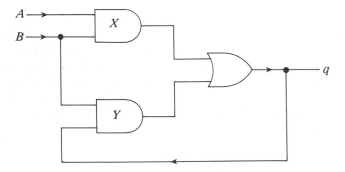

Fig. 15.1 An Asynchronous Sequential Circuit

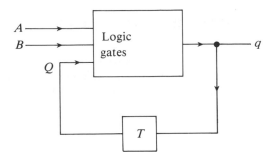

Fig. 15.2   Feedback Delay

Primary, Secondary and Output Variables

From the foregoing it can be seen that a circuit such as Fig. 15.1 has two types of input variable:

(a) A and B – the external inputs – which are called *primaries*;
(b) Q – the feedback variable – which 'remembers' the state of the circuit: the fact that $Q = 1$ or 0 tells us whether AB was 11 or 00 before becoming 01. A variable such as Q is called a *secondary*.

In Fig. 15.1 not only is q fed back to become a secondary input, but it is also the circuit output. However, combinational logic can be provided using primary and secondary variables as inputs to form outputs such as Z_1 and Z_2 in Fig. 15.3, which is a general model for an asynchronous sequential circuit.

The delay is not shown in this circuit diagram but is implied by using q as the output of the gate forming the secondary and Q as an input used in combination with the primaries.

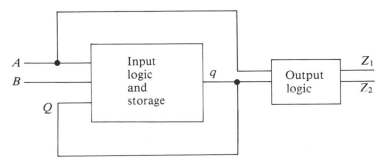

Fig. 15.3   General Model of an Asynchronous Sequential Circuit

15.2 ANALYSIS OF AN ASYNCHRONOUS SEQUENTIAL CIRCUIT

It is sometimes necessary to determine the specification of an asynchronous sequential circuit, i.e. to determine the value of the output variable(s) for all combinations of primary values occurring in every possible sequence. A procedure for finding the specification, in other words, analyzing such a circuit, can be described by example.

☐ **Example 15.1**

It is required to specify the behavior of Fig. 15.4 in which the primaries A and B can only change one at a time. In this particular circuit the secondary variable q is also the output, but the next example will deal with the problem of analyzing a circuit with separate output logic.

Excitation Map: The first step in analyzing Fig. 15.4 is to suppose that the feedback connection is broken at X, thus separating the secondary as an output q and as an input Q. Then

$$q = A\bar{B} + AQ + BQ$$

This function q is represented on the K map of Fig. 15.5 which is called, in this context, an excitation map. It is convenient to enter the primaries A and B on this map as the column headings and the secondary Q on the rows. We can now consider the concept of stable and transient (unstable) states. Consider first the square $\bar{A}\bar{B}\bar{Q}$, i.e. $ABQ = 000$. In this state $Q = 0$ and $q = 0$, so if the feedback connection were remade the circuit would remain in this state. In contrast, consider the square $A\bar{B}\bar{Q}$, i.e. $ABQ = 100$. Here $Q = 0$ but $q = 1$, so if the feedback were reconnected one of these variables would have to change. Note that in any particular case the primaries as

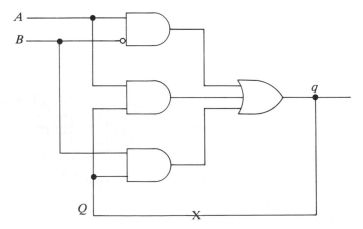

Fig. 15.4 Circuit for Example 15.1

Q \ AB	00	01	11	10
0	0	0	0	1
1	0	1	1	1

q

Fig. 15.5   Excitation Map for Example 15.1

the external inputs are fixed, so the only change which can occur is a vertical shift, i.e. a change in Q, on the excitation map. Therefore the circuit changes from $ABQ = 100$ to 101 and $q = Q = 1$. The state 100 is said to be transient. The sequence of events could be that with $ABQ = 000$ the circuit is stable ($q = Q = 0$) and then A changes to 1. Now the circuit is unstable because q has changed to 1, so the circuit cannot remain in this state and Q also changes to 1: the circuit has switched out of its transient state 100 into a stable state 101, where $Q = q = 1$.

On the excitation map stable states are those where $q = Q$, i.e. $q = 0$ on the $Q = 0$ row and $q = 1$ on the $Q = 1$ row. It should be emphasized that transient states are essential to the operation of a sequential circuit; otherwise the output could never change between 0 and 1.

Flow Table: To continue the analysis we now draw another K map – a flow table (Fig. 15.6) – on which both stable and transient states are entered by arbitrarily assigned state numbers. First the stable states are entered, i.e, those in which $q = Q$. These are easily identified on the excitation map of Fig. 15.5 because, as noted, they will be the states in which $q = 0$ in the $Q = 0$ row and in which $q = 1$ in the $Q = 1$ row. Clearly there are six stable states. To indicate that they are stable they are entered as ringed numbers, as shown in Fig. 15.6. The transient states are given the same number as the stable states in the vertically adjacent squares, but they are not ringed. This aspect of the numbering is logical because suppose the circuit goes into transient state 6 (100); then it will rapidly be forced into stable state ⑥ (101) as Q changes from 0 to 1.

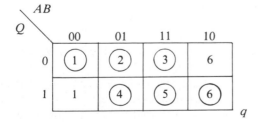

Fig. 15.6   Flow Table for Example 15.1

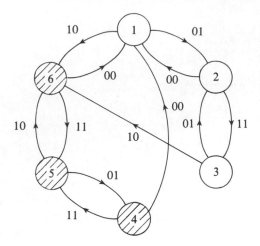

Fig. 15.7   Total State Diagram for Example 15.1

Total State Diagram: The flow table (Fig. 15.6) enables us to construct a total state diagram (Fig. 15.7) which summarizes the circuit behavior. Each ring represents a stable state, so we can begin drawing the diagram by sketching a circle of numbered rings – in this case there will be six. Lines are then drawn between states to represent transitions. For example, if the circuit is in state ① and the primaries change from 00 to 01, it can be seen from the flow table that the circuit changes to ②. A line with an arrow on the total state diagram shows the direction of the transition, and also, 01 is marked against the line to indicate the primary values needed to effect the transition. If the primaries, instead of changing from 00 to 01, had changed from 00 to 10, then the flow table shows that the circuit goes from stable state ① to stable state ⑥ via transient state 6. This is simply shown as a line from ① to ⑥ on the total state diagram with 10 against it – the diagram does not show transient states.

There cannot be any other lines *from* state ① on the total state diagram because the only other possibility is for both primaries to change – AB to go from 00 to 11 – and this is not permitted. It follows that there will be two lines *from* every state because a change in one of the primaries always takes the circuit to a different state. Another convention is to shade any ring for a state which gives a 1 output from the circuit. This happens to be where $q = 1$ in this circuit, i.e. in states ④, ⑤ and ⑥.

Waveform Diagram and Specification: A circuit specification can be deduced from a total state diagram, but it is common practice to draw a waveform diagram (Fig. 15.8) first. Ideally, a waveform diagram shows every possible sequence of changes in the primaries. There are many possible sequences of changes for two variables and to sketch them all is burdensome. By sketching a representative set of changes, such as those shown for A and B in Fig. 15.8, most eventualities are catered for, and in drawing up the specification, which shows the sequences of A

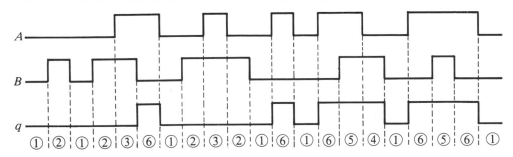

Fig. 15.8   Waveform Diagram for Example 15.1

and B leading to a 1 at the output, it will be evident if a significant sequence has been omitted. The primary waveforms used in Fig. 15.8 first show B going through the sequence 0–1–0 with A held at 0. Then B goes to 1 followed by A, and then B goes to 0 before A. The sequence is then repeated, except that A goes to 0 before B. The whole sequence is then repeated but with A and B changing place with each other. Observing the values of the primaries as they change, the value of the output and the number of the state can be deduced from the total state diagram and entered on the waveform diagram.

To draw up the specification it is necessary to write down the circumstances in which the output will be 1. From the waveform diagram of Fig. 15.8, the circuit goes into state ⑥ when B goes to 0 after both A and B were 1 (state ③). It next goes to ⑥ when A goes to 1 after $AB = 00$ (①). The next significant state to note is ④ when only $B = 1$ following $AB = 11$ (⑤), which in turn followed $AB = 10$ (⑥). This can be summarized by saying that the circuit first gives an output 1 only if $A = 1$ when $B = 0$. However, once the output becomes 1 it will remain at 1 as long as A or B or both are 1. If B goes to 1 before A, then the output cannot become 1 until B goes to 0.

In conclusion it is immediately evident from the total state diagram of Fig. 15.7 that the circuit is sequential. If it were not, all adjacent states would be linked by two lines showing that whenever a variable is changed and then returned, it always returns to the previous state. In Fig. 15.7 when the circuit is in state ③ ($AB = 11$) and B goes to 0, the circuit goes to state (⑥): restoring B to 1 does not take the circuit back to ③ but to ⑤ instead. Similarly, beginning in state ④ with $AB = 01$, the sequence 00 to 01 takes the circuit to state ① but then to ② rather than back to ④. The existence of a single line between states on a total state diagram indicates that the circuit is sequential. □

15.3   ASYNCHRONOUS SEQUENTIAL CIRCUIT WITH OUTPUT LOGIC

In Fig. 15.9 the part of the circuit which forms the secondary variable q is the same as Fig. 15.4 but the addition of an AND gate with inputs B and $\bar{Q}$ gives an output

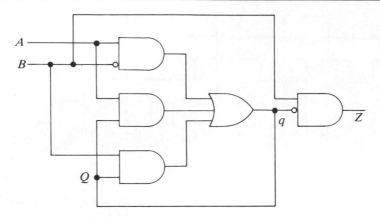

Fig. 15.9   An Asynchronous Sequential Circuit with Output Logic

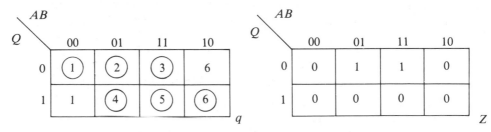

Fig. 15.10   Flow Table (Fig. 15.6 Repeated) and Output Map

Z which is the output of the entire circuit. The analysis follows similar lines to Example 15.1 because, irrespective of whether there is output logic, it is necessary to examine the secondary variable q in order to determine stable and transient states of the circuit. Thus having derived the flow table as before, another K map – the output map – is drawn adjacent to it, as in Fig. 15.10. The function $Z = B\bar{Q}$ is entered on the output map and comparing the two maps it can be seen that the output of the circuit Z will be 1 in states ② and ③. The total state diagram will be the same as in Fig. 15.7 except that now, states ② and ③ will be shaded. It is now only necessary to note on the waveform diagram (Fig. 15.8) that $Z = 1$ in states ② and ③ and to draw up the specification, which will be: Z will only become 1 if B is 1 and A is 0 and will remain at 1 whatever happens to A but will fall back to 0 when B falls to 0. If A goes to 1 first, then Z cannot become 1 until B returns to 0.

15.4   DESIGN OF AN ASYNCHRONOUS SEQUENTIAL CIRCUIT USING GATES

The specification of an asynchronous sequential circuit can either be realized entirely by interconnecting gates as in the circuit of Fig. 15.4, or by using gates plus an SR

flip flop to store the secondary variable. In the following sections the design of each type of circuit will be described by example.

☐ **Example 15.2**

A circuit with inputs A and B is required to give an output $Z = 1$ when $AB = 11$ but only if A becomes 1 before B.

The design procedure is, to an extent, the reverse of analysis, but at the outset we shall not know if output logic is required or whether Z is simply equal to the secondary variable q.

Waveform Diagram: Using the same sequence of A and B changes used in Fig. 15.8 and observing the specification, the output waveform Z can be drawn as in Fig. 15.11. There are five different combinations of ABZ and these are arbitrarily given state numbers with only state ⑤ having $Z = 1$.

Total State Diagram: From the waveform diagram we can construct a total state diagram (Fig. 15.12) in which the states are linked by lines marked with the input (primary) combination required to effect the change from one state to the other. State ⑤ is shaded to indicate that $Z = 1$ in this state.

Primitive Flow Table: In reversing the analysis procedure, the next step in design would be the production of a flow table. This cannot be taken directly from the total state diagram because we do not know how many secondary variables are required and therefore how may rows are needed in the flow table. The procedure is to construct what is known as a *primitive flow table,* in which it is assumed that in moving between stable states the circuit always passes through a transient state. The first two rows of such a table would be initiated as in Fig. 15.13a. Notice in the total state diagram of Fig. 15.12 that in stable state ①, $AB = 00$ so ① is entered in row 1 under $AB = 00$. The total state diagram shows that when the circuit is in state ①, making $AB = 01$ takes the circuit to state ②. This is entered on the primitive flow table by assuming first that the circuit enters transient state 2

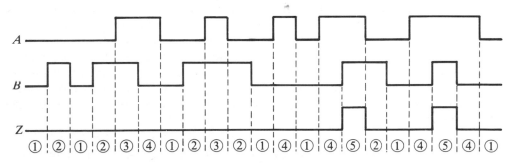

Fig. 15.11   Waveform Diagram for Example 15.2

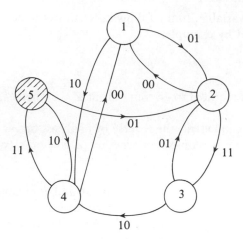

Fig. 15.12   Total State Diagram for Example 15.2

(a)

		AB		
		00	01	11	10
1	①	2		
2	1	②		

(b)

		AB		
		00	01	11	10
1	①	2	—	4
2	1	②	3	—
3	—	2	③	4
4	1	—	5	④
5	—	2	⑤	4

Fig. 15.13   Primitive Flow Table for Example 15.2 (a) First Two Rows (b) Complete Table

(horizontal shift) and then to stable ② (vertical shift). Restoring AB to 00 takes the circuit first to transient state 1 and then to stable ①.

The complete primitive flow table (Fig. 15.13b) has as many rows as there are stable states, and each stable state is in the corresponding row under the required AB value. Either side of each stable state, a change of one of the primaries takes the circuit to the transient state (horizontal shift) and then to the stable state (vertical shift) dictated by the total state diagram. Since both inputs cannot change simultaneously, each row has a 'can't happen' state at a distance of two columns from the stable state (remember that the end columns in a K map are considered to be adjacent).

Merger Diagram: The primitive flow table of Fig. 15.13b could be directly realized by logic gates but the five rows imply the need for three secondary variables,

	AB			
(a)	00	01	11	10
	①	②	3	4

	AB			
(b)	00	01	11	10
	①	②	③	4

Fig. 15.14 Row Merging (a) Merging the First Two Rows of Fig. 15.13b (b) Merging the First Three Rows of Fig. 15.13b

with consequent complexity and high cost. Although a circuit must have some transient states, it is not always necessary for a transition between stable states to take place via a transient state. It can be seen in Fig. 15.13b that, for permitted primary values, some rows contain the same state numbers albeit that some are stable and others transient. Thus, by eliminating transient states 1 and 2 in rows 2 and 1, the two rows could be merged to form Fig. 15.14a. The only difference in implementation is that in passing between states ① and ② the circuit will not enter a transient condition, i.e. the secondary variable will not change value. This new row could, in turn, be merged with row 3 of the primitive flow table, forming Fig. 15.14b which allows the circuit to move between state ①, ② and ③ without entering a transient state. Neither row 2 nor row 3, with state ③ under $AB = 11$, could be merged with row 4 or row 5, which have transient state 5 and stable state ⑤ respectively in that column.

To effect the merger, one technique is to draw a diagram – the merger diagram of Fig. 15.15 – in which states which can be merged are linked. Taking row 1 of Fig. 15.13b first, we note that 1 and 2, 1 and 3, 1 and 4, and 1 and 5 all contain the same states in so far as permitted primary changes allow these states. Therefore lines linking 1–2, 1–3, 1–4 and 1–5 are drawn in Fig. 15.15. Then we consider row 2 and note that 2 and 3 can be merged, so the 2–3 link is drawn. Row 3 has no succeeding row that it can be merged with, but rows 4 and 5 can be merged so the 4–5 link is drawn. The complete merger diagram consists of two closed triangles ①, ②, ③ and ①, ④, ⑤.

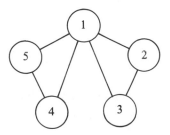

Fig. 15.15 Merger Diagram

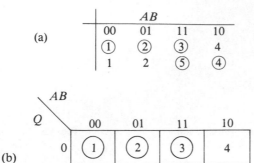

Fig. 15.16 (a) Result of Merging Rows 1, 2 and 3 and Rows 4 and 5 (b) One Flow Table Which Can Be Obtained from Merging Rows

Having completed the merger diagram, suppose that rows 1, 2 and 3 were merged into one row, and rows 4 and 5 into another, resulting in Fig. 15.16a. Comparing this with Fig. 15.6, it can be seen that it is basically a flow table. It remains to allocate the two rows to the two states of the secondary variable, which we shall again label Q. Choosing, arbitrarily, $Q = 0$ for row 1 and $Q = 1$ for row 2, we have the flow table of Fig. 15.16b.

The choice of mergers − we could have merged rows 1, 4 and 5, and 2 and 3 − and the allocation of the rows to $Q = 0$ and $Q = 1$ will all lead to a circuit which meets the specification. The final choice will probably rest with the solution leading to the greatest economy in gates. Very often, if the allocation of $Q = 1$ can be made to embrace the states giving an output $Z = 1$ − in this case state ⑤ − this does lead to an efficient implementation.

Excitation and Output Maps: Continuing to reverse the analysis procedure, an excitation map can be developed from the flow table in which $q = \bar{Q}$ for the transient states. The resulting map is shown in Fig. 15.17, and from it,

$$q = A\bar{B} + AQ$$
$$= A(\bar{B} + Q)$$

The output Z will be 1 in state ⑤ and this can, if desired, be represented on an output map as in Fig. 15.18, which gives

$$Z = ABQ$$

Although this fulfills the specification, simplification is possible by including an extra 1 on the output map at $ABQ = 011$ as shown in Fig. 15.19. This can be used because $Z = 1$ when $ABQ = 111$ and when the circuit enters transient state 2 as A changes to 0, the output merely remains at 1 for a split second before going to stable

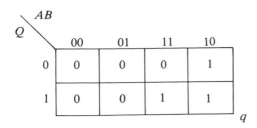

Fig. 15.17   Excitation Map for Example 15.2

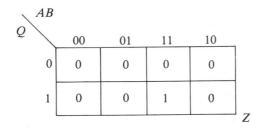

Fig. 15.18   Output Map for Example 15.2

state ②. This does not cause a malfunction but only a brief delay. It allows us to implement Z as

$$Z = BQ,$$

and thus reduce the output gate from a 3 input AND gate to a 2 input AND gate. The complete circuit to meet the specification is given in Fig. 15.20.

It is instructive to return to the merger diagram of Fig. 15.15 and to use different mergers and/or allocation of the resulting rows to the values of the secondary variable Q. Merging of rows 2 and 3, and 1, 4 and 5, and allocating the former to $Q = 0$ results in

$$q = \bar{B} + AQ$$
$$Z = BQ$$

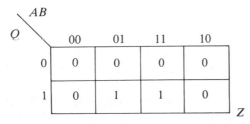

Fig. 15.19   Output Map in which Transient State is Included

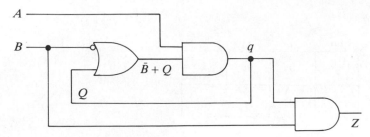

Fig. 15.20   A Circuit which Meets the Specification of Example 15.2

from which we can see that the implementation of q is different although, in this case, no simpler than before. □

15.5 DESIGN OF AN ASYNCHRONOUS SEQUENTIAL CIRCUIT USING AN *SR* FLIP FLOP

By using an *SR* flip flop in conjunction with gates, an asynchronous sequential circuit can be designed which results in a secondary variable free from static hazards (Section 12.10). The basic layout of such a circuit is shown in Fig. 15.21. Combinational logic is used to generate S (set) and R (reset) signals for the flip flop whose ouput is the secondary variable Q. If necessary, combinational logic may be used to derive the output Z. As before, the design procedure will be described using an example.

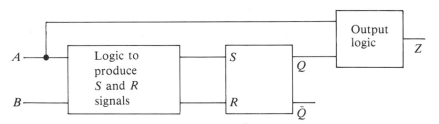

Fig. 15.21   Layout of Circuit Using an *SR* Flip Flop in an Asynchronous Sequential Circuit

□ **Example 15.3**

A circuit with primary inputs A and B is required to produce an output Z which will become 1 when A becomes 1 if B is already 1. Once $Z = 1$ it will remain so until A goes to 0.

The preliminary work of drawing a waveform diagram, total state diagram, primitive flow table, merger diagram and flow chart is carried out in the same manner as in Example 15.2, and for this example is shown in Figs. 15.22a to e. One point to observe, which would have been equally true for a wholly gate design, is that with these six states the abbreviated form of waveform diagram we have been using does not go from state ④ when $AB = 10$ to $AB = 11$. However, it is evident from the specification that Z remains at 1, i.e. the circuit returns to state ③, and this can be entered on the total state diagram. Fig. 15.22d shows that only by merging

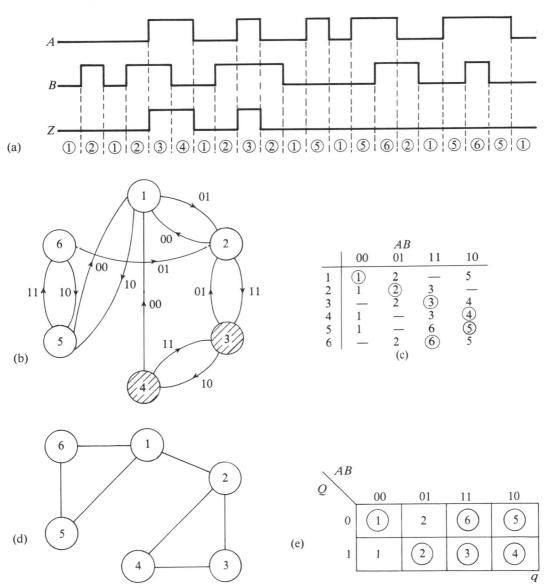

Fig. 15.22 Diagrams for Example 15.2 (a) Waveform Diagram (b) Total State Diagram (c) Primitive Flow Table (d) Merger Diagram (e) Flow Table

rows 1, 5 and 6, and 2, 3 and 4, can the circuit be reduced to two secondary states. Since $Z = 1$ in states ③ and ④, then the first group is allocated to $Q = 0$ and the second to $Q = 1$ on the flow table.

To find the logic to drive the flip flop we first draw two K maps (Figs. 15.23a and b), one for S and one for R. For each square there are three possibilities: 0, 1 and 'doesn't matter', which will be marked X.† Filling the corresponding squares in each map together, first consider the row $Q = 0$. If a change in the primaries causes the circuit to move between the three stable states (①, ⑤ and ⑥) then the circuit must remain reset. Therefore, for these three squares S must be 0 to prevent the flip flop from being set to 1 but R doesn't matter. Hence $S = 0$, $R = X$. When entering transient state 2, $Q = 0$ but is required to become 1. Therefore $S = 1$ but since $SR = 11$ is forbidden for an SR flip flop, $R = 0$. In the $Q = 1$ row, when entering a stable state (②, ③ or ④), S doesn't matter because the flip flop is already set but R must be 0 to prevent a reset. To make the remaining square (1) transient the circuit must be reset so $R = 1$, $S = 0$.

Looping 1s with any adjacent 'doesn't matter' squares to find the minimal logic for S and R, we have

$$S = \bar{A}B; \qquad R = \bar{A}\bar{B}$$

Having previously determined that $Z = 1$ in states ③ and ④, then referring back to the flow table of Fig. 15.22e, we can deduce that

$$Z = AQ$$

The required circuit is shown in Fig. 15.24.   □

It is instructive to return to the flow table of Fig. 15.22e and to consider the result of designing with gates. The excitation map for the secondary is given in Fig. 15.25 which has two adjacent prime implicant loops and therefore the circuit will have a static hazard. To eliminate the hazard an additional loop would have to be included, giving

$$q = \bar{A}B + AQ + BQ$$

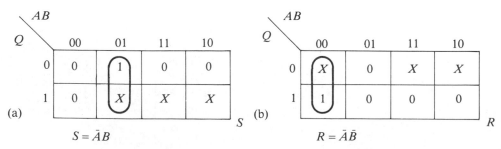

Fig. 15.23   *S* and *R* Maps for Example 15.3

† An alternative method of labeling a K map for S and R is given in the Appendix at the end of this chapter.

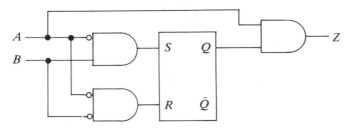

Fig. 15.24   Circuit for Example 15.3

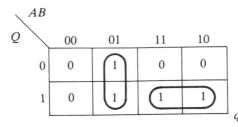

Fig. 15.25   Secondary Map Developed from Flow Chart of Fig. 15.22e Showing Existence of Static Hazard

15.6   DESIGN OF AN ASYNCHRONOUS SEQUENTIAL CIRCUIT WITH TWO SECONDARY VARIABLES

Previous design examples have been chosen so that the merging process reduced the problem to the implementation of a flow table with two rows. Therefore only a single secondary variable Q with values 0 and 1 was required. If the merging process does not reduce the flow table to two rows, then an extra secondary variable is needed. The following example gives the implementation of a two secondary variable design with gates although it could equally well be effected with SR flip flops using methodology similar to that of Example 15.3.

☐   **Example 15.4**

A system with input primaries A and B, which cannot change simultaneously, is required to give a 1 at its output Z when $A = 1$ and $B = 0$ but only if preceded by $AB = 11$; once $Z = 1$ it remains at 1 as long as $A = 1$ whatever happens to B.

As noted it is not always necessary to draw a waveform diagram which gives every possible sequence of two variables. However, having drawn the sequence used in previous examples (Fig. 15.26), it can be seen on the total state diagram (Fig. 15.27) that having got the circuit into state ④ ($ABZ = 101$), the waveform does not provide the sequence in which B returns to 1 and back to 0. This sequence is added by providing the last 6 states on the waveform diagram which reveals the existence

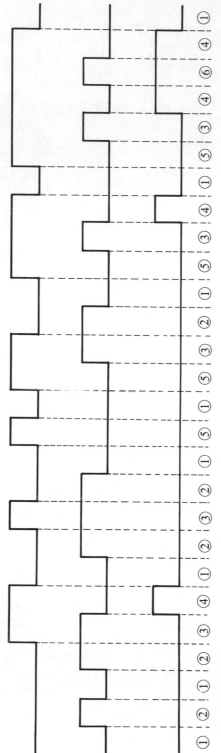

Fig. 15.26 Waveform Diagram for Example 15.4

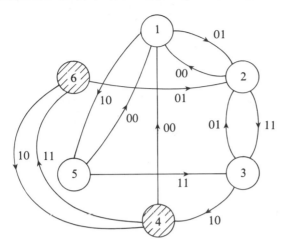

Fig. 15.27 Total State Diagram for Example 15.4

of state ⑥ ($ABZ = 111$). Strictly speaking, the waveform diagram should be extended further to see what happens from state ⑥ when A goes to 0. However, it is obvious from the specification that this makes $Z = 0$ so the circuit must go back to state ②. We can now complete the total state diagram (Fig. 15.27).

Construction of the primitive flow table (Fig. 15.28) and the merger diagram (Fig. 15.29) indicates that there must be at least three secondary states and therefore two secondary variables which will be labeled q_1 and q_2. Noting that $Z = 1$ in states ④ and ⑥, these may be assigned to two rows in which either of the secondary variables is 1; Q_1 has been chosen for this design (Fig. 15.30). One of these rows could have been left as a 'can't happen' condition because only three secondary states are needed, and it is a worthwhile exercise to work out the design based on this assignment. However, the choice made does simplify the design, so we shall proceed and complete the flow table by allocating rows 1 and 5 on the primitive flow table to $Q_1Q_2 = 00$ and rows 2 and 3 to $Q_1Q_2 = 01$.

	AB			
	00	01	11	10
1	①	2	—	5
2	1	②	3	—
3	—	2	③	4
4	1	—	6	④
5	1	—	3	⑤
6	—	2	⑥	4

Fig. 15.28 Primitive Flow Table for Example 15.4

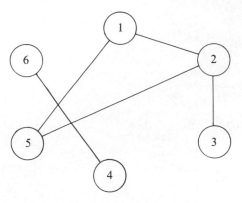

Fig. 15.29   Merger Diagram for Example 15.4

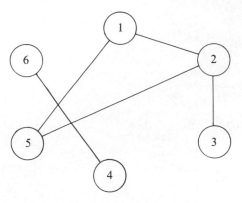

Fig. 15.30   Flow Table for Example 15.4

To construct the excitation map for both q_1 and q_2 on a single map (Fig. 15.31), note that each stable state will have $q_1q_2 = Q_1Q_2$, e.g. for states ① and ⑤ $q_1q_2 =$ 00. The value of q_1q_2 for a transient state will be equal to the value of Q_1Q_2 which the circuit must take up to enter the corresponding stable state. For example, for transient state 6 in row $Q_1Q_2 = 00$ then q_1q_2 must be made 10 to switch the circuit to state ⑥. The 'can't happen' states of the unmerged rows 4 and 5 are also entered and from the completed excitation map q_1 and q_2 can be read off separately, giving

$$q_1 = AQ_1 + A\bar{B}Q_2$$
$$q_2 = \bar{A}B + B\bar{Q}_1 + A\bar{Q}_1Q_2 + A\bar{B}Q_1 + A\bar{B}Q_2$$

The term is included to eliminate static hazards (section 12.10).

It remains to find the logic for Z by first drawing the output map (Fig. 15.32). First, for the stable states ④ and ⑥ which are required to produce a 1 at the output, then 1 is entered in the corresponding squares of the output map. Similarly, 0s are entered for states ①, ②, ③ and ⑤. Next we must avoid the hazard of pro-

AB

Q_1Q_2	00	01	11	10
00	00	01	01	00
01	00	01	01	11
11	00	X	10	11
10	X	01	10	11

q_1q_2

Fig. 15.31   Excitation Map for Example 15.4

ducing a momentary unwanted transition in the output as the circuit passes through a transient state in between two stable states which have the same output value. For example, if transient state 2 in row $Q_1Q_2 = 00$ is made 1, then in passing from state ① to state ② via transient state 2 the circuit will give a momentary output 1 even though it should remain at 0. Therefore transient state 2 in row $Q_1Q_2 = 00$ must be 0. However, when the circuit passes through a transient state between stable states which have opposite output values, then the value of Z in the transient state doesn't matter. For example, if $Z = 1$ in transient state 2 of row $Q_1Q_2 = 10$, then in passing from state ② to state ⑥, which requires Z to change from 0 to 1, Z will become 1 momentarily sooner than if we have made $Z = 0$ in the transient state. This is of no consequence, so transient state 2 in row $Q_1Q_2 = 10$ doesn't matter and is accordingly marked X, which may simplify the output logic. In fact, the resulting Z map gives the simple solution

$Z = Q_1$.

AB

Q_1Q_2	00	01	11	10
00	0	0	0	0
01	0	0	0	X
11	X	X	1	1
10	X	X	1	1

Z

Fig. 15.32   Output Map for Example 15.4

15.7 PRIMARY PLANE METHOD FOR ASYNCHRONOUS SEQUENTIAL CIRCUIT DESIGN

For some problems this method produces a solution quite readily. Describing the method by example, consider the problem of designing a circuit which will produce an output $Z = 1$ when the primary inputs $AB = 01$ but only if preceded by $AB = 11$.

The stable states of the circuit are set out in Fig. 15.33, which is known as the primary plane. Note that it is not necessary to mark the primary values on the transition lines because these are explicitly represented on the two-variable map forming the primary plane. It will be noted on the primary plane that the square $AB = 01$ contains two stable states, ③ and ⑤ (this implies that the circuit is sequential). We then divide the primary plane into the two secondary planes of Fig. 15.34, ensuring that:

(a) states occupying the same square on the primary plane ③ and ⑤ are separated on the secondary planes;
(b) transition between the states on a secondary plane can be effected simply by primary changes;
(c) each secondary plane contains at least one empty square which represents a transient state to enable the circuit to switch between the two planes.

Each secondary plane is associated with the state of a secondary variable such as

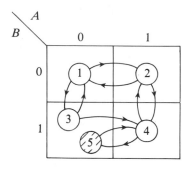

Fig. 15.33   Primary Plane

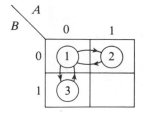
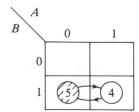

Fig. 15.34   Secondary Planes with Transient States Indicated by Empty Squares

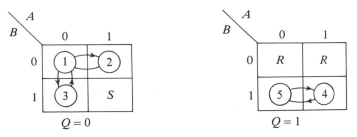

Fig. 15.35 Secondary Planes Allocated to the Two States of an *SR* Flip Flop

the output Q of an SR flip flop. The function of the flip flop (FF) is to 'remember' in which state the circuit resides. In particular, for this design problem, the FF must 'remember', when $AB = 01$, whether the previous state of the circuit was $AB = 00$ (1) or $AB = 11$ (4): it can then determine, with $AB = 01$, whether the circuit should be in state ③ or ⑤.

Allocating the secondary planes, arbitrarily, to $Q = 0$ and $Q = 1$ as shown in Fig. 15.35, then $Q = 1$ for states ①, ② and ③ and $Q = 0$ for states ④ and ⑤. If, with $Q = 0$, the circuit goes into the transient state ($AB = 11$) then S indicates that the FF must be set to $Q = 1$ thus changing over to the $Q = 1$ secondary plane and into state ④. Likewise, the transient states on the $Q = 1$ plane are marked R to show that in these two states the FF must be reset to $Q = 0$. To illustrate the way in which the circuit is required to operate, suppose it is in state ③ ($ABQ = 010$) and A changes to 1. The circuit must move to the transient square S and then to ④ on the $Q = 1$ map. If, now, A returns to 0, the FF will not be reset because the circuit simply moves to state ⑤ on the $Q = 1$ plane.

To evaluate the logic, the FF inputs can be deduced from the secondary planes of Fig. 15.35 as

$$S = AB; \qquad R = \bar{B}.$$

The output Z must be 1 in state 5, i.e. when $Q = 1$ and $AB = 01$, and therefore

$$Z = \bar{A}BQ.$$

However, by using the transient square $AB = 00$ when $Q = 1$, the output logic can be simplified to

$$Z = \bar{A}Q.$$

Alternative solutions can be found by:

(a) reversing the allocation of $Q = 0$ and $Q = 1$ to the secondary planes; or

(b) by allocating states ① and ③ to one secondary plane and ③, ④ and ⑤ to the other.

Logic should be evaluated for each of these propositions to find the most economical design.

SUMMARY

One of the distinguishing features of asynchronous sequential circuits is that of making the inputs a combination of input primary variables and fed back secondary variables. The latter give the circuits the capability of 'memorizing' previous states but their introduction makes the circuits conceptually more challenging. However, a routine of logical steps will always lead to the solutions of analyzing an existing circuit or designing from a specification. A danger to be avoided, particularly in design, is to lose sight of the physical significance of the various maps and charts being produced.

APPENDIX

Alternative K Map labeling for the S and R inputs of a flip flop used in an Asynchronous Sequential Circuit Design

Having produced the flow table of Fig. 15.22e for Example 15.3, it is then necessary to derive the S and R logic for the flip flop. An alternative to the two K maps of Fig. 15.23 is Fig. 15.36, the entries on which are found as follows:

(a) For stable states in the $Q = 0$ row of Fig. 15.23e, the flip flop must remain reset. Therefore the S input must be 0 but it is immaterial whether R is 0 or 1. Enter r to indicate optional reset.
(b) For a transient state in the $Q = 0$ row the flip flop must be set to 1. Enter S to show compulsory Set.
(c) For stable states in the $Q = 1$ row the flip flop must remain set, so $R = 0$. The value of S is optional so we enter s.
(d) For transient state, in the $Q = 1$ row the flip flop must be reset so R is entered to show a compulsory Reset.

Minimal logic for the R input must include all compulsory resets (R) and may

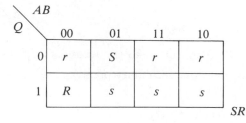

Fig. 15.36   Alternative *SR* Map for Example 15.3

include optional resets (r). Therefore, from Fig. 15.36,

$$R = \bar{A}\bar{B}.$$

Similarly,

$$S = \bar{A}B.$$

REVISION QUESTIONS

Some of the design problems have several solutions.

15.1 Analyze the operation of the asynchronous sequential circuit shown in Fig. Q.15.1. Draw a waveform diagram showing the complete operation of the circuit and hence write a specification for the function of the circuit.

($q = 1$ when $AB = 11$ and also, when $AB = 01$ if preceded by $AB = 11$)

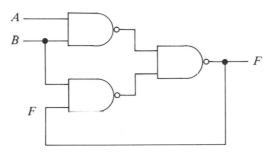

Fig. Q.15.1

15.2 Analyze the circuit of Fig. Q.15.2 to produce a total state diagram. Hence show that the circuit is unnecessarily complex and could be replaced by three NOR gates assuming A and B can never change together.

$$(\bar{q} = \overline{\overline{A + B} + \overline{A} + \bar{Q}}; z = \bar{q})$$

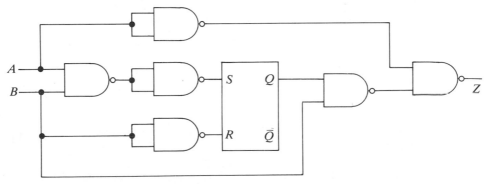

Fig. Q.15.2

15.3 Analyze the asynchronous sequential circuit of Fig. Q.15.3 so that its operation can be shown on a total state diagram and write a specification which describes the action of the circuit.

Identify the static hazard in this circuit and explain how it can cause an unwanted transition at q.

Show how the hazard can be eliminated by the addition of one gate.

($z = 1$ when $AB = 11$ if preceded by $AB = 10$. Add AQ in function q)

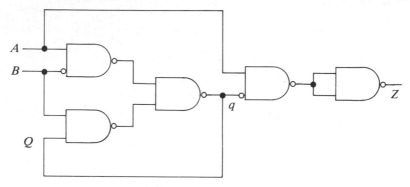

Fig. Q.15.3

15.4 Analyze the circuit of Fig. Q.15.4. drawing state and waveform diagrams to show the complete operation of the circuit. ($Z = 1$ when $X_1 X_2 = 11$ if preceded by $X_1 X_2 = 10$. It remains at 1 as long as $X_2 = 1$ regardless of the value of X_1.)

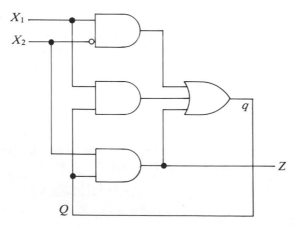

Fig. Q.15.4

15.5 In the circuit shown in Fig. Q.15.5 the binary inputs A and B can never change at the same time. Analyze the circuit so that its operation can be displayed on a total state diagram and a waveform diagram. Hence write a specification which describes the circuit action. Explain the function of Gate *a*.

(Z goes to 1 when B goes to 1 if $A = 0$, and remains at 1 when A also becomes 1. If A goes to 1 first, Z cannot become 1 until A returns to 0.)

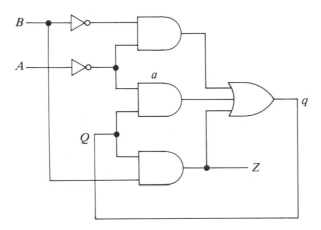

Fig. Q.15.5

15.6 The circuit diagram in Fig. Q.15.6 shows an asynchronous sequential circuit. The binary inputs A and B can never change together. Analyze this circuit so that its operation can be displayed on a total state diagram. Hence redesign the circuit in a simpler form, still using NAND gates.

$$(q = \overline{\overline{A} \cdot \overline{BQ}}; \; Z = q)$$

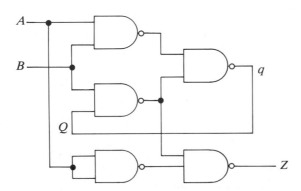

Fig. Q.15.6

15.7   Analyze the asynchronous sequential circuit of Fig. Q.15.7, drawing state waveform diagrams to show the complete operation of the circuit.
($Z = 1$ when $AB = 11$ and also, if $AB = 10$ but only if preceded by $AB = 11$).

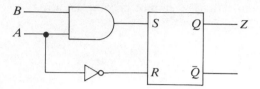

Fig. Q.15.7

15.8   Analyze the circuit of Fig. Q.15.8 giving state and waveform diagrams to show the operation of the circuit. Hence, or otherwise, redesign the circuit using AND, OR and NOT gates only.

(Z goes to 1 if A goes to 1 when $B = 0$ or if B goes to 1 when $A = 0$.
It remains 1 when $A = B = 1$ but falls to 0 if B falls to 0.
$(q = B + AQ; Z = B + A\bar{Q})$

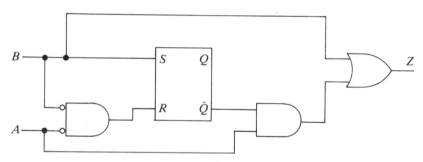

Fig. Q.15.8

15.9   An asynchronous sequential circuit has two binary inputs C and G which can never change together. C consists of clock pulses with mark/space ratio of $1:1$ and G is a gating signal of variable duration. The clock pulses are to appear at the circuit output Z when the gate input is at logic '1' and the output should be '0' when the gate input is at logic '0'. Nevertheless, all the output pulses must be full length clock pulses. Design a suitable circuit using only NOR gates, and which uses an SR flip flop to generate the secondary variable.

$$(S = \overline{C + \bar{G}}; R = \overline{C + G}; Z = \overline{\bar{C} + \bar{Q}})$$

15.10   A circuit has two binary inputs A and B, only one of which can change at any time. The output rises to 1 when A rises to 1 or when B rises to 1 provided in each case that the other input is already 1. The output must return 0 when A returns to 0 regardless of the state of B. Design a suitable circuit using gates.

$$(q = AB + AQ; Z = Q)$$

15.11   A circuit has two binary inputs A and B, only one of which can change at any time. The output rises when A rises if B is already 1 or when B rises if A is already 1. The output falls to 0 when B falls to 0. Design a circuit using an SR flip flop to generate the secondary variable.

$$(S = AB; R = \bar{B}; Z = Q)$$

15.12   A circuit has two binary inputs A and B, only one of which can change at any time. The output must take the value of A when $B = 1$; when $B = 0$, the output remains fixed at its last value prior to B becoming 0. Design a suitable circuit, which has no static hazards.

$$(S = AB; R = \bar{A}B; Z = Q, \text{ or } q = AB + \bar{B}Q + AQ; Z = Q)$$

15.13   A detector receives pulse trains on two separate lines A and B. The detector is required to produce a 1 output when coincident pulses are present on both lines but only if the pulse on A arrives before a pulse on B.

Design a suitable asynchronous sequential circuit for this purpose which uses NAND gates only.

State whether or not your design is free from hazards and briefly justify your statement.

$$(F = \overline{\overline{A\bar{B}}\ \overline{AQ}}; Z = \overline{\overline{BQ}})$$

15.14   The conditions for powering a transmitter ($P = 1$) are that the cooling system is operating ($C = 1$) and the filament switch is closed ($F = 1$) but only if $C = 1$ first. If F is switched to 1 before C, it must be returned to 0 and the correct sequence initiated before the transmitter can be powered. Design an asynchronous sequential circuit which meets these conditions which uses an RS flip flop to generate the secondary variable. The whole system including the flip flop must be implemented with NOR gates.

$$(R = \bar{B}; S = \overline{A + \bar{B}}; Z = \overline{\bar{B} + Q})$$

15.15   Two sensors A and B, spaced 10 cm apart as shown in Fig. Q.15.15, are used to select bars of 12 cm length from a batch on a moving conveyor belt carrying bars of 8 cm and 12 cm length. Each sensor produces a logic '1' while a bar is passing it. Although the bars may come in any order, they are always separated by a distance greater than 12 cm and the belt cannot reverse direction.

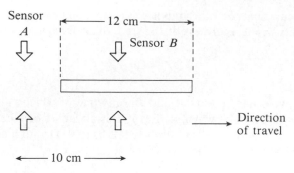

Fig. Q.15.15

Design an asynchronous sequential circuit using NOR gates only so that an output logic '1' indicates the selection of each 12 cm bar. This output should start when it is certain that a 12 cm bar is passing and stop when the bar has passed sensor B. The output should be logic '0' for all other conditions.

$$(q = \overline{\bar{B} + \overline{A + Q}}; Z = Q)$$

15.16 A circuit has two binary inputs A and B, only one of which can change at any time. The output must be 0 whenever $A = 0$. The first change in B occurring while $A = 1$, causes the output to become 1. Thereafter the output remains 1 until A returns to 0.

$$(q_1 = AQ_1 + A\bar{B}; q_2 = AB + AQ_2; Z = AQ_1)$$

15.17 Two sensors, A and B, are used to monitor the direction of rotation of a slowly rotating remote shaft and produce binary outputs which can never change at the same time. If the direction of rotation is clockwise, the output of A leads that of B and if it is anticlockwise, the output of B leads that of A. Sample waveforms are shown in Fig. Q.15.17 and the direction of rotation can reverse at any time.

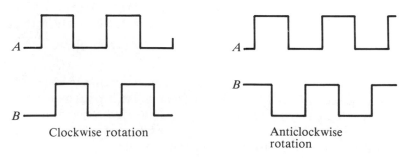

Clockwise rotation   Anticlockwise rotation

Fig. Q.15.17

Derive logic equations for an asynchronous sequential circuit which indicates by a logic '1' at its output Z when rotation is clockwise and by a logic '0' when it is anticlockwise.

$$q_1 = AB + BQ_1 + AQ_1; q_2 = A\bar{B} + AQ_2 + \bar{B}Q_2$$
$$Z = \bar{A}\bar{B}\bar{Q}_2 + \bar{A}BQ_1 + ABQ_2 + A\bar{B}\bar{Q}_1$$
$$+ \bar{A}B\bar{Q}_2 + BQ_1Q_2 + A\bar{Q}_1Q_2$$

16

Digital IC Families

Chapter Objectives
16.1   Properties of Digital ICs
16.2   Bipolar Logic ICs
16.3   MOS Digital ICs
16.4   Wired Logic
16.5   Interfacing TTL and CMOS ICs
16.6   Tristate Outputs
        Summary
        Revision Questions

At an early stage in the development of transistor logic circuits it was recognized that the complexity associated with the requirements of such systems as digital computers would stimulate a demand for high-density packaging and miniaturization. Originally this was met by simply encapsulating discrete components in epoxy resin to form packages of gates with leads brought out from the resin. However, the potential of this arrangement was severely limited. It was followed by the invention of the silicon integrated circuit which was truly a watershed in the development of electronic engineering. Packing densities increased by spectacular orders of magnitude until we now have the capability of making ICs containing millions of transistors.

The first ICs used bipolar technology, the most prominent development being the invention of TTL (transistor transistor logic) by Texas Instruments: it became, and remains, an industry standard. TTL is regarded as a high-speed logic family but is limited in this respect because its transistors have to be saturated when switched on. This limitation was counteracted by the invention of ECL (emitter coupled logic) which is an unsaturated bipolar logic and which remains the fastest logic family available. However, even though bipolar ICs can be more complex than packaged discrete components by several orders of magnitude, the number of gates that can be fabricated on a single chip (the packing density) is limited. This limitation was overcome by the development of the MOSFET, which made large-scale integration

(LSI) and very large-scale integration (VLSI) possible. MOS ICs are limited in operating speed, so apart from increasing packing densities, a major research activity is that of making circuits faster – if MOS technology could be made to operate at BJT speeds then we should have the best of both worlds. In this context a material which offers the potential of greatly increased operating speed is gallium arsenide (GaAs) (Section 1.13) but the extent to which it will replace silicon is difficult to predict.

In this chapter we shall first define the main properties of digital ICs and then describe the main logic families. It will be seen that there are sometimes trade-offs between the properties for some IC families. Some of the bipolar logic IC families, e.g. RTL, were invented many years ago but are still being made.

16.1 PROPERTIES OF DIGITAL ICs

In this chapter positive logic is assumed, i.e. 1 and 0 represent high and low voltages respectively.

16.1.1 Fan In

This is defined as the number of logic inputs a circuit can accommodate. Thus, an AND gate, which can have up to 4 input variables, has a fan in of 4. In some logic families (e.g. RTL) the operating speed falls as the number of inputs is increased so fan in may be determined by first specifying an acceptable speed. In many cases it is simply a practical matter such as the number of pins available on the package.

16.1.2 Fan Out

Fan out is the number of logic gate inputs of the same family that a gate output can drive before its performance is seriously impaired. For example, with DTL it is necessary for the driving transistor to be saturated for the gate output to be '0' but once the number of loads it has to drive reaches a certain figure this cannot be guaranteed.

16.1.3 Propagation Delay Time t_{PD}

Using the inverter of Fig. 16.1 as an illustrative example, when v_i is going from 0 to 1 it reaches a point at which the fall in v_o is initiated. Then t_{PHL} is the time taken for v_o to complete its transition from 1 to 0, i.e. to reach its *low* voltage.†

† Some texts use the time taken to complete 50% of the transition.

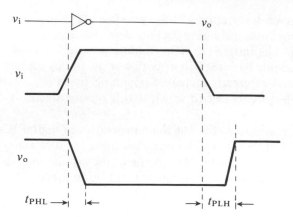

Fig. 16.1 Illustrating Propagation Delay

Figure 16.1 also illustrates the corresponding definition for t_{PLH}. Propagation delay time t_{PD} is defined as

$$t_{PD} = \tfrac{1}{2}(t_{PHL} + t_{PLH})$$

Propagation delay time for a digital IC will depend on such factors as the type of transistors used and the manner in which they are operated. Bipolar transistors are faster than MOSFETs and their speed of operation is increased if they are not saturated in the ON condition. As noted, ECL (unsaturated bipolar logic) is faster than TTL (saturated bipolar logic) which in turn is faster than MOS logic, although CMOS is challenging TTL in this respect.

16.1.4 Noise Margin

Manufacturers specify voltage limits to represent logical '0' and '1'. These voltages may not be the same for gate inputs and outputs. Where noise which is produced by, for example, switching spikes or mains hum, causes a logic voltage at a gate input to be outside the specified limits, then the circuit is not guaranteed to perform its logical function. Since this topic is actually concerned with voltage levels then, rather than use 1 and 0 to represent the logic levels, it is normal practice to refer to the high and low values of input and output voltages as V_{IH}, V_{IL}, V_{OH} and V_{OL}.

Consider the logic circuits of Fig. 16.2 with the output of A connected to the

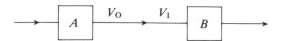

Fig. 16.2 Noise on the Link Between Gates A and B can Cause V_I to be Outside Logic Voltage Specification for Gate B

input to B. When the output of A is low, V_{OL}, the input to B, should also be low, V_{IL}, but there is a danger than noise on the link might cause the input to B to rise to a level above the maximum permissible value of V_{IL} ($V_{IL(max)}$) and be interpreted as a logical '1'. This is most likely to occur when V_{OL}, for gate A, is at its highest permissible value, which is termed $V_{OL(max)}$. For TTL, $V_{OL(max)} = 0.4$ V, and $V_{IL(max)} = 0.8$ V. Then as long as V_{IL} does not rise above 0.8 V the manufacturer guarantees that this voltage will be interpreted as a logical '0'. Therefore provided the noise on the link between A and B does not exceed 0.4 V and thus, when superimposed on the 0.4 V out of A, cause the input to B to exceed 0.8 V, the circuit will operate correctly. In general then, the low state noise margin (NML) is defined as

$$NML = V_{IL(max)} - V_{OL(max)}$$

which for TTL is 0.8 V $-$ 0.4 V $=$ 0.4 V.

In a similar manner, the high state noise margin (NMH) can be defined but in this case the quantities which are important are $V_{IH(min)}$ and $V_{OH(min)}$. Again using TTL as an example,

$$V_{OH(min)} = 2.4 \text{ V}, \ V_{IH(min)} = 2.0 \text{ V, and}$$
$$NMH \ = V_{OH(min)} - V_{IH(min)}$$
$$= 2.4 - 2.0 = 0.4 \text{ V}$$

High and low state noise margins are not equal for all logic families.

16.1.5 Power Dissipation

Even though the power dissipated by logic gates is, at most, a matter of milliwatts (about 40 mW for an ECL gate is one of the highest), it becomes a considerable factor when dealing with thousands of gates. Apart from the usual problems associated with power supplies, cooling, etc., one of the requirements of high-density packaging is low power consumption/gate. To achieve high switching speeds, relatively low value resistors are used in TTL and power consumption is 10 to 20 mW/gate. MOS logic ICs use MOSFETs as load resistors which can be operated in a manner resulting in low power consumption. For example, CMOS can be run at less than 1 mW/gate.

16.1.6 Packing Density

The number of gates that can be fabricated on a single IC has risen dramatically over the last decade, the most spectacular development occurring in MOS technology which has led to the production of such systems as the microprocessor. Packing density is often defined in terms of 'scales of integration'. Texas Instruments use, as the basis of their definition, a 'gate or circuit of similar complexity' and a summary of their scales of integration is as follows:

Small-scale integration (SSI) –   a single microcircuit containing less than 12 gates

Medium-scale integration (MSI) –   a single microcircuit containing 12 to 100 gates

Large-scale integration (LSI) –   a single microcircuit containing 100 to 1000 gates

Very large-scale integration (VLSI) – a single microcircuit containing over 1000 gates

VLSI is only feasible with MOS technology, the others can be bipolar or MOS.

16.2 BIPOLAR LOGIC ICs

16.2.1 BJT as a Switch

Unlike analog circuits, BJTs in binary logic circuits are operated in only two states. Their operating conditions can be deduced from Figs. 16.3a and b, for which the two states are:

(a) OFF defined by $I_B = 0$ and therefore $I_C = 0$.
   Then $V_{CE} = V_{CC} = 5$ V using the figures marked on the diagram.
(b) ON, in which case the transistor is saturated so that $V_{CE(sat)}$ approaches 0 V and therefore $I_C = V_{CC}/R = 2.5$ mA.

In practice $V_{CE(sat)}$ must be finite but for a good switching transistor it will be as low as 0.1 V to 0.2 V. In this chapter it will be assumed that $V_{CE(sat)} = 0.2$ V and that for this purpose $V_{BE(ON)} = 0.7$ V unless stated differently. However, the 0.2 V is not usually significant so we shall use the approximation that $I_{C(ON)} = V_{CC}/R$. To ensure saturation, then taking the case of using a transistor with the lowest current gain, for a particular type, as $h_{FE(min)}$, then,

$$I_{B(ON)} \geqslant \frac{I_{C(ON)}}{h_{FE(min)}}$$

For example, if $h_{FE(min)} = 40$, then using the figures previously quoted,

$$I_{B(ON)} \geqslant \frac{2.5 \text{ mA}}{40} = 62 \ \mu A$$

One of the disadvantages of saturating transistors in the ON state is that having fixed $I_{B(ON)}$ for the worst case – $h_{FE(min)}$ – then for the average case this value of $I_{B(ON)}$ is excessive. This will not affect $V_{CE(sat)}$ but when it is required to switch the transistor OFF the excess charge stored in the base must be extracted before

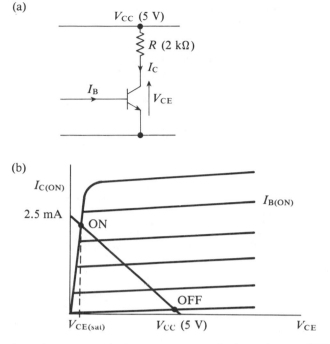

Fig. 16.3 (a) BJT and (b) Its Load Line Showing ON and OFF States

collector current stops flowing which lengthens t_{PLH}. An exception to this is emitter coupled logic (ECL) which is designed not to saturate in the ON state, thereby gaining in speed.

16.2.2 Resistor Transistor Logic (RTL)

Figure 16.4 is an RTL gate which has two logic inputs and which feeds three similar gates. If these are the maximum numbers of inputs and outputs then fan in is 2 and fan out 3. Let us assume that the logic inputs are fed from the outputs (collectors) of two similar gates, so that a '0' is represented by a saturated transistor and may be taken as 0.2 V, and a '1' by a nonconducting transistor and will be $+5$ V. With A and B both 0, the input voltages will be too low to bring Q into conduction so $F = 1$. If either A or B (or both) is 1, Q will conduct and the gate resistors can be chosen for Q to be saturated so $F = 0$. This, as the truth table (Table 16.1) shows, corresponds to NOR logic. A careful study of this circuit reveals several general points.

(a) When Q is OFF, $F = 1$, and to bring the transistor of the following gate(s) ON the circuit must supply current to them. This is called *current sourcing*. With Q OFF current flows from the 5 V supply through R and into the bases of the

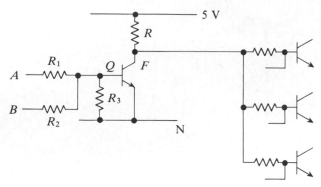

Fig. 16.4 RTL Gate

Table 16.1 Truth Table for RTL Circuits

A	B	F
0	0	1
0	1	0
1	0	0
1	1	0

following gates. This means that the '1' inputs to the following gates will be somewhat less than the 5 V assumed.

(b) When $A = 1$, Q must be saturated even if $B = 0$. If, additionally, $B = 1$, then the base current of Q will be approximately doubled. When it is required to turn Q OFF then excess charge must be extracted from the base and the time taken to do this will be the major factor determining t_{PLH}. Therefore, as fan in is increased, propagation delay time is also increased so there has to be a trade-off between t_{PD} and fan in.

(c) Connecting the base to a negative supply rail as in Fig. 16.5 has two benefits:
   (i) In the OFF condition V_{BE} is negative as opposed to 0 V (approx.). It will

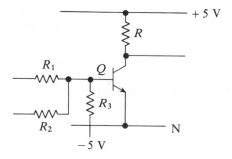

Fig. 16.5 RTL Gate with the Base Connected to a Negative Supply

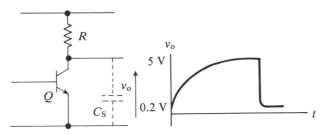

Fig. 16.6 Showing That Rise Time is Larger Than Fall Time Because for the Former C_s Charges via R but for the Latter C_s Discharges via Q

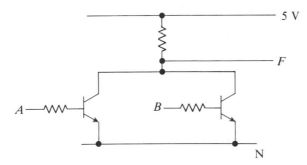

Fig. 16.7 An Alternative Form of RTL

therefore take a larger noise voltage to bring Q ON, so noise immunity is improved.

(ii) When Q is being switched OFF the circuit can be designed so that a substantial current is drawn through R_3 to shorten t_{PLH}.

(d) There is inevitable stray capacitance across the transistor output (Fig. 16.6). When Q is turned OFF, v_o rises relatively slowly because C_s must charge from 0.2 V to 5 V via R. When Q is turned ON, C_s discharges from 5 V to 0.2 V via Q which will have a much lower resistance than R. Therefore $t_{PLH} > t_{PHL}$.

An alternative form of RTL is shown in Fig. 16.7. In this circuit when either (or both) input is 1 then the corresponding transistor will conduct and make $F = 0$. Again, this is NOR logic. It has the advantage that the number of inputs at 1 does not affect base current for any one transistor and therefore t_{PLH} will be shorter than for the more traditional RTL. This form of RTL has been made by Ferranti using collector diffusion isolation (CDI). Its performance is better than NMOS as regards speed and is comparable in packing density.

16.2.3 Diode Transistor Logic (DTL)

Although DTL is not popular nowadays, it was the precursor of TTL and the action of the two circuits is similar. A brief description of DTL is justified because it will

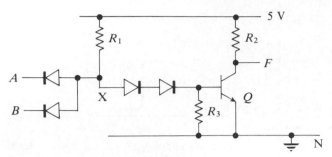

Fig. 16.8 DTL Positive NAND Gate

Table 16.2 Truth Table for DTL Circuit

A	B	F
0	0	1
0	1	1
1	0	1
1	1	0

show how its shortcomings in regard to propagation delay led to the development of TTL.

Given that $V_{CE(sat)} = 0.2$ V in Fig. 16.8, then a logical '0' at either of the inputs A and B will also be represented by 0.2 V if fed from a similar circuit. This will cause the associated diode to conduct by current flowing from the $+5$ V rail through R_1 to the input. With a 0.7 V across the diode the voltage at X wrt N will be 0.9 V. This is greater than the voltage which would be needed to make a single diode connected between X and the base of Q conduct. However, two diodes need 1.4 V, so Q stays OFF. Thus if either or both of the inputs is 0, $F = 1$. If both inputs are 1 $(+5$ V), the input diodes will be reverse biased so the voltage at X rises and the resistors can be chosen to make Q saturate. In this condition $F = 0$ and, as the truth table (Table 16.2) for the circuit shows, the DTL circuit is a positive NAND gate.

Note that when Q is saturated the base current is not affected by the number of inputs so therefore, unlike the single transistor version of RTL, propagation delay is not a function of fan in.

When $F = 0$ the function of Q is to *sink* current from the input diodes of the gates that it feeds, as shown in Fig. 16.9. To ensure that Q is saturated, it should be noted that there are two components of collector current:

(a) the current through $R_2 \doteqdot 5$ V$/R_2$;
(b) NI

where N is the fan out and I the current which must be sunk by Q from each gate

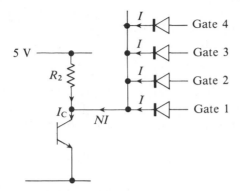

Fig. 16.9 Current Sinking

that it feeds. Therefore the level of base current must be high enough to ensure saturation for the worst case of maximum fan out and minimum h_{FE}. For the average case the base of Q will contain excess charge and when F is required to go to 1, the excess charge must be removed. Therefore a consequence of high fan out is prolonged t_{PLH}. Deterioration of t_{PLH} also occurs because of the same effect which was noted with RTL, viz. that when Q is cut off, shunt capacitance C_s must be charged through R_2.

16.2.4 Transistor Transistor Logic (TTL)

TTL was the first type of logic circuit specifically designed to be manufactured in IC form; DTL was originally conceived as a discrete component circuit. Referring to the basic circuit of Fig. 16.10, the multiple emitter transistor Q_1 stands out as a novel feature. In DTL (Fig. 16.8) the path between inputs and the transistor base is via diodes connected back to back. In TTL each E–B–C path from an input to Q_2 base is similar, so it follows that the logic function of TTL is the same as DTL, i.e. positive NAND. With any input at 0 (0.2 V) Q_1 is saturated by base current

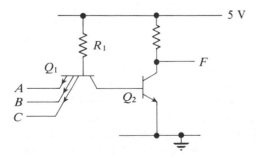

Fig. 16.10 Basic TTL

through R_1 so the voltage at Q_2 base is 0.4 V, which is too low to bring Q_2 ON. With all inputs 1 ($+5$ V) all E–B junctions of Q_1 are reverse biassed so its B–C junction becomes forward biassed and Q_2 saturates with base current flowing from the $+5$ V rail through R_1 and the B–C junction of Q_1.

As noted, the propagation delay of DTL is prolonged by the time taken to remove excess charge stored in the transistor base and because C_s has to be charged through the collector load resistance. The basic TTL circuit only shortens the time to remove excess charge in the base. When an input falls to 0.2 V, $V_{BE2} = 0.7$ V as long as it is saturated, and therefore $V_{CE1} = 0.5$ V. This means that Q_1 is not saturated but is in its active region. Therefore any current flowing *out of B_2* is the collector current of Q_1. This will be an amplified version of I_{B1} and the gate will be designed for this current to ensure that excess charge in Q_2 base is removed rapidly.

Unused Inputs

Any input which is left unconnected acts as a 1 input because both have the same effect in leaving the E–B junction nonconducting. It is better, however, to connect unused inputs to $+5$ V through resistors of 1 kΩ to avoid pick up.

Totem Pole TTL

The most common form of TTL is shown in Figs. 16.11a and b which are identical but are drawn separately to show significant voltages in the '0' and '1' output states. Output transistors Q_3 and Q_4 form what is known as a totem pole arrangement which has a low output resistance for both '0' and '1' output conditions: Q_3 and Q_4 are ON for the 1 and 0 conditions respectively. This makes t_{PLH} shorter than for DTL because stray capacitance charges via Q_3 which has a much lower resistance than the collector load of a DTL gate.

To confirm the output states for different input conditions, consider first Fig. 16.11a with all inputs 1 (nominally $+5$ V but as we shall see, rather less in practice). All E–B junctions of Q_1 are now reverse biassed but the B–C junction is forward biased by current from the $+5$ V rail through R_1. Working backwards from Q_4 emitter and taking 0.7 V as the voltage across each of the three conducting junctions (BC1, BE2 and BE4), we have the voltages shown in Fig. 16.11a, Then with $V_{CE2(sat)} = 0.2$ V, $V_{BN3} = 0.9$ V. It requires 1.4 V across V_{BE3} and D to make them conduct, so these two devices are OFF. For any one or more inputs at 0 (0.2 V), Q_1 will be ON and the voltages will be as shown in Fig. 16.11b. With $V_{CE1(sat)} = 0.2$ V, there will be 0.4 V at Q_2 base which is too low to make Q_2 and Q_4 conduct. Therefore $V_{BN3} = 5$ V and with 0.7 V across BE3 and across D the output is 3.6 V (if there is an output load), which is nominally 1 for TTL.

We can now relate the TTL noise margin figures quoted in Section 16.1.4 to Figs. 16.11a and b. For the low noise margin (NML), a worst-case figure of $V_{CE(sat)} = 0.4$ V for Q_4 in Fig. 16.11a gives $V_{OL(max)} = 0.4$ V. In Fig. 16.11b with

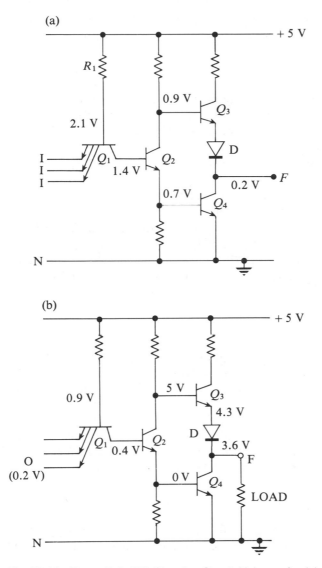

Fig. 16.11   Totem Pole TTL Showing Circuit Voltages for (a) Output O, (b) Output I

$V_{IL} = 0.8$ V, V_{BN2} can be as high as 1.2 V but this is still less than the 1.4 V required to make Q_2 and Q_4 conduct. Therefore $V_{IL(max)} = 0.8$ V and

NML $= 0.8$ V $- 0.4$ V $= 0.4$ V

For NMH, $V_{OH(min)}$ is given the conservative figure of 2.4 V. Reference to Fig. 16.11b shows that this worst-case figure would only come into consideration with a reduced power supply (4.75 V is regarded as the minimum acceptable) and with maximum pn junction voltages. In Fig. 16.11a, if $V_{IH} = 2$ V this is not quite low enough to bring the E–B junctions of Q_1 into conduction. Therefore $V_{IH(min)} = 2$ V and

NMH = 2.4 − 2 = 0.4 V.

TTL circuits are made by the planar process. This results in the E–B junction having a reverse breakdown voltage of 7 V so the inputs should not exceed this value. The supply voltage must be in the range 4.75 V to 5.25 V. Fan in is fixed in manufacture but can be as high as 8.

Calculations of fan out are based on the unit load (UL). For TTL one UL is an input to one TTL gate. When the output is 1 a gate has to supply only 40 μA to a following gate for standard TTL, and this is not a significant factor in determining fan out. However, when the output is 0 a gate output has to sink 1.6 mA from each following gate input. Reference to data sheets for standard TTL shows that the maximum permissible output current I_{OL} is 16 mA. If this figure is exceeded by connecting too many ULs across the output then, because of a gate's finite output resistance, V_{OL} will rise above its specified maximum value of 0.4 V. Therefore, with one UL = 1.6 mA and $I_{OL(max)} = 16$ mA, fan out = 10.

Schottky TTL

Aluminum can act as a p type impurity and when brought into contact with n type silicon during the process of forming IC terminals it forms a pn junction. This is known as a Schottky barrier diode and it can be introduced into the fabrication process to produce transistors with the configuration shown in Fig. 16.12a which are given the distinguishing symbol of Fig. 16.12b. The forward voltage of a Schottky diode is only 0.25 V but this is sufficient to prevent the transistor from saturating. As V_{CE} tries to fall to 0.2 V, D conducts and diverts excess current away from the base. The result is a significant reduction in propagation delay.

High and Low Power TTL

The resistor values used in TTL ICs can be selected to meet different specifications. Low resistor values are used for fast response and are often used in Schottky TTL

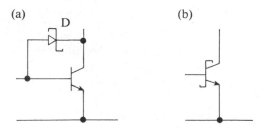

Fig. 16.12   (a) Schottky Barrier Diode which Prevents a Transistor from Saturating (b) Symbol for a Transistor with a Schottky Diode

Table 16.3 Comparison of Different TTL Families

Type No.	Description	t_{PD}(ns)	P_d(mW)
74	Standard	10	10
74L	Low power	33	1
74H	High speed	6	22
74S	Schottky	3	19
74LS	Low power Schottky	9.5	2
74ALS	Advanced Low power Schottky	5	1.25

to make a further increase in speed. However, they incur the penalty of high power dissipation. Low power TTL ICs use larger resistor values which increase propagation delay. Table 16.3 summarizes typical values for various types of TTL.

16.2.5 Emitter Coupled Logic (ECL)

ECL is a nonsaturating, bipolar logic family with emitter follower output drive: these features combine to make ECL the fastest of all logic ICs, with t_{PD} typically 2 ns.

Manufacturers give many of the values of voltage, current and resistance for their ECL circuits to 3 significant figures. In the following example only 2 significant figures are used but they are representative for a typical ECL gate. The circuit is based on the long-tailed pair (Section 7.4) of Fig. 16.13 which has one input held at a dc reference of -1.3 V. Any one of the three supply points can be earthed but manufacturers claim that earthing the collector rail gives the best noise immunity. Taking -1.6 V and -0.8 V as representing logic inputs 0 and 1 respectively, and

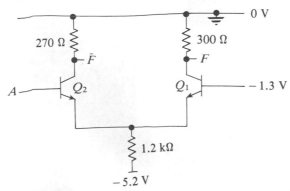

Fig. 16.13 A Long-tailed Pair Used as the Basic Circuit of an ECL Gate

taking $V_{BE(ON)} = 0.8$ V (characteristic of ECL transistors), then the following deductions can be made for Fig. 16.13.

When $A = 0 (-1.6$ V); Q_2 is OFF, Q_1 is ON and $V_E = -2.1$ V (wrt the collector rail).

Therefore $I_E = \dfrac{(-2.1 + 5.2) \text{ V}}{1.2 \text{ k}\Omega} = 2.6 \text{ mA} \doteq I_{C1}$.

Therefore $V_{C1} = -I_{C1} \times 300 \ \Omega = -0.8$ V, and $V_{C2} = 0$ V.

When $A = 1 (-0.8$ V); Q_1 is OFF, Q_2 is ON and $V_E = -1.6$ V.

Therefore $I_E = \dfrac{(-1.6 + 5.2) \text{ V}}{1.2 \text{ k}\Omega} = 3 \text{ mA} \doteq I_{C2}$

Therefore $V_{C2} = -I_{C1} \times 270 \ \Omega = -0.8$ V, and $V_{C1} = 0$ V.

Note that in both cases, V_{CE} for the ON transistor exceeds 0.2 V (1.3 V and 0.8 V) so the transistor is not saturated.

Taking V_{C1} as the circuit output F and V_{C2} as its complement $\bar{F}$, the logic specification of the circuit is summarized in Table 16.4. Although input and output logic voltages are different, nevertheless.

$F = A$ and therefore $\bar{F} = \bar{A}$.

Now by adding extra transistors, as in Fig. 16.14, then if any of the inputs is 1, F

Table 16.4   Truth Table for ECL Circuit of Fig. 16.13

A	F (V_{C1})	$\bar{F}$ (V_{C2})
0 (-1.6 V)	0 (-0.8 V)	1 (0 V)
1 (-0.8 V)	1 (0 V)	0 (-0.8 V)

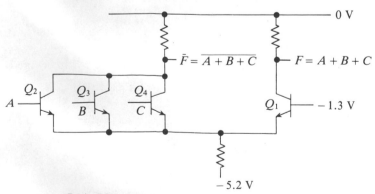

Fig. 16.14   Basic ECL Gate

o not cause severe switching spikes on power
han 20 because of the high current-driving cap
ts. Finally, the choice of complementary OR and
ner's task.

ction Logic (I²L)

high speed of operation characteristic of bipolar logic circuits
ch higher packing density than, for example, TTL because it
ication stages. In Fig. 16.16 it can be seen that to provide a 0 at
, and for a 1, Q must be OFF. However, the circuit would work
thout R, the main requirement being that Q must be a current sink
0. Figure 16.17 is an I²L inverter which uses this principle. A current
from a source external to the chip through a pnp common base
ansistor Q_1 to point A, which is the logic input to the gate. It will be
e collector output of a similar gate. When this feeder stage is ON, i.e.
is 0 and therefore $A = 0$, it sinks the current I away from Q_2 base so Q_2
nd Q_2 output(s) is 1. When the feeder stage is OFF (1) the injection current
go to Q_2 base, so Q_2 will be ON and its output(s) 0.

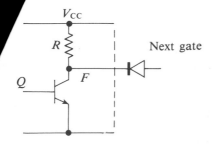

Fig. 16.16 Current Sink

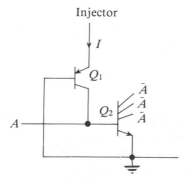

Fig. 16.17 I²L Inverter

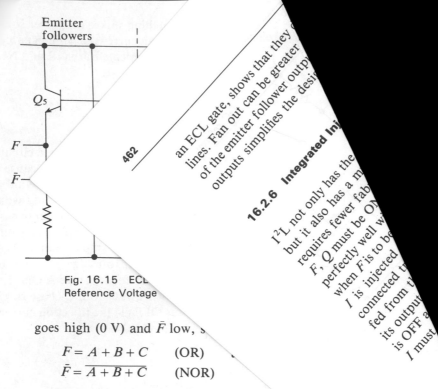

Emitter followers

Q_5

F

$\bar{F}$

Fig. 16.15 ECL
Reference Voltage

462

an ECL gate, shows that they
lines. Fan out can be greater
of the emitter follower outp
outputs simplifies the desi

16.2.6 Integrated Inj

I^2L not only has the
but it also has a m
requires fewer fab
F, Q must be ON
perfectly well w
when F is to be
I is injected
connected t
fed from t
its outpu
is OFF a
I must

goes high (0 V) and $\bar{F}$ low,

$$F = A + B + C \qquad \text{(OR)}$$
$$\bar{F} = \overline{A + B + C} \qquad \text{(NOR)}$$

Figure 16.15 shows the addition of tw
Q_5 and Q_6 which are fed from Q_1 and Q_2 to
follower stages perform two functions:

(a) They reduce the logic output voltage levels by one
the 1 and 0 output voltages are reduced from 0 V a.
-1.6 V respectively, which makes them the same as .
simplifies the cascading of gates.
(b) They provide a lower output resistance drive to charge anc
capacitance rapidly.

Figure 16.15 also shows a temperature-stabilized dc reference sour
base of Q_1.

Advantages and Disadvantages of ECL

As may be inferred from the relatively small voltage between logic voltages, noise
immunity of ECL is poor. Further, the consequence of relatively low resistor values
is high power dissipation. However, the unsaturated transistor operation results in
switching speeds which are faster even than Schottky TTL. Also, the fact that there
is little difference in the current drawn from the supply between the two states of

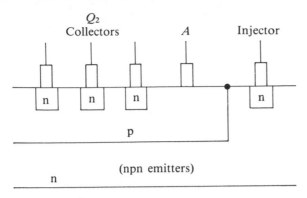

Fig. 16.18 I^2L Layout

The capability for high packing density can be explained by reference to the layout of an inverter on a chip, as shown in Fig. 16.18. All npn transistors such as Q_2 have a common grounded emitter, which also forms the base of the injector transistor. A single diffusion forms the base of Q_2 and the collector of Q_1 and one further diffusion forms the collector(s) of Q_2. No resistors are needed and the layout does not require isolation diffusion to separate the transistors nor metal interconnections. These features make I^2L the most suitable bipolar logic for LSI.

There is also a Schottky version of I^2L which has a propagation delay of about 10 ns compared with 25 ns for the standard version.

16.3   MOS DIGITAL ICs

Although it was recognized that NMOS (N enhancement MOSFETs) had superior characteristics to PMOS for the implementation of digital ICs, there were difficulties associated with NMOS fabrication so the first MOS digital ICs were PMOS. However, NMOS has now superseded PMOS because: (a) its packing density is higher for reasons associated with electrons being more mobile than holes; (b) its speed is faster as a consequence of reduced size; and (c) its gate and drain voltages are positive which make it compatible with TTL. Complementary P and NMOS – CMOS – is a further addition to the range of digital ICs. It is faster than NMOS and has lower power dissipation, but it has lower packing density and higher cost.

16.3.1   MOSFETs as Switches

MOS logic circuits usually contain n channel enhancement MOSFETs. They sometimes also include n channel depletion MOSFETs or p channel enhancement MOSFETs. For an n channel enhancement MOSFET (Fig. 16.19), in which the

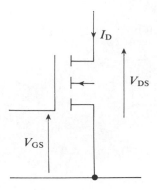

Fig. 16.19   N Channel Enhancement MOSFET

substrate may be grounded or taken to a negative supply, V_{GS} must be positive and at least equal to the threshold voltage V_T before it conducts, where V_T may be 1 to 2 V.

Therefore if $V_{GS} \geqslant V_T$ the MOSFET is ON,
            if $V_{GS} < V_T$ the MOSFET is OFF.

N channel depletion MOSFETs are usually used as drain load resistors for an enhancement device. They will conduct when $V_{GS} = 0$ V so usually the only condition which it is necessary to observe for practical purposes is that $V_{GS} \geqslant 0$ V.

A p channel enhancement MOSFET is complementary to the n channel and therefore, to conduct, V_{GS} must be more negative than the threshold voltage V_T which will typically be -1 to -2 V.

16.3.2   NMOS Inverter

NMOS operation is exemplified by the inverter of Fig. 16.20. Assume logic levels of 0 V (0) and $+5$ V (1) with $V_{DD} = +5$ V. These voltages make the circuit compatible with TTL although, unlike TTL, significantly higher operating voltages can safely be used with NMOS. Q_2 acts as the drain load for Q_1 and is held ON since its gate is at $+V_{DD}$ and is therefore positive wrt source. Q_1 only comes ON when $A = 1$, i.e. when its gate is positive. The considerations governing the operation of Q_2 as a nonlinear resistance, the value of which is a function of its terminal voltages, are complex. However, if we first note that $I_{D1} = I_{D2} = I_D$, then if $A = 0$ (0 V), $V_{GS1} = 0$ V and Q_1 is OFF. Since, as noted, Q_2 is ON at all times, then its resistance will be much lower than that of Q_1 and therefore the voltage at F will approach V_{DD}, i.e. $F = 1$. When $A = 1$, Q_1 conducts and I_D rises. If the two MOSFETs were identical then the voltage at F would be $V_{DD}/2$, but by making the channel length : width ratio of Q_2 at least 10 times that of Q_1, then Q_2 will have a much higher resistance than Q_1 so the voltage at F will approach 0 V, i.e. $F = 0$. Therefore

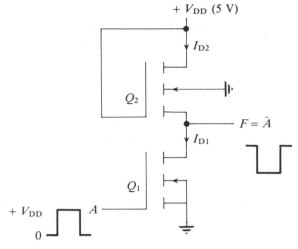

Fig. 16.20   NMOS Inverter

Fig. 16.21   Simplified MOSFET Symbol

the circuit is an inverter. Figure 16.20 shows the substrate being grounded, but by making it 2 or 3 V negative wrt ground, interelectrode capacitances are reduced, which increases operating speed.

Usually data books simply draw MOSFETs as in Fig. 16.21. The lack of an arrow to indicate N or P MOS is justified because the logic family under consideration will be known. The fact that the gate terminal is centralized emphasizes the symmetry of the device in which case drain and source are interchangeable. Substrate connections are not shown: they are known to be commoned and, as noted, taken to a negative supply (NMOS) or grounded.

Use of Depletion MOSFET as Load

We have assumed that by connecting the gate of Q_2 in Fig. 16.20 to V_{DD}, the device will remain in conduction and act like an amplifier load resistor. However, an enhancement MOSFET will not conduct below its threshold voltage V_T which is the minimum value of V_{GS} required to bring it into conduction. Therefore when the output is logical 1 which ideally will be V_{DD}, it will be less than this value by V_T which may be 1 to 2 V. This drawback can be eliminated by using a depletion MOSFET for the load because these devices conduct when $V_{GS} = 0$ V. It is also claimed that depletion MOSFETs have better transient characteristics than enhancement devices. Further, enhancement/depletion combinations lend themselves to efficient use of chip area and are therefore favored for large-scale integration.

NMOS NOR Gate

The inverter of Fig. 16.20 is easily extended to a NOR gate by adding extra MOSFETs as shown in Fig. 16.22. The output F will be 0 when any input is 1.

NMOS NAND Gate

A NAND gate is shown in Fig. 16.23. For the output to be 0 all MOSFETs must be conducting and this will only be the case when all inputs are 1.

NMOS Characteristics

Capacitance between gate and the other two electrodes combined with high output resistance results in NMOS being a relatively slow logic. High input resistance allows an NMOS gate to drive many gates of the same family so fan out can be high but input capacitance introduces a speed penalty so there must be a trade-off. The effective current drain in the OFF state of an NMOS gate is virtually zero. In the ON state it will still be much lower than for any bipolar logic and averaging the two states yields a power dissipation of about 150 μW/gate.

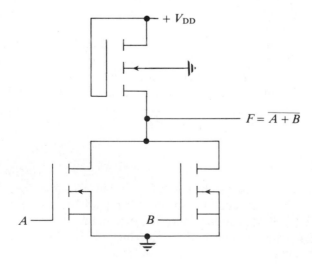

Fig. 16.22   NMOS NOR Gate

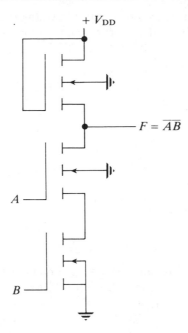

Fig. 16.23   NMOS NAND Gate

16.3.3   Complementary MOS (CMOS)

CMOS gates are fabricated by forming NMOS and PMOS on the same IC. For the inverter shown in Fig. 16.24 the two gates are commoned to form an input and the two drains to form an output. Then with $A = 0$ (0 V) Q_1 will be OFF but Q_2 will

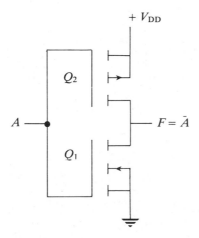

Fig. 16.24   CMOS Inverter

be ON because its gate is 0 V and its substrate positive: thus the voltage between gate and substrate is negative, which is the polarity for conduction with PMOS. Therefore $F = 1$. With $A = 1$, Q_1 will be ON and Q_2 OFF, so $F = 0$ which demonstrates that the circuit is an inverter.

CMOS NOR Gate

In Fig. 16.25

(a) when $A = 1$, Q_1 is ON and Q_3 OFF so $F = 0$;
(b) when $B = 1$, Q_2 is ON and Q_4 OFF so $F = 0$;
(c) when $A = B = 0$, Q_3 and Q_4 are ON so $F = 1$.

Therefore the circuit is a NOR gate.

CMOS NAND Gate

In Fig. 16.26, when $A = B = 1$, Q_1 and Q_2 are both ON and Q_3 and Q_4 are both OFF so $F = 0$. If either (or both) input, say A, is 0, then Q_1 will be OFF and Q_3 ON so $F = 1$. These are the NAND function conditions.

CMOS Characteristics

Fabrication of two types of MOSFET on one IC is inefficient in the use of silicon,

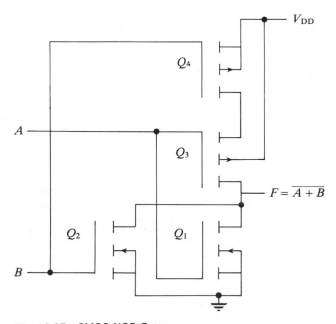

Fig. 16.25   CMOS NOR Gate

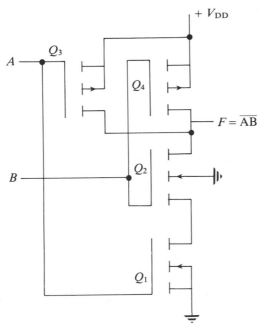

Fig. 16.26   CMOS NAND Gate

so packing density for CMOS is lower than for NMOS. In both states one type of MOSFET will be OFF but in transition load capacitance has to be charged. Therefore in the static condition, power dissipation is virtually zero but it rises with switching frequency. Switching speed is faster than for NMOS because, although gate capacitance values are similar, the fact that one of the driving MOSFETs is ON results in faster charge and discharge. The complementary nature of the family determines the range of input voltages that can be used for 0 and 1, so that

$$V_{IL(max)} = 0.3 V_{DD}$$
$$V_{IH(min)} = 0.7 V_{DD}$$

which, for a 5 V supply gives 1.5 V (0) and 3.5 V (1).

With higher switching speed than NMOS and lower power dissipation than TTL, CMOS is attractive for many traditional TTL applications. For this reason TTL and CMOS have identical pin outs and similar numbers. For example, 7400 is a TTL quad 2 input NAND gate. 74C00 is a CMOS quad 2 input NAND gate.

16.4   WIRED LOGIC

When the outputs of two TTL NAND gates are connected as in Fig. 16.27, the resulting logic can be deduced as follows. If either gate produces a 0 then F will be

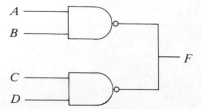

Fig. 16.27   Two NAND Gates with Outputs Wired Together

0 because the output transistor (Q_4 in Fig. 16.11) will hold the output down even if the other output, in isolation, would have produced a 1. This is equivalent to saying that F is the AND function of the two outputs $\overline{AB}$ and $\overline{CD}$, i.e.

$$F = \overline{AB} \cdot \overline{CD}$$

Hence the arrangement is called WIRED-AND, (oddly enough it is often known as WIRED-OR). The advantage of the arrangement is that it saves an AND gate if that particular logic function is required.

Other families, for example, ECL can operate with wired outputs. With ECL the output is 1 when an output transistor is OFF. Therefore, for the OR output,

$$F = (A + B) + (C + D)$$

and for the NOR output,

$$F = \overline{A + B} + \overline{C + D}$$

16.4.1   TTL Open Collector Gates

TTL totem pole outputs should not be WIRE-ANDed. In Fig. 16.28 if only the left-hand gate is at 0 output then Q_4 must not only sink current from the load but

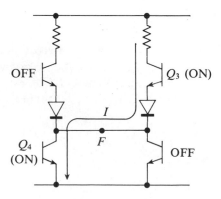

Fig. 16.28   A Totem Pole Output Sinking Current From a Gate to which it is Wired

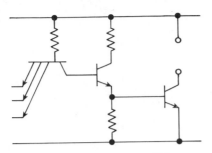

Fig. 16.29   Open Collector TTL

it must also sink current from Q_3 of the other gate. This is likely to overload Q_4. The solution is to use 'open collector' versions of TTL, the circuit of which is shown in Fig. 16.29. When a number of these are wired together, a single external pull up resistor must be provided whose value will depend on the number of gates wired together.

16.5   INTERFACING TTL AND CMOS ICs

TTL is always operated from a nominal $+5$ V supply while CMOS will operate from anything from $+3$ V to $+15$ V. Therefore, when a designer wishes to capitalize on the advantages of both circuits in a single system they can be operated from a common $+5$ V rail.

CMOS Driving TTL

When the CMOS gate output of Fig. 16.30 is 1 it only has to supply 40 μA to each TTL gate it feeds; but when it is 0 it has to sink 1.6 mA from each gate input. With $V_{IL(max)}$ for TTL at 0.8 V, then to maintain NML at 0.4 V, V_{OL} of the CMOS gate must be no more than 0.4 V. Data sheets for the CMOS gate being used will specify how much current it can sink before the output rises to 0.4 V and on this basis, the number of 1.6 mA loads (ULs) from which it can sink current can be calculated to give the fan out. Buffers are available where direct connection is unsatisfactory.

TTL Driving CMOS

With $V_{OL(max)}$ for TTL at 0.4 V and $V_{IL(max)}$ for CMOS at 1.5 V (0.3 times the supply voltage), there is no incompatibility with regard to the low-voltage state. Further

Fig. 16.30   CMOS Driving TTL

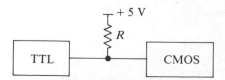

Fig. 16.31 TTL Driving CMOS with Pull Up Resistor Loading Open Collector Output to Make $V_{OH} = 5V$

CMOS has a very high input resistance so the current requirement is negligible. For the high state, $V_{OH(min)} = 2.4$ V for TTL and for CMOS $V_{IH(min)} = 3.5$ V (0.7 times the rail voltage) which is clearly unsatisfactory since the CMOS gate may well not accept 2.4 V as a 1 input. The solution, as shown in Fig. 16.31, is to use open collector TTL with a pull up resistor to raise V_{OH} to about 5 V.

16.6 TRISTATE OUTPUTS

In binary logic circuits one becomes accustomed to the concept of two output states, 0 and 1, which may be represented, for example by 0 V and $+5$ V. In bus systems where several ICs may be connected to a common bus, a 0 V output from TTL, for example, will hold a bus line at 0 V even though the requirement is that a different IC sets the voltage. For this reason some gates have a third output state which is an open circuit. In this state a gate output will have no effect on a bus line it is connected to. A simplified circuit showing how this is done with a totem pole TTL gate is shown in Fig. 16.32. When the control pulse is $+5$ V, D_1 and D_2 are OFF and the gate operates in the usual way. When the control pulse is at 0 V, the diodes are ON and both output transistors are held OFF, leaving the output open circuited.

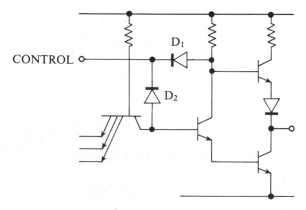

Fig. 16.32 Tristate TTL Gate. When CONTROL = 0 V Output Transistors will be OFF and Therefore Output Open-circuited

Table 16.5   Typical Characteristics of Digital ICs

	RTL (CDI)	*DTL*	*TTL (standard)*	*ECL*	*I^2L*	*NMOS*	*CMOS*
Propagation delay (ns)	15	25	9	2	20	50	25
Noise immunity (V)	0.3	0.8	0.8	0.2	0.2	1	1.5
Power dissipation (mW)	12	10	10	25	0.05	0.1	0.001†
Packing density	LSI	MSI	MSI	SSI	LSI	VLSI	LSI
Fan in	5	10	8	15			
Fan out	4	10	10	25		>50	>50

† Static – increases with clock frequency.

SUMMARY

Basically, bipolar logic ICs are chosen for fast operating speed and MOS for high packing density and low power dissipation/gate. However, this is a generalization and Table 16.5 shows typical values for the important characteristics of the major IC families.

REVISION QUESTIONS

16.1   An ECL circuit is quoted as having the following logic voltage:

	Max (V)	Min (V)
V_{IL}	−1.33	−5
V_{IH}	−0.7	−1.03
V_{OL}	−1.5	−1.8
V_{OH}	−0.7	−0.85

Calculate NML and NMH.   (0.17 V; 0.18 V)

16.2   What must be the minimum value of h_{FE} for the BJT in the inverter of Fig. Q.16.2 for it to saturate when the input A is 1 (+5 V)? Assume $V_{BE(ON)} = 0.7$ V; $V_{CE(sat)} = 0$ V.

(31)

16.3   If Fig. Q.16.2 is converted into an RTL gate by simply adding another 33 kΩ resistor connected to the base, what effect does it have on t_{PHL} and t_{PLH}?

16.4   Explain why shunt capacitance across the output of Fig. Q.16.2 does not extend t_{PHL} and t_{PLH} equally.

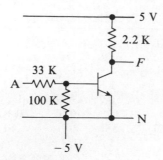

Fig. Q.16.2

16.5 Use RTL and TTL to explain the difference between current sourcing and current sinking.

16.6 Describe two features of TTL which make it faster than DTL. How is Schottky TTL made faster than standard TTL?

16.7 Would you expect low-power TTL to be faster or slower than standard TTL?

16.8 Compare the advantages and disadvantages of MOS logic over bipolar logic.

16.9 What are the main advantages of MOS logic over bipolar logic?

16.10 Explain the operation of the enhancement NMOS gate of Fig. Q.16.10. What logic function does it realize? What must be the difference in the geometry of input and load MOSFETs? What are the approximate output logic voltage levels?

(0 V, +4 V)

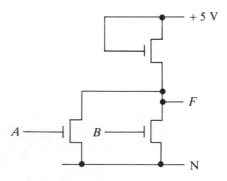

Fig. Q.16.10

16.11 Why is a depletion MOSFET often used as the load device for Fig. Q.16.10?

16.12 What are the advantages of CMOS (a) over NMOS; (b) over TTL?

16.13 Explain the statement that it is only when static that CMOS has lower power dissipation than NMOS.

16.14 Explain why the outputs of standard TTL ICs should not be wired together. Under what condition is wiring permissible?

16.15 Two TTL gates have inputs (1) $\bar{A}$ and B, (2) C and $\bar{D}$. What will be the output logic function when the gates are wired together?

16.16 Why is it necessary to use a pull up resistor when TTL is used to drive CMOS? Under what circumstances is this operation successful?

16.17 Why are tristate circuits necessary for certain applications?

17

Semiconductor Memories

Chapter Objectives

17.1 Read Only Memory (ROM)
17.2 ROM Applications
17.3 Programmable Read Only Memory (PROM)
17.4 Erasable and Programmable Read Only Memory (EPROM)
17.5 Electrically Erasable and Programmable Read Only Memory (EEPROM)
17.6 Random Access Memory (RAM)
17.7 RAM Memory Cells
17.8 Programmable Logic Array (PLA)
17.9 Programmable Array Logic (PAL)
17.10 Uncommited Logic Array (ULA)
17.11 Memory Speeds and Logic
       Summary
       Revision Questions

Memory device technology has advanced rapidly over the last decade or so. Magnetic cores used to be prevalent but have been superseded by electronic memories. A system such as a digital computer will have some mass storage medium such as magnetic disk or tape for storing user software and files, but for processing it will use electronic memories which are the subject of this chapter. They may be classified into ROMs (read only memories) and RAMs (random access memories). Once programmed, a ROM, as its name suggests, can only be read from – data cannot be written into it. Data can be written into a RAM as well as being read from it, and since the writing aspect is the feature which distinguishes a RAM from a ROM, then arguably a better name would have been READ/WRITE memory. It could also be argued that the use of the word 'random' here is stretching its meaning somewhat because it is simply meant to indicate that any memory location can be accessed by choice. But this is equally true of a ROM. However, the names ROM and RAM are firmly established.

Before describing ROMs and RAMs, some other terms associated with memory should be defined:

(a) *Static* means that data are retained permanently.
(b) *Dynamic* means that data are not stored permanently and must therefore be refreshed periodically.
(c) *Destructive* means that data are erased from memory when they are read.
(d) *Volatile* means that power supplies must be kept on for data to be retained.

ROMs are static, nondestructive and nonvolatile. RAMs are volatile and may be static or dynamic, destructive or nondestructive.

17.1 READ ONLY MEMORY (ROM)

A ROM – more fully a mask programmable ROM – is an LSI device in the form of a matrix in which each crossover point may be joined by a semiconductor device during the fabrication process. Figure 17.1 is a 4×2 ROM in which four of the eight crossovers are joined by diodes. In practice the joins are more likely to be made by BJTs or MOSFETs. The four rows A_0 to A_3 can be regarded as inputs or better, addresses. The two columns D_0 and D_1 are the outputs.

One way to operate a ROM is that when an address is selected a logical 1 (high voltage) would be connected to the selected line. Thus a 1 at A_2 in Fig. 17.1 brings on diode X. With the other address lines at 0 the remaining diodes will be OFF. The 1 at A_2 then, is conducted through diode X to give a 1 at D_1. With both diodes of column D_0 OFF, D_0 will be 0. In this arrangement then, the presence of an interconnection at a junction results in a 1 at the output when the particular row is

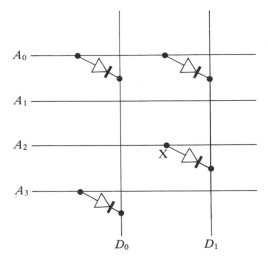

Fig. 17.1 4×2 ROM

addressed. In Fig. 17.1 four addresses can access four stored words as follows:

$$A_0 = 1 \qquad D_0 D_1 = 11 \qquad\qquad A_2 = 1 \qquad D_0 D_1 = 01$$
$$A_1 = 1 \qquad D_0 D_1 = 00 \qquad\qquad A_3 = 1 \qquad D_0 D_1 = 10$$

ROM Address Decoding

In practice ROMs are much larger than 4×2 and for for a device such as a 1024×4 ROM the idea of having 1024 address lines is not a practical proposition. It is therefore customary to have an address decoder (Section 13.4) as in Fig. 17.2. With 10 input address lines A_0 to A_9, any one of $2^{10} = 1024$ rows in the ROM can be selected, so that any one of 1024, 4-bit words will appear at D_0 to D_3.

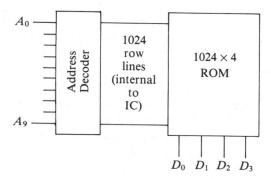

Fig. 17.2   ROM Address Decoding

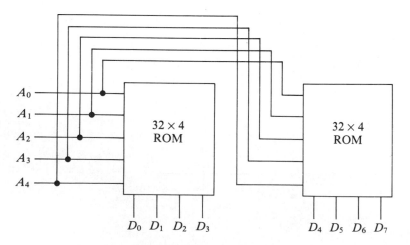

Fig. 17.3a   Two 32×4 ROMs Connected to Make a 32×8 ROM

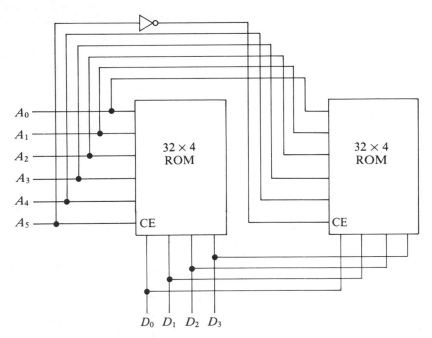

Fig. 17.3b   Two 32 × 4 ROMs Connected to Make a 64 × 4 ROM

Memory Expansion

Identical ROMs can be cascaded to provide an expanded memory. The possibilities are:

(a) Increased word length.
(b) Increased number of addresses.
(c) Increase in both word length and number of addresses.

Figure 17.3a shows two 32×4 ROMs − in which the address decoders are included in the ROM block − being used to create a 32×8 memory. Corresponding address lines on the two ROMs are commoned so that any address activates each row of the same number in each ROM and produces an 8-bit output word D_0 to D_7.

Two 32×4 ROMs are connected in Fig. 17.3b to create a 64×4 memory. Each ROM has a chip enable (CE) input which activates the ROM when it is at, say, logical 1. Then with $A_5 = 1$ the left-hand ROM is activated, while with $A_5 = 0$ it will be the right-hand ROM. Thus the 32 addresses 000000 to 011111 produce output words D_0 to D_3 from the right-hand ROM, and addresses 100000 to 111111 from the left.

To expand both word length and number of locations, note the arrangement of Fig. 17.3c which combines the two previous configurations to create a 64×8 ROM.

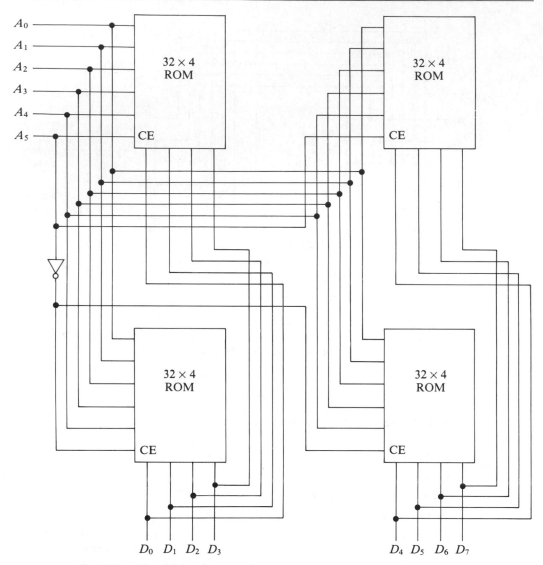

Fig. 17.3c   Four 32 × 4 ROMs Connected to Make a 64 × 8 ROM

17.2   ROM APPLICATION

Generating Boolean Functions

Instead of wiring up individual gates to produce Boolean sums of products, they can be generated by a ROM as illustrated in Fig. 17.4. The 8 × 4 ROM used here has the capability of producing four words (expressions) W_0 to W_3, each containing any

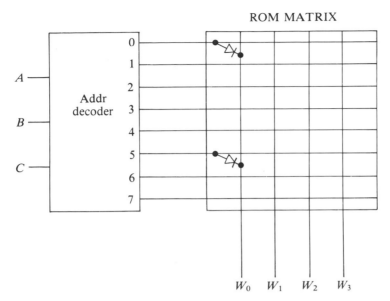

Fig. 17.4 Using a ROM to Generate Boolean Sums of Products

or all of the eight minterms associated with three input variables A, B and C. For example, when $ABC = 000$, $\bar{A}\bar{B}\bar{C} = 1$, and row 0, exclusively, will be at logical 1. With a diode at the crossover of row 0 and column W_0, then $W_0 = 1$ when $ABC = 000$. Similarly, there is a diode on row 5 of column W_0 so when $ABC = 101$, i.e. $A\bar{B}C = 1$, then $W_0 = 1$. Therefore,

$$W_0 = \bar{A}\bar{B}\bar{C} + A\bar{B}C$$

In a similar manner W_1, W_2 and W_3 can also be made into sums of minterms. Although the address decoder does contain gates, the saving for a large array can be considerable. Further, any change in logic requirements only involves a new ROM. Although this is expensive for a mask programmable ROM, the cost with a PROM (see Section 17.3) will be quite low.

Code Conversion

Conversion from one binary code to another by using gates was described in Section 13.5. It can also be achieved using ROMs as illustrated in the direct binary to excess 3 converter of Fig. 17.5. Using ten rows of a 16×4 ROM, then when row 0 is activated by $B_3B_2B_1B_0 = 0000$, $E_3E_2E_1E_0$, the excess 3 code output, must be 0011. This will be achieved by the two diodes at row 0 in columns E_0 and E_1 and by the absence of diodes in columns E_2 and E_3. Checking each row will show that the ROM output is the binary input converted into excess 3 code.

Fig. 17.8   Floating Gate MOSFET

17.4   ERASABLE AND PROGRAMMABLE READ ONLY MEMORY (EPROM)

Erasure of data stored in a ROM has been made possible by the floating gate MOSFET (Fig. 17.8) which is basically a p or n channel MOSFET with no connection on the gate. A reverse voltage between drain and source breaks down the drain–substrate junction and causes charge to be stored on the gate. The effect is to make the resistance between drain and source low – for all practical purposes a short-circuit. Storage time is long, perhaps a decade or more, so the device is regarded as static. Exposure to UV light through a window in the EPROM removes the stored charge by causing the flow of a photoelectric current. Total erasure of stored data can be effected in tens of minutes. Although stored data are regarded as static, exposure to normal lighting does reduce the duration of storage considerably. The disadvantage of the EPROM is that the deliberate erasure by UV light cannot be applied to locations selectively – the erasure is total.

17.5   ELECTRICALLY ERASABLE AND PROGRAMMABLE READ ONLY MEMORY (EEPROM)

Sometimes called electrically alterable ROMs (EAROMs), these devices can either be based on floating gate MOSFETs or on MNOSFETs where N stands for a silicon nitride layer between gate and oxide. Current pulses of appropriate polarity can be used to program or erase individual locations.

17.6   RANDOM ACCESS MEMORY (RAM)

Data written into memory devices such as EPROMs are retained until the device is to be reprogrammed for some different purpose. A RAM, on the other hand, is designed for data to be written into it and read from it throughout an operation so that it can be used as a temporary memory. For example, while a ROM might be used as a look-up table the analogous role for a RAM would be that of a scratch pad. RAMs are based on MOS or bipolar technology depending on whether storage size or speed is the paramount consideration.

The operation of a RAM can be appreciated from Fig. 17.9. Whatever data exist on the four data input lines D_{i0} to D_{i3} they will not be *written* into the four D flip flops (pp. 377–9) until the *memory write* line MW is set to 1 (assuming positive edge trigger). Further, the data held in the four flip flops cannot be read out at D_{o0} to D_{o3} until a 1 on the MR (*memory read*) line enables the four AND gates.

By modern standards a small RAM is the 64 bit Motorola MCM 14505 whose block diagram is shown in Fig. 17.10. Although the internal organization is a 16 row × 4 column arrangement, it is sufficient to note that the 6-bit address $A_0 \ldots A_5$ can address any one of the 64 CMOS memory cells. To read out of memory R/$\overline{\text{W}}$ must be 1 and for writing into memory R/$\overline{\text{W}}$ must be 0. For the pur-

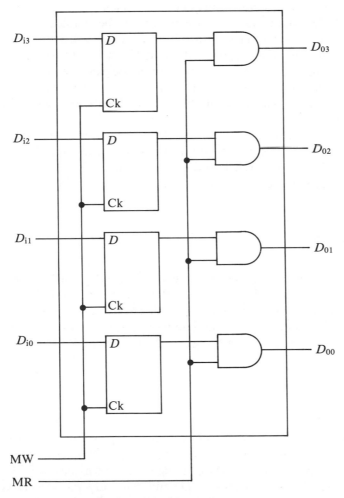

Fig. 17.9 Basic Clocked Parallel Register Operating as a RAM

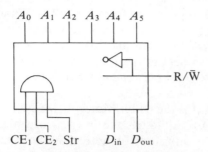

Fig. 17.10   MCM 14505 64-bit RAM (Courtesy Motorola)

pose of combining ICs into larger memory arrangements there are two chip enable inputs CE_1 and CE_2. Both must be 1 for the chip to be enabled. There is a strobe line which must also be 1 to enable the chip: this is provided to ensure that only when data being written in at D_{in} or read out at D_{out} are stable (valid) will the chip operate. By using four chips, the total memory can be expanded to 64×4: it is only necessary to common the corresponding address lines of the four chips. Then with four separate D_{in} and D_{out} lines all four bits of a 4-bit word can be read in or written out simultaneously.

To create a 256×1 memory from four 64×1 chips the chip enable inputs CE_1 and CE_2 can be used as extra address lines as shown in Fig. 17.11. Then with $A_7 A_6 = 00$, chip 0 only is enabled, and so on. For example, for address 123 (0111 1011) the two MSBs, 01, enable IC_1 so that only bit 59 of chip 1 is activated.

17.7 RAM MEMORY CELLS

The capability of being able to write new data into a RAM during operation is the feature which distinguishes it from the various types of ROM. Writing involves setting memory cells to 1 or resetting them to 0. The devices used – MOS or bipolar – and the circuit techniques used determine such factors as the potential RAM storage capability, operating speed, cost, and whether the RAM is to be static or dynamic.

17.7.1 Static RAM Cells

Static RAM cells are basically flip flops in the form of cross-coupled gates (see Section 14.1). The fastest type of RAM is based on ECL gates but the maximum ECL RAM size is small. TTL RAMs, although not quite as fast as ECL, are nevertheless regarded as being high-speed logic and made in somewhat larger memory arrays than ECL. RAM cells consisting of MOS and CMOS flip flops can be made

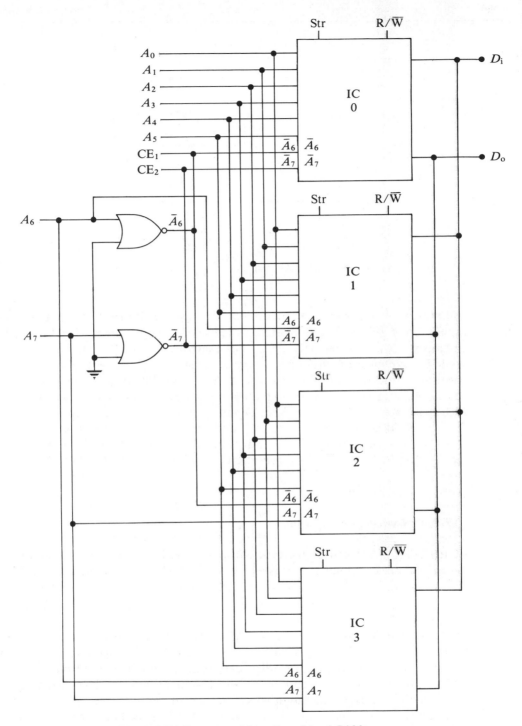

Fig. 17.11   256 × 1 RAM Constructed from Four 64 × 1 RAMs

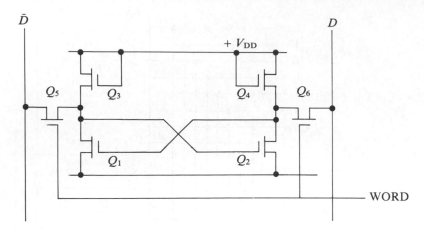

Fig. 17.12   NMOS Static RAM Cell

into larger arrays than TTL but are correspondingly slower. Even so, the fact that static RAM cells are flip flops entails the use of several transistors/cell which limits memory size even using MOSFETs. For example, Fig. 17.12 is an NMOS memory cell. The circuit configuration is that of a transistor multivibrator (Chapter 11) with Q_3 and Q_4 acting as drain load resistances. Using MOSFETs in this way is more economical in terms of chip area than laying down a resistive track. The memory cell is isolated from the data lines D and $\bar{D}$ until the cell is addressed by a voltage on the WORD line which brings Q_5 and Q_6 into conduction. Then, with a read operation, the complementary voltages on the drains of Q_1 and Q_2 are connected to the data lines. With a write operation the cell takes up the logic value on the data lines.

17.7.2   Dynamic RAM Cells

Each static RAM cell of the type shown in Fig. 17.12 has six MOSFETs, which clearly limits the number of cells which can be packaged into one IC. The dynamic RAM cell of Fig. 17.13 has only one MOSFET which is switched on when the ROW line is selected. Each column has a sensing amplifier, and in a READ operation the charge on C is sensed when the row is selected. For a write operation data on the COL/DATA line charges C when the MOSFET is opened by the row being selected. There are two disadvantages with this circuit. First, a read operation is destructive, in other words the action of reading data from a cell erases it. Second, even if the cell is not selected, the charge on C is lost through leakage. The leakage problem is solved by a refresh operation which consists of reading evey cell once every few milliseconds and rewriting back into each cell the data read out. In a similar manner the destructive readout is followed by a rewrite back into the cell.

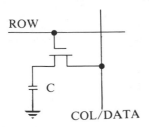

Fig. 17.13 Dynamic RAM Cell

17.8 PROGRAMMABLE LOGIC ARRAY (PLA)

As the number of variables increases the economics of using a ROM for a few output words becomes questionable. For example, with 16 input variables a ROM would have to have a capacity of 512 kbits to produce 8 words. The complexity arises because a ROM must be capable of producing 2^N minterms each of N variables: in this case, N is 16. Where, in principle, minimization could be used to reduce the number of variables in each Boolean product, implementation by ROM does appear to be a clumsy technique.

To see how a PLA reduces the complexity for some functions, consider a system with four variables A, B, C, D from which the following words are required:

$$W_1 = A\bar{B} + \bar{B}C\bar{D} + BD$$
$$W_2 = A\bar{B} + \bar{B}\bar{C} + C$$

One product $A\bar{B}$ appears in both words so only five terms are required. For this purpose the PLA would require five 8-input AND gates, the 8 arising so that each variable and its complement is available at the AND gate input. These are shown in Fig. 17.14 where the variables required at each input are connected in manufacture. The gate P has A and $\bar{B}$ connected, gate Q has $\bar{B}$, C and $\bar{D}$ connected, and so on. The five product lines can then be interconnected to the word lines. Each OR gate has four input lines so each word can have up to four terms. Figure 17.14 shows the interconnections made for W_1 and W_2.

Programming of AND functions and words is carried out by a manufacturer on the basis of a table completed by the user for standard PLA chips. Field programmable logic arrays (FPLAs) are also available in which the PROM technique of blowing fusible links is carried out with a PROM programmer.

17.9 PROGRAMMABLE ARRAY LOGIC (PAL)

A PAL has a fixed array of OR gates which is fed by a set of AND gates whose inputs are programmable. A simple form of PAL is shown in Fig. 17.15, in which

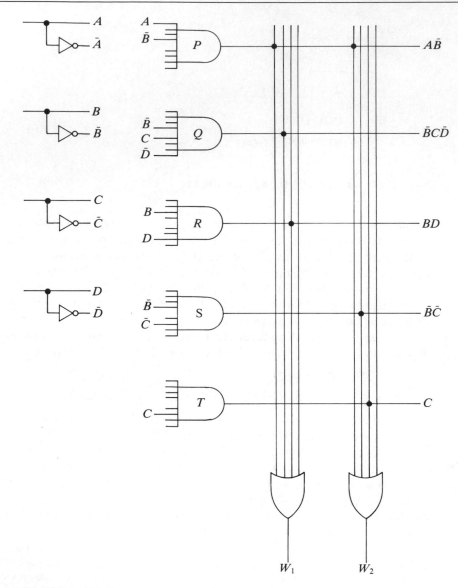

Fig. 17.14   PLA Layout

one OR gate is driven by four AND gates so that the output W can be the sum of four products. With three variables each AND gate can have up to six inputs, the actual choice being made by maintaining connection at the crossover points. PALs are field programmable using a programmer to blow fusible links.

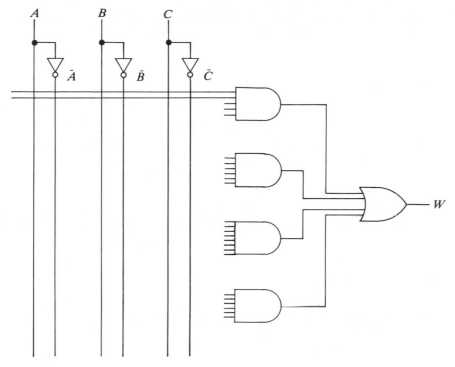

Fig. 17.15   PAL Layout

17.10 UNCOMMITTED LOGIC ARRAY (ULA)

Where there is a requirement for a complete system on a single chip – full custom design – the cost is high and the design and manufacture time long. A ULA – sometimes called a gate array – can fulfill many specifications. ULAs contain a large number of passive and active components and only require one or two layers of aluminum interconnection to become what is known as semicustom design. They are available in MOS and bipolar technologies and the different types available contain from several hundred to several thousand gates.

17.11 MEMORY SPEEDS AND LOGIC

Memories have several timing parameters, the most significant being access time. For a read operation this may be specified as the time it takes valid data to occur following the establishment of an address. For bipolar memories access time can be as low as 25 ns and for MOS memories as high as 450 ns. However, MOS devices can be manufactured in large memory sizes with ROMs available up to 1 Mbit, static

RAMs up to 4 Mbit and dynamic RAMs up to 16 Mbit. These figures are almost certain to be revised upwards as the technology develops.

SUMMARY

ROMs and RAMs are widely available in bipolar and MOS technologies and in various sizes and operating speeds. They can be used in cascade to increase storage size. There are also field programmable ROMs such as PROMs and EPROMs. ROMs are usually static but RAMs can be static or dynamic, the advantage of the dynamic RAM (DRAM) being its large storage capability: this, however, must be offset against its having to be refreshed. More recently memory devices such as PLAs, PALs and ULAs have been provided to improve design potential.

REVISION QUESTIONS

17.1 What do the following terms mean when applied to electronic memories?

Volatile
Static
Dynamic
Destructive

17.2 State the full names of the following memory devices and describe very briefly the function of each one.

ROM, PROM, EPROM, EAROM, RAM, PLA, PAL, ULA.

17.3 Describe the methods of storing data in ROMs, PROMs and EPROMs. What are their relative merits?

17.4 State the relative merits of static and dynamic RAMs.

17.5 Show the connection of diodes in the ROM matrix of Fig. 17.4 so that

$$W_1 = A\bar{B}C + AB\bar{C} + ABC$$

17.6 Redraw the matrix of Fig. 17.5 for the device to act as an excess 3 to binary converter.

17.7 In Fig. 17.11 state the logic levels on CE_1 and CE_2 when the input address is 170. Hence state which ICs are enabled and which disabled.

18

Analog-to-digital and Digital-to-analog Conversion

Chapter Objectives
18.1 Representation of an Analog Signal in Digital Form
18.2 Digital-to-analog Converters
18.3 Analog-to-digital Converters
18.4 An 8-bit DAC/ADC
      Summary
      Revision Questions

Although most of the chapters in this book are concerned either with analog or digital topics, it is wrong to think of electronic systems as being exclusively either analog or digital. Many systems exploit the advantages of both in the different contexts of various parts of a system, so there is often a need to convert signals from one form to the other. As technology changes, more and more signal processing is carried out by first converting an analog signal into digital form. Signals from transducers such as microphones, heat sensors, strain gauges, medical monitoring devices, and so on, are usually in analog form. Digital computers and micro-processors can only accept signals in digital form but the advantage of these devices in terms of reliability, adaptability and range of facility is considerable. Therefore analog-to-digital conversion at a computer input is a common requirement. Equally, when a signal is to be taken from a computer for control, reproduction or whatever, it may be necessary to convert it from digital to analog form.

It is convenient to describe digital-to-analog converters (DACs) before analog-to-digital converters (ADCs) because some types of ADC actually use a DAC as part of the complete system. First, however, it is necessary to describe some fundamental factors associated with converting an analog signal into digital form.

18.1 REPRESENTATION OF AN ANALOG SIGNAL IN DIGITAL FORM

Analog Signal Voltage Range

The analog signal voltage range is illustrated in Fig. 18.1 as the maximum peak-to-peak voltage excursion of an analog signal. It may have negative and positive limits (bipolar) or it may have 0 V as one of its limits with either a positive or a negative voltage defining the analog signal voltage range (unipolar).

Quantizing and Sampling

In Fig. 18.2 the analog signal voltage range is divided into 2^N – in this case 8 – equal voltage bands or quanta. (In transmission systems the bands may not be equal but such specialist applications are outside the scope of this book.) When the analog-to-

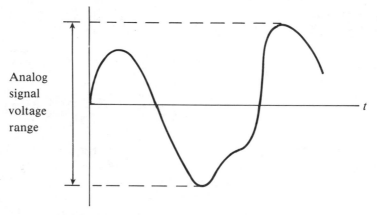

Fig. 18.1   Illustrating Analog Signal Voltage Range

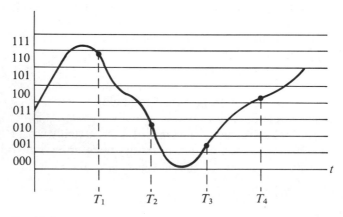

Fig. 18.2   Quantizing and Sampling

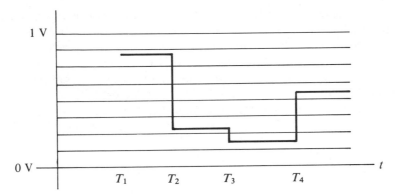

Fig. 18.3   A Digital Signal Reconstructed in Analog Form

digital (A/D) conversion is being carried out, the analog signal waveform is continually sampled at equal time intervals marked at T_1, T_2, T_3 and T_4 in Fig. 18.2. When the signal voltage is being sampled, it is held at this value until the next sample so that, in effect, the signal takes up the waveshape shown in Fig. 18.3. Having sampled the signal, the ADC generates a binary code which represents the voltage level in which the sample lies. Since the voltage range is divided into 2^3 bands, the binary code generated will have 3 bits/sample. In the most straightforward of ADCs the representation will be in direct binary code, so for the samples of Fig. 18.3 the code will be

110, 010, 001 and 100

If the signal range is 0 to 1 V then each voltage band will be $\frac{1}{8}$ V so these four binary numbers indicate that the samples are in the following bands:

110    $\frac{6}{8}$ V to $\frac{7}{8}$ V

010    $\frac{2}{8}$ V to $\frac{3}{8}$ V

001    $\frac{1}{8}$ V to $\frac{2}{8}$ V

100    $\frac{4}{8}$ V to $\frac{5}{8}$ V

At the reverse D/A conversion process all that happens is that each binary number is converted back into a nominal level so that the reconstructed analog signal will be of the sampled form shown in Fig. 18.3.

Quantizing Error – Resolution

Clearly the waveform of Fig. 18.3 is a distorted version of the original analog signal of Fig. 18.2 although the basic similarity between the two can be discerned. Passing the reconstructed waveform through a lowpass filter does round off the corners to improve the likeness without necessarily removing any high-frequency component contained in the original analog signal. However, an inherent error introduced by the A/D conversion process is that dividing the voltage range into quantum voltage bands implies that on D/A conversion the exact value of the original signal within a quantum band cannot be known. This is called a quantizing error and on transmission systems the subjective distortion introduced is called quantizing noise.

Quantizing error can be reduced by generating more bits/sample, each extra bit implying the doubling of the number of voltage bands, with each band occupying only half as much of the voltage range and therefore halving the quantizing error. Using the previous example of a 0 to 1 V voltage range, then with 4 bits/sample there will be 16 voltage bands, each of $\frac{1}{16}$ V. At the present state of the art, 14-bit ADCs are common, which implies 16 384 voltage bands. The quality of an ADC in this respect is called its resolution so that, for a 14-bit ADC, the resolution is 1/16384. High resolution does create a considerable design problem in that accurate discrimination between very small voltage levels is required.

Sampling Frequency

The other fundamental form of distortion in A/D conversion can be caused by sampling at too low a frequency. It can be avoided by sampling at a frequency a little higher than twice the highest signal component frequency f_h contained in the analog signal: $2f_h$ is known as the Nyquist frequency.

18.2 DIGITAL-TO-ANALOG CONVERTERS

Binary Weighted Resistor DAC

Figure 18.4 illustrates the most basic form of DAC. The version shown is only for 3 bits but to extend it simply means adding resistors in the progression $8R$, $16R$, and so on. In Fig. 18.4, $B_2B_1B_0$ is a 3-bit binary word in which B_2 is the MSB. For each bit a 1 is represented by a voltage V and a 0 by 0 V. These voltages are the outputs

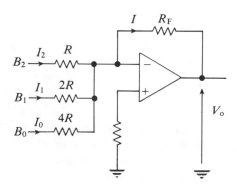

Fig. 18.4   Binary Weighted 3-bit Resistor DAC

of transistor switches actuated by the digits of the binary signals to be converted to analog form: an example of the type of electronic switch used in this operation will be described at the end of this section. Referring to Section 9.4 on op amps, the currents through the input resistors when 1s occur will be,

$$I_2 = \frac{V}{R}; \; I_1 = \frac{V}{2R}; \; I_0 = \frac{V}{4R}$$

$$V_0 = -IR_F = -\frac{R_F}{R}\left[V + \frac{V}{2} + \frac{V}{4}\right]$$

From this last equation it is evident that the analog voltage is weighted in accordance with the significance of the input bits as required. Accuracy of conversion is dependent on the ratio of the input resistors being accurately realized and for an 8-bit DAC the ratio of the largest to the smallest will be 128:1. High accuracy of resistance at high ratios is difficult to achieve.

Ladder Type (*R/2R* Network) DAC

In Fig. 18.5 the four switches represent the digital input of a 4-bit DAC: as shown, the switch positions represent an input 1101. Remembering that the inverting input

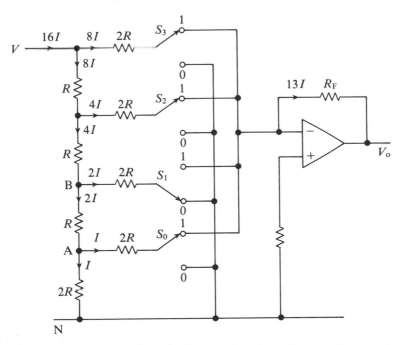

Fig. 18.5   *R/2R* Ladder Network DAC Showing a Digital Input 1101 producing a Current 13*I* into the Op Amp

to the op amp is a virtual earth then, irrespective of its position, a switch is always at earth potential. Therefore the two $2R$ resistors are in parallel and the resistance between A and N is R. A current such as $2I$ entering A will divide equally between the two resistors, as shown. The resistor R between B and A will, when added to the resistance R between A and N, give a total resistance of $2R$. This in turn is in parallel with the $2R$ resistor connected to S_2, so again there are two $2R$ resistances in parallel, this time between B and N, giving a total resistance R. Any current entering B will divide equally between the two paths. The principle can be extended for a ladder of as many bits as are necessary.

Now consider a reference voltage V providing a constant current $16I$ into the ladder. It can be seen to divide $8I$ into S_3, $4I$ into S_2 and so on. These currents will pass into the op amp resistor R_F and produce an output voltage when the corresponding switch is at 1; otherwise the current goes to ground. Therefore, when set to 1, S_3 provides twice as much current as S_2, four times as much as S_1 and so on, and results in the required analog output voltage. The advantage of this circuit is that it is only necessary to provide accurately matched resistors in the ratio $2R:R$, i.e. $2:1$.

Analog Switches

Bit trains do not occur at DAC inputs, as indicated in Fig. 18.5, as mechanical switches operating between a dc supply and ground. They occur as pulse trains, often appearing more rapidly than a mechanical switch could be made to operate. Further, the bits in a pulse train are often unsuitable for connecting direct to DAC inputs because their voltage levels have wide tolerance. For example, a 1 from a TTL circuit might have any voltage from 2.4 V to 3.6 V, and this is clearly unsuitable for

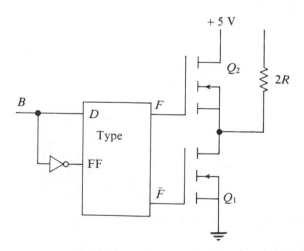

Fig. 18.6   NMOS Switch which Connects $2R$ Resistor to 0 V when $B = 0$ and to $+5$ V when $B = 1$

accurate conversion as both of the DACs described have been predicated on the basis of a 1 being represented by a switch set to a constant voltage V.

DACs use a variety of electronic switches for converting the 0s and 1s of an incoming bit train into stable low and high voltages: a simple example is shown in Fig. 18.6. The bit train B is fed into a D type flip flop, so that if $B = 1$, $F = 1$ and $\bar{F} = 0$, and vice versa if $B = 0$. The flip flop outputs are connected to two NMOS devices whose common terminal is connected to the DAC input, in this case a $2R$ resistor of a $R/2R$ ladder. When $B = 1$, $F = 1$, $\bar{F} = 0$. Therefore Q_2 is ON, Q_1 is OFF and $+5$ V is connected to the $2R$ resistor. When $B = 0$, only Q_1 will be ON and the $2R$ resistor will be connected to ground as required in Fig. 18.5.

18.3 ANALOG-TO-DIGITAL CONVERTERS

Counter (Staircase or Single-ramp) ADC

In Fig. 18.7, V_A is a sample of an analog signal to be converted into digital form. The comparator, which may simply be an op amp, has two inputs, V_A and V_D − the output of a DAC. As long as $V_A > V_D$ the comparator output will be a logic 1 which will keep the AND gate enabled for clock pulses to pass into the counter. At the beginning of each sample conversion, the counter is cleared so its output $B_3B_2B_1B_0 = 0000$ and therefore the DAC output $= 0$ V. Therefore $V_A > V_D$ and the counter will start counting clock pulses. This causes the binary number $B_3B_2B_1B_0$ to increase so the DAC output rises until $V_D > V_A$ which closes the AND gate and stops the count. At this point $B_3B_2B_1B_0$ is a binary number which represents the analog sample V_A.

Figure 18.8 shows the DAC output V_D climbing as a staircase waveform with

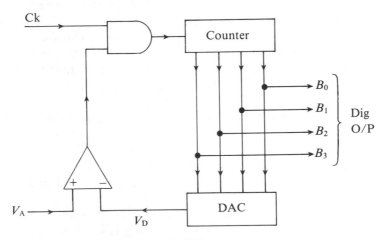

Fig. 18.7　Counter (Single-ramp) ADC

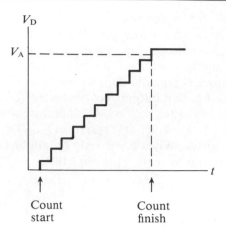

Fig. 18.8 Staircase (Single-ramp) Waveform from DAC Part of ADC of Fig. 18.7

each step being a quantizing step occurring for each advance in the count. The staircase waveform is, in effect, a quantized ramp and the circuit is sometimes called a single-ramp ADC.

To convert the maximum amplitude of an analog sample the count will reach $B_3B_2B_1B_0 = 1111$ which takes $2^4 = 16$ clock pulses. In general 2^N clock pulses must be allowed to convert a sample which makes this type of ADC relatively slow.

Successive Approximation ADC

This type of ADC (Fig. 18.9) works on the principle of setting the programmable register bits one at a time, beginning with the MSB, and comparing its reconverted analog value with the analog sample to determine whether the setting of a bit was correct. For example, suppose that the correct output for a particular sample is 1011. On a 0 to 1 V analog range this would correspond to an analog voltage of between $\frac{11}{16}$ V and $\frac{12}{16}$ V. At the first clock pulse the MSB is set to 1 and all other bits to 0. In this condition $V_D < V_A$ so the comparator confirms that $B_3 = 1$ and the next clock pulse sets $B_2 = 1$. Now $V_D > V_A$ so the comparator output causes B_2 to be reset to 0 and the next clock pulse sets $B_1 = 1$. This results in $V_D < V_A$ so $B_1 = 1$ is maintained and B_0 set to 1. As V_A lies between $\frac{11}{16}$ V and $\frac{12}{16}$ V, $B_0 = 1$ is confirmed and the conversion is complete. Essentially the 4 bits define 16 voltage bands with 1011 defining the band $\frac{11}{16}$ V to $\frac{12}{16}$ V. At 1000 the comparator indicates that $V_A > \frac{1}{2} V$. At 1100 it indicates that it is less than $\frac{3}{4}$ V. At 1010 it indicates that $V_A > \frac{5}{8}$ V. At 1011 it indicates that $V_A > \frac{11}{16}$ V but since the second check showed that $V_A < \frac{3}{4}$ V, then 1011 is the correct result.

Note that the conversion only takes the time of as many clock pulses as there are bits in the digital output so the successive approximation ADC is much faster than the counter ADC.

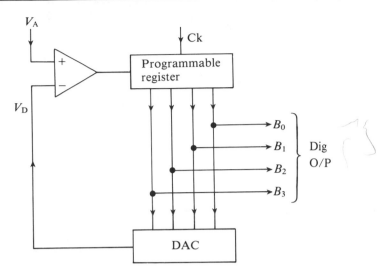

Fig. 18.9 Successive Approximation ADC

Parallel Comparator (Flash) ADC

In Fig. 18.10 a constant voltage V – the upper limit of the analog voltage V_A – is potentially divided across 8 equal resistors for a 3-bit conversion. Thus the non-inverting inputs of the 7 comparators have voltages equally separated from $\frac{1}{8}V$ to $\frac{7}{8}V$ as shown. Suppose, when sampled, that V_A is between $\frac{3}{8}V$ and $\frac{4}{8}V$. Then the comparator outputs will be

$$C_7C_6C_5C_4C_3C_2C_1 = 0\ 0\ 0\ 0\ 1\ 1\ 1.$$

Larger values of V_A cause 0s to be replaced by 1s until $V_A > \frac{7}{8}V$ when all comparators will be 1.

The 8 combinations of comparator outputs are coded into the 3-bit binary number required. The process is very fast and does not involve counting clock pulses. Its disadvantage is that of hardware complexity. An 8-bit ADC, which defines 128 analog voltage levels, would require 128 resistors and 127 comparators, so flash converters are usually restricted to small numbers of bits/sample.

Dual Ramp ADC

In Fig. 18.11, S represents an analog switch of the type described in Section 18.2. It is controlled by the carry out C_0 of the counter which, at the beginning of a conversion of an analog sample V_A will be set to 0 with the counter set to 0000. Assuming that V_A is positive then v_s will be a negative-going ramp as illustrated in the period 0 to T_A in Fig. 18.12. Here it is important to appreciate that the rate of fall of v_s is proportional to the value of the analog sample V_A. As soon as the downward ramp v_s begins, it causes the comparator to give a 1 output which enables

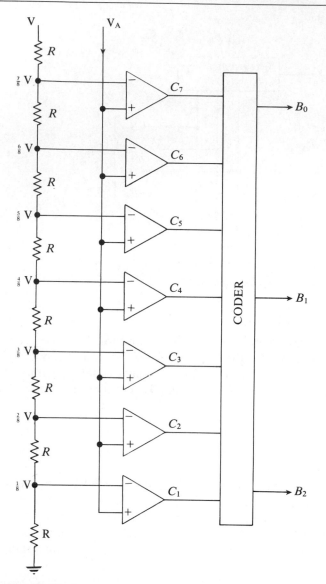

Fig. 18.10   Parallel Comparator (Flash) ADC

the AND gate. Clock pulses are fed into the counter which runs through the complete count 0000 to 1111. The time T_A taken to complete this part of the operation is constant. However, the voltage $-v_S$ reached by the ramp in this time will be proportional to V_A. At the next clock pulse the counter is set to 0000 and C_0 to 1 which causes S to switch to a constant reference voltage $-V_R$. Now, as illustrated in Fig. 18.12, v_s ramps back from $-v_S$ to 0 V. The rate at which this reverse ramp takes

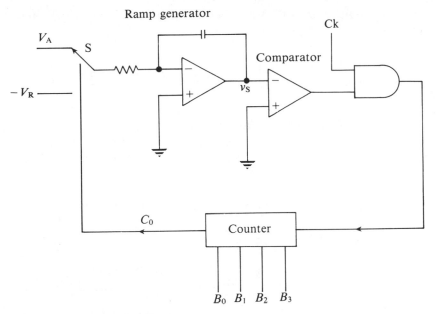

Fig. 18.11   Dual Ramp ADC

place is constant and fixed by $-V_R$. However, the time it takes, the period from T_A to T_R, is fixed by $-v_S$ which, as noted, is proportional to V_A. During the period from T_A to T_R the counter counts clock pulses and when v_s reaches 0 V the comparator output changes to 0, closes the AND gate, and the count stops. Although the second ramp is at a constant rate, the voltage through which it passes, $-v_S$ to 0 V, was fixed by the original value of V_A, so the number of clock pulses passing through the gate from T_A to T_R, and therefore the final counter setting, is proportional to the analog sample V_A.

Any nonlinearity in the first ramp is compensated for by the same effect producing a compensating error in the second. Although the process is slow, it is accurate

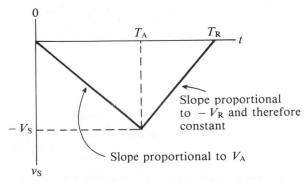

Fig. 18.12   Output of Ramp Generator in Dual Ramp ADC

and commonly used in DVMs: it is slow because a double clock pulse count is an inherent part of the process.

18.4   AN 8-BIT DAC/ADC

Figure 18.13 shows that the layout of the ZN 425E which is an 8-bit dual mode DAC/ADC. It contains an 8-bit DAC consisting of an $R/2R$ ladder network and an array of 8 bipolar switches. It also contains an 8-bit binary counter so that, in the

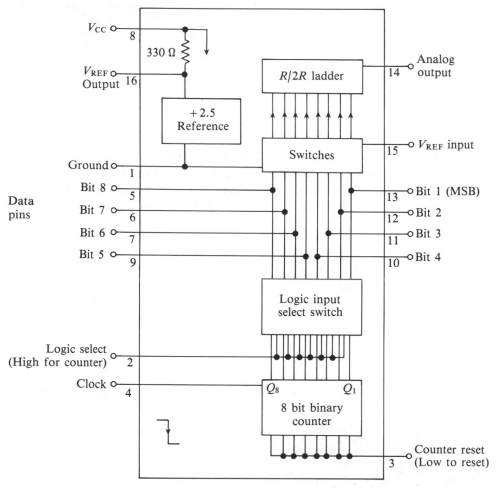

Fig. 18.13   Block Diagram of ZN 425E Dual Mode DAC/ADC (Courtesy RS Components Ltd)

ADC mode, clock pulses can be counted to establish a digital output. However, for A/D conversion, additional external circuitry is required. D/A conversion is quite straightforward in that a 0 on pin 2 ('Logic Select') disconnects the counter output from the 8 data pins. Then, inputs to the 8 data pins are converted by the $R/2R$ network to an analog output at pin 14.

The chip contains an internal 2.5 V dc reference for setting logic levels. Pin 16 makes this available as an output. Pin 15 permits an external reference to be used instead of the internal one.

A/D Conversion – Handshaking

For A/D conversion a 1 is required on pin 2 which enables the counter output to appear on the 8 data pins. The counter is also connected to the $R/2R$ ladder network to convert the counter state back to analog form so that the technique noted in Section 18.3 of comparing the reconverted signal with the incoming analog sample can be used to stop the counter when it reaches the correct value. The additional external circuitry required for A/D conversion is shown in Fig. 18.14 and consists of a comparator and three NAND gates, two of which are cross-coupled to form an *SR* flip flop. Figure 18.14 also shows the 8 data lines being connected to the I/O (input/output) port of a device such as a microprocessor and two control lines which connect what is known as a 'handshake' to control the passage of data between the two devices.

The handshaking process is complicated but revealing. The following account should be read in conjunction with the handshake waveforms of Fig. 18.15. The two handshake lines are:

(a) Start conversion (SC) which is an output from the microprocessor to the ADC. When it is 1 the microprocessor is requesting the ADC to convert the analog sample at the input to the comparator into a digital signal and to send it as 8 data bits along the 8 data lines.

(b) Not Data Valid ($\overline{DV}$) which is an output from the ADC to the microprocessor. When $\overline{DV} = 0$ the ADC is informing the microprocessor that the data on the 8 data lines are valid, i.e. a conversion has been completed. The fact that $\overline{DV}$ is in the complemented form is a convenience of circuitry and is not, in itself, significant.

At the completion of the previous sample the analog output would have reached the analog sample voltage so the comparator output $D = 0$. This makes $A_0 = 1$. An $SC = 1$ command would be maintained so $B_0 (= \overline{DV}) = 0$ indicating that the data are valid. With $B_0 = 0$, gate C is disabled so clock pulses cannot pass into the counter. To begin the conversion of a new sample the microprocessor sends a 0 on SC. This is connected to counter reset (pin 3) which zeroes the counter output and prevents it from counting. The 0 on SC also makes $B_0 = 1$. This makes $\overline{DV} = 1$ showing that the data are not valid. Making $B_0 = 1$ also enables C but although

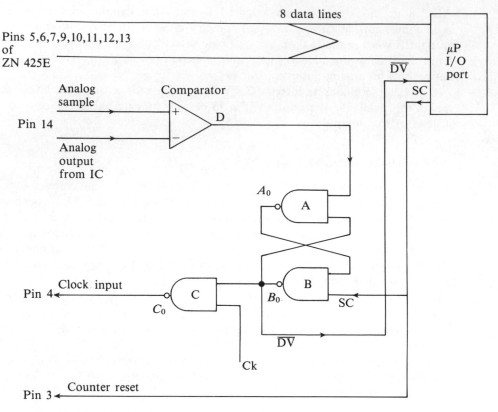

Fig. 18.14   Additional Circuitry for ZN 425E for A/D Conversion

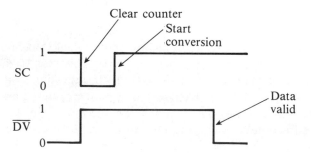

Fig. 18.15   Handshake Signals for A/D Conversion

clock pulses pass through it they will not be counted with SC = 0. With the counter reset to 0 the analog output (pin 14) is also at 0 V so $D = 1$.

The next step is that the microprocessor makes SC = 1 to start a conversion. This enables the counter which starts counting clock pulses so the analog output reconverts the counter state and ramps up in steps. The data lines follow the counter but as long as $\overline{DV} = 1$ the microprocessor will know that the data are not valid. When the analog output reaches the value of the analog sample input, D goes to 0. This makes the flip flop change over so B_0 ($\equiv \overline{DV}$) goes to 0, which:

(a) disables gate C thus stopping the flow of clock pulses into the counter;
(b) makes $\overline{DV} = 0$ indicating to the microprocessor that the data are valid.

SUMMARY

A/D and D/A conversions have become important aspects of electronic engineering. Many signals are originated in analog form but technology now exists for their processing and control, with advantage, in digital form. Reduction in signal distortion is related to sampling frequency and the number of bits/sample: any increase in these factors has a bearing on cost. Increasing the number of bits/sample to reduce quantizing error has implications for circuit design with the tolerance and stability of components becoming more critical. There are several different types of ADC, the choice being influenced by the fact that fast conversion involves increased circuit complexity and therefore higher cost.

REVISION QUESTIONS

18.1 An 8-bit ADC has an analog voltage input range from -5 V to $+5$ V. What is the value of each quantum voltage step?

(0.039 V)

18.2 A 4-bit ADC with an analog input range from -5 V to $+5$ V gives an output 0101. What are the minimum and maximum values of input voltage sample?

(-1.875 V; -1.25 V)

18.3 An ADC has an analog input voltage range of 0 to $+2.5$ V. If the resolution is to be no worse than 1/1000, how many bits/sample are required and what would be the resolution and quantum voltage step value?

(10; 1/1024; 2.4 mV)

18.4 What is the advantage of the $R/2R$ ladder network DAC over the binary weighted resistor version?

18.5 Draw an $R/2R$ ladder network for a 4-bit $(B_3B_2B_1B_0)$ DAC and show, separately, how, when B_0 and B_1 are 1, the currents from their dc sources divide throughout the network.

18.6 List the advantages and disadvantages of single ramp, dual ramp, flash and successive approximation ADCs.

18.7 A 4-bit single ramp ADC has an analog input voltage range of 0 to $+5$ V. With an input sample of $+3$ V, sketch the output of the DAC section from the beginning to the end of the conversion.

18.8 A 4-bit successive approximation ADC has an analog input voltage range of -2 V to $+2$ V. If the sample value is -1 V, draw up a table showing the counter values throughout the conversion period.

18.9 A dual ramp ADC has analog samples:

(a) half full voltage range;
(b) full voltage range.

Sketch the dual ramp for each case, showing the relative timing and voltage levels for each sample. Hence justify the relative ADC output values.

19

DC Power Supplies and Other Diode Applications

Chapter Objectives

Rectification, as part of the process of deriving a source of dc power from the ac mains, is a common application of diodes. Although the process of stabilizing the output voltage of dc supplies within narrow limits requires rather more in the way of electronic device sophistication than merely diodes, particularly with the present emphasis on switch mode power supplies, nevertheless the inclusion of this process as an extension of rectification can be justified in that it confines the topic of dc power supplies within a chapter. Having developed the topic from introductory material on diodes as circuit elements, the inclusion of other applications of diodes within a single chapter forms a coherent grouping for an important class of circuits.

19.1 SILICON DIODES AS CIRCUIT ELEMENTS

Under reverse bias conditions ($V_{AK} \leqslant 0$ V in Fig. 19.1), leakage current in silicon diodes can safely be ignored at normal working temperatures and therefore, as a circuit element, a silicon pn diode can be treated as an open circuit.

Judicious use of approximation simplifies the analysis of diode circuits under

509

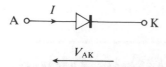

Fig. 19.1   Voltage and Current Labels for a Diode

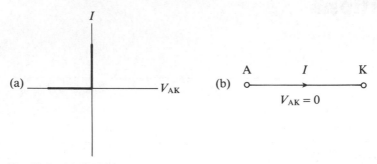

Fig. 19.2   (a) I/V Characteristic of an Ideal Diode (b) Its Representation at a Circuit Element when Forward Biased

forward bias conditions. An ideal diode has the I/V characteristic shown in Fig. 19.2a which imples that, when forward biased, a diode acts like a short circuit as in Fig. 19.2b.

In practice, diode current I is insignificant until V_{AK} is at least 0.5 V. Once conduction is established, I increases very steeply with V_{AK}, so for many purposes the I/V characteristic of Fig. 19.3a is an approximation which is more acceptable than Fig. 19.2a. Typically, once $V_{AK} = 0.7$ V, I increases without any further increase in V_{AK}. This leads to Fig. 19.3b as the representational circuit element. It should be noted that the figure 0.7 V is typical rather than universal, and figures between 0.5 V and 1.0 V are used for various diodes and operating conditions.

For some applications this forward bias characteristic is over-simple: the fact that V_{AK} does increase with I should be taken into account so that the I/V characteristic of Fig. 19.4a and the equivalent circuit element of Fig. 19.4b are more realistic.

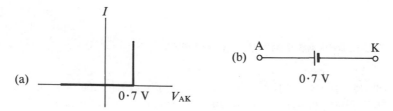

Fig. 19.3   (a) I/V Characteristic of an Ideal Diode Delineating the Forward Conducting Voltage (b) The Simplified Equivalent Circuit

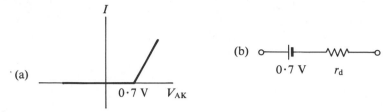

Fig. 19.4 (a) *I/V* Characteristic of a Diode with Forward Conducting Voltage and Finite Forward Resistance (b) The Implied Equivalent Circuit

In general, the 0.7 V of Figs. 19.3b and 19.4b need only be used when significant compared with other circuit voltages. Similarly, the resistance r_d of Fig. 19.4b should only be included when the pd developed across it is significant. For the rectifier circuits of the following section, Fig. 19.2b is acceptable.

19.2 DC POWER SUPPLIES

General Considerations

Conversion of ac, usually the 240 V, 50 Hz mains, into a dc supply to be used, for example, for electronic circuits is a common requirement. Figure 19.5 shows the first three stages of the process. These are common to most dc power supplies and, in many cases, all that is required. First, the ac mains input is transformed, usually downwards, to a voltage near the output voltage level required. Second, the alternate positive and negative half-cycles are rectified to produce a waveform of half sine waves of one polarity. This is, in fact, dc and is all that is required for a simple application such as a battery charger. Where a steady dc is required, the alternating components of the rectified waveform are removed at the third stage, a lowpass filter, which delivers a smoothed output voltage. For many purposes it is acceptable for this voltage to vary as mains voltage and/or load current change but when output voltage is specified to be constant within narrow limits, a fourth process called regulation, to be described in Section 19.5, is necessary.

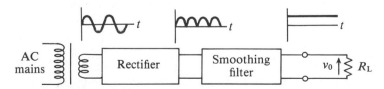

Fig. 19.5 Main Components of a dc Power Supply. Sometimes a Regulator is Provided Between the Smoothing Filter and the Load

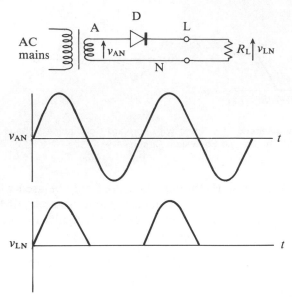

Fig. 19.6   Half-wave Rectification

19.3   RECTIFICATION

Figure 19.6 is a half-wave rectifier circuit with its associated waveforms. Diode D only conducts when v_{AN} is positive so that output v_{LN}, neglecting any volt drop across the diode when conducting, consists of the positive half-cycles of input only. Reversal of D would result in v_{LN} being the negative half-cycles of v_{AN}.

Figure 19.7 illustrates full-wave rectification. The two sine waves v_{AN} and v_{BN} are in antiphase so D_1 and D_2 conduct on alternate input half-cycles. The output half-cycles which make up v_{LN} and the corresponding transformer secondary waveforms from where they originate are labeled.

A full-wave rectifier which does not require a center-tapped transformer secondary winding is the bridge rectifier of Fig. 19.8. When v_{AB} is positive D_1 and D_4 will be reverse biased and D_2 and D_3 forward biased. Current flows $A \rightarrow D_2 \rightarrow L \rightarrow N \rightarrow D_3 \rightarrow B$. When v_{AB} is negative the diodes exchange roles and current flows $B \rightarrow D_4 \rightarrow L \rightarrow N \rightarrow D_1 \rightarrow A$. For both half-cycles of input, current flows from L to N so the output is a fully rectified waveform as illustrated.

19.4   SMOOTHING

Rectified waveforms must be smoothed before they can be used for applications such as dc supplies to electronic circuits. A rectified sine wave consists of a dc com-

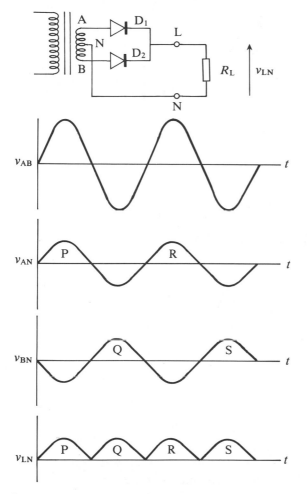

Fig. 19.7   Full-wave Rectification

ponent and harmonics of the supply frequency. Therefore, a smoothing circuit is basically a lowpass filter which passes the dc component and attenuates the ac components. By inspection of Figs. 19.6 and 19.7, the ratio of the ac to dc components of a half-wave rectified waveform is higher than for the full wave and it is therefore more difficult to smooth. Full-wave rectifiers are supplied as packaged components and are almost universally used for dc supplies.

The first two terms of the Fourier series for a fully rectified sine wave are

$$\frac{2}{\pi} V_m - \frac{4}{3\pi} V_m \cos 2\omega t$$

i.e. the dc component is equal to the average value of the sine wave and the lowest

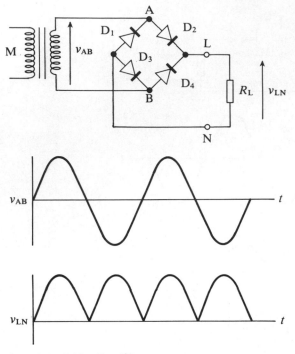

Fig. 19.8   Bridge Rectifier

frequency to be attenuated for a 50 Hz supply is the second harmonic, 100 Hz. In this context the figure of merit for a smoothing filter is the ripple factor r, which may be defined as

$$r = \frac{\text{RMS value of ac component of the output waveform}}{\text{Nominal dc value of the output waveform}}$$

It can be equally well expressed in terms of voltage or current.

Capacitor Filter

Connection of a capacitor C (Fig. 19.9) across the load resistance R_L forms the simplest of smoothing circuits. For either a half- (as shown) or a full-wave rectifier,

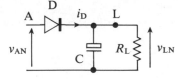

Fig. 19.9   Capacitor Smoothing Filter

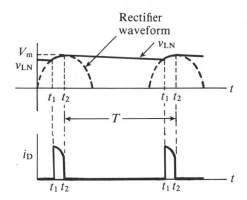

Fig. 19.10 Voltage Ouput and Diode Current Waveforms for a Half-wave Rectifier Smoothed with a Capacitor Filter

C charges up to V_m, the peak value of ac input to the rectifier during diode conduction. Ideally, C would maintain V_m across the load, but in delivering current to R_L it partly discharges, as shown in Fig. 19.10. Its lost charge is replenished when the ac input to D, v_{AN}, rises above the load voltage v_{LN}. Then, in the periods t_1 to t_2, D conducts, C is recharged and the output voltage is restored to V_m. There are two particular effects to observe. First, the output voltage ripple: in designing the circuit, CR_L must be chosen to limit the fall in v_{LN} – which constitutes the ripple – to an acceptable minimum. Therefore, for most purposes, $CR_L \gg T$, where T is the period of the waveform and will be 20 ms for the 50 Hz mains. Second is the fact that the diode current i_D which charges C consists of a series of short spikes. No current flows in D until v_{AN} rises to the voltage to which v_{LN} has fallen. This will be at the instants marked t_1 on Fig. 19.10. Output voltage v_{LN} now follows v_{AN} until they both reach V_m at instants t_2. Now v_{AN} falls below v_{LN} so D cuts off. In the relatively

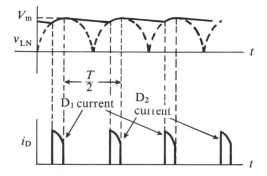

Fig. 19.11 Voltage and Current Waveforms for a Full-wave Rectifier Smoothed with a Capacitor Filter

short periods t_1 to t_2, C must regain all of the charge lost over the complete cycle. Consequently the diode current i_D has the shape shown in Fig. 19.10.

For a full-wave rectifier, C has only half of the time to discharge as illustrated in Fig. 19.11. Therefore, for a given value of CR_L, the ripple will be smaller. Charge lost by C is replenished on every half-cycle but note that one diode is pulsed for positive half-cycles of input and the other for negative.

Ripple Factor of a Capacitor Filter

Derivation of ripple factors is an aspect of network theory and for the most part it is only necessary to quote the expressions. However, it is worth deriving the ripple factor for a capacitor filter, which is the simplest of smoothing circuits, to illustrate the concepts involved.

For a full-wave rectifier Fig. 19.12 shows the output waveform v_{LN}, load resistance R_L and smoothing capacitor C across which it is developed. The instant at which v_{LN} reaches V_m is arbitrarily designated as $t = 0$. From $t = 0$, diode conduction ceases and as C discharges through R_L,

$$v_{LN} = V_m \exp(-t/CR_L)$$

Output voltage v_{LN} continues to fall until diode conduction resumes at $t = t_1$. If the ripple is to be small then the total decay time t_1 is almost equal to $T/2$ where T is the period of the waveform and is 20 ms for a 50 Hz mains. Then the minimum value of v_{LN} is given by

$$v_{LN}(min) = V_m \exp(-T/2CR_L)$$

The peak-to-peak ripple is the difference between the maximum and minimum values of v_{LN}, so therefore

$$\text{peak-to-peak ripple} = V_m - V_m \exp(-T/2CR_L)$$
$$= V_m[1 - \exp(-T/2CR_L)] \tag{19.1}$$

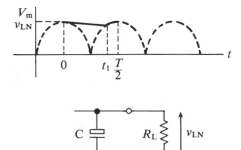

Fig. 19.12   Illustrating that if Ripple is Small the Discharge Time t_1 of *C* is Almost Equal to the Period of a Half-cycle of Input $T/2$

Again, if ripple is to be small,

$$CR_L \gg \frac{T}{2} \quad \text{and} \quad \frac{T}{2CR_L} \ll 1.$$

When $x \ll 1$ in the exponential series

$$e^{-x} = 1 - x + \frac{x^2}{2!} \cdots,$$

only the first two terms are significant. Therefore, equation (19.1) approximates to

$$\text{peak-to-peak ripple } V_r = V_m \left[1 - \left(1 - \frac{T}{2CR_L} \right) \right] \tag{19.2}$$

$$= \frac{V_m T}{2CR_L} = \frac{V_m}{2fCR_L} = \frac{V_m \times \pi \times X_c}{R_L} \tag{19.3}$$

$$\text{Fractional ripple } \frac{V_r}{V_m} = \frac{\pi X_c}{R_L}$$

For a half-wave rectifier the discharge time will be for a whole period T and the peak-to-peak ripple will be

$$V_r = \frac{V_m}{fCR_L} = \frac{V_m \times 2\pi \times X_c}{R_L}$$

$$\text{Fractional ripple} = \frac{2\pi X_c}{R_L}$$

An exponential discharge waveform, which only decays for a small fraction of its fully charged voltage, may be regarded as linear, a statement which is implied by using the approximation equation (19.2). Therefore the waveshape of v_{LN} is approximately triangular (Fig. 19.13). The rms value of a triangular waveform is given by

$$V_{RMS} = \frac{V_{pk-pk}}{2\sqrt{3}} \tag{19.4}$$

Fig. 19.13   Illustrating that if Ripple is Small the ac Component of the Output Voltage is Approximately Triangular

and therefore, using the definition of ripple factor r,

$$r = \frac{\text{RMS value of ripple}}{\text{Nominal output voltage}}$$

$$r = \frac{(V_m/2fCR_L)(1/2\sqrt{3})}{V_m} = \frac{1}{4\sqrt{3}fCR_L} \tag{19.5}$$

For a half-wave rectifier, r will be doubled.

☐ **Example 19.1**

A power supply using a full-wave rectifier is to provide 150 mA at 20 V with no more than 1% ripple. Calculate the minimum value of smoothing capacitor.

$$R_L = \frac{20 \text{ V}}{0.15 \text{ A}} = 133 \text{ } \Omega$$

Using equation (19.5),

$$C = \frac{1}{4\sqrt{3} \times 50 \times 133 \times 0.01}$$

$$= 2200 \text{ } \mu\text{F}$$

☐

Regulation of a Capacitor Filter

As noted, the diode current in a rectifier smoothed by a capacitor filter consists of a series of short spikes rather than being a continuous current flow. These spikes are drawn not only through the forward resistance of the diode (which at high current values will be significant) but also through the transformer windings. The volt drop which occurs in the circuit resistance because of these current spikes is greater than that which would occur due to a steady current capable of delivering the same charge to the capacitor over the whole cycle of input. Therefore the output voltage V_{LN} falls quite significantly as load current increases, as indicated in Fig. 19.14: in other words, the regulation of a dc supply with capacitor smoothing is poor.

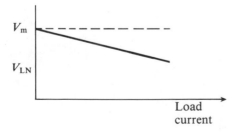

Fig. 19.14   Regulation Graph of a dc Supply with Capacitor Smoothing

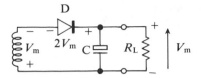

Fig. 19.15 Illustrating That When the Half-wave Rectifier Input Voltage is Maximum Negative, the Reverse Voltage Across D is $2V_m$

Diode Ratings

Figure 19.15 illustrates the instant in a smoothed half-wave rectifier circuit when the reverse voltage across D is at its maximum. Assuming that the output voltage is constant at V_m, then when the input is at its maximum negative voltage $-V_m$, the voltage across D will be $2V_m$. Therefore the peak inverse voltage (PIV) rating of D must be at least twice the nominal output voltage. It is easy to deduce that, for a full-wave rectifier with center-tapped secondary, the same PIV rating is necessary for each diode. For a bridge rectifier, each pair of nonconducting diodes divides the reverse voltage between them, so theoretically the PIV requirement is halved. In practice, since we cannot rely on diodes being identical the reverse voltage may not be evenly divided so it is better not to choose diodes with PIV down at the theoretical value of V_m.

Current ratings are more difficult to determine because of the complex shape of the diode current waveform. It is not difficult to calculate the average diode current during its conduction period.

☐ **Example 19.2**

A capacitor filter has to provide 100 mA at 2% ripple from a full-wave rectifier. Calculate the average diode current while it is conducting.

The rms value of the ripple will be 2 mA and using equation (19.4), the peak-to-peak value will be

$$V_{RMS} \times 2\sqrt{3} = 7 \text{ mA}$$

Now refer to Fig. 19.16 and note that diode conduction begins when the output current i_L has fallen to 93 mA at an angle θ given by

$$93 = 100 \sin \theta$$

from which $\theta = 68°$. Therefore current flows in the diode for $22°$ out of $180°$ for the half-cycle. Then, using simple proportion, the average diode current $I_{D(AV)}$ during the conduction period will be

$$I_{D(AV)} = \frac{180}{22} \times 100 \text{ mA} = 0.82 \text{ A}$$

The peak value of the diode current will be higher than the average because the

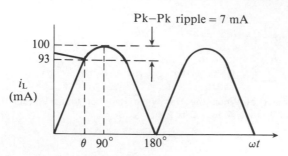

Fig. 19.16   Diagram for Example 19.2

charge current is not constant during the spike but is more like the shape indicated in Fig. 19.11, and so the diode current rating should be uprated by several times. □

Generally the two related factors of poor regulation and large peak diode current lead to the conclusion that the capacitor filter should not be used for large load currents.

L Section Filter

Both of the reactive components of the L section filter illustrated in Fig. 19.17 contribute to the attenuation of the ac component of the rectified waveform: the series inductance L offers high reactance to ac but none to dc; the shunt capacitance bypasses ac components but offers no path for dc. Another advantage of this filter is that the inductance tends to maintain a continuous charging current. The nature of inductance is that, if the current through it changes, an emf is induced across it which opposes the change and tends to maintain the current. Therefore the 'spiky' nature of the diode current characteristic of a capacitor filter can be avoided and regulation improved. The smaller the value of L, the less effective it will be in maintaining continuous current flow, so a minimum value of inductance L(min) has to be specified. Further, the smaller the load current, the more likely it is that the charging current will fall to zero during cycles of input. Therefore the value of L(min) specified is given for a maximum value of load resistance R(max). The implication of this is that there must be a minimum 'bleed' current if charging current is to be continuous. For a 50 Hz full-wave rectifier the relationship is

$$L(\text{min}) = \frac{R(\text{max})}{950} \tag{19.6}$$

with L in henries and R in ohms.

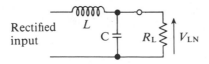

Fig. 19.17   L Section Filter

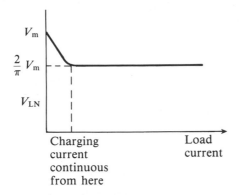

Fig. 19.18   Regulation Curve for L Section Filter

Figure 19.18 is the regulation curve of an L section filter. It falls sharply from V_m to $2V_m/\pi$ – the average value of the fully rectified waveform. Up to this point the load current is too small for the inductance to sustain continuous current flow so regulation is correspondingly poor. From this point on the regulation curve is far superior to that of a capacitor filter in equivalent circumstances. Thus the nominal output voltage for an L section filter is $2V_m/\pi$ compared with V_m for a capacitor filter.

Finally, quoting the ripple factor for an L section filter as

$$r = \frac{1.2}{LC} \tag{19.7}$$

with L in henries and C in μF, we can design a circuit as follows.

☐ **Example 19.3**

It is required to supply 0.5 A at 15 V with no more than 2% ripple. Choose suitable components:

(a) assuming that the load current is to be constant at 0.5 A,
(b) assuming that the load current can have any value from 0 to 0.5 A.

(a) $R_L = \dfrac{15\ \text{V}}{0.5\ \text{A}} = 30\ \Omega$

Using equation (19.6),

$$L(\text{min}) = \frac{30}{950} = 31\ \text{mH}$$

Using equation (19.7),

$$LC = \frac{1.2}{r} = 60$$

Therefore $C = \dfrac{60}{0.031} = 2000 \; \mu F$

(b) Choosing, arbitrarily, a 'bleed' current of 15 mA,

$$R_B = \frac{15 \text{ V}}{15 \text{ mA}} = 1 \text{ k}\Omega$$

Therefore, $L(\min) = \dfrac{1000}{950} = 1.05 \text{ H}$

and $C = \dfrac{60}{1.05} = 57 \; \mu F$

If we were prepared to double the bleed current then the inductance could be halved to advantage: the fact that C would have to be doubled is of little consequence. $\square$

π Section Filter (LC)

Excellent smoothing can be provided by the π filter of Fig. 19.19 but, as with the capacitor filter, regulation is poor because C_1 will be charged by current spikes in the same way as the capacitor filter. However, it is not necessary to provide a no-load 'bleed' resistor and the no-load output voltage will be equal to the peak value of the rectified waveform. Ripple factor r is given by

$$r = \frac{5700}{R_L L C_1 C_2}$$

at 50 Hz, with R_L in ohms, L in Henries, and C_1 and C_2 in μF.

π Section Filter (RC)

Inductors are bulky and expensive. By replacing the inductance of Fig. 19.19 with the resistance R of Fig. 19.20, attenuation of the ripple component will be achieved in a similar manner. Unlike an inductance, however, the resistance will dissipate power. For example, consider supplying a 100 Ω load at 10 V using a resistance R in the filter instead of a 0.5 H inductance. To be equally effective in the attenuation of ripple, R must be equal to X_L at 100 Hz. Therefore,

$$R = 2\pi 100 \times 0.5 = 314 \; \Omega$$

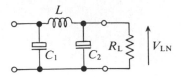

Fig. 19.19 π Section (LC) Filter

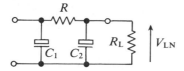

Fig. 19.20 π Section (RC) Filter

Since the load current flows through R, the voltage across it will be

$$V_R = 100 \text{ mA} \times 314 \ \Omega = 31.4 \text{ V}$$

Therefore the voltage across R is over 3 times the load voltage and correspondingly poor in power terms. In conclusion, π section RC filters are not recommended for high load current applications.

19.5 REGULATION AND STABILIZATION

Rectification and smoothing of an ac input give an output voltage V_O which is free from ripple. However, there are two effects which make the output vary. First, if the amplitude of the ac input voltage changes then V_O will, over a few cycles of input, change in proportion. Mains voltage fluctuations of 5% to 10% are not uncommon, particularly under winter loading conditions. Second, as the load current increases, more voltage is dropped across the internal resistance of the rectifier, so V_O falls. This effect is particularly pronounced with a capacitor filter because the current drawn through the transformer and diodes is, as noted, in the form of short spikes and therefore produces greater volt drops than the equivalent steady current of an L section filter. Maintaining V_O against mains voltage change is called stabilization whereas maintaining V_O against load current change is called regulation. However, the name 'regulated power supply' is given to a circuit designed to combat both sources of output voltage change.

Zener Diode as a Reference Voltage

Regulated power supplies invariably contain an internal reference voltage. They are used as a reference against which V_O is continuously checked and corrected as appropriate. In some aspects batteries are ideal as reference voltages but they are inconvenient. More common is to use a zener diode operating in its reverse voltage breakdown condition (Fig. 19.21). With the ideal zener diode the I/V characteristic is vertical once the zener diode voltage V_Z is attained. In other words, once the zener diode has broken down, the voltage across it will be V_Z regardless of the current I passing through it. In practice the characteristic is not absolutely vertical, i.e. it has

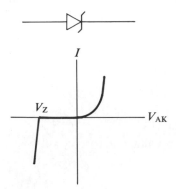

Fig. 19.21   Zener Diode Symbol and I/V Characteristic

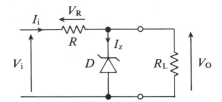

Fig. 19.22   Simple Zener Diode Regulator

an incremental resistance r_z, where

$$r_z = \frac{\Delta V_{AK}}{\Delta I} \quad \text{for } V_{AK} \geqslant V_Z$$

Therefore, for good regulation, the variation of current through a zener diode used as a reference should be as small as possible.

Zener Regulator

Figure 19.22 is the simplest of voltage regulators. In this, and other regulators to be described in this chapter, V_i is the smoothed but unregulated output from the rectifier. The regulator, in this case the series resistor R and zener diode D, are connected between the rectifier and the load represented by resistance R_L. In Fig. 19.22, the zener diode voltage $V_Z = V_O$, the required output voltage. Then, as the input voltage V_i, say, increases, the input current rises and the increase of V_i is dropped across the series resistor R while the excess current which accompanies it flows through D. The action of the circuit can be illustrated by example.

☐   **Example 19.4**

In Fig. 19.22 let $V_O = V_Z = 10$ V, and to ensure that the zener diode is well into its breakdown region, let $I_z(min) = 10$ mA. The input voltage $V_i = 15 \pm 20\%$ and $R_L = 100\ \Omega$. Find a suitable value for R and calculate the maximum value of I_z.

$$I_L = \frac{10\text{ V}}{100\ \Omega} = 100\text{ mA}.$$

$V_i(min) = 15$ V $- 20\% = 12$ V. At this voltage, $V_R = 2$ V.

$I_i = I_L + I_z(min) = 110$ mA

Therefore $R = \dfrac{2\text{ V}}{110\text{ mA}} = 18\ \Omega$

$V_i(max) = 15$ V $+ 20\% = 18$ V

Assuming that the zener diode holds V_O stable at 10 V, $V_R = 8$ V.

Therefore $I_i = \dfrac{V_R}{R} = \dfrac{8\text{ V}}{18\ \Omega} = 440$ mA.

But I_L remains at 100 mA.

Therefore $I_z(max) = 440 - 100 = 340$ mA. $\qquad\qquad\qquad\qquad$ □

The assumption that V_O remains constant at 10 V only holds if the zener diode is ideal under breakdown conditions. If, however, $r_z = 2\ \Omega$, then an increase of current of from 10 mA to 340 mA would cause V_O to increase by

$$\Delta V_O = 0.33 \times 2 = 0.66\text{ V}$$

which is not an impressive result for a regulated power supply.

The other disadvantage of the circuit is inefficiency. All of the input current flows through R and the excess current, as noted, flows through the diode. Input power under maximum voltage condition $= 18 \times 0.44 = 7.9$ W. Neglecting the output voltage rise, output power $= 10 \times 0.1 = 1$ W, which demonstrates that the circuit is inefficient. For regulation against varying load current a similar picture emerges. Using the previous example, i.e. $V_O = 10$ V, $R = 18\ \Omega$, but with V_i constant at 15 V, let R_L have any value from 100 Ω to ∞. Then I_L can have any value from 100 mA to 0.

With $V_i = 15$ V, $I_i = \dfrac{V_R}{18\ \Omega} = 280$ mA

and will remain constant for all values of R_L. Then with $R_L = 100\ \Omega$, $I_L = 100$ mA, $I_z = 180$ mA; and with $R_L = \infty$, $I_L = 0$, $I_z = 280$ mA, so again V_O will vary because of the wide range of values of I_z, and the circuit will be inefficient because of power loss in the series resistor and in the zener diode.

Series (Emitter Follower) Regulator

In the series (emitter follower) regulator of Fig. 19.23, $V_O = V_Z - V_{BE}$. Therefore, if the voltage changes across the diode and B–E junction are less than for com-

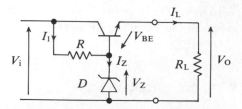

Fig. 19.23   Series (Emitter Follower) Regulator

parable circumstances with the simple zener diode regulator, then V_O will be more stable. Any increase in V_i is simply absorbed by an increase in V_{CE} – transistor action being such that this does not involve any significant increase in I_C and I_E. There will be an increase in the current through R and D but since, as we shall show by example, this will be less than for the corresponding case with the zener regulator, V_O will be less affected. This stabilizing of V_O against input voltage change not only improves performance against mains voltage fluctuation, but also will reduce any remaining ripple in the unregulated input. Further, the input current does not have to be fixed at a value required to provide the largest specified value of R_L. Instead, as the load draws more current, V_O will tend to fall, but as D holds the base at V_Z, V_{BE} tends to rise so emitter follower action causes more current to be drawn through the collector and emitter, and into the load.

☐   **Example 19.5**

Using a similar specification to the one used for the simple zener regulator of Example 19.4, i.e. $V_i = 15$ V $\pm$ 20%, $V_O = 10$ V, $R_L = 100$ Ω, choose a value for R for the circuit of Fig. 19.23 and determine the maximum variation in V_o given

$$r_Z = 2\ \Omega, h_{FE} = 50.$$

Take $I_z(\text{min}) = 10$ mA, $V_{BE} = 0.7$ V.

$$I_E = I_L = \frac{V_O}{R_L} = \frac{10\ \text{V}}{100\ \Omega} = 100\ \text{mA}$$

Therefore $I_B \doteqdot \dfrac{I_E}{h_{FE}} = 2$ mA.

$$V_Z = V_O + V_{BE} = 10.7\ \text{V}$$

With V_i at its minimum value of 12 V,

$$V_R = 1.3\ \text{V}$$

In this condition, I_Z should be minimum at 10 mA.
Then $I_1 = I_Z + I_B = 12$ mA

Therefore $R = \dfrac{V_R}{I_1} = \dfrac{1.3\ \text{V}}{12\ \text{mA}} = 108\ \Omega$

When V_i rises to its maximum value of 18 V,

$$V_R = 7.3 \text{ V}$$

Then $I_1 = \dfrac{7.3 \text{ V}}{108 \ \Omega} = 67 \text{ mA}$

But with I_L constant at 100 mA, I_B will remain at 2 mA.

Therefore $I_Z = 65$ mA, i.e. it increases by 55 mA

Therefore V_Z will rise by $\Delta I_Z \times r_z = 55 \text{ mA} \times 2 \ \Omega = 0.11$ V

V_O will increase by the same voltage, which is clearly more satisfactory than the simple zener stabilizer whose output increased by 0.66 V. □

Consider the same circuit with the load resistance being increased to 10 kΩ – unlike the simple zener regulator, current must be drawn through the B – E junction to maintain the output voltage V_O.

Then $I_E = I_L = \dfrac{10 \text{ V}}{10 \text{ k}\Omega} = 1$ mA

and $I_B = 0.02$ mA

With $V_1 = 15$ V,

$$I_i = \dfrac{4.3 \text{ V}}{108 \ \Omega} = 40 \text{ mA}$$

and with $I_B = 0.02$ mA,

$$I_z \doteq 40 \text{ mA}$$

However with V_i constant, I_1 will remain constant so with I_B falling by 1.8 mA, I_z will rise by 1.8 mA and increase V_z by only 3.6 mV. The difficulty, however, is that in I_B falling by 1.8 mA, V_{BE} will fall by an amount determined by the I_B/V_{BE} characteristic of the transistor. In practice this may be as much as 0.2 V and V_O will rise correspondingly.

The circuit is more efficient than the zener regulator because there is no requirement to sink large quantities of unwanted current through the diode.

Feedback (Op Amp) Series Regulator

In this circuit a high-gain amplifier – the op amp shown in Fig. 19.24 is ideal for this purpose – amplifies any change in V_O and feeds it back to restore V_O to its specified value. For the op amp the noninverting input voltage $V_A = V_z$, the zener reference voltage. The inverting input voltage

$$V_B = \dfrac{R_2}{R_1 + R_2} \ V_O \text{ and therefore } V_O = \left(1 + \dfrac{R_1}{R_2}\right) V_B.$$

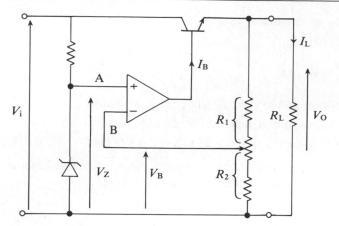

Fig. 19.24   Feedback (Op Amp) Series Voltage Regulator

In the steady state $V_A = V_B$ and therefore

$$V_O = \left(1 + \frac{R_1}{R_2}\right) V_z$$

(less V_{BE} which may be disregarded).

To follow the regulating action of the circuit, suppose that V_O tends to fall, either because the unregulated input voltage V_i falls or because the load current I_L rises. Then V_B will fall so the op amp increases I_B until V_O is restored. With R_1 and R_2 being made up of three resistors of which the middle one is variable, the circuit has an effective output voltage control.

For a large output current the op amp may not be able to meet the transistor base current requirements, in which case a Darlington pair can be used as illustrated in Fig. 19.25.

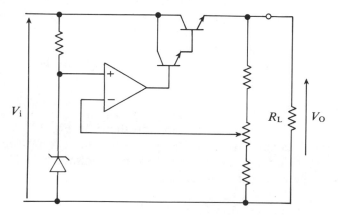

Fig. 19.25   Use of Darlington Pair in Feedback Voltage Regulator

Switching Regulators

In a series regulator the series transistor carries the load current. Since the series transistor is the means by which the output voltage is controlled, there must always be voltage across it and therefore there will always be a power loss in it. Another disdvantage of the series regulator is that the output voltage V_O will also be less than the input voltage V_i. Switching regulators are better in both respects. Efficiency is higher because a transistor operated as a switch ideally has zero voltage across it when ON and zero current through it when OFF, so it dissipates no power in either state. Further, a switching regulator can be connected to either step up or to step down the unregulated input voltage.

In principle a switching regulator operates as in Fig. 19.26a in which an electronic switch represented by S, typically operating between 200 Hz and 20 kHz, controls current i through an LC smoothing circuit and load resistance R_L. With S at A the current rises, and with S at B it falls. However, i changes at a slow rate because the effect of inductance is to oppose current change. As indicated in Fig. 19.26b, i consists of a relatively small triangular waveform superimposed on a larger direct current. The charge on C, and therefore V_O, will increase with the average value of i, which in turn will increase with the proportion of time that the switch is at A. However, note that the primary function of L and C is to produce a smooth output voltage, a task which is eased by the switch being operated above mains frequency.

The basic action of a step-down switching regulator can be followed with Fig. 19.27. Transistor Q is alternately switched ON and OFF by the waveform driving its base. With Q ON, i flows through L and C, the voltage developed across L holding D OFF. When Q is switched OFF, the direction of i will be maintained by the current-sustaining effect of the inductance (Fig. 19.27b) so the current path will be around the loop L, C and D, with i falling to give the polarity of voltage shown.

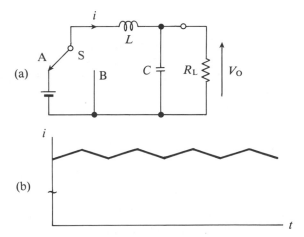

Fig. 19.26   (a) Basic Circuit of a Switching Regulator (b) Its Current Waveform

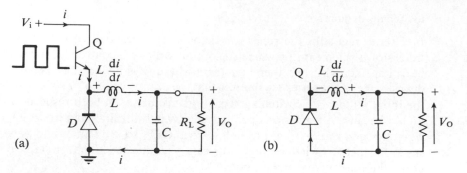

Fig. 19.27 Step-down Switching Regulator (a) Q ON, i Rising; (b) Q OFF, i Falling

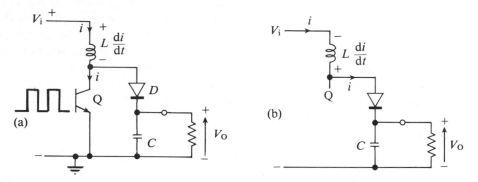

Fig. 19.28 Step-up Switching Regulator (a) Q ON, i Rising; (b) Q OFF, i Falling

From Fig. 19.27a,

$$V_i = L \frac{di}{dt} + V_O$$

Therefore $V_O < V_i$.

To make $V_O > V_i$ the circuit is modified as in Fig. 19.28a. With Q ON, the output voltage reverse biases D, so i flows through L and Q with i increasing and with the polarity of voltage across L as shown. When Q is switched OFF (Fig. 19.28b) the direction of i will be maintained so D is turned ON. This can only happen if $L \, di/dt$ has the polarity shown, which implies that i is falling. From Fig. 19.28b,

$$V_O = V_i + L \frac{di}{dt} > V_i$$

Switching Regulator 78S40

Figure 19.29 shows the essential elements of the 78S40 switching regulator. The manufacturer provides a table showing the values of L, C_1, and C_2 (which sets the waveform generator frequency) required to achieve specified values of output

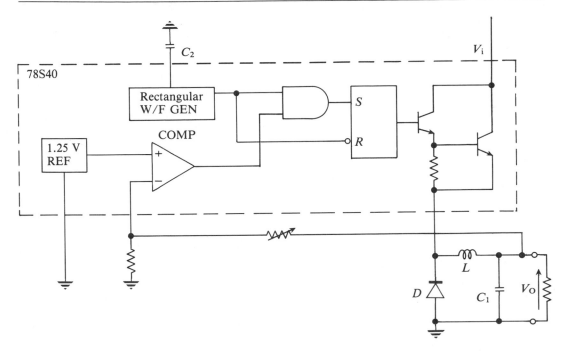

Fig. 19.29   Switching Regulator 78S40 in Step-down Operation (Courtesy RS Components Ltd)

voltage V_O, percentage ripple and maximum output current. The components which are external to the 78S40 IC in Fig. 19.29 are connected for step-down operation, but step-up is also available with the connection configuration of Fig. 19.28.

When V_O rises above its specified voltage, its attenuated value is compared with the 1.25 V reference which causes the comparator output to go negative. This disables the AND gate so the *SR* flip flop remains reset and the Darlington switch stays OFF. Therefore V_O falls until the comparator output goes positive. This allows the *SR* flip flop to be set and reset by the waveform generator which, in turn, opens and closes the Darlington switch and this enables V_i to sustain V_O.

Switching Regulator – Pulse Width Modulation Operation

An alternative mode of operation for a switching regulator such as the 78S40 is to drive the switch with a waveform whose duty cycle falls as V_O rises; in other words, the switch drive waveform is pulse width modulated. In Fig. 19.30 an auxiliary output from the generator is integrated to produce a ramp of constant frequency which is connected to the inverting input of the comparator. The op amp is connected as an analog negative feedback amplifier whose output rises as V_O rises. It will therefore delay the instant at which the ramp makes the comparator output

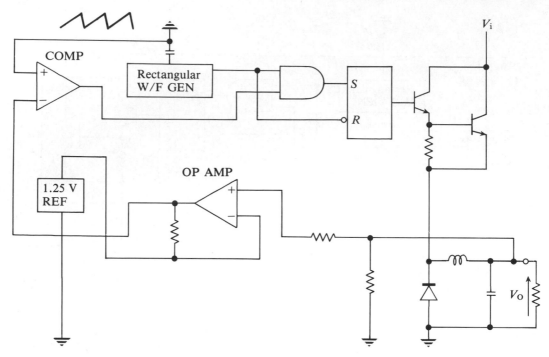

Fig. 19.30 Switching Regulator 78S40 in Pulse Width Modulation Operation

go positive. Figure 19.31 illustrates the effect with V_O rising to delay the setting of the flip flop and therefore the instant at which the Darlington switch is turned ON. As the flip flop is always reset when the ramp flyback is initiated, then it is evident that the flip flop output pulse width falls as V_O rises. Therefore the charging current falls to counteract the rise in V_O.

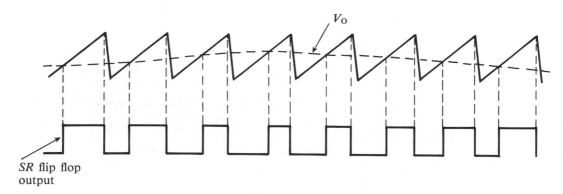

Fig. 19.31 Illustrating that Rise in V_O Shortens Flip Flop Output Pulse

19.6 PRECISION RECTIFIER

It is acceptable, as noted, to disregard the volt drop across a conducting diode when dealing with large voltages such as those encountered in rectifiers for power supplies. In many applications, particularly low-level signal processing, where millivolts rather than volts may be the order of amplitude under consideration, then not only is the approximation invalid, the signals may well not be large enough to bring a silicon diode into conduction at all. In these circumstances the precision rectifier of Fig. 19.32 uses the virtually infinite gain of an op amp to good effect. When v_s is negative, v_C is negative and D is cut off. Therefore $v_o = 0$. When v_s is positive, v_C will tend to rise to $+15$ V but D conducts and the circuit is essentially the op amp voltage follower of Fig. 9.7 whose gain is $+1$. In other words, v_o rises until $v_B = v_s$ and since $v_o = v_B$, then $v_o = v_s$. There will still be a forward volt drop of, say, 0.7 V across D, but it is of no consequence because op amp action dictates that the voltage differences between inverting and noninverting inputs must be virtually zero. As this only applies when D is ON the circuit rectifies, giving in this case positive-going

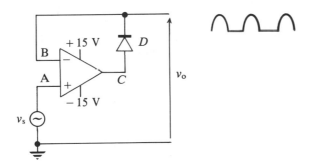

Fig. 19.32   Precision Rectifier

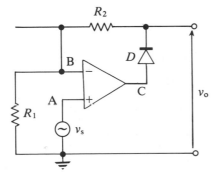

Fig. 19.33   Precision Rectifier with Voltage Gain A_v

signals at the output equal, in amplitude, to the positive components of v_s. Reversal of D would result in the circuit giving only negative-going signals.

Voltage gain greater than 1 can be provided by adding R_1 and R_2 as in Fig. 19.33. With this noninverting input circuit,

$$A_v = 1 + \frac{R_2}{R_1}$$

but again, only for positive input voltages.

19.7 DIODE DETECTOR

A circuit which is similar in operation to a half-wave rectifier is the diode detector of Fig. 19.34a. The function of the circuit is to produce, from the amplitude-modulated (AM) wave at its input (Fig. 19.34b), the relatively slowly varying shape of the waveform which is the modulating signal of frequency f_m, and to reject the high-frequency (f_c) carrier waves which are the individual waves which make up the waveform. Figure 19.34c shows the required output. As the input wave builds up from A to B, each successively larger positive half-cycle of carrier makes D conduct and charges C up to the peak value of the half-cycle. In between positive peaks D cuts off because C remains charged – the time constant CR is such that the degree of discharge between cycles of carrier is negligible. After B the waveshape falls and by choosing $CR < 1/f_m$, then v_o will decay with the envelope and therefore have the required waveshape of Fig. 19.34c. For small signals it would be necessary to use a circuit such as the precision rectifier described in Section 19.6.

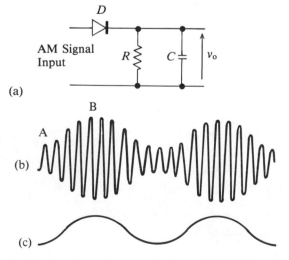

Fig. 19.34   Diode Detector (a) Circuit (b) AM Signal Input (c) Modulating Signal Output

19.8 DIODE CLIPPERS AND DIODE CLAMPS

A clipper is a circuit which prevents a waveform exceeding some positive or negative voltage by slicing off the waveform extremity at the clipping level. Some clippers slice off both extremities. In the following circuits the diode models of Figs. 19.2b, 19.3b and 19.4b will simply be referred to as models A, B and C respectively. In all cases it will be assumed that the input voltage v_i is a sine wave of peak value 10 V, although clippers and clamps are equally useful with nonsinusoidal waveforms.

Shunt Diode Clipper

Assuming diode model A for Fig. 19.35a, when D is ON it acts as a short circuit across the output so $v_o = 0$. When D is OFF it acts as an open circuit so $v_o = v_i$. D will only be ON when v_i is positive so v_o will be as illustrated in Fig. 19.35b. Had diode model B been used then D would not have conducted until $v_i = 0.7$ V, so v_o would result in Fig. 19.35c. Reversal of D would result in Fig. 19.35d.

Adding a dc voltage in series with D (Fig. 19.36a) causes a shifting of the clipping level. One technique for analyzing these circuits is first to consider whether D is conducting when $v_i = 0$. If D is ON, then using model A, v_o is equal to the clipping voltage, in this case $+3$ V. If D is OFF then $v_o = v_i$ until v_i reaches a voltage which brings D ON. In Fig. 19.36a the 3 V source reverse biases D so it is OFF with $v_i = 0$ V. When v_i reaches $+3$ V, D comes ON and the circuit clips. Figure 19.36b is the output waveform. With model B, D will be OFF until $v_i = +3.7$ V, from which point the circuit behaves as in Fig. 19.37a. Up to this point the shunt arm is open circuit so the overall effect is the output waveform of Fig. 19.37b.

Diode model C only needs to be used if r_d, the diode forward resistance, is comparable with R. It does not affect the input voltage at which clipping occurs but once D is ON the three voltages v_i, $+3$ V and $+0.7$ V divide between R and r_d. Suppose that in the clipper of Fig. 19.36a, $R = 1$ kΩ and $r_d = 100$ Ω. Then once D is ON the equivalent circuit will be as shown in Fig. 19.38a, and using superposition,

$$v_o = \frac{100}{1100} v_i + \frac{1000}{1100} 3.7 \text{ V}$$

from which we can develop the waveform of Fig. 19.38b. Thus, when v_i is at its peak value of $+10$ V, $v_o = 4.3$ V.

A double clipping action — sometimes called slicing — is achieved by the circuit of Fig. 19.39a. Using diode model A, D_1 will clip at all voltages above 3 V and D_2 at all voltages less (i.e. more negative) than -2 V. Figure 19.39b is the output waveform.

Figure 19.40a gives an output which is a slice of the negative part of the input only. D_1 is ON for all voltages down to -2 V. For $-2 \text{ V} > v_i > -5 \text{ V}$, neither diode is ON so $v_o = v_i$. When $v_i < -5$ V, D_2 conducts so the circuit clips at -5 V. Figure 19.40b is the output waveform. A waveform can be sliced using two zener

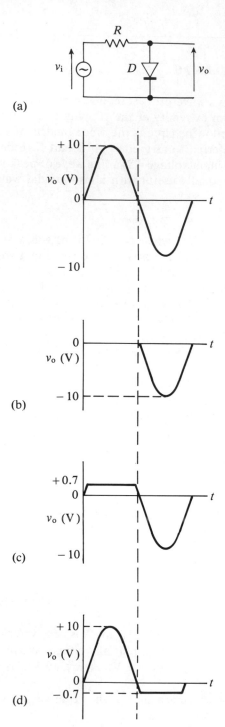

Fig. 19.35 (a) Shunt Diode Clipper and Input Waveform (b) Clipper Output Using Diode Model A (c) Clipper Output Using Diode Model B (d) Clipper Output with Diode Reversed

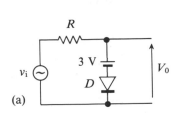 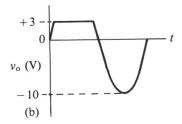

(a) (b)

Fig. 19.36 (a) Diode Clipper with dc Source to Set the Clipping Level (b) Clipper Output with Diode Model A

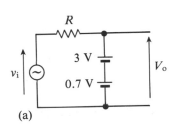 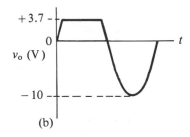

(a) (b)

Fig. 19.37 (a) Equivalent Circuit of Clipper when $v_i \geqslant 3.7$ V (b) Clipper Output with Diode Model B

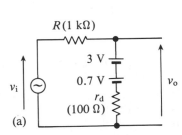 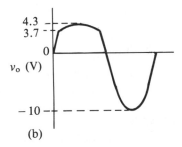

(a) (b)

Fig. 19.38 (a) Equivalent Circuit of Clipper with Diode Model C when $v_i \geqslant 3.7$ V (b) Clipper Output

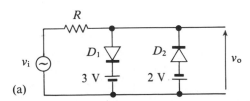 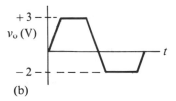

(a) (b)

Fig. 19.39 (a) Slicing Circuit (b) Slicing Circuit Output

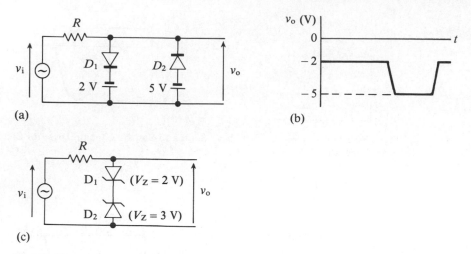

Fig. 19.40   (a) Circuit for Producing Negative Slice (b) Negative Slice of Output (c) Zener Diode Slicing Circuit

diodes across the output as illustrated in Fig. 19.40c. Remembering that when forward biased a zener diode acts like a standard silicon diode then, when $v_i \geq 3$ V, and D_2 breaks down, D_1 will be forward biassed, so assuming model A, v_o clips at $+3$ V. When $v_i \leqslant -2$ V, D_1 breaks down so the circuit clips at -2 V. With 3 V $> v_i > -2$ V, neither diode is broken down but one will be reversed biased so the shunt arm is open circuit. The output waveform will be the same as Fig. 19.39b.

Series Diode Clippers

By contrast with the shunt diode clippers which clip when a diode is turned ON, the series diode clipper of Fig. 19.41a clips when D is turned OFF. This simplest of clippers will be recognized as the half-wave rectifier which, assuming diode model A, produces an output v_o consisting of positive half sine waves of v_i, which is the condition for D to be ON, but clips off the negative half sine waves, as shown in Fig. 19.41b, because D will be acting as an open circuit isolating v_i from the output.

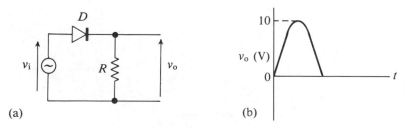

Fig. 19.41   (a) Series Diode Clipper (b) Clipper Output Waveform

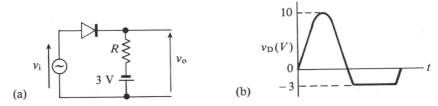

Fig. 19.42 (a) Shifting the Clipping Level in a Series Clipper (b) Clipper Output Waveform

Shifting the clipping levels requires the inclusion of a dc source (Fig. 19.42a). When $v_i = 0$, the 3 V source makes D conduct so, assuming model A, $v_o = v_i$ until $v_i < -3$ V, at which point D is cut off. Now, no current flows around the circuit so $v_o = -3$ V and the output waveform is as in Fig. 19.42b.

With diode model B, D cuts off when $v_i < -2.3$ V. Until then the circuit output can be deduced from Fig. 19.43a, from which $v_o = v_i - 0.7$ V, and which leads to the output waveform of Fig. 19.43b. The input waveform has been included on this diagram so that it can be observed that:

(a) v_o is always 0.7 V below v_i when D is ON; and

(b) although the voltage at which D cuts off is $v_i = -2.3$ V, the clipping level of the output waveform is -3 V.

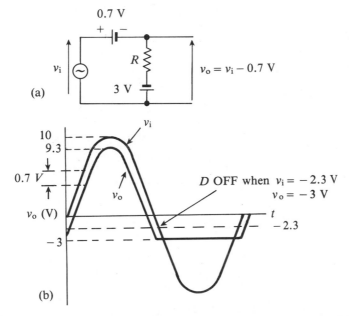

Fig. 19.43 (a) Equivalent Circuit of Series Clipper when D is ON Using Diode Model B (b) Clipper Output Waveform

Diode Clamp

A clamp is a circuit which is designed to fix one extremity of a waveform to a certain voltage level regardless of changes in the input waveform. For example, the waveform of Fig. 19.44a has had its positive extremity fixed at -2 V in Fig. 19.44b.

In the circuit of Fig. 19.45a, D conducts in the first quarter-cycle of v_i (0 to t_1 in Fig. 19.45b) and, assuming diode model A, C charges up to $+10$ V. With D conducting $v_o = 0$ V. During the next two quarter-cycles (t_1 to t_2), v_i falls by 20 V from $+10$ V to -10 V, and since there is no path for C to discharge, both sides of it must fall by 20 V so v_o falls from 0 V to -20 V. Now v_i can rise from -10 V to $+10$ V and v_o can follow it from -20 V to 0 V without bringing D into conduction. Therefore the output v_o has its positive extremity clamped at 0 V.

With D reversed (Fig. 19.46a), then given the same starting condition, the first half-cycle (0 to t_1 in Fig. 19.46b) produces no current flow so v_o follows v_i. For the next quarter-cycle (t_1 to t_2) current flows and C is charged to -10 V (Fig. 19.46c) with v_o being clamped at 0 V (Fig. 19.46d). From t_2 to t_3, v_i rises by 20 V but as there is no current path for C to discharge, v_o also rises by 20 V, i.e. from 0 V to $+20$ V. Therefore v_o has its negative extremity clamped at 0 V.

If v_i, in Fig. 19.46a, falls to, say, 5 V peak, then the negative extremity will become clamped at $+5$ V because C retains its -10 V charge. By connecting a discharge resistor across C (Fig. 19.47) with $CR > T$ – the period of the input

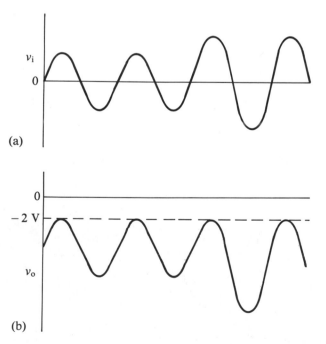

Fig. 19.44   An Input Waveform (a) with its Positive Extremity Clamped (b) at -2 V

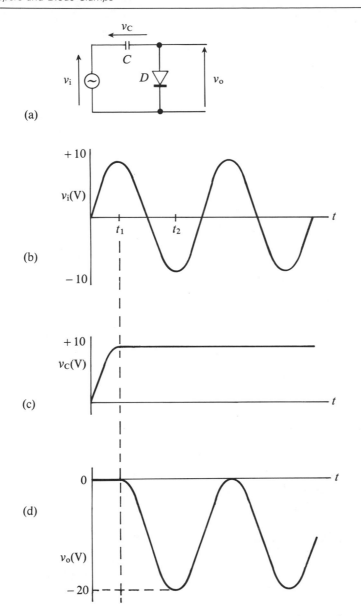

Fig. 19.45 Diode Clamp with Positive Extremities of Output Clamped at 0 V

waveform − then, when the input signal level does fall, C will discharge over a few cycles down to a voltage which will bring the clamping level back to 0 V. Over each cycle the discharge resistor will cause a slight decay even when the signal level is not changing, but the lost charge will be replenished by a brief period of diode conduction when the waveform is at the clamping level.

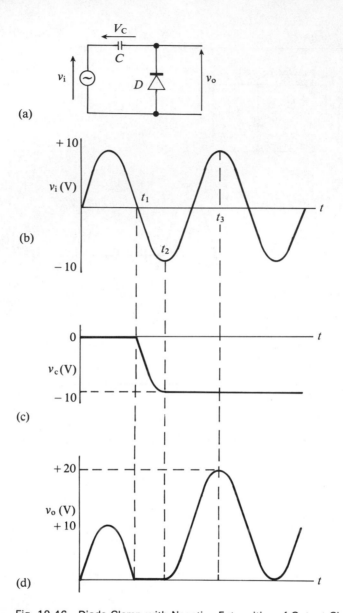

(a)

(b)

(c)

(d)

Fig. 19.46   Diode Clamp with Negative Extremities of Output Clamped at 0 V

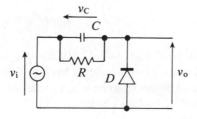

Fig. 19.47   Diode Clamp with Discharge Resistor to Accommodate Varying Input Signal Levels

The clamping level can be shifted from 0 V by including a dc supply in series with the diode as shown in Fig. 19.48a. In this circuit the 3 V supply prevents D from conducting until $v_i > 3$ V on the first quarter-cycle. Therefore, from 0 to t_1 of v_i (Fig. 19.48b) v_o rises from 0 to 3 V (Fig. 19.48d) but v_c remains at 0 V (Fig. 19.48c). From t_i to t_2, D conducts so C charges up to $+7$ V. From t_2 to t_3, v_i falls by 20 V and v_o follows it from $+3$ V to -17 V. Therefore the positive extremity is clamped at $+3$ V.

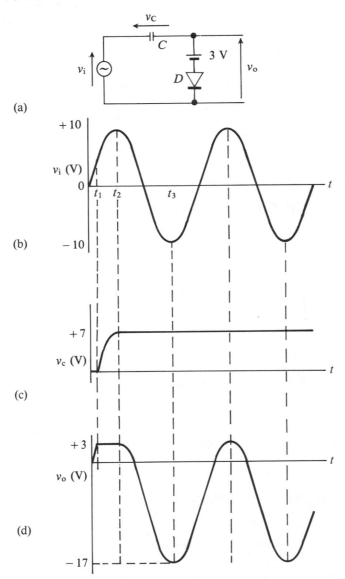

Fig. 19.48   Diode Clamp with Positive Extremities of Output Clamped at $+3$ V

Diode Clamp as a DC Restorer

Clamps are widely used in television circuits where it is necessary to retain a reference voltage to represent black level. In the two waveforms of Fig. 19.49, T_p are the periods of picture signal and T_s the periods for synchronizing for two consecutive lines in each waveform. The synchronizing pulses have been stripped off, leaving picture signals only with 0.3 V representing black level and 1.0 V representing peak white, which is the maximum picture voltage for a black-and-white signal. Figure 19.49a would represent a rather dark scene and Fig. 19.49b a much brighter one. Clearly the latter has the much higher dc component of the two, and this dc component must be retained if overall scene brightness is to be properly represented on the displayed picture. If these signals are passed through, for example, a *CR* coupling, the dc component will be lost and the two waveforms will appear as in Fig. 19.50a and b. By clamping these signals with the negative extremity at 0.3 V the waveforms will be restored to their former levels of Figs. 19.49. Used in this manner the circuit is known as a dc restorer.

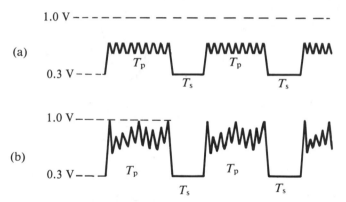

Fig. 19.49   Low Brightness (a) and High Brightness (b) TV Picture Signals with Black Level (0.3 V) and Peak White (1.0 V) Correctly Established

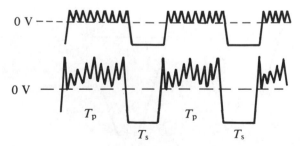

Fig. 19.50   TV Picture Signals with Black Level, Peak White and dc Component Lost after Passing Through a CR Coupling

SUMMARY

Conversion of ac to dc power is a long-standing application of electronics but the introduction of digital techniques has led to the switch mode power supply and, as with many other aspects of electronics, the results have been beneficial. Nevertheless, the traditional techniques of rectification and smoothing are essential aspects of power supplies generally and should therefore be understood. Clipping and clamping circuits are deceptively simple and experience suggests that students have difficulty in mastering them. Problems are eased when it is legitimate to use the simpler diode models for analysis.

REVISION QUESTIONS

19.1   A half-wave rectifier is fed from the 240 V ac mains via a 12 : 1 step-down transformer. Calculate the output voltage and minimum PIV required for the diode if the output is smoothed.

(28 V; 56 V)

19.2   A full-wave rectifier is required to supply 50 mA dc to a load of 150 Ω. Calculate the turns ratio of the transformer primary to each secondary half winding and the minimum diode PIV if the input is to be fed from the ac mains.

(45 : 1; 15 V)

19.3   A full-wave rectifier with capacitor smoothing feeds a 50 Ω load at 12 V. If current flows for one-tenth of the period of each half-cycle, calculate the diode current rating on the assumption that when current does flow it is constant.

(2.4 A)

19.4   A rectified power supply is to have an output of 24 V (max) at 500 mA, with a ripple of 1 V peak-to-peak, derived from ac mains of frequency 50 Hz.
        Draw the circuit of a suitable bridge rectifier power supply with capacitive smoothing.
        Determine:

(a) the capacitance and peak voltage rating of the capacitor.

(5,000 μF; 24 V)

(b) the current and peak inverse voltage ratings of the diodes, assuming constant current flow for one-tenth of each half-cycle;

(5 A; 24 V)

(c) the transformer rms secondary voltage.

(17 V)

19.5 A full-wave bridge rectifier supplies a small current to a resistive load which is shunted by a large capacitance. Sketch, indicating relative amplitudes, the current waveforms in (a) the load, (b) the capacitor, and (c) one arm of the bridge.

Develop an expression for the approximate magnitude of the peak-to-peak ripple voltage across the load in terms of load current I, capacitance C and supply frequency f.

Given that the supply frequency is 50 Hz and the current supplied to the load is 6 mA, estimate the value of capacitance required to reduce the peak-to-peak ripple voltage to 30 mV.

($C = 2000\ \mu F$)

19.6 A full-wave rectifier with capacitor smoothing has 2% ripple. Calculate the proportion of a half-cycle of input for which current flows.

(6.4%)

19.7 Compare the properties of capacitor and L section filters.

19.8 Why is it desirable to maintain a bleed current with an L section filter?

19.9 A full-wave rectifier with an L section filter provides 12 V to a fixed load of 50 Ω. If the ripple is to be 1% maximum, calculate the minimum value of filter components.

(53 mH; 2300 μF)

19.10 If the load resistance in Q.19.9 can have any value from 50 Ω to 2000 Ω, recalculate the filter components.

(2.1 H; 57 μF)

19.11 A zener stabilizer has a dc input supply of 30 V ± 20%. It is required to feed a 50 Ω load at 20 V. Calculate the maximum series resistance and maximum zener diode current if the minimum zener diode current is to be 20 mA.

(9.5 Ω; 1.28 A)

19.12 A series stabilizer is required to feed a resistive load which can vary between 100 Ω and 400 Ω. The stabilizer input is fixed at 20 V and the output is to be 15 V. Calculate the maximum stabilizer resistance if the minimum zener current is 10 mA. Calculate the maximum zener current.

(31 Ω; 123 mA)

19.13 For Fig. 19.23 it is required that $V_o = 5$ V with $V_i = 7$ V ± 1 V. The load current is to be 250 mA. Given that $I_z(\text{min}) = 5$ mA; $h_{FE} = 50$; $r_z = 3\ \Omega$; $V_{BE(ON)} = 0.6$ V,
(a) calculate the maximum value of R;
(b) calculate the change in V_o as V_i rises from 7 V to 9 V.

$$(40\ \Omega;\ +0.15\text{V})$$

19.14 What is the advantage of the feedback series regulator over the emitter follower series regulator?

19.15 What are the advantages of switching regulators over nonswitching regulators?

19.16 Describe the action of switching regulators in

(a) stepping down
(b) stepping up

input voltage.

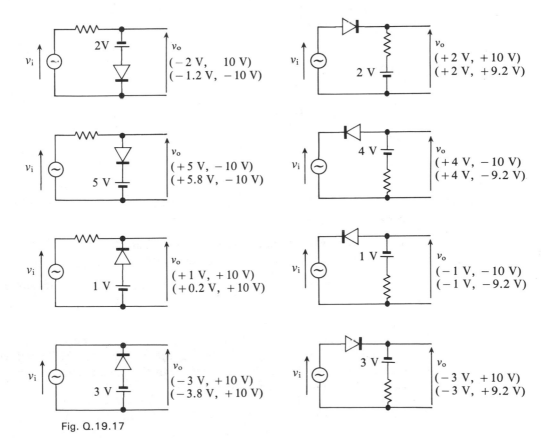

Fig. Q.19.17

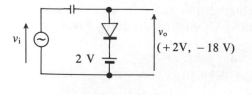

$(+2V, -18\,V)$

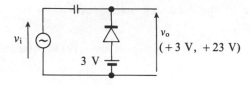

$(+3\,V, +23\,V)$

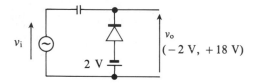

$(-2\,V, +18\,V)$

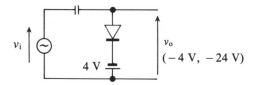

$(-4\,V, -24\,V)$

Fig. Q.19.18

19.17   For each of the circuits in Fig. Q.19.17, $v_i = 10 \sin \omega t$ (V). Sketch v_o showing the voltages of the waveform extremities for:

(a) diode acting as short circuit when forward biased,
(b) diode acting as 0.8 V pd when forward biased.

(In the answers shown next to each circuit, the first voltage is the clipping level.)

19.18   For each of the clamping circuits in Fig. Q.19.18, v_i is a sine wave of 10 V peak. Sketch v_o in each case, showing the voltage of the waveform extremities. Assume that diodes are ideal. (In the answers shown next to each circuit, the first voltage is clamping level.)

20

A Microprocessor System and its Instruction Set

Chapter Objectives

20.1   Constituents of a Microprocessor System
20.2   Memory
20.3   Arithmetic/Logic Unit (ALU)
20.4   Control Section
20.5   Two-phase (2ϕ) Clock
20.6   Microprocessor System
20.7   Addressing Modes
20.8   Instruction Set
20.9   Condition Code Register
20.10   Branching − Relative Addressing
20.11   Subroutine
20.12   Delay Loops
        Summary
        Appendix
        Revision Programs

Invention of the MOSFET led to very large-scale integration (VLSI) technology with the result that the central processing unit (CPU) of a digital computer could be fabricated on a single IC. Such an IC, a *microprocessor*, when interconnected with other components has permitted the realization of a digital computer on a single PCB − a microprocessor system. Although MOSFET technology has a low operating speed this has not prevented the introduction of microprocessors in a wide range of industrial applications and in low-cost computer systems − microcomputers. The purpose of this chapter is to give a brief description of the architecture of a microprocessor system and to explain the nature of microprocessor instructions and how they are carried out. It is directed specifically towards the Motorola M 6800 microprocessor system but the basic principles covered are universal. The important topic of input/output (I/O) is given a chapter of its own.

20.1 CONSTITUENTS OF A MICROPROCESSOR SYSTEM

A microprocessor system or indeed, any digital computer, can be regarded as being made up of four essential elements:

Memory
Arithmetic/Logic Unit (ALU)⎫
Control Section ⎬ CPU
Input/Output ⎭

The ALU and the control section are collectively known as the central processing unit (CPU) and can be fabricated on a single IC called a microprocessor unit (MPU). With increasing chip complexity an MPU may also contain a useful amount of memory. An MPU mounted on a PCB with memory and I/O forms a microprocessor system.

20.2 MEMORY

The sections of a microprocessor system are interconnected in such a manner that there is no logical starting point, but one advantage of describing memory first is that it introduces the data and address buses which are the principal highways of communication between sections. The term '8-bit machine'† (of which the Motorola M 6800 is an example) implies that the registers processing data in the ALU are, to invoke the standard jargon, 8 bits wide. Therefore instructions are produced as groups of 8 bits − 8 bits being known as a byte. For convenience a byte is usually written as two hex numbers (see Section 13.1), e.g. $4E \equiv 0100\ 1110$. Commonly, 8-bit microprocessor systems have provision for up to 65 536 memory locations each capable of storing 1 byte. ($65\ 536 = 2^{16} = 64 \times 2^{10} = 64 \times 1024$, usually expressed as 64 K of memory). Therefore to address (i.e. select) any individual location a 16-bit word is required: a 16-bit word is the same as two bytes and can be represented by four hex numbers. The general idea is illustrated in Fig. 20.1. Memory location $A001 contains $4F. Therefore if we were reading data from memory, the 16-bit pattern corresponding to $A001 is put on to the address bus and the 8 bits corresponding to $4F are output from memory on to the data bus. Note that the address bus at memory is unidirectional but that the data bus is bidirectional. Therefore, to write a number into a memory location, say $A008, the 16 bits corresponding to $A008 are put on to the address bus and the 8-bit pattern to be stored is put on to the data bus by some other component in the microprocessor system and stored in location $A008 because that is the number on the address bus.

† Older machines were predominantly 4-bit processors, commonly used in toys. In recent years 16-bit machines have become more widely used.

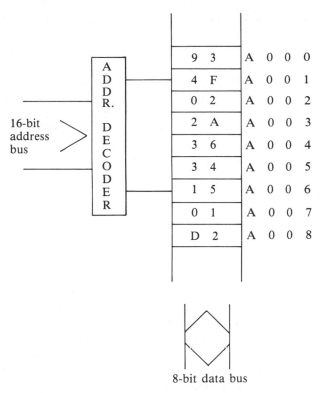

8-bit data bus

Fig. 20.1 Address and Data Buses at Memory

20.3 ARITHMETIC/LOGIC UNIT (ALU)

An ALU may be regarded as a device which contains a number of working registers called accumulators, and circuitry to carry out a range of arithmetic and logic functions on data which has been loaded into them from memory. For instance, the contents of an accumulator could be added to, subtracted from, or ANDed with the contents of another accumulator or storage location.

20.4 CONTROL SECTION

A program consists of a sequence of instructions. Usually, each instruction consists of an operator and an operand. Thus, with an instruction such as 'load accumulator A with the number $27', 'load accumulator A' is the operator and '$27' the operand. These would appear in memory and, when required, be put out on to the data bus as two bytes, $8B and $27, where $8B is known as the op code which represents the operator. In other words, a program consists of nothing but numbers in the form of consecutive 8-bit bytes. One function of the control section is to know whether program bytes are operators or operands. For example, modifying the previous instruction into loading $8B into accumulator A, the two bytes would be $8B and $8B. The control section must also develop binary patterns from instructions which, in effect, direct the ALU to carry out the required operations. It also has a sequencing function in that it must ensure that having set the ALU to carry out a particular operation (e.g. ADD), it is next provided with the data on which to carry it out. Not all microprocessor instructions conform to this pattern. For example, in an instruction such as 'increment accumulator A', the directive to 'add 1' is inherently contained in the instruction. Therefore all that is required is the program byte which represents the instruction 'increment accumulator A': in this case it is $4C.

20.5   TWO-PHASE (2ϕ) CLOCK

The reference for timing microprocessor operations, such as the sequencing of operators and operands, is known as the two-phase (2ϕ) clock which consists of two nonoverlapping waveforms as illustrated in Fig. 20.2. Given this and the components previously mentioned, we can build up a simplified block diagram of a microprocessor system which can be used to show how the various parts of the system interact when running a program.

20.6   MICROPROCESSOR SYSTEM

Only two control section registers are shown in the simplified microprocessor system of Fig. 20.3. However, these registers, the program counter (PC) and the instruction register (IR), merit emphasis in an explanation of the processing of microprocessor instructions. The program counter is a 16-bit register which puts on to the address bus the address of the next op code or operand to be executed. It is therefore usually,

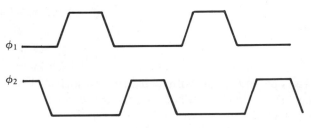

Fig. 20.2   Two-phase Clock

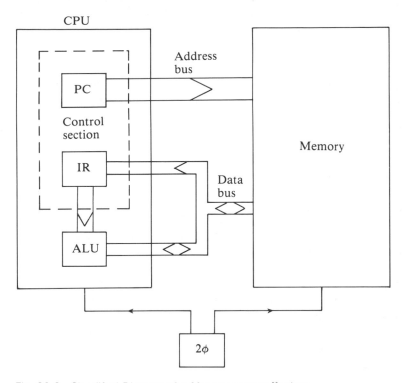

Fig. 20.3   Simplified Diagram of a Microprocessor System

but by no means always, simply incremented as program bytes are put on to the data bus in the sequence with which they occur in memory. When an op code is output from memory on to the data bus, it is latched into the instruction register which generates a bit pattern instructing the ALU to carry out the required operation.

Suppose that the program in the following example is stored in memory of the microprocessor system of Fig. 20.3.

☐   **Example 20.1**

The first column of Table 20.1 is a description of each instruction. The second consists of mnemonics called assembly language — if a microprocessor system has an assembler it can be programmed in assembly language. Ultimately, however, the program actually stored will consist of bytes which can be represented, as in the third column, by hex numbers. Thus, if we refer to the M 6800 Instruction Set (see Appendix on p. 558), we see that INCA is represented by the op code $4C. This is the first instruction and will be stored at the location chosen for the start of the program, $0200. As noted, the data is self-contained in this particular instruction so the operation requires only one program byte which can be stored in a single

Table 20.1

Operation	Assembly Mnemonic	Machine Code	Location in Memory
Increment Accumulator A	INCA	$4C	$0200
Load Accumulator B with the number $A6	LDAB#$A6	$C6A6	$0201/2
Add to Accumulator A the contents of location $341E	ADDA$341E	$BB341E	$0203/4/5

memory location. When the program is run the sequence will be as follows and should be read in conjunction with the memory table of Fig. 20.4.

At A on the 2ϕ clock (Fig. 20.5), the PC puts the start address $0200 on to the address bus where it is latched.

At B, the PC is incremented to $0201 but $0200 remains latched on the address bus.

At C the op code $4C is output from $0200 on to the data bus – the instruction is said to have been *fetched*.

At D, the instruction register (IR) decodes op code $4C and produces the control word which causes the ALU to increment accumulator A – the instruction has been *executed*.

The op code $4C contains the information which, in effect, tells the control section not to expect data because the data required is inherent in the increment instruction.

Addr.	
0200	4C
0201	C6
0202	A6
0203	BB
0204	34
0205	1E

Fig. 20.4   Memory Contents for Example 20.1

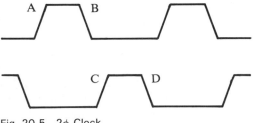

Fig. 20.5 2ϕ Clock

The fetch/execute cycle is a feature of the way that a computer deals with both operators and operands.

The next instruction LDAB # $A6 is made up of two program bytes, $C6 being the op code and $A6 the operand. The # ('hash') sign simply indicates that it is the number $A6 which is to be loaded – without the # sign the data is whatever number is in location number $A6. More will be said on this in the next section. During the previous instruction, once $0200 had been latched on to the address bus the PC was incremented to $0201. For this operation the sequence is similar.

At A on the 2ϕ clock, the PC puts $0201 on to the address bus where it is latched.
At B, the PC goes to $0202.
At C, the op code $C6 is fetched from $0201.
At D, $C6 is decoded in the IR which produces the bit pattern instructing the ALU to prepare to load accumulator B.

This time the op code $C6 contains the information that there will be a second program byte which will be the data for this instruction. The sequence will be:

At A, the PC puts $0202 on to the address bus.
At B, the PC is incremented to $0203.
At C, the data $A6 is fetched from memory.
At D, the ALU takes the data which is on the data bus and loads it into accumulator B in accordance with the instruction decoded from the previous op code.

The sequence for the final instruction is similar but the differences should be noted. First the machine code for the whole operation consists of the bytes $BB341E so the control section must provide for three fetch/execute cycles. Second, the data, which will be one byte, is found by fetching the contents of a location having a two-byte address $341E. Had the address been no higher than $00FF, the zeros would be omitted and the operation reduced to two bytes. □

20.7 ADDRESSING MODES

In the previous section we drew the distinction between an instruction such as LDAA#$A6 and one such as LDAA$A6. We have also noted that an instruction such as INCA has an op code but no operand. Instructions such as INCA or COMA (complement accumulator A) are said to have an *implied* addressing mode. For an instruction such as LDAA#$A6, the # sign indicates that it is the actual number $A6 which is to be added: the addressing mode is called *immediate*. When the # sign is omitted the data is to be found in location $00A6. Thus if a group of locations had contents as shown in Fig. 20.6 then the instruction LDAA$A6 would result in accumulator A being loaded with $F3. The operation LDAA#$A6 would lead to accumulator A being loaded with $A6. The addressing mode for LDAA$A6 is called *direct*. For addresses higher than $00FF the operation runs to three bytes and the addressing mode is said to be *extended*: it is, in essence, the same as direct addressing.

Indexed Addressing

An instruction such as

　　LDAB $ 04, X

is more complex than those previously described. One of the 16-bit registers contained in the M 6800 microprocessor is the index register. It can be loaded by immediate, direct or extended addressing. Although it can be loaded with numbers up to $FFFF, let us simply load it with a one byte number as follows:

　　LDX # $ 36

Now, if we write an instruction such as the one at the beginning of the paragraph,

　　LDAB $ 04, X

then to find the contents of accumulator B we must add to the offset $04 the number

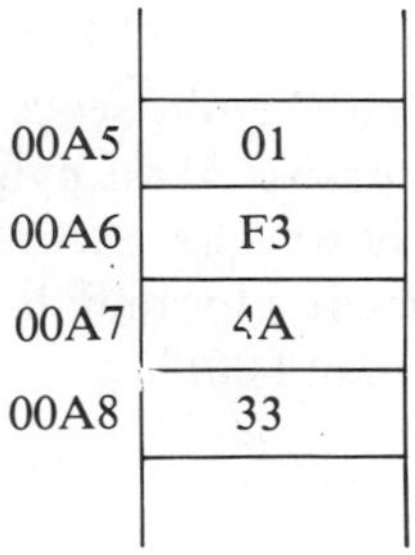

00A5	01
00A6	F3
00A7	4A
00A8	33

Fig. 20.6   A Group of Locations Holding Data

0036	69
0037	8A
0038	CC
0039	B0
003A	5F
003B	12

0069	44
006A	F0
006B	1D
006C	2D
006D	33

Fig. 20.7   Two Groups of Locations Holding Data

in the index register, which is $36. Therefore accumulator B is now loaded with the contents of location $003A which, as shown in Fig. 20.7, is $5F.

If instead the index register loading had been direct, e.g.

LDX $ 36

then its contents would be $69. The result of the instruction

LDAB $ 04, X

leads to accumulator B being loaded with the contents of location ($0069 + $04) which is location $006D. Therefore accumulator B will be loaded with the number $33.

Indexed addressing is often useful when dealing with a group of consecutive locations. This is illustrated in Example 20.3 on p. 529.

20.8   INSTRUCTION SET

Manufacturers provide an 'Instruction Set' for each of their microprocessors such as the one shown in the Appendix on pp. 558–62 for the Motorola M 6800.

Instruction Column: This briefly describes the instruction, e.g. 'Load accumulator', 'Decrement index register', and so on.

Mnemonic Column: These mnemonics are the formal assembly language instructions. As such they will be recognized by an assembler and converted into machine code.

Addressing Modes: These six columns indicate the various addressing modes in which an instruction can be used. It will be noted that no instruction can be addressed in all modes. For example, load and store instructions must have an operand and therefore cannot have an entry in the 'Implied' column. On the other hand, Decrement A means 'subtract 1 from accumulator A' and therefore an operand is inappropriate, so there will only be one entry under 'Implied' for instructions such as Decrement A and Increment B.

Op Codes (OP), Program Bytes (PB) and Machine Cycles (MC): Within each addressing mode column there are three supplementary columns. The first shows the op code (OP) for each instruction. Thus for Load Accumulator A Immediate the op code is $86, while for Load Accumulator A Direct the op code is $96.

The number given in the column labeled MC is the number of machine cycles it takes to complete an operation. Since the 2ϕ clock runs at 1 MHz, each machine cycle lasts 1 μs. For example, an operation such as ADDA direct requires 3 machine cycles and therefore takes 3 μs to complete. Some instructions are particularly long, e.g. JSR extended, which takes 9 machine cycles.

The next column gives the number of program bytes (PB) associated with each operation. An instruction such as DECA has no operand and therefore only one program byte ($4A) is required. An immediate or direct load operation requires two bytes. For example, for LDAA $2E the machine code is $962E. An extended load or store operation requires three bytes. Thus, the machine code for STAB $431D is $F7431D.

Boolean/Arithmetic Operation: Each entry in this column is a logical statement of a particular operation. Where the letter M is used it stands for the operand. The diagrams used to illustrate SHIFT instructions illustrate important differences between some instructions. For instance the diagram for ASRA shows that the value of bit 7 is retained but with LSRA a 0 is moved into bit 7.

20.9 CONDITION CODE REGISTER

This section describes the purpose of the final column 'Condition Reg.' in the Instruction Set. If a microprocessor system were restricted to arithmetic operations such as ADD and DEC or to logic operations such as AND and OR, its usefulness for control purposes, where a decision or action is to be taken based on the state of a process, would be limited. For example, when the temperature of a process reaches a certain level there may be a requirement to reduce the heat input: this will involve the frequent checking of temperature and, when it reaches a certain level, giving out an instruction to reduce the heat. A microprocessor system can use branching instructions, i.e. alternative courses of action based on incoming data and the state of a special register called the condition code register. Basically, this device uses six of its locations as *flag* bits, a term intended to indicate whether the result of an instruction is, or is not, for example, negative. The six flag bits are:

N – Negative
Z – Zero
C – Carry
V – Overflow
H – Half carry from bit 3 to bit 4
I – Interrupt

For the purpose of this book it is sufficient to consider only the first 3 flag bits, N, Z and C. In the examples which follow it is important to realize that the condition code register is updated after every instruction.

Z Flag

The Z flag is set ($Z = 1$) whenever the result of an instruction is zero. To take a trivial example consider the instructions

```
CLRA
LDAB # $04
```

The first concerns accumulator A and results in its contents becoming zero. Therefore $Z = 1$. The second concerns accumulator B and results in its contents being something other than zero. Therefore $Z = 0$.

N Flag

The N (negative) flag simply treats bit 7 as a sign bit and will be set to 1 if bit 7 is 1. Therefore, for any number up to and including 7F, $N = 0$ and for all other numbers $N = 1$. Taking two instructions in sequence and noting the Z and N flags, we have

```
CLRB    Z = 1    N = 0
DECB    Z = 0    N = 1
```

C Flag

This carry/borrow flag, to give it its full name, will be set to 1 whenever the result leads to a carry out of bit 7 or a borrow into bit 7. For example,

```
LDAA # $FF    C = 0    Z = 0    N = 1
ADDA # $01    C = 1    Z = 1    N = 0
CLRA          C = 0    Z = 1    N = 0
SUBA # $01    C = 1    Z = 0    N = 1
```

C will also be set to 1 in certain shift instructions. Thus,

```
LDAA # $FF C = 0
ASLA         C = 1
```

which can be confirmed from the instruction set. The C flag would also have been set by using LSRA instead of ASLA.

20.10 BRANCHING – RELATIVE ADDRESSING

As noted, it is sometimes necessary to choose between two alternative courses of action. The choice is usually based on examining the contents of a register, in the process of which flags will be set or reset. The action taken can be determined by using one of the branching instructions listed in the Instruction Set (Appendix, pp. 558–62). It will be noted that the addressing mode is called 'Relative' which simply means that, when branching, the next instruction is found by moving a predetermined number of steps relative to the branching instruction.

☐   **Example 20.2**

Given numbers in location $0500 and $0501, put the larger in $0500.

```
LDAA $0500
CMPA$0501     * IF THE NUMBER IN $0501 IS THE LARGER THE N FLAG
BPL FINISH    * WILL BE SET AND THEREFORE THIS NUMBER
LDAA $0501    * MUST BE PUT IN $0500, OTHERWISE THIS PART
STAA $0500    * OF THE PROGRAM MUST BE OMITTED
              * BY BRANCHING

         *
FINISH SWI
```

The following points should be noted from this short program:

(a) The process of branching is achieved by making the program counter jump out of its incrementing sequence to the address of the branching destination. When programming in machine code it is necessary to calculate the number of steps by which the program counter must jump. With the prevalence of assemblers in microprocessor systems this is unnecessary because, using the simple expedient of a label such as FINISH, the assembler calculates the number of steps.

(b) Note the use of comments – these make the program readable.

(c) The program is terminated with SWI, which stands for Software Interrupt. This instruction stops the program counter. Without it the microprocessor will continue to step through memory whose contents are random. ☐

The following example illustrates the use of branching and indexed addressing.

☐ **Example 20.3**

Find the largest number in 10 locations beginning at $B000 and store it in $B000.

```
        LDX # $B000 * FIRST ADDR IN X
        CLRB *B USED FOR COUNTING NUMBER OF LOCNS
        *
SWOP LDAA $00,X * NUMBER LOADED
NEXT INX * ADDRESS INCREMENTED
        INCB * LOCN COUNTER INCREMENTED
        CMPB # $09 *
        BEQ FINISH * BRANCHES TO FINISH WHEN 10 LOCNS CHECKED
        *
        CMPA 00, X * X HAS BEEN INCREMENTED SO COMPARISON IS WITH
        NEXT LOCN
        *
        BMI SWOP * IF ACC A HAS SMALLER NUMBER BRANCH WILL CAUSE
                 * LARGER TO BE LOADED BECAUSE X HAS
                 * BEEN INCREMENTED
        *
        BRA NEXT * THIS BRANCH ONLY OCCURS IF PREVIOUS BRANCH IS
        PASSED
        *
        STAA $B000
FINISH SWI                                                           ☐
```

20.11 SUBROUTINE

A subroutine is a self-contained program segment which can be used within a program for a specified task. For example, it might be necessary at some stage in a program to find the average of two numbers. If there is a subroutine which will carry out this task then the programmer merely has to incorporate into the program an instruction to jump to the subroutine which finds the average. When this is accomplished the subroutine will be terminated with an instruction to return to the main program.

There are several advantages of using subroutines:

(a) Because they are self-contained they can be looked at in isolation from the rest of the program, which improves readability and simplifies debugging.

(b) A library of subroutines can be produced which reduces development time.

(c) A subroutine can be used repeatedly in a program without extra memory cost.

In assembly language the procedure is quite simple: one simply writes the instruction for 'Jump to subroutine' accompanied by a label which identifies the subroutine. For example,

JSR AVERAGE

This instruction causes the program counter to jump to a memory address which will have been designated for the subroutine by the assembler: the subroutine itself will be labeled AVERAGE and it will end with the instruction RTS − return from subroutine. A segment of program using a subroutine might appear as follows:

```
        STAA $01A7
        JSR AVERAGE
        LDAB # $04
        DECA
          .
          .
AVERAGE LDAA $A100  * THIS SUBROUTINE FINDS
        LDAB $A101  * THE AVERAGE OF 2 NUMBERS
        ABA         * BY ADDING AND DIVIDING
        ASRA        * BY 2
        RTS
```

Briefly, when the instruction JSR occurs, the contents of working registers, in particular, the program counter, are saved on a section of memory called the *stack*. In this way, the results of processes contained in the working registers up to the point at which the subroutine is called will be preserved. Since one of these is the program counter then, when RTS occurs, the original PC contents are retrieved from the stack and put back into the PC so that the program resumes at the next instruction after JSR.

In the M6800 the section of memory used for the stack must be designated by the programmer. For this purpose there is a 16-bit register called the stack pointer. By loading this with a number corresponding to an address in RAM, which is unlikely to be overwritten by the program, the first address in the stack is designated, e.g.,

LDS # $3FFF.

The following example illustrates the program structure to be used with a subroutine.

☐ **Example 20.4**

Whenever two numbers are stored in locations $A100 and $A101 a subroutine can be called so that the larger number is always in $A100. Write a program which uses this subroutine to take numbers in $F510 and $F512 and puts the larger in $F510.

```
                LDS # 3FFF   * NUMBER IN STACK POINTER
                *
                LDAA $F510   * INSTRUCTIONS FOR STORING NUMBERS TO
                STAA $A100   * BE PUT IN ORDER
                LDAA $F512   *
                STAA $A101   *
                *
                JSR ORDER    *
                LDAA $A100   * AFTER SUBROUTINE $A100 CONTAINS THE
                STAA $F510   * LARGER NUMBER
                LDAA $A101   *
                STAA $F512   *
                *
ORDER           LDAA $A100   * IF $A100 CONTAINS LARGER NUMBER THE
                CMPA $A101   * SUBROUTINE WILL NOT BRANCH
                BMI SWOP     *
                RTS          *
                *
SWOP            LDAB $A101   * THIS SECTION WILL ONLY BE USED IF NUMBERS
                STAA $A101   * HAVE TO BE EXCHANGED
                STAB $A100   *
                RTS
                SWI
```

20.12 DELAY LOOPS

As noted, the M 6800 microprocessor is synchronized by a 2ϕ clock. This runs at 1 MHz and therefore makes each machine cycle (MC) last 1 μs. Accurate delays can be created by writing a segment of program known as a delay loop which is repeated a predetermined number of times. The technique is illustrated in the following example.

☐   **Example 20.5**

Set bit 2 of location $F512 alternating between 0 and 1 with 1 ms between each change.

```
                                            MC
BEGIN     LDAA # $04    * THESE 2 INSTRS.    2  ┐
          STAA $F512    * SET BIT 2 TO 1     4  │
          JSR DELAY                          9  ┘
          CLRA          * THESE 2 INSTRS.    2
          STAA $F512    * SET BIT 2 TO 0     4
          JSR DELAY                          9
          BRA BEGIN                          4
DELAY     LDAB # $XY                         2      ┐
LOOP      DECB          * THESE 2 INSTRS.    2  ┐   │
          BNE LOOP      * ARE REPEATED       4  ┘XY │
                        * XY TIMES               │
          RTS                                5  ┘
```

Bit 2 will remain at 1 for the time in μs shown by the square brackets. The curly bracket section in the subroutine will be repeated XY times until accumulator B is decremented down to 0, so XY must be calculated for the required delay. In this program the first three instructions last for 15 μs and the subroutine lasts,

$$2 + (2 + 4)XY + 5 \; \mu s$$

Therefore, for a 1 ms delay XY can be calculated from,

$$15 + 2 + 6XY + 5 = 1000$$

all numbers being decimal, from which

$$XY = 163_{10} = \$A3$$

The output will be at 0 for an additional 4 μs because of the BRA BEGIN instruction, but the two periods could be equalized by adding NOP (no operation) instructions into the program. For many purposes it is sufficient simply to calculate the subroutine delay or even just take the loop time 6(XY). ☐

Nested loops

The type of subroutine used in Example 20.5 is limited to a maximum delay of $6(FF) = 1530 \; \mu$s. Longer delays can be generated by using the index register but it may be required for other purposes and in any event, the delay is still limited. A more elegant approach is to use two or more storage locations as demonstrated in the following example.

☐ **Example 20.6**

```
                                                       MC
              CLR $A000                                 6
    OUTER     CLR $A001                                 6    ⎫
    INNER     DEC $A001    * FIRST DECR. MAKES          6    ⎬
                           * CONTENTS = $FF                  ⎪
              BNE INNER                                 4  ⎭
              DEC $A000                                 6
              BNE OUTER    * FIRST TIME AROUND          4
                           * THIS MAKES CONTENTS
                           * OF $A000 = $FF. THE
                           * INNER LOOP WILL THEN
                           * BE REPEATED UNTIL
                           * $A000 CONTENT IS $00.
```

In this program segment the inner loop (curly bracket) is repeated $FF times and then the outer loop (square bracket) is also repeated FF times which includes the FF repetitions of the inner loop. Therefore the total delay (in decimal) is

$$6 + 255[6 + 255\{6 + 4\} + 6 + 4]$$
$$= 255 \times 255 \times 10 = 0.65 \; s$$

The delay can be varied by loading numbers other than zero into the locations and can be extended by using more storage locations. □

SUMMARY

Having read this chapter students should be able to write simple M6800 assembly language programs using various addressing modes. Subroutines should be used where appropriate. Programs should also be made readable by including adequate comments. More challenging programs will be developed in the next chapter when we enter the true province of the engineer in dealing with input/output.

APPENDIX: M 6800 INSTRUCTION SET (COURTESY MOTOROLA)

Instruction	Mnemonic	Implied OP MC PB	Immediate OP MC PB	Direct OP MC PB	Extended OP MC PB	Indexed OP MC PB	Relative OP MC PB	Boolean/Arith Operation	H (5)	I (4)	N (3)	Z (2)	V (1)	C (0)
Load accumulator	LDAA		86 2 2	96 3 2	B6 4 3	A6 5 2		$M \to A$	•	•	↕	↕	R	•
Load accumulator	LDAB		C6 2 2	D6 3 2	F6 4 3	E6 5 2		$M \to B$	•	•	↕	↕	R	•
Load stack pointer	LDS		8E 3 3	9E 4 2	BE 5 3	AE 6 2		$M \to SP_H.(M+1) \to SP_L$	•	•	9	↕	R	•
Load index register	LDX		E 3 3	DE 4 2	FE 5 3	EE 6 2		$M \to X_H.(M+1) \to X_L$	•	•	9	↕	R	•
Store accumulator	STAA			97 4 2	B7 5 3	A7 6 2		$A \to M$	•	•	↕	↕	R	•
Store accumulator	STAB			D7 4 2	F7 5 3	E7 6 2		$B \to M$	•	•	↕	↕	R	•
Store stack pointer	STS			9F 5 2	BF 6 3	AF 7 2		$SP_H \to M.SP_L \to (M+1)$	•	•	9	↕	R	•
Store index register	STX			DF 5 2	FF 6 3	EF 7 2		$X_H \to M.X_L \to (M+1)$	•	•	9	↕	R	•
Transfer accumulators	TAB	16 2 1						$A \to B$	•	•	↕	↕	R	•
Transfer accumulators	TBA	17 2 1						$B \to A$	•	•	↕	↕	R	•
Transfer Acc. to cond. reg.	TAP	06 2 1						$A \to CCR$			Note 12			
Transfer cond. reg. to Acc.	TPA	07 2 1						$CCR \to A$	•	•	•	•	•	•
Transfer stck ptr to index	TSX	30 4 1						$SP + 1 \to X$	•	•	•	•	•	•
Transfer index to stck ptr	TXS	35 4 1						$X - 1 \to SP$	•	•	•	•	•	•
Pull data	PULA	32 4 1						$SP + 1 \to SP.M_{SP} \to A$	•	•	•	•	•	•
Pull data	PULB	33 4 1						$SP + 1 \to SP.M_{SP} \to B$	•	•	•	•	•	•
Push data	PSHA	36 4 1						$A \to M_{SP}.SP - 1 \to SP$	•	•	•	•	•	•
Push data	PSHB	37 4 1						$B \to M_{SP}.SP - 1 \to SP$	•	•	•	•	•	•
Add accumulators	ABA	1B 2 1						$A + B \to A$	↕	↕	↕	↕	↕	↕
Add	ADDA		8B 2 2	9B 3 2	BB 4 3	AB 5 2		$A + M \to A$	↕	↕	↕	↕	↕	↕
Add	ADDB		CB 2 2	DB 3 2	FB 4 3	EB 5 2		$B + M \to B$	↕	↕	↕	↕	↕	↕
Add with carry	ADCA		89 2 2	99 3 2	B9 4 3	A9 5 2		$A + M + C \to A$	↕	↕	↕	↕	↕	↕
Add with carry	ADCB		C9 2 2	D9 3 2	F9 4 3	E9 5 2		$B + M + C \to B$	↕	↕	↕	↕	↕	↕
Subtract accumulators	SBA	10 2 1						$A - B \to A$	•	•	↕	↕	↕	↕
Subtract	SUBA		80 2 2	90 3 2	B0 4 3	A0 5 2		$A - M \to A$	•	•	↕	↕	↕	↕
Subtract	SUBB		C0 2 2	D0 3 2	F0 4 3	E0 5 2		$B - M \to B$	•	•	↕	↕	↕	↕
Subtract with carry	SBCA		82 2 2	92 3 2	B2 4 3	A2 5 2		$A - M - C \to A$	•	•	↕	↕	↕	↕
Subtract with carry	SBCB		C2 2 2	D2 3 2	F2 4 3	E2 5 2		$B - M - C \to B$	•	•	↕	↕	↕	↕

Condition Reg: 5 = H, 4 = I, 3 = N, 2 = Z, 1 = V, 0 = C

Appendix (Continued)

The following table continues the microprocessor (MC6800-type) instruction set summary. Addressing modes are grouped as IMMED, DIRECT, INDEX, EXTND, and IMPLIED. For each mode the columns give OP (Operation Code), ~ (Number of MPU Cycles = MC), and # (Number of Program Bytes = PB).

Operation	Mnemonic	IMMED OP	~	#	DIRECT OP	~	#	INDEX OP	~	#	EXTND OP	~	#	IMPLIED OP	~	#	Boolean/Arithmetic Operation	H	I	N	Z	V	C
Increment	INCA													4C	2	1	$A + 1 \rightarrow A$	●	●	↕	↕	5	●
	INCB													5C	2	1	$B + 1 \rightarrow B$	●	●	↕	↕	5	●
	INC							6C	7	2	7C	6	3				$M + 1 \rightarrow M$	●	●	↕	↕	5	●
Increment stack pointer	INS													31	4	1	$SP + 1 \rightarrow SP$	●	●	●	●	●	●
Increment index reg.	INX													08	4	1	$X + 1 \rightarrow X$	●	●	●	↕	●	●
Decrement	DECA													4A	2	1	$A - 1 \rightarrow A$	●	●	↕	↕	4	●
	DECB													5A	2	1	$B - 1 \rightarrow B$	●	●	↕	↕	4	●
	DEC							6A	7	2	7A	6	3				$M - 1 \rightarrow M$	●	●	↕	↕	4	●
Decrement stack pointer	DES													34	4	1	$SP - 1 \rightarrow SP$	●	●	●	●	●	●
Decrement index register	DEX													09	4	1	$X - 1 \rightarrow X$	●	●	●	↕	●	●
Complement (1's)	COMA													43	2	1	$\overline{A} \rightarrow A$	●	●	↕	↕	R	S
	COMB													53	2	1	$\overline{B} \rightarrow B$	●	●	↕	↕	R	S
	COM							63	7	2	73	6	3				$\overline{M} \rightarrow M$	●	●	↕	↕	R	S
Complement (2's)	NEGA													40	2	1	$00 - A \rightarrow A$	●	●	↕	↕	1	2
	NEGB													50	2	1	$00 - B \rightarrow B$	●	●	↕	↕	1	2
	NEG							60	7	2	70	6	3				$00 - M \rightarrow M$	●	●	↕	↕	1	2
Decimal adjust accumulator	DAA													19	2	1	(see text)	●	●	↕	↕	3	3
Logical and	ANDA	84	2	2	94	3	2	A4	5	2	B4	4	3				$A \bullet M \rightarrow A$	●	●	↕	↕	R	●
	ANDB	C4	2	2	D4	3	2	E4	5	2	F4	4	3				$B \bullet M \rightarrow B$	●	●	↕	↕	R	●
Inclusive or	ORAA	8A	2	2	9A	3	2	AA	5	2	BA	4	3				$A + M \rightarrow A$	●	●	↕	↕	R	●
	ORAB	CA	2	2	DA	3	2	EA	5	2	FA	4	3				$B + M \rightarrow B$	●	●	↕	↕	R	●
Exclusive or	EORA	88	2	2	98	3	2	A8	5	2	B8	4	3				$A \oplus M \rightarrow A$	●	●	↕	↕	R	●
	EORB	C8	2	2	D8	3	2	E8	5	2	F8	4	3				$B \oplus M \rightarrow B$	●	●	↕	↕	R	●
Shift left arithmetic	ASLA													48	2	1		●	●	↕	↕	6	↕
	ASLB													58	2	1		●	●	↕	↕	6	↕
	ASL							68	7	2	78	6	3					●	●	↕	↕	6	↕
Shift right arithmetic	ASRA													47	2	1		●	●	↕	↕	6	↕
	ASRB													57	2	1		●	●	↕	↕	6	↕
	ASR							67	7	2	77	6	3					●	●	↕	↕	6	↕
Shift right logical	LSRA													44	2	1		●	●	R	↕	6	↕
	LSRB													54	2	1		●	●	R	↕	6	↕
	LSR							64	7	2	74	6	3					●	●	R	↕	6	↕
Rotate left	ROLA													49	2	1		●	●	↕	↕	6	↕
	ROLB													59	2	1		●	●	↕	↕	6	↕
	ROL							69	7	2	79	6	3					●	●	↕	↕	6	↕
Rotate right	RORA													46	2	1		●	●	↕	↕	6	↕
	RORB													56	2	1		●	●	↕	↕	6	↕
	ROR							66	7	2	76	6	3					●	●	↕	↕	6	↕

The Boolean/Arithmetic operation for the shift and rotate instructions is shown by bit-shift diagrams operating on A, B, or M with the carry bit C and bit positions b7 … b0.

Instruction	Mnemonic	Implied OP MC PB	Immediate OP MC PB	Direct OP MC PB	Extended OP MC PB	Indexed OP MC PB	Relative OP MC PB	Boolean/Arith Operation	H (5)	I (4)	N (3)	Z (2)	V (1)	C (0)
Compare accumulators	CBA	11 2 1						$A - B$	●	●	↕	↕	↕	↕
Compare	CMPA		81 2 2	91 3 2	B1 4 3	A1 5 2		$A - M$	●	●	↕	↕	↕	↕
	CMPB		C1 2 2	DI 3 2	F1 4 3	E1 5 2		$B - M$	●	●	↕	↕	↕	↕
Compare index register	CPX		8C 3 3	9C 4 2	BC 5 3	AC 6 2		$X_H - M.X_L - (M+1)$	●	●	7	↕	8	●
Test (zero or minus)	TSTA	4D 2 1						$A - 00$	●	●	↕	↕	R	R
	TSTB	5D 2 1						$B - 00$	●	●	↕	↕	R	R
	TST				7D 6 3	6D 7 2		$M - 00$	●	●	↕	↕	R	R
Bit test	BITA		85 2 2	95 3 2	B5 4 3	A5 5 2		$A \bullet M$	●	●	↕	↕	R	●
	BITB		C5 2 2	D5 3 2	F5 4 3	E5 5 2		$B \bullet M$	●	●	↕	↕	R	●
								TEST						
Branch	BRA						20 4 2		●	●	●	●	●	●
Branch if carry clear	BCC						24 4 2	$C = 0$	●	●	●	●	●	●
Branch if carry set	BCS						25 4 2	$C = 1$	●	●	●	●	●	●
Branch if overflow clear	BVC						28 4 2	$V = 0$	●	●	●	●	●	●
Branch if overflow set	BVS						29 4 2	$V = 1$	●	●	●	●	●	●
Branch if equal to zero	BEQ						27 4 2	$Z = 1$	●	●	●	●	●	●
Branch if greater or equal to zero	BGE						2C 4 2	$N \oplus V = 0$	●	●	●	●	●	●
Branch if greater than zero	BGT						2E 4 2	$Z + (N \oplus V) = 0$	●	●	●	●	●	●
Branch if less than zero	BLT						2D 4 2	$N \oplus V = 1$	●	●	●	●	●	●
Branch if less than or equal to zero	BLE						2F 4 2	$Z + (N \oplus V) = 1$	●	●	●	●	●	●
Branch if not equal to zero	BNE						26 4 2	$Z = 0$	●	●	●	●	●	●
Branch if minus	BMI						2B 4 2	$N = 1$	●	●	●	●	●	●
Branch if plus	BPL						2A 4 2	$N = 0$	●	●	●	●	●	●
Branch if higher	BHI						22 4 2	$C + Z = 0$	●	●	●	●	●	●
Branch if lower or same	BLS						23 4 2	$C + Z = 1$	●	●	●	●	●	●

Addressing Mode / Condition Reg

	Mnemonic	Immediate/Direct			Index			Extend			Implied			Special Operations	Condition Codes
Branch to subroutine	BSR										8D	8	2		
Jump to subroutine	JSR				AD	8	2	BD	9	3					
Jump	JMP				6E	4	2	7E	3	3					
Return from subroutine	RTS										39	5	1		
Return from interrupt	RTI										3B	10	1		Note 10
Software interrupt	SWI										3F	12	1		
Wait for interrupt	WAI										3E	9	1		
No operation	NOP										02	2	1	$PC + 1 \rightarrow PC$	
Clear	CLRA										4F	2	1	$00 \rightarrow A$	
	CLRB										5F	2	1	$00 \rightarrow B$	
	CLR				6F	7	2	7F	6	3				$00 \rightarrow M$	
Clear carry	CLC										0C	2	1	$0 \rightarrow C$	
Clear interrupt mask	CLI										0E	2	1	$0 \rightarrow I$	
Clear overflow	CLV										0A	2	1	$0 \rightarrow V$	
Set carry	SEC										0D	2	1	$1 \rightarrow C$	
Set interrupt mask	SEI										0F	2	1	$1 \rightarrow I$	
Set overflow	SEV										0B	2	1	$1 \rightarrow V$	

CONDITION CODE SYMBOLS:

H Half-carry from bit 3;
I Interrupt mask
N Negative (sign bit)
Z Zero (byte)
V Overflow, 2's complement
C Carry from bit 7
R Reset Always
S Set Always
‡ Test and set if true, cleared otherwise
● Not Affected

LEGEND:

OP Operation Code (Hexadecimal);
MC Number of MPU Cycles;
PB Number of Program Bytes;
+ Arithmetic Plus;
− Arithmetic Minus;
● Boolean AND;
M_{SP} Contents of memory location pointed to be Stack Pointer;
+ Boolean Inclusive OR;
⊕ Boolean Exclusive OR;
$\overline{M}$ Complement of M;
→ Transfer into;
0 Bit = Zero;
00 Byte = Zero;

Note—Accumulator addressing mode instructions are included in the IMPLIED addressing.

CONDITION CODE REGISTER NOTES:

(Bit set if test is true and cleared othewise)

1	(Bit V)	Test: Result = 10000000?
2	(Bit C)	Test: Result = 00000000?
3	(Bit C)	Test: Decimal value of most significant BCD Character greater than nine? (Not cleared if previously set.)
4	(Bit V)	Test: Operand = 10000000 prior to execution?
5	(Bit V)	Test: Operand = 01111111 prior to execution?
6	(Bit V)	Test: Set equal to result of $N \bullet C$ after shift has occurred.
7	(Bit N)	Test: Sign bit of most significant (MS) byte = 1?
8	(Bit V)	Test: 2's complement overflow from subtraction of MS bytes?
9	(Bit N)	Test: Result less than zero? (Bit 15 = 1)
10	(All)	Load Condition Code Register from Stack. (See Special Operations)
11	(Bit I)	Set when interrupt occurs, if previously set, a Non-Maskable Interrupt is required to exit the wait state.
12	(ALL)	Set according to the contents of Accumulator A

REVISION PROGRAMS

Write programs in 6800 assembly language for the following.

20.1 Add $25 to $16 and store the result in $0E00.

20.2 Add $34 to the contents of $A05B and store the result in $00FF.

20.3 Find $(05_{10} + 15_{10} - 07_{10})$ and store the result in $0A00.

20.4 Carry out the operation $54 AND $07 OR $88 store the result in $00C0.

20.5 Add the contents of locations $01F0 and $E7 and OR the result with the number $5A.

20.6 Compare the numbers in locations $0100 and $0101. If they are equal, clear accumulator B. If they are unequal, load accumulator B with the number $22.

20.7 Add the numbers $00 to $0A and store the results in $B100 (*Hint:* keep on incrementing and adding and storing the contents of an accumulator until its contents are $0A.)

20.8 Put the numbers in locations $0B00, $0B01 and $0B02 so that the largest is in $0B00 and the smallest in $0B02.

20.9 Increment the contents of locations $0200 until its contents exceed $FF.

20.10 Whenever bit 7 of location $B100 is 1, use a subroutine to check the number in $B101. If it is even, make bit 6 of $B100 a 0; otherwise make it 1. (*Hint:* bit 7 should be continually checked by testing N flag. A number will be even if bit 0 is 0 – shift right and check carry).

20.11 Use time delay subroutines to continuously output a 0 for 20 ms and a 1 for 30 ms on bit 4 of location $0BA8.

21

Microprocessor Input/Output – the Peripheral Interface Adapter

Chapter Objectives
21.1   Peripheral Interface Adapter (PIA)
21.2   Addressing Data and Data Direction Registers
21.3   Data Direction Register
21.4   Initializing the PIA
21.5   Flow Charts
21.6   Microprocessor as a Controller
21.7   Control Register
21.8   Software Handshake
         Summary
         Revision Programs

For engineering applications it is important for a microprocessor system to be able to communicate with peripherals. It can then be used to receive data from peripherals such as VDU keyboards and measurement transducers and to send out data for communication (e.g. displays) and control purposes. Data generated by peripherals and by microprocessors do not usually occur in forms which are compatible. It is therefore necessary to provide an interface between them so that they can communicate with each other. This chapter does no more than introduce the peripheral interface adapter (PIA) which is one of the parallel interfaces designed for M 6800 systems.

21.1 PERIPHERAL INTERFACE ADAPTER (PIA)

A PIA is an IC (M6821) with pins for address and data which make it compatible with the M6800 microprocessor. It has two virtually identical sections, each of which can be connected to a set of eight external lines, thus forming two input/output (I/O) ports, A and B, as shown in Fig. 21.1. The lines are numbered PA0 to PA7 and PB0

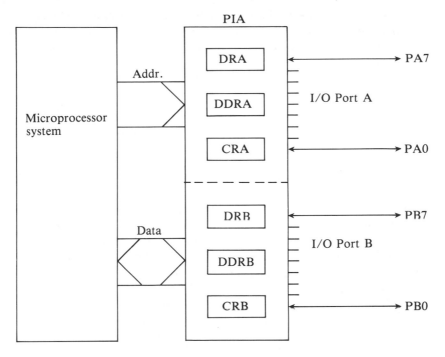

Fig. 21.1   PIA Layout

to PB7. Each side of the PIA consists of three 8-bit registers:

Data Register (DRA)
Data Direction Register (DDRA)
Control Register (CRA)

with the bracketed abbreviations being labeled for the A side.

Figure 21.1 also shows the PIA being linked to the microprocessor system via the system address and data buses. In fact, the PIA registers are treated as memory locations which can be written to or read from when addressed. For each side of the PIA the data and data direction registers share an address in a manner to be described; each control register has its own address. Throughout this chapter the addresses will be as follows:

DRA and DDRA $F510
CRA $F511
DRB and DDRB $F512
CRB $F513

It should be noted, however, that not all systems use these addresses for the PIA.

A data register is essentially a buffer between the eight external lines of an I/O port and the system data bus. Each bit in a data register can either be used as an input into which it accepts data from a line and puts it onto the bus, or it takes data from the system bus and outputs it to the corresponding line. The direction of each bit in a *data register* is determined by programming the corresponding *data direction register*. The control register has several functions, but it is basically concerned with managing the flow of data through the PIA.

21.2 ADDRESSING DATA AND DATA DIRECTION REGISTERS

As the two sides of the PIA are almost identical, descriptions will be written in terms of the A side to avoid repetition. To understand PIA operation it is important to appreciate the interdependence of the three types of register. Of these the CRA is the most complicated but at this stage it is only necessary to deal with CRA2 (control register A, bit 2). As noted, DRA and DDRA share the address $F510. If CRA2 = 0 then DDRA will be activated when $F510 is addressed. When CRA2 = 1 then address $F510 applies to DRA. Thus the instruction

CLR $F511

clears all CRA bits including CRA2 and subsequently any instruction involving $F510 applies to DDRA. The instructions

LDAA # $04
STAA $F511

make CRA2 = 1 so that address $F510 applies to DRA.

21.3 DATA DIRECTION REGISTER

The eight bits of each DDR determine whether the corresponding DR bits are inputs or outputs. A '0' on a DDR bit makes the corresponding DR bit an input, and vice versa if the DDR bit is a '1'. The eight bits can be individually programmed to produce any pattern of inputs and outputs. Thus if CRA2 = 0, then address $F510 applies to DDRA. If $F510 is being addressed and the number on the data bus is $A7 (1010 0111) then bits 6, 4 and 3 of the DRA will be inputs (because these bits in the DDRA have been made '0') and the other five bits of the DRA will be outputs.

21.4 INITIALIZING THE PIA

Initialization refers to instructions, usually near the beginning of a program, which establish the operating conditions at the two I/O ports required for the application in hand. For example, suppose that all eight bits of DRA are to be inputs, and bits 0 and 1 of DRB are to be outputs with bits 2 to 7 inputs. For each side of the PIA two 8-bit words are required, one for DDR and the other for CR. Since the index register is 16 bits wide it can be used to program both registers for one side of PIA simultaneously.

Consider the instructions:

```
CLR $F511 * CRA2 = 0 SO THAT $F510 IS DDRA ADDRESS
LDX # $0004 * 00 TO $F510 (DDRA) MAKES ALL DRA BITS
STX $F510    * INPUTS. 04 TO $F511 MAKES CRA2 = 1 SO
             * THAT $F510 BECOMES DRA ADDRESS
```

By clearing $F511, $CRA2 = 0$ so that DDRA will be addressed by $F510. Then the first $00 puts eight 0s into DDRA so that PA0 to PA7 will all be inputs. The $04 goes to $F511 so that $CRA2 = 1$ and from then on $F510 is the address of DRA.

The next instructions will be:

```
CLR $F513    * CRB2 = 0
LDX # $0304  * 03 (0000 0011) MAKES
STX $F512    * PBO-1 OUTPUTS AND PB2-7 INPUTS
```

☐ **Example 21.1**

To illustrate the operation of a PIA, consider the problem of using a delay loop (Section 20.12) to send out a 50 Hz square wave on PA3 when a signal fed in on PA0 changes from 0 to 1 (Fig. 21.2). To make PA0 an input and PA3 an output, then DDRA must be loaded with any number which makes bit $0 = 0$ and bit $3 = 1$, e.g. $08 (0000 1000).

To initialize the PIA use the program segment

```
CLR $F511    * PIA INITIALIZATION
LDX # $0804 *
STX $F510    *
```

To detect a 1 on PA0 it can be shifted right and will set CCR carry flag $C = 1$ when it is 1:

```
CHECK LDAA $F510 * THIS LOOP WILL BE REPEATED UNTIL PAO = 1
      ASRA       *
      BCC CHECK  *
```

To send out the waveform:

```
REPEAT CLRA         * SENDS A ZERO ON ALL BITS INCL PA3
       STAA $F510 *
       JSR DELAY  *
       LDAA # $08 * SENDS A ONE ON PA3
       STAA $F510 *
       JSR DELAY  *
       BRA REPEAT *
```

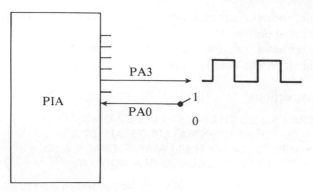

Fig. 21.2   Diagram for Example 21.1

It is left as an exercise to write the delay subroutine on the lines suggested in Section 20.12.   □

21.5   FLOW CHARTS

For more complex problems it is often helpful to construct a flow chart which, in effect, divides the problem into sections, each of which can be accomplished by a segment of program. Consider the following.

□   **Example 21.2**

Samples of the temperature of a chemical process are represented in binary code by an 8-bit number fed in at I/O port A (Fig. 21.3). When the system is ready to accept

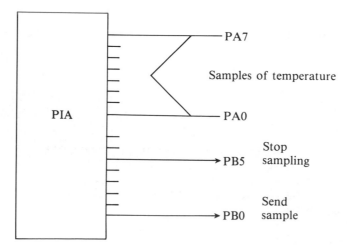

Fig. 21.3   Diagram for Example 21.2

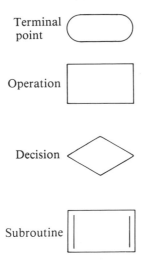

Terminal point

Operation

Decision

Subroutine

Fig. 21.4 Flow Chart Symbols

a sample it sends out a '1' on PB0. When it has checked the sample temperatures and counted 100 in the range $80 to $A0, it must send out a '1' on PB5. It is also required to count the total number of samples received in reaching 100 samples in the specified range and to store this total number in $A000.

Figure 21.4 shows some of the accepted symbols used in flow charts. Figure 21.5 is a flow chart for this particular problem. In the following program, accumulator B is used to count the samples in the specified range.

```
        CLRA $F511    * PIA INITIALIZATION MAKING PA0-7 INPUTS
        CLRB $F513    * AND PB0 AND PB5 OUTPUTS
        LDX = $0004 *
        STX $F510     *
        LDX # $2104 *
        STX $F512     *
        *
        CLR $A000     * SAMPLE COUNT
        CLRB          * VALID SAMPLE COUNT
        *
NXTSAMP LDAA # $01    * SENDS 1 ON PB0 REQUESTING A SAMPLE
        STAA $F512    *
        INC $A000     * COUNTS SAMPLE
        LDAA $F510    * READS SAMPLE
        *
        CMPA #$80     * IS SAMPLE > $80
        BMI NXTSAMP *
        *
        CMPA # $A0  * IS SAMPLE < $80
        BPL NXTSAMP *
```

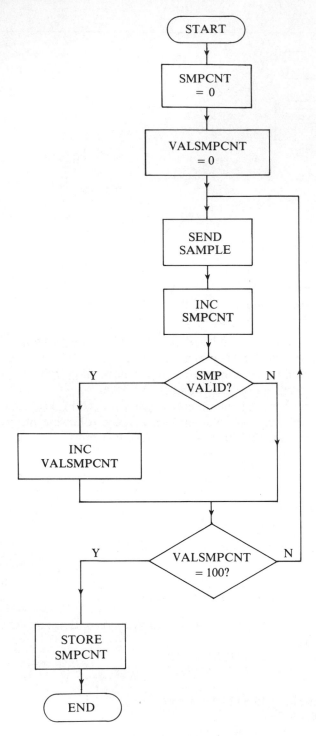

Fig. 21.5   Flow Chart for Example 21.2

```
INCB * PASSING PREVIOUS BRANCH POINTS INDICATES VALID
SAMPLE
*
CMPB # $64  * CHECKS FOR 100 VALID SAMPLES
BNE
NXTSAMP     *
SWI         * THE NUMBER OF SAMPLES IS IN $A000        □
```

21.6 MICROPROCESSOR AS A CONTROLLER

The application of a microprocessor to exercise a control function is illustrated by the following example.

□ **Example 21.3**

A single byte input on I/O port A indicates, as a direct binary number, the level of fluid in a tank, so that $00 shows it to be empty and $FF full. It is required to control the fluid supply to the tank so that it is turned off when the tank reaches 3/4 full ($C0) and turned on when it has fallen to 1/4 full ($40). The supply is controlled from PB7, a 0 turning it OFF and a 1 turning it ON.

Figure 21.6 is a flow chart for the problem which yields the following program.

```
        CLR $F511    * PIA INITIALIZATION MAKING PA0-7 INPUTS
        CLR $F513    * AND PB7 OUTPUT
        LDX # $0004  *
        STX $F510    *
        LDX # $8004  *
        STX $F512    *
        *
LOSAMP  LDAA $F510   * CHECK FOR 1/4 FULL
        CMPA # $40   *
        BPL OFF
        *
ON      LDAB # $80   * SUPPLY ON WITH '1' ON PB7
        STAB $F510   *
        *
        LDAA $F510   * CHECKS FOR 3/4 FULL
        CMPA # $C0   *
        BMI ON
        *
OFF     CLRB         * SUPPLY OFF WITH '0' ON PB7
        STAB $F512   *
        BRA LOSAMP                                        □
```

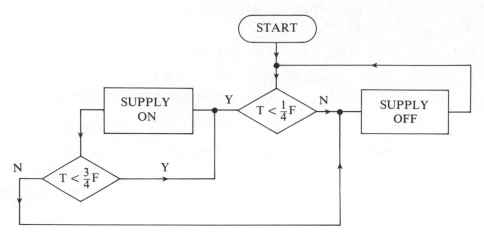

Fig. 21.6   Flow Chart for Example 21.3

21.7   CONTROL REGISTER

For the preceding programs involving I/O we have only used bit 2 of the control registers (CRA2 and CRB2). For applications such as software handshakes it is necessary to explain some of the other control register functions. First, however, we should note that as well as having eight data lines, each side of the PIA has two control lines. These are shown as CA1 and CA2 in Fig. 21.7 for the A side and, as indicated, CA1 is always an input to the PIA but CA2 may be used either as an input or as an output: the choice is exercised by programming the control register (CRA). The B side control lines have corresponding roles.

Let us now describe the function of each control register bit in roughly ascending order of difficulty. The descriptions are written in terms of the A side, which is almost identical to the B side.

(a) CRA2 – as noted, when CRA2 = 0, $F510 is DDRA address and when CRA2 = 1, $F510 is DRA address.
(b) CRA5 is used to determine whether CA2 is to be an input or an output. Consistent with DDR/DR, when CRA5 = 0, CA2 is an input.
(c) When CA2 is an output, then if CRA4 = 1, CA2 = CRA3. To illustrate the interdependence of CRA3, 4 and 5 consider the following program segment.

     LDAA # $3C
     STAA $F511

Figure 21.8 notes the function of the four 1s in the middle of the byte stored in CRA. The logical sequence for examining the effect of each bit is CRA2, 5, 4, 3. If the requirement had been to make CA2 = 0 then only CRA3 would be

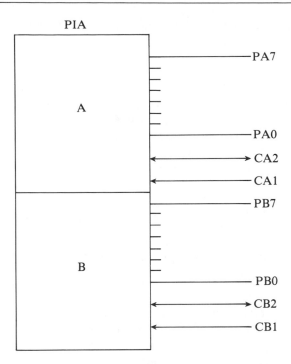

Fig. 21.7   I/O Ports A and B Showing Control Lines

changed and this would be done by storing $38 instead of $3C in CRA. (There will be a brief description of CRA3 and 4 when CA2 is an input in due course.)

(d) CRA6 and 7 are called, respectively, CA2 and CA1 flags or sometimes, interrupt flags. CRA6 is only used when CA2 is an input. A peripheral signaling for attention will be connected to one of the control lines and will set the corresponding flag bit. Peripherals either signal with a 0-to-1 or with a 1-to-0 transition. For CA1 flag (CRA7), CRA1 enables the programmer to choose which polarity of transition will set CRA7: by setting CRA1 = 1 the flag will be set by a 0–1 transition on CA1. For CA2 the same choice can be exercised by programming CRA4

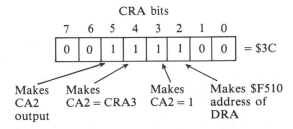

Fig. 21.8   Effect of $3C Being Loaded into Control Register A

when CA2 is an input (in which case CRA5 must be 0). Both flags are reset when the data register is read by an instruction such as

    LDAA $F510

(e) This leaves CRA0, and CRA3 when CA2 is an input. They involve some understanding of interrupts, any elaboration of which is beyond the scope of this text. Briefly, provision is made in a microprocessor system to interrupt a program, but for this to happen two conditions must obtain. First, when initializing the PIA then CRA0 for CA1 or CRA3 for CA2 must be set to 1. When either of these flags is set, $\overline{\text{IRQA}}$, which is a PIA output connected to the microprocessor, goes to 0 indicating a request to interrupt. Second, the interrupt will only occur when one of the interrupt instructions (see the instruction set of Fig. 20.4) has been used to set the I flag in the condition code register.

The foregoing items (a) to (e) do not exhaust all possibilities for the control registers, but they provide the basis for some useful applications. First, however, consider the following PIA initialization requirements:

1. All A side data lines output.
   CA1 flag to be set with high input and providing interrupt request.
   CA2 – an input – with flag set on high input but without interrupt request.
   For this specification all DDRA bits must be 1 to make the data lines outputs.
   Figure 21.9 indicates the control register requirements.
   Therefore the initialization program will be

       CLR $F511
       LDX # $FF17
       STX $F510

2. I/O port A – all data lines inputs.
   CA1 flag to be set on low input – no interrupt.
   CA2 to be an output with CA2 = CRA3.
   I/O port B – PB0–3 outputs, PB4–7 inputs.
   CB1 flag to be set on high input with interrupt.
   CB2 flag to be an input with flag set on high input but without interrupt.

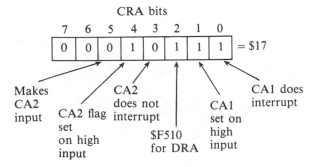

Fig. 21.9   Effect of $17 Being Loaded into Control Register A

For this specification the register contents will be

```
DDRA    $00
CRA     $34
DDRB    $0F
CRB     $17
```

and the initialization program will be

```
CLR $F511
CLR $F513
LDX # $0034
STX $F510
LDX #$0F17
STX $F512
```

21.8 SOFTWARE HANDSHAKE

In Section 18.4 we described an analog-to-digital converter (ADC) with a handshake facility. The ADC lines and handshake waveforms are shown in Fig. 21.10. Briefly repeating the waveform sequence, at the end of a conversion the SC (start conversion) signal from the microprocessor system will be held at 1 and the $\overline{DV}$ (not data

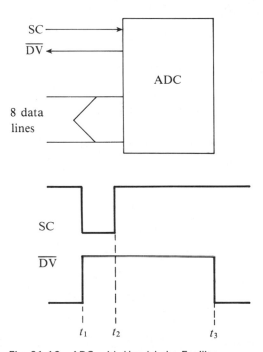

Fig. 21.10   ADC with Handshake Facility

valid) signal which the ADC sends to the microprocessor will be 0, indicating that the data on the 8 data lines were valid. To begin a conversion the microprocessor makes $SC = 0$ (at t_1) which instructs the ADC to clear its contents. The ADC responds by making $\overline{DV} = 1$, thus indicating that the data on the 8 data lines are not valid. At t_2 the microprocessor makes $SC = 1$, thus instructing the ADC to 'start conversion'. When the conversion has been completed (at t_3) the ADC informs the microprocessor that the data are valid by making $\overline{DV} = 0$. An application of the process is illustrated in the following example.

☐   **Example 21.4**

A program is required which takes 100 samples from an ADC connected to I/O port A and stores them in consecutive locations beginning at $B000.

Figure 21.11 shows the ADC to PIA connection. Of the two control lines only CA2 can be used as an output and must therefore be connected to SC. For control register initialization Fig. 21.12 denotes the bits which are significant. CRA7, the CA1 flag, will be looking for a 0 to show that the sample is valid so CRA1 = 0. CRA3 = 1 thus making CA2 = 1 because the ADC SC terminal requires a 1-to-0 transition to clear the converter. When the program is running, CRA3 can be reprogrammed for start or clear converter as required.

```
              CLR $F511      * PIA INITIALIZATION MAKES
              LDX # $003C    * PA0-7 INPUTS CA2 OUTPUT
              STX $F510      * AND CA1 FLAG SET ON ZERO.
              *
              LDX # $B000    * FIRST LOCN FOR DATA
              *
NXTSAMP  LDAA # $34    * CLEAR ADC BY MAKING SC = 0
              STAA $F511     *
              *
              LDAA # $3C     * START CONVERSION BY MAKING SC = 1
              STAA $F511     *
              *
```

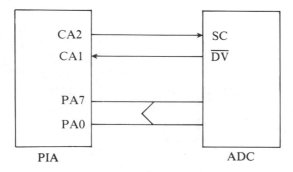

Fig. 21.11   Diagram for Example 21.4

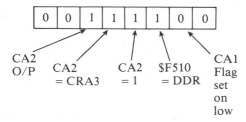

Fig. 21.12   Control Register Initialization for Example 21.4

```
CHECK      LDAA $F511    * CHECKS CA1 FLAG AND BRANCHES UNTIL IT IS SET
           BMI CHECK     *
           *
           LDAA $F510    * STORES SAMPLE
           STAA $00,X    *
           INX
           CPX # $B064   * CHECKS IF 100 SAMPLES HAVE BEEN STORED
           *
           BNE NXTSAMP
           SWI                                                        □
```

SUMMARY

Like any other electronic device, a microprocessor is only of value if it can be interfaced in the context of some particular application. As a digital computer it has a fixed operating speed and it processes data in fixed numbers of bits determined by the size of its operating registers, this being 8 in the case of M 6800. To communicate with peripheral devices special interface chips such as the PIA have been developed. Although the PIA is a complex device, its operation is a particularly revealing aspect of digital computer organization and application. The chapter concludes with an example of interfacing an analog system to microprocessor using an ADC and a PIA.

REVISION PROGRAMS

Rather than simply provide a number of revision questions, the following problems have been provided with solutions. In the spirit of competition, readers may wish to produce programs using fewer machine cycles.

21.1 A motor is to be controlled by switches A and B connected as shown in Fig. Q.21.1a.

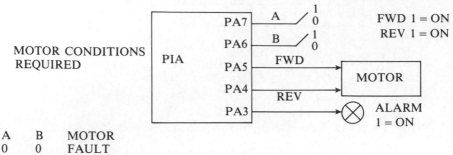

MOTOR CONDITIONS REQUIRED

A	B	MOTOR
0	0	FAULT
0	1	FORWARD
1	0	REVERSE
1	1	STOP

Fig. Q.21.1a

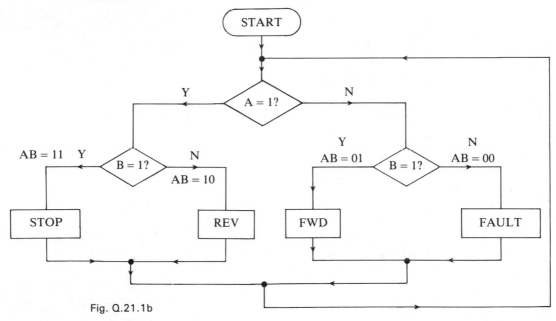

Fig. Q.21.1b

Figure Q.21.1b is a flow chart for the problem and results in the following program.

```
        CLR $F511
        LDX # $3804    * PA3, 4, 5 OUTPUTS, PA6, 7 INPUTS
        STX $F510      *
BEGIN LDAA $F510
        BMI AONE       * BRA IF PA7 = 1
        ASLA
        BMI FWD        * BRA IF PA6 = 1
        LDAB # $08     * THIS STEP WILL ONLY BE REACHED IN FAULT
        STAB $ F510    * CONDITION
        *
```

```
AONE  ASLA            * PA6 TO ACC A 7
      BMI STOP
      LDAB # $10       * REVERSE BY PA4 = 1
      STAB $F510       *
      BRA BEGIN
STOP  CLRB             * STOP BY PA4, 5 = 00
      STAB $F510       *
      BRA BEGIN
FWD   LDAB # $20       * FORWARD BY PA4, 5 = 01
      STAB $F510       *
      BRA BEGIN
```

21.2 Before processing data on PA0–3, a device requires a high to low pulse of at least 20 μs from CA2. When it has processed that data which it returns to the PIA on PA4–7 it signals the fact with a low to high pulse on CA1. The process is to run continuously. Having written the program, calculate the number of operations per second it will run.

```
                                              MACHINE CYCLES
         CLR $F511        * PIA INITIALIZATION
         LDX # $FO3E      *
         STX $F510        *
NXTSAMP  LDAA # $ 36      *                          2
         STAA $F511       * CA2 = 0                   5
         *
         LDAB # $03       * DELAY LOOP TO MAKE CA2 = 0   2
LOOP     DECB             * FOR 20 MICROSEC              2 ⎫
         BNE LOOP         *                             4 ⎬ × 3
         *                                                ⎭
         LDAA # $3E       *CA2 = 1                    2
         STAA $F511       *                           5
TEST     LDAA $F511       * THIS LOOP CONTINUES       4
         BPL TEST         * UNTIL CA1 = 1             4
         BRA NXTSAMP                                  4
```

If the program does not have to branch back on the CA1 TEST loop it takes 46 μs and can therefore be repeated 21 739 times a second.

21.3 If the byte on PA0–7 has even parity (an even number of 1s including no 1s) then a 1 is to be sent out on CA2.

Figure Q.21.3 is a flow chart for this problem. By shifting LSB to carry we can use BCC or BCS to check it. The 1s can be accumulated and after 8 shifts the accumulated number will be even if its LSB = 0. The solution uses accumulator B to store the number of bits, and $A000 to store the number of 1s.

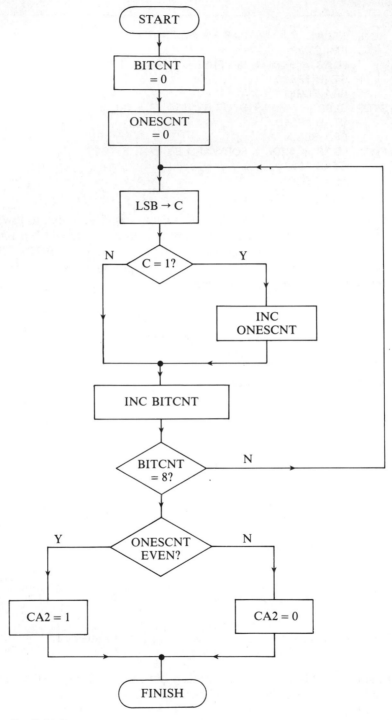

Fig. Q.21.3

```
              CLR $F511      * PIA INITIALIZATION
              LDX # $0034    *
              STX $F510      *
              CLRB           * BITCOUNT
              CLR $A000      * ONESCOUNT
              LDAA $F510
NXTDIG        LSRA           * CHECK LSB
              BCC BITCOUNT
              INC $A000      * LSB = 1 COUNTED
BITCOUNT INCB
              CMPB # $08     * HAVE 8 BITS BEEN CHECKED
              BNE NXTDIG
              LDAA $A000     * CHECK ONESCOUNT
              LSRA           *
              BCC EVEN       * PARITY EVEN IF LSB = 0
              *
              CLRB           * '0' ON CA2 FOR ODD
              BRA PARITY
EVEN          LDAB # $08     * '1' ON CA2 FOR EVEN
PARITY        STAB $F511
              SWI
```

21.4 For Fig. Q.21.4 write a program which displays a number stored in location $0150 on two 7-segment displays when $S = 1$. The displays are to be enabled so that the least significant hex digit is displayed on L and the most significant hex digit on II. A display is enabled with a '0'.

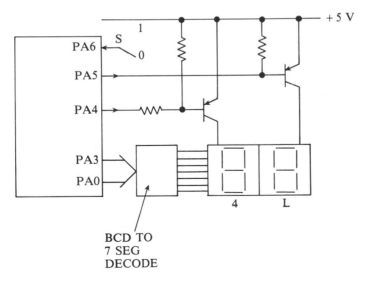

Fig. Q.21.4

In the following program note that ANDA # $0F is used to mask off bits 4–7 in accumulator A so that its contents do not enable the displays. Then ORAA # $10 makes PA5 = 0 and ORAA # $20 makes PA4 = 0.

```
          CLR $F11      * PIA INITIALIZATION
          LDX # $3F04   *
          STX $F510     *
          *
TEST LDAA $F510
          ALSA          * CHECK FOR PA6 = 1
          BPL TEST      *
          *
          LDAA $0150
          ANDA # $0F    * MASKS BITS 4–7
          ORAA # $10    * ENABLES DISPLAY L WITH PA5 = 0
          STAA $F510    * LSD TO L
          *
          LSRA          * PUTS MSD IN BITS 0–3 OF ACC A AND MAKES BITS
          LSRA          * 4–7 = 0
          LSRA          *
          LSRA          *
          *
          ORAA # $20    * ENABLES DISPLAYS H WITH PA4 = 0
          *
          STAA $F510    * MSD TO H
          SWI
```

22

The Intel 8088 Microprocessor

Chapter Objectives
22.1   Constituents of a Microprocessor System
22.2   Addressing Modes
22.3   Instruction Set
22.4   Flags Register
22.5   Subroutines
22.6   Software Delays
22.7   Microprocessor Input/Output – The Programmable Peripheral Interface
22.8   Flow Charts
        Appendix
        Revision Questions

There is a tendency for technology to become not only more complex but also cheaper in terms of cost per function. When the first edition of this book was written most undergraduate students of electronics worked on 8-bit microprocessors such as the MC6800 described in the last two chapters. However, 16-bit microprocessors are now commonplace. In writing about microprocessors within a general electronics text an author has to choose the particular device which she/he is going to use as the context for his work. In this case the choice was made relatively easy because of the ubiquity of IBM Personal Computers (PCs) and the remarkable number of clones. Although technology has advanced to ATs, PS2s and doubtless further developments which will be available very soon, the IBM PCXT is widely used for both microprocessor applications and desktop computing. It is based on the Intel 8088 microprocessor which will be the focal point of this chapter.

Types of Programming Language

Programs can be written in:

(a) Machine (or object) code.

591

(b) Assembly language.

(c) High-level language such as Pascal.

☐ **Machine Code**

All programs have eventually to be converted into machine code which is the representation of instructions in binary code, although it is normal to write the code as hex numbers. Thus the instruction, 'Add to the lower byte of register AX the number 6AH' would, in 8088 machine code, be 046A. Then when 04 is output from memory the computer responds by setting up the ALU to add to AL the next number (6A) appearing on the data bus.

☐ **Assembly Language**

In assembly language the instruction used in the last section would be written,

ADD AL, 6AH

Instructions are written as mnemonics such as ADD, SUB (subtract), MOV (move), etc. and as such are an easier method of writing programs in that they are nearer to everyday language than machine code. However, the computer system must have a program called an assembler, such as MASM (Microsoft Macro Assembler), which converts the assembly language program into machine code before it can be run on the system – the assembly language program is said to have been assembled.

☐ **High-level Language**

This gives a much easier method of writing programs. A line of high-level language program such as,

IF A < B THEN A := A + 1;

would be equivalent to several assembly language instructions. Again the high-level language must be converted into machine code and for this there are two types of program; interpreters and compilers. An interpreter simply takes each high-level language statement and converts it into machine code as the program is run. Therefore the interpreter has to reside in the computer memory while the program runs. It also results in slow operation especially when a loop occurs because each statement has to be converted into machine code by the interpreter every time it is run. A compiler, on the other hand, converts the complete program into machine code which is similar to using an assembler in an assembly language program.

Some of the more general aspects of computer engineering are described both here and in the introduction to Chapter 20. However, it is better to illustrate the technology with reference to the particular microprocessor under consideration. To some extent the descriptions are influenced by the author's own working environment.

For example, using the assembler MASM, hexadecimal numbers have to be written 26H rather than $26 as in Chapter 20. Most of the programs are in assembly language but there is a trend towards the use of programming microprocessors in high-level languages and some examples using Pascal programs are provided and in these the dollar sign is used for hexadecimal numbers.

22.1 CONSTITUENTS OF A MICROPROCESSOR SYSTEM

22.1.1 Memory – Data and Address Buses

A microprocessor system is made up of several interdependent sections. The fact that they are interdependent implies that there is no uniquely logical starting point, but one advantage of beginning with memory is that of establishing the difference between data and addresses and the functions of the two highways, better known as buses, which carry the data and address signals between the computer sections.

Although the 8088 is a 16-bit microprocessor, its memory is made up of 8-bit (one byte) locations. (The relevant terminology is that 1 bit = 1 binary digit, 1 nibble = 4 bits, 1 byte = 8 bits, 2 bytes (i.e. 1 word) = 16 bits.) The total number of 8 bit memory locations that the microprocessor can access is $2^{20} = 1\,048\,576$ bytes. This is approximately one million so the memory is said to be one megabyte (MB). If each one of the 2^{20} locations is to be numbered, i.e. to have an address expressed as a binary number, then the number would have to have 20 bits.

Expressing numbers of this length in bits is tiresome so it is normal to number the addresses in hexadecimal code: five hex numbers can be used to identify the individual single byte locations. However each location does need to be activated individually so an address decoder is used as shown in Fig. 22.1. The system works as follows. Memory location A0004H contains the byte of data expressed in hex as 3A. Then if we are reading data from memory, the 20 bits corresponding to A0004 are placed on the address bus, the corresponding line out of the decoder is activated and the 8 bits of data corresponding to 3AH are output from memory on to the data bus. Note that the address bus is unidirectional whereas the data bus is bidirectional.

Therefore, to write a byte of data, say 2FH, into location A0009, first A0009 is put on to the address bus and when some other component of the system puts 2F on the data bus it will be read into memory location A0009. A complication of the 8088 is that its memory is organized into segments.

Segmented Memory

As noted, the 8088 has the capacity for 1 MB of memory which requires a 20 bit (5 nibble) address. The 1 MB can be divided into sixteen 64 K blocks beginning at address 00000 and ending at FFFFFH. For addressing, the difference between adja-

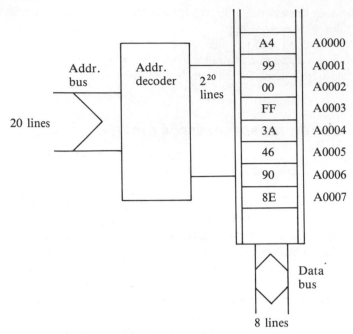

Fig. 22.1   Address and Data Buses at Memory

cent 64 K blocks resides only in the most significant nibble and gives the following arrangement:

Block 0   00000   to 0FFFFH
Block 1   10000H to 1FFFFH
   ⋮
Block 15 F0000H to FFFFFH

Four 64 K blocks of memory are designated:

Code segment − used for the program being executed
Data segment − for program data
Stack segment − for such applications as holding provisional data and addresses while subroutines are being serviced,
Extra segment − for storing additional data

Four corresponding registers CS, DS, SS and ES, each of two bytes, are used to point to the first address of each segment, the fifth nibble being provided by adding four 0 bits at the least significant end of the segment register address. For example, if the CS register contains A000H then the first address of the code segment will be A0000H. The segments are not necessarily coincident with 64 K blocks; for example if the DS register contains B432H then the data segment would be from B4320H to C431FH.

Physical Addresses and Offsets

When an instruction specifies an address it will always use a 2 byte number which indicates its location within a segment. For example, if CS contains A000H then the code segment consists of addresses from A0000H to AFFFFH. An instruction recognized by the microprocessor as being in the code segment and using address B432H will automatically point to address AB432H. Thus AB432H is the physical address and B432H is the offset from the beginning of the segment. To specify an address within a segment the notation used would be CS: B432H.

One advantage of this arrangement is that the physical location of a program within memory does not have to be fixed. Whenever it is run it is only necessary to specify the segment register contents to place the program in memory. This is particularly useful in a multi-tasking operation where a program can be interrupted, temporarily stored on disc, and then re-located as convenient.

22.1.2 Arithmetic/Logic Unit (ALU) and General Registers

An 8088 microprocessor system can carry out a large repertoire of instructions. These include arithmetic operations such as addition and multiplication, and Boolean logic operations such as ANDing and ORing bytes or words of data. These operations are carried out by the ALU, frequently by first fetching data from memory and putting it into one of four word length general-purpose registers labelled AX, BX, CX or DX. Each can be divided into two separate one-byte registers which can be used quite independently of one another for byte length rather than word length operations. The single byte half-registers are labeled AH and AL, etc. to signify high byte and low byte. As well as being used for general-purpose arithmetic and logic operations each of these registers separately has special functions some of which will be described. There are also several other registers such as the flags register and the instruction pointer to be covered later in the chapter.

22.1.3 Microprocessor System – Fetch/Execute Sequence

Having described the functions of memory and ALU we are now in a position to look at the layout of a stored program computer. The model drawn in Fig. 22.2 is illustrative of digital computers generally and can be used to show how instructions are fetched and executed. The following description uses a particular instruction to illustrate the operation of the system.

Consider the instruction 'Add to the existing contents of register AL (the lower byte of AX) the hex number 85'. Most of the programs in this chapter will be written in 8088 assembly language in which this instruction would appear as,

ADD AL, 85H

An instruction usually has two parts: an operator (in this case ADD AL) and an operand (here 85H). We should note the arrangement that the destination – register AL – is always stated before the data (or source of data) – the number 85H – and that the two must always be separated by a comma. Suppose that the instruction is held in two consecutive memory locations at addresses B3100H and B3101H (Fig. 22.2). The byte contained in B3100H is known as an op code. The computer will use the op code to set up the operation 'Add to the existing contents of AL'. In this case the op code is 02. The following sequence now occurs. The instruction pointer (IP), which always holds the offset address of the next instruction to be carried out, causes B3100H to be put on to the address bus where it is latched. Now the instruction pointer is incremented in readiness for the next part of the instruction. With B3100H on the address bus, memory outputs op code 02 on to the data bus. This op code is routed into the instruction register which uses the 8 bits representing the hex number 02 to set up the ALU for an 'add' operation. With the instruction pointer already set, B3101H is latched on the address bus so memory outputs the operand 85H on to the data bus. The complete instruction has been *fetched* from memory. Because the op code has already set up the ALU for addition then with the operand now available the instruction can be executed.

22.1.4 Clock

This fetch/execute sequence characterises the processing of all instructions. Each part of the process is synchronized by a clock pulse generator or just clock (Ck) which typically runs at 8 MHz and therefore each clock pulse lasts for 0.125 μs. Each instruction takes a specified number of clocks for completion. The previous ADD instruction takes four clocks ie $4 \times 0.125 = 0.5$ μs.

The number of clocks differs widely between the various instructions taking from two for an increment (INC) instruction up to 80 for divide (DIV). The number will also depend on whether the instruction involves byte or word length data.

22.1.5 Control Section

The foregoing describes the bare essentials of a computer system including memory, address and data buses, and the execution of instructions by the ALU. There is also an organizational aspect of the system. For example, when the instruction pointer causes an address to be put on the address bus what determines whether the memory is 'written into' or 'read from'? Also, when two separate bytes appear on the data bus, one has to be treated as an operator and the other as an operand. If the previous example were altered to ADD AL, 02 then what would appear on the data bus would be the op code 02 followed by the operand 02. The fact that these two bytes are treated quite differently is because of the control section of the system.

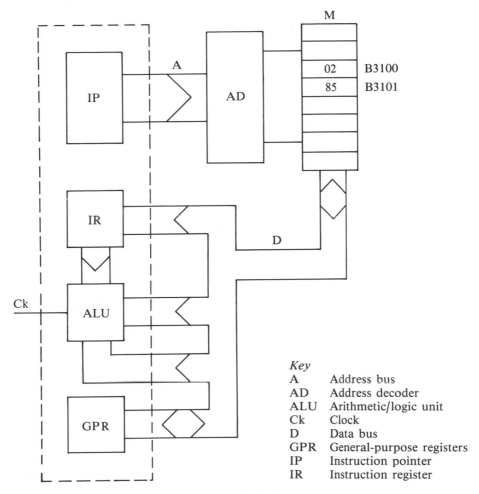

Key
A   Address bus
AD   Address decoder
ALU   Arithmetic/logic unit
Ck   Clock
D   Data bus
GPR   General-purpose registers
IP   Instruction pointer
IR   Instruction register

Fig. 22.2   Generalized Model of a Stored Program Computer

Not shown in Fig. 22.2 is a third bus called the control bus. It includes lines such as 'memory read' and 'memory write'. When the instruction pointer causes an address such as B3100H in the previous example to be output the CPU also activates the memory read line which causes the op code 02 to be put on the data bus. The instruction register, which is part of the control section, causes the ALU to take the contents of AL and set up the addition process. For example, using an adder/subtractor such as the one in Fig. 13.5 it makes SUB = 0. When the next memory read operation is initiated the ALU is expecting a byte of data and when the second 02 appears on the data bus it will be added to the contents of AL. In this way the op code 02 and the operand 02 are treated quite differently in fetching and executing the instruction ADD AL, 02.

22.2 ADDRESSING MODES

The 8088 microprocessor has several addressing modes of which three will be described. They are exemplified by the following instructions

INC BL   – implied addressing
ADD BL,46H   – immediate addressing
ADD BL,[0800H]   – direct addressing

22.2.1 Implied Addressing

The instruction INC BL means add one to the existing contents of register BL (the low byte of BX). Therefore the operand, the number, is not stated but implied. There would be no carry to BH in the event of an overflow (although there would be an effect on the flags register as we shall see in Section 22.4). For example, if BX contains 23FFH then INC BL results in BX containing 2300H. On the other hand the instruction INC BX would lead to BX containing 2400H.

22.2.2 Immediate Addressing

This is the addressing mode used in the example of Section 22.1.3. The instruction ADD BL,46H means add to the existing contents of BL the hex number 46. Thus if BL contained 7CH before the instruction then afterwards it would contain C2H. Again word length and byte length instructions have to be treated differently.

For example, suppose BX contains 45FAH. Then ADD BX,1234H results in BX containing 582EH. However, ADD BL,34H would result in BX containing 452EH, the carry not being propagated when BL only is specified.

22.2.3 Direct Addressing

The instruction ADD BL,[0800H] means add to the existing contents of register BL *the contents of memory location 0800H*. (Note that the address specified is the offset.)

Consider the small block of five memory locations in Fig. 22.3. Now also assume that BX contains 4F29H i.e. BH contains 4FH and BL contains 29H. The instruction ADD BL,[0800H] means that the contents of location 0800H,

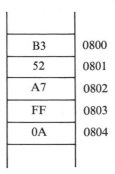

B3	0800
52	0801
A7	0802
FF	0803
0A	0804

Fig. 22.3   Five Memory Locations and Contents

(i.e. B3H), must be added to the existing contents of BL, (i.e. 29H). Therefore Bl will now contain the number DCH. If the instruction were such as to cause BL to overflow then the carry would not propagate into BH.

22.2.4 Low Byte, High Byte Principle

If a direct mode instruction specifies a word length register then two consecutive memory bytes will be involved. For example, using Fig. 22.3 again, consider the instruction, MOV AX, [0802H]. This will cause the contents of location 0802H to be put into AL and those of 0803H into AH i.e. AX will contain FFA7H.

Direct addressing can also be used to load the contents of a register into memory. For example, now that AX contains FFA7H the instruction, MOV [A100H], AX puts A7H into location A100H (lower byte) and FFH into A101H. This principle of accessing the lower byte of the destination first is not universal.

22.3 INSTRUCTION SET

Microprocessor manufacturers such as INTEL provide a document called an instruction set (see the appendix on pp. 612–13) which summarizes, in terms requiring a certain level of sophistication for its interpretation, the functions of their particular system. The 8088 has over eighty instructions many of which can be used with several addressing modes. This section is devoted to a description of some of these instructions. It includes flag settings, the significance of which will be described in Section 22.4, input/output (IN/OUT) instructions, to be covered in Section 22.5, and subroutines (CALL/RET), the subject of Section 22.6.

22.3.1 Data Transfer Instructions

Mnemonic	*Description*	*Examples*	*Flags*†
			Z S C

MOV	Moves data into and between registers and memory	MOV DX,B3144H
		MOV DL,F3H
		MOV AL,[A300H]
		MOV [A300H],AL
		MOV BX,SAMPLE	

XCHG	Exchanges data between registers and memory	XCHG AL,BL XCHG AL,[A046H]	
IN/OUT	Inputting and Outputting data between I/O ports and AX register	IN AL,DX OUT DX,AX	

22.3.2 Arithmetic Instructions

Mnemonic	*Description*	*Examples*	*Flags*		
			Z	S	C

ADD
SUB
CMP
	Adds, subtracts and compares numbers and data in registers and memory. For add/ subtract data in destination (first named locations) is affected by the instruction	ADD BL,A3H
		ADD BL,AL
		ADD BL,[0400H]
		SUB AL,A3H
		SUB [A000H],AL
		SUB AL,TEMP
		CMP AL,4
		CMP AL,[A1000H]
		CMP BX,SAMPLE
			*
			*
			*
			*
			*
			*
			*
			*
			*
				*
				*
				*
				*
				*
				*
				*
				*
				*
					*
					*
					*
					*
					*
					*
					*
					*
					*

INC DEC	Adds or subtracts 1 from register or memory	INC AL DEC ACCOUNT	* *	* *	* *

22.3.3 Logic Instructions

Mnemonic	*Description*	*Examples*	*Flags*		
			Z	S	C
AND OR XOR	ANDs, ORs or XORs contents of registers, memory and numbers. Only destination affected	AND AL,BL OR CX,[A100H] AND DL,8 XOR BX,CX OR AL,POR7B	* * * * *	* * * * *	

† The presence of an asterisk in a column indicates that the flag is affected by the instruction.

22.3.4 Shift and Rotate Instructions

Mnemonic	*Description*	*Examples*	*Flags* Z S C
ROR	Rotate right register or memory contents LSB to MSB and CF	ROR AL,1	*

ROL	Rotate left. Similar to ROR but MSB to LSB and to CF	ROL [A100H],1 ROL SAMPLE,2	* *
SAL/SHL	Shift arithmetic left and shift logical left are synonymous. Has the effect of multiplying by 2, but bit next to sign bit must be 0 for positive and 1 for negative otherwise the sign will not be preserved and there will be overflow		

Mnemonic	*Description*		*Examples*	*Flags*
				Z S C
SAR	Shift arithmetic right, sign bit retained, LSB to CF. Divides by 2			

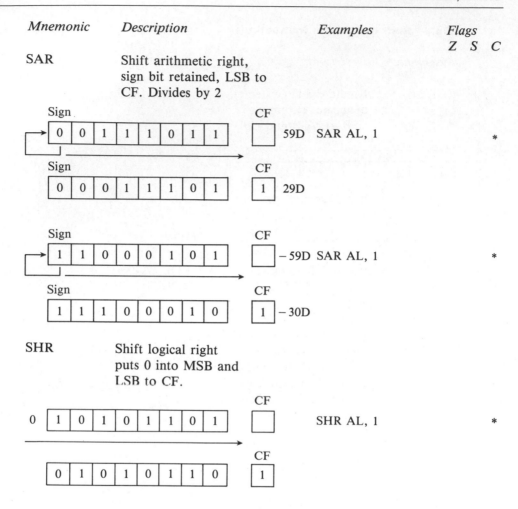

| SHR | Shift logical right puts 0 into MSB and LSB to CF. | |

22.3.5 Control Transfer Instructions

Mnemonic	*Description*	*Examples*	*Flags*
			Z S C
CALL	Call a subroutine	CALL DELAY	
RET	Return from subroutine	RET	
JB/JNAE	Jump on below/not above or equal	JB LESS	
JBE/JNA	Jump on below or equal/ not above	JBE PARITY	
JE/JZ	Jump on equal/zero	JZ TEMP	
JL/JNGE	Jump on less/not greater or equal	JL UNDERWEIGHT	

Mnemonic	*Description*	*Examples*	*Flags*		
			Z	*S*	*C*
JLE/JNG	Jump on less or equal/ not greater	JLE OLDSAMPLE			
JS	Jump on sign	JS NEGSAMPLE			
JMP	Jump always (unconditional)	JMP BEGIN			

22.4 FLAGS REGISTER

This section explains the 'flags' column of the instruction set. If a microprocessor were restricted to arithmetic operations such as DEC and ADD or to logic operations such as AND and OR its usefulness for control purposes, where a decision or action has to be taken based on the state of a process, would be limited. For example, when the temperature of a process reaches a certain level there may be a requirement to reduce the heat input: this will involve frequent checking of the temperature and, when it reaches a certain level, giving out an instruction to reduce the heat. The 8088 microprocessor has a number of conditional 'jump' instructions, i.e. alternative courses of action based on the state of the flags register. This is a 16-bit register of which 9 are used, each associated with a separate flag. The three flags which are the subject of this section are:

Z – zero flag
S – sign flag
C – carry/borrow flag

The setting of these flags can be controlled by data to determine which of two courses of action a microprocessor will take. For example, as long as there is fluid in a tank its contents are not zero and therefore the Z flag can be under program control so that it is not set, i.e. $Z = 0$ and the process of emptying the tank can be continued. As soon as it is empty $Z = 1$ and the program 'jumps' to an alternative set of instructions which causes the tank to be refilled.

It is important to realize that the flags register is updated after every instruction although an instruction will only control flags marked with an asterisk in the instruction set. The three flags under consideration operate as follows.

Z (Zero Flag)

The Z flag is set ($Z = 1$) when the result of an instruction is zero. To take a trivial example consider the instructions

```
              Z
MOV AL,4      ?
INC AL        0
SUB AL,5      1
```

The first instruction is in the 'data transfer group' which do not affect the flags, therefore the Z flag is unaffected; as we were not given the Z flag setting before the first instruction the state of Z is unknown after it. The second instruction increases the contents of AL to 5: the flags register examines AL after this instruction and since the contents are not zero then $Z = 0$. The final instruction reduces the contents of AL to 0 so $Z = 1$.

S (Sign) Flag

The S flag simply regards the MSB (bit 7 for byte operation and bit 15 for word operation) as a sign bit and will be set to 1 when MSB is 1. The previous program segment supplemented by a decrement instruction produces the following flag settings:

```
              Z     S
MOV AL,4      ?     ?
INC AL        0     0
SUB AL,5      1     0
DEC AL        0     1
```

C (Carry/Borrower) Flag

The C flag will be set ($C = 1$) when the result leads to a carry out or a borrow into the MSB. For example

```
                                              Z  S  C
MOV AL,7FH   ; (FLAGS UNAFFECTED BY DATA      ?  ?  ?
             ; (TRANSFER INSTRUCTION
             ;
INC AL       ; (INCREASES CONTENTS OF AL      0  1  0
             ; (TO 80H. THEREFORE MSB = 1
             ; (AND S FLAG = 1
             ;
ADD AL,80H   ; (THIS CAUSES THE CONTENTS      1  0  1
             ; (OF AL TO BECOME ZERO PLUS
             ; (A CARRY OUT
```

One point of note in this short program, apart from the flag settings, is the use of comments to make the program readable. A short program such as this is easy to follow, but one consisting of several hundred instructions would be extremely difficult to follow without comments.

Reference to the instruction set shows that carry flag can be set or reset with any of the shift or rotate instructions. For example MOV AL, FFH makes all bits of $AL = 1$. Then any shift or rotate, left or right instruction makes $C = 1$.

The following two examples show how flags can be used to direct a program to one of two alternative courses of action.

☐ **Example 22.1**

It is required to check the number in register AL and store it in memory location A100 if it is even and in A101 if it is odd.

This problem can be approached by recognizing that a binary number is even if its LSB is 0; otherwise it is odd. A rotate right instruction puts the LSB into the carry position and will set the carry flag if the number is odd.

```
        ROR AL,1        ; PUTS LSB TO CARRY
        JC ODD          ; JUMPS IF NUMBER IS ODD
        ROL AL,1        ; RESTORES EVEN NUMBER
        MOV [A100H],AL  ; STORES EVEN NUMBER
        JUMP FINISH     ; JUMPS TO END OF PROG
ODD     ROL AL,1        ; NUMBER ODD AND RESTORED
        MOV [A101H],AL  ; STORES ODD NUMBER
FINISH RET              ; END OF ROUTINE
```
☐

☐ **Example 22.2**

Operate a continuous check on the contents of register BL. Whenever its contents are decimal 23 make bit 3 of register BH = 1, otherwise 0.

```
CHECK   CMP BL,17H  ; COMPARES BL WITH 23D
        JNZ FALSE   ; (WHEN BL<>23 PROG JUMPS
                    ; (TO FALSE
        OR BH,8     ; (MAKES BH3 = 1 WITHOUT AFFECTING
                    ; (OTHER BITS
        JMP CHECK   ; JUMPS BACK TO RECHECK BL
FALSE   AND BH,F7H  ; MAKES BH3 = 0 WITHOUT AFFECTING
                    ; OTHER BITS
        JMP CHECK
```

The use of AND and OR functions for setting individual bits in a register should be noted. To make BH3 = 1 then OR BH, 8 applies an OR to each bit of BH, except bit 3, with 0 and therefore has no effect. However, BH3 is ORed with 1 and therefore set to 1. To set BH3 to 0 then AND BH, F7H ANDs each bit except BH3 with 1 and therefore has no effect. ANDing BH3 with 0 resets it to 0. ☐

22.5 SUBROUTINES

A subroutine is a self-contained program segment which can be used within a program for a specified task. For example, it may be necessary at some stage in a program to find the average of two numbers. If there is a subroutine to carry out this task then the programmer simply has to incorporate into the program an instruction which calls up the subroutine. At the end of the subroutine there will be an instruction to return to the main program.

There are several advantages in using subroutines:

(a) They are self-contained and can therefore be looked at in isolation from the rest of the program: this improves program readability and simplifies debugging.
(b) A library of subroutines can be produced which reduces development time.
(c) A subroutine can be used repeatedly in a program without extra memory cost.

In assembly language the procedure is simple: one simply writes the instruction for 'Call the subroutine' accompanied by a label which identifies the subroutine. For example:

    CALL AVERAGE

This instruction causes the instruction pointer to jump to a memory address which will have been designated for the subroutine by the assembler: the subroutine itself will be labeled AVERAGE and it will end with the instruction RET − return to the main program from the subroutine. A segment of program using a subroutine might appear as follows:

```
                MOV DX,1A47H
                CALL AVERAGE
                MOV CL,47H
                DEC CH
                :
AVERAGE         MOV AX,[A000H]    ; (THIS SUBROUTINE
                MOV BL,[A001H]    ; (FINDS AVERAGE OF
                ADD AX,BL         ; (TWO NUMBERS
                SHR AX,1          ; (BY ADDING THEM
                RET               ; (AND DIVIDING BY 2
```

Briefly, when the CALL instruction occurs, the contents of the registers, in particular, the instruction pointer, are saved in a section of memory called the *stack*. In this way the results of processes up to the point at which the subroutine is called will be preserved. Since one of these is the instruction pointer contents then when RET occurs the original instruction pointer contents are retrieved from the stack and put back into the instruction pointer so that the program resumes at the next instruction after CALL.

22.6 SOFTWARE DELAYS

As an alternative to using programmable delay ICs, delays can be introduced by writing a program segment and forcing it to be repeated by the number of times necessary to produce the required delay. Assuming, as previously, an 8088 in which each machine cycle (MC) lasts for $0.125 \mu s$, then given the number of machine cycles taken up by each instruction the delay can be calculated.

☐ **Example 22.3**

Calculate the time taken for the following program segment to be executed

```
                                               MC
        MOV AL,85H                              4
DELAY   DEC AL      ; (THESE TWO INSTRUCTIONS   3
        JNZ DELAY   ; (WILL BE EXECUTED         4
                    ; (85 H TIMES
```

To calculate delay first note that $85H = 133D$. Register AL will have to be decremented 133 times to reduce its contents to zero. Therefore if we include the 4 machine cycles (MC) for the MOV instruction the total number of MC for this program segment will be:

$$4 + 133(3 + 4) = 935$$

Therefore delay $= 935 \times 0.125 = 117 \mu s$. ☐

For delays longer than can be produced by loading a register up to its maximum before decrementing, a technique known as nesting of registers or memory locations can be used.

☐ **Example 22.4**

Calculate the delay produced by the following program segment.

```
                                                          MC
            MOV [B100H],8  ;                              10
OUTER_LOOP  MOV [B101H],0  ;                              10
INNER_LOOP  DEC [B101H]    ; (THE FIRST DECREMENT   ⎫     15
            JNZ INNER_LOOP ; (REDUCES B101 CONTENT  ⎬256  4
                           ; (TO FF SO IT WILL TAKE ⎭
                           ; (256 TURNS ROUND THIS        8
                           ; (LOOP BEFORE PROCEEDING
                           ;
            DEC [B100H]    ; (THIS LOOP WILL BE            15
            JNZ OUTER_LOOP ; (REPEATED 8 TIMES BUT         4
                           ; (EACH TIME THE PROCESS
                           ; (OF GOING ROUND THE
                           ; (INNER LOOP 256 TIMES
                           ; (WILL BE REPEATED
```

Total number of MC will be:

$$10 + 8[10 + 256\{15 + 4\} + 15 + 4] = 10[29 + 256 \times 19] = 48930$$

Therefore delay = 6.2 ms. □

The number in B100 is a multiplier and can be chosen to produce different delays. More locations can be used to extend the delay.

22.7 MICROPROCESSOR INPUT/OUTPUT – THE PROGRAMMABLE PERIPHERAL INTERFACE

It is necessary for microprocessors such as the 8088 to be able to pass data to and from computer peripherals such as keyboards and displays, and with devices used for engineering purposes such as instruments and machine tools. The 8088 microprocessor is one of a family of compatible ICs designed for many purposes including interfacing. This section is an introduction to the Intel 8255 Programmable Peripheral Interface (PPI) which has been designed for parallel input/output (I/O) applications.

22.7.1 Programmable Peripheral Interface 8255

The Intel 8255 is an IC which can be used to interface the 8088 to 3 ports, each of 8 bits (Fig. 22.4). Data passing between ports and the 8088 must pass through a register in the PPI, each port having its own register. Each register has an address and in this section we shall use:

1F0H   Port A
1F1H   Port B
1F2H   Port C

Although it is unlikely that the system would be used to its full capacity it can, by using a sufficient number of PPIs, accommodate up to 64 K I/O ports at different addresses.

The addresses are similar to those used in the 8088 system memory and are accessed on the same address bus but there are two instructions peculiar to I/O operations namely IN and OUT. When one of these instructions is used then it causes the PPI to be addressed rather than system memory.

Program instructions can be written so that *all* eight lines of port A (A_0-A_7) are either inputs or outputs and likewise for port B. Port C can be divided into two sets of four lines, CL (CL_0-CL_3) and CH (CH_4-CH_7) and each set can either be a group of inputs or outputs.

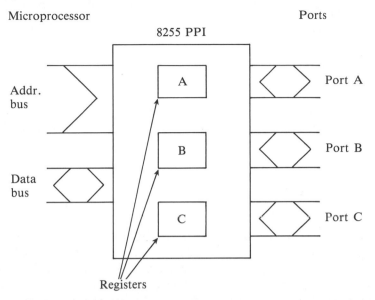

Fig. 22.4   8255 PPI Interfacing Between Three I/O Ports and an 8088 Microprocessor

22.7.2 Control Register

The PPI also includes a fourth 8-bit register – the control register. It has its own address, in this case 1F3H. It is used to determine, among other matters, which of the ports (A, B, CH and CL) will be inputs and which outputs. It is also used to determine in which of three available modes a PPI is operated. The three modes are:

Mode 0   Basic I/O
Mode 1   I/O with handshaking signals
Mode 2   Bidirectional I/O with handshaking signals

As an introduction to microprocessor I/O this chapter will deal only with basic I/O operations which are carried out in mode 0.

The 8 bits of the control register (Fig. 22.5) are used as follows. The operating mode is determined by bits 6, 5 and 2: for mode 0 these must all be 0. To set the mode bit 7 must be 1. This leaves us with bits 4, 3, 1 and 0 which are used to determine whether the ports will be inputs or outputs. A bit set to 1 or 0 makes the sets of lines inputs or outputs respectively. They are designated as follows:

Bit 4   Port A
Bit 3   Port CH
Bit 1   Port B
Bit 0   Port CL

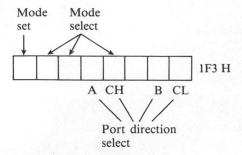

Fig. 22.5   Control Register

☐   **Example 22.5**

What hex number must be written into the control register (address 1F3H) to set
the PPI in mode 0 with ports A and CL inputs and ports B and CH outputs?
As noted, and at all times, to set the PPI in mode 0, bits 7, 6, 5 and 2 must be set
1000 (Fig. 22.6). To make A and CL inputs bits 4 and 0 must be 1. To make B and
CH outputs bits 3 and 1 must be 0. Figure 22.6 shows that the bit pattern in the
control register must be 1001 0001 = 91H. Therefore to set up the control register
as required we must write into the PPI address (1F3H) the number 91H. This procedure is covered in the next section.   ☐

22.7.3   IN and OUT Instructions

The use of an IN or an OUT instruction with an address will activate a PPI register
rather than system memory. For an 8088 microprocessor to input from (IN) or output to (OUT) a PPI register the data must pass through register AX of the 8088.
Since we shall be dealing only with byte length rather than word length instructions

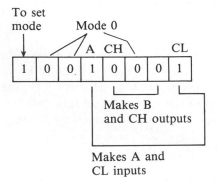

Fig. 22.6   Diagram for Example 22.5

then the data will have to pass through AL, the lower byte of AX. Also to input (IN) or output (OUT), the port's register address must be held in the 8088 DX register unless it is only a byte length address. For example if the PPI port A register were at address EOH then to input data one would simply use the instruction,

IN AL,EOH

However, PPI addresses are more commonly word length and for the remainder of this chapter the PPI addresses will be as previously stated, i.e.:

Port A	1F0H
Port B	1F1H
Port C	1F2H
Control register	1F3II

To use any of the registers its address is first put into 8088 register DX and the data then moved through AL. Assuming that port A has been programmed to be an input then to input data the instruction would be:

```
MOV DX,1F0H   ; PUTS PORT A ADDR TO DX
IN AL,DX       ; PUTS PORT A DATA TO AL
```

If port B were programmed as an output then having put the data into AL the instructions would be:

```
MOV DX,1F1H   ; PORT B ADDR TO DX
OUT DX,AL      ; DATA IN AL TO PORT B
```

It will be observed that IN and OUT instructions embody the general principle of destination preceding source, e.g.:

```
IN AL,DX   ; PORT CONTENTS INTO AL
OUT DX,AL; AL CONTENTS OUT TO PORT
```

22.7.4 PPI Initialization

To set up the PPI control register for a particular mode and I/O pattern it is only necessary to output from register AL to the control register address (1F3H) the appropriate number. Thus to implement Example 22.5 where the requirement was to write to 1F3H (control register address) the number 91H to give mode 0 with A and CL inputs and B and CH outputs the instruction will be:

```
MOV DX,1F3H   ; CONTROL REG ADDR TO DX
MOV AL,91H     ; (PUTS 1001 0001 TO
OUT DX,AL      ; (CONTROL REG VIA AL
```

Familiarity with the instruction set and basic I/O puts us in a position to implement a wide range of microprocessor applications.

☐ **Example 22.6**

When bit 0 of PPI port CL is 1, send out a 50 Hz square wave on bit 7 of port A (Fig. 22.7a).

PPI initialization is required to make port A an output and port CL an input. Putting 0s in the control register for ports B and CH then Fig. 22.7b shows that 81H should be written into the control register (addr. 1F3H). Once the 1 has been detected on CL0 then A7 must alternate between 0 and 1 with 10 ms between transactions: a subroutine can be used for the delay. The program could be as follows:

```
        MOV DX,1F3H   ; CONTROL REG ADDR TO DX
        MOV AL,81H    ; (SETS PORT A AS OUTPUT
        OUT DX,AL     ; (AND PORT CL AS INPUT
                      ;
        MOV DX,1F2H   ; (PORT CL
TEST    IN AL,DX      ; (INTO REGISTER AL
                      ;
        SHR AL,1      ; (CLO TO CARRY, JUMPS
        JNC TEST      ; (BACK IF CL0 = 0
                      ;
        MOV DX,1F0H   ; PORT A TO DX
SQUARE  MOV AL,0      ; (MAKES
        OUT DX,AL     ; (A7 = 0
        CALL DELAY    ; 10 MS DELAY
        MOV AL,80H    ; (MAKES
        OUT DX,AL     ; (A7 = 1
        CALL DELAY    ; 10 MS DELAY
        JMP SQUARE    ;
```

This completes the program except for writing the delay subroutine as previously described.

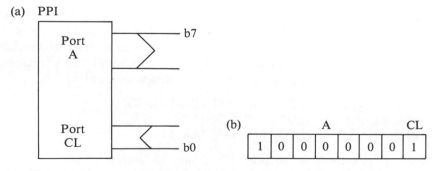

Fig. 22.7   (a) Diagram and (b) Control Register Contents for Example 22.6   ☐

22.8 FLOW CHARTS

For complex problems it is helpful to construct a flow chart which, in effect, divides the problem into sections each of which can be accomplished by a program segment or subroutine. Consider the following example.

☐ **Example 22.7**

Samples of the temperature of a chemical process are represented in binary code by an 8-bit number received on PPI port A (Fig. 22.8a). When the microprocessor system is ready to accept a sample it sends out a 1 followed by a 0 on CLO. When, having checked a series of samples and counted 25 in the range 80H to A0H it must send out a 1 on CH4. It is also required to count the total number of samples checked in reaching 25 in the specified range and then send this number out on port B.

Figure 22.9 shows some of the symbols accepted for use in flow charts. Figure 22.8c is a flow chart for this particular problem. Figure 22.8b shows the bit pattern to be written into the control register to make port A an input and ports B, CH and CL outputs: the number required is 90H.

For this problem microprocessor register BL will be used to count samples in the range 80H to A0H and BH to count the total number of samples. The program could be as follows:

```
MOV DX,1F3H      ; (PPI INITIALIZATION
MOV AL,90H       ; (FOR A AS INPUT AND
OUT DX,AL        ; (B, CH, CL AS OUTPUTS
                 ;
MOV BX,0         ; (BL AND BH SET TO ZERO
                 ; (BEFORE COUNTING SAMPLES
                 ;
```

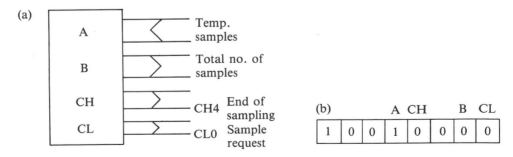

Fig. 22.8   (a) Diagram, (b) Control Register Contents for Example 22.7

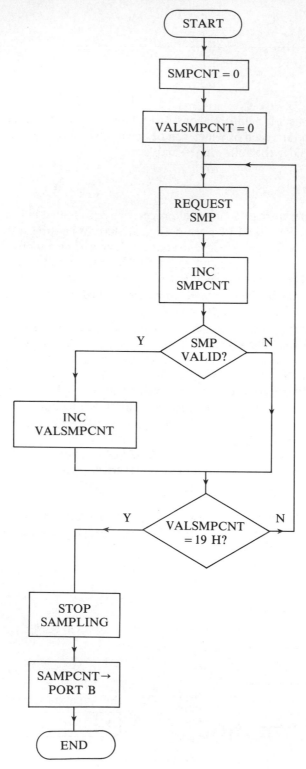

Fig. 22.8   (c) Flow chart for Example 22.7

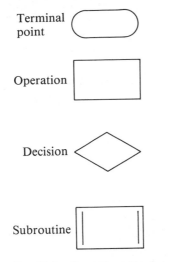

Terminal point

Operation

Decision

Subroutine

Fig. 22.9 Flow Chart Symbols

```
REQSMP        MOV DX,1F2H     ; (SENDS1
              MOV AL,1        ; (FOLLOWED BY 0
              OUT DX,AL       ; (OUT ON CLO
              MOV AL,0        ; (REQUESTING
              OUT DX,AL       ; (SAMPLE
                              ;
              INC BH          ; (INCREMENTS TOTAL NO OF
                              ; (SAMPLES CHECKED
              MOV DX,1F0H     ; (INPUTS SAMPLE
              IN AL,DX        ; (AND JUMPS PAST
              CMP AL,80H      ; (INCREASING VALID SAMPLE
              JL VALSMPCNT    ; (COUNT UNLESS IN
              CMP AL,A0H      ; (RANGE
              JG VALSMPCNT    ; (80H to A0H
                              ;
              INC BL          ; (INCREMENTS VALID SAMPLE COUNT
VALSMPCNT     CMP BL,17H      ; (REQUESTS ANOTHER SAMPLE
              JNZ REQSMP      ; (UNLESS THERE ARE 25 IN RANGE
                              ;
              MOV DX,1F3H     ; (SENDS 1 OUT
              MOV AL,10H      ; (ON CH4 TO
              OUT DX,AL       ; (STOP SAMPLING
                              ;
              MOV DX,1F2      ; (SEND VALSMPCNT
              OUT DX,BL       ; (OUT ON PORT B
                              ;
```

☐   **Example 22.8**

This problem is virtually identical to revision program 21.1. However the motor condition required has been detected by bit masking in this case but the flow chart of Fig. 21.1b could be used as an alternative approach.

A motor is to be controlled by switches connected to port A lines 0 and 1 as shown in Fig. 22.10a. A flow chart for the problem is given in Fig. 22.10b.

```
              MOV DX,1F3H   ; (INITIALIZATION
              MOV AL,90H    ; (PORT A INPUT
              OUT DX,AL     ; (PORT B OUTPUT
                            ;
BEGIN         MOV DX,1F0H   ; (INPUTS MOTOR
              IN AL,DX      ; (CONDITION SWITCHES
                            ;
              AND AL,3      ; (MASKS OFF ALL BITS
                            ; (EXCEPT A1 A0
              CMP AL,0      ; (ONLY JUMPS TO ALARM
              JZ ALARM      ; (IF A1 A0 = 00
                            ;
              CMP AL,1      ; (ONLY JUMPS TO FORWARD
              JZ FORWARD    ; (IF A1 A0 = 01
                            ;
              CMP AL,2      ; (ONLY JUMPS TO REVERSE
              JZ REVERSE    ; (IF A1 A0 = 10
                            ;
              MOV DX,1F1H   ; (ONLY REACHES THIS POINT
              MOV AL,0      ; (IF A1A0 = 11. THEREFORE
              OUT DX,AL     ; (STOP CONDITION EXECUTED
              JMP BEGIN     ;
                            ;
ALARM         MOV DX,1F1H   ; (ALARM
              MOV AL,4      ; (INSTRUCTIONS
              OUT DX,AL     ; (
              JMP BEGIN     ;
                            ;
FORWARD       MOV DX,1F1H   ; (MOTOR
              MOV AL,2      ; (FORWARD
              OUT DX,AL     ; (INSTRUCTIONS
              JMP BEGIN     ;
                            ;
REVERSE       MOV DX,1F1H   ; (MOTOR
              MOV AL,1      ; (REVERSE
              OUT DX,AL     ; (INSTRUCTIONS
              JMP BEGIN     ;
```

(a)

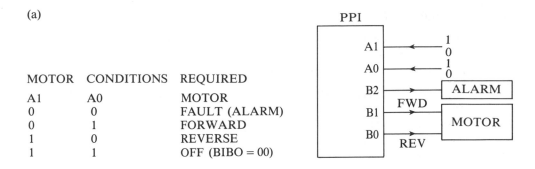

MOTOR   CONDITIONS   REQUIRED

A1	A0	MOTOR
0	0	FAULT (ALARM)
0	1	FORWARD
1	0	REVERSE
1	1	OFF (BIBO = 00)

(b)

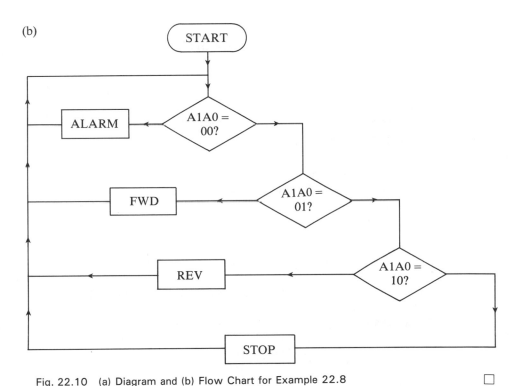

Fig. 22.10   (a) Diagram and (b) Flow Chart for Example 22.8

22.9 PROGRAMMING THE 8088 IN PASCAL

Turbo Pascal is a version of the language which is suitable for programming 8088-based systems particularly where there is a need for I/O applications. The relative simplicity of using Pascal compared with assembly language can be demonstrated by a simple example.

☐ **Example 22.9**

When a 1 is input on CLO (Fig. 22.11a) feed the complement of the number being input on port A out on port B. Figure 22.11b shows that the control register ($1F3) must be set to $91. A suitable program is as follows:

```
PROGRAM COMPLEMENT;
CONST
  PORTA = $1F0 ; (*THESE ASSIGNMENTS*)
  PORTB = $1F1 ; (*PERMIT THE USE OF*)
  PORTC = $1F2 ; (*DESCRIPTORS WHEN*)
  CONTROL = $1F3 ; (*REFERRING TO PORTS*)

VAR
  SAMPLE, INVERSE : BYTE; (*TWO BYTE*)
                          (*LENGTH VARIABLES*)

BEGIN
  REPEAT
    IF PORT[PORTC] AND $01 = $01
    THEN PORT[PORTA] := SAMPLE;
      BEGIN
        INVERSE := $FF - SAMPLE;
        PORTB := PORT[INVERSE];
      END;
  UNTIL KEYPRESSED;

END.
```

Notes
1. The statement SAMPLE := PORT[PORTC] is a means of inputting data from a port to a variable.
2. PORT[INVERSE] := PORTB is a means of outputting a variable to·a port.

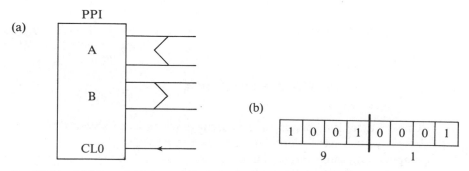

Fig. 22.11 (a) PPI and (b) Control Register Contents for Example 22.9

3. IF PORT[PORTC] AND $01 = $01 means that all bits except CLO can be disregarded and that the statement is only true when CLO = 1.
4. REPEAT UNTIL KEYPRESSED is a means of running a program continuously until any key on the microprocessor system is pressed. □

□ **Example 22.11**

The temperature sampling problem (Example 22.7) could be programmed as follows:

```
PROGRAM SAMPLING;

CONST
   A = $1F0;
   B = $1F1;
   C = $1F2;
   CONTROL_WORD = $1F3;
   (*THESE CONSTANTS ASSIGN LABELS TO THE PORT ADDRESSES*)
VAR
   SMPCNT, VALSMPCNT, SAMPLE : BYTE;
BEGIN
   PORT[CONTROL_WORD] := $90; (*PPI INITIALIZATION*)
   SMPCNT := 0   ;
VALSMPCNT := 0;
(*VARIABLES SET TO ZERO BEFORE COUNT*)
   REPEAT
      PORT[C] := PORT[C] OR $01;
      PORT[C] := PORT[C] AND $FE;
      (*SETS CLO TO '1' THEN '0' WITHOUT AFFECTING OTHER BITS*)
      SAMPLE := PORT[A];     (*INPUTS A SAMPLE*)
      SMPCNT := SMPCNT + 1; (*INCREMENT SAMPCNT*)

      IF (SAMPLE > $80) AND (SAMPLE < $100) THEN
      VALSMPCNT := VALSMPCNT + 1;
      (*INCREMENTS VALSMPCNT IF SAMPLE IN RANGE*)
   UNTIL VALSMPCNT = $19;
   PORT[C] := PORT[C] OR $10;
   (*PUTS '1' OUT ON CH4*)
   PORT [B] := SMPCNT; (*OUTPUTS SMPCNT*)
END.
```
□

APPENDIX: INSTRUCTION SET SUMMARY

DATA TRANSFER

MOV = Move:

	76543210	76543210	76543210	76543210
Register/memory to/from register	1 0 0 0 1 0 d w	mod reg r/m		
Immediate to register/memory	1 1 0 0 0 1 1 w	mod 0 0 0 r/m	data	data if w 1
Immediate to register	1 0 1 1 w reg	data	data if w 1	
Memory to accumulator	1 0 1 0 0 0 0 w	addr-low	addr-high	
Accumulator to memory	1 0 1 0 0 0 1 w	addr-low	addr-high	
Register/memory to segment register	1 0 0 0 1 1 1 0	mod 0 reg r/m		
Segment register to register/memory	1 0 0 0 1 1 0 0	mod 0 reg r/m		

PUSH = Push:

Register/memory	1 1 1 1 1 1 1 1	mod 1 1 0 r/m
Register	0 1 0 1 0 reg	
Segment register	0 0 0 reg 1 1 0	

POP = Pop:

Register/memory	1 0 0 0 1 1 1 1	mod 0 0 0 r/m
Register	0 1 0 1 1 reg	
Segment register	0 0 0 reg 1 1 1	

XCHG = Exchange:

Register/memory with register	1 0 0 0 0 1 1 w	mod reg r/m
Register with accumulator	1 0 0 1 0 reg	

IN = Input from:

Fixed port	1 1 1 0 0 1 0 w	port
Variable port	1 1 1 0 1 1 0 w	

OUT = Output to:

Fixed port	1 1 1 0 0 1 1 w	port
Variable port	1 1 1 0 1 1 1 w	
XLAT = Translate byte to AL	1 1 0 1 0 1 1 1	
LEA = Load EA to register	1 0 0 0 1 1 0 1	mod reg r/m
LDS = Load pointer to DS	1 1 0 0 0 1 0 1	mod reg r/m
LES = Load pointer to ES	1 1 0 0 0 1 0 0	mod reg r/m
LAHF = Load AH with flags	1 0 0 1 1 1 1 1	
SAHF = Store AH into flags	1 0 0 1 1 1 1 0	
PUSHF = Push flags	1 0 0 1 1 1 0 0	
POPF = Pop flags	1 0 0 1 1 1 0 1	

ARITHMETIC

ADD = Add:

Reg/memory with register to either	0 0 0 0 0 0 d w	mod reg r/m		
Immediate to register/memory	1 0 0 0 0 0 s w	mod 0 0 0 r/m	data	data if s w 01
Immediate to accumulator	0 0 0 0 0 1 0 w	data	data if w 1	

ADC = Add with carry:

Reg/memory with register to either	0 0 0 1 0 0 d w	mod reg r/m		
Immediate to register/memory	1 0 0 0 0 0 s w	mod 0 1 0 r/m	data	data if s w 01
Immediate to accumulator	0 0 0 1 0 1 0 w	data	data if w 1	

INC = Increment:

Register/memory	1 1 1 1 1 1 1 w	mod 0 0 0 r/m
Register	0 1 0 0 0 reg	
AAA = ASCII adjust for add	0 0 1 1 0 1 1 1	
DAA = Decimal adjust for add	0 0 1 0 0 1 1 1	

SUB = Subtract:

Reg/memory and register to either	0 0 1 0 1 0 d w	mod reg r/m		
Immediate from register/memory	1 0 0 0 0 0 s w	mod 1 0 1 r/m	data	data if s w 01
Immediate from accumulator	0 0 1 0 1 1 0 w	data	data if w 1	

SBB = Subtract with borrow:

Reg/memory and register to either	0 0 0 1 1 0 d w	mod reg r/m		
Immediate from register/memory	1 0 0 0 0 0 s w	mod 0 1 1 r/m	data	data if s w 01
Immediate from accumulator	0 0 0 1 1 1 0 w	data	data if w 1	

DEC Decrement:

	76543210	76543210	76543210	76543210
Register/memory	1 1 1 1 1 1 1 w	mod 0 0 1 r/m		
Register	0 1 0 0 1 reg			
NEG Change sign	1 1 1 1 0 1 1 w	mod 0 1 1 r/m		

CMP = Compare:

Register/memory and register	0 0 1 1 1 0 d w	mod reg r/m		
Immediate with register/memory	1 0 0 0 0 0 s w	mod 1 1 1 r/m	data	data if s w 01
Immediate with accumulator	0 0 1 1 1 1 0 w	data	data if w 1	
AAS ASCII adjust for subtract	0 0 1 1 1 1 1 1			
DAS Decimal adjust for subtract	0 0 1 0 1 1 1 1			
MUL Multiply (unsigned)	1 1 1 1 0 1 1 w	mod 1 0 0 r/m		
IMUL Integer multiply (signed)	1 1 1 1 0 1 1 w	mod 1 0 1 r/m		
AAM ASCII adjust for multiply	1 1 0 1 0 1 0 0	0 0 0 0 1 0 1 0		
DIV Divide (unsigned)	1 1 1 1 0 1 1 w	mod 1 1 0 r/m		
IDIV Integer divide (signed)	1 1 1 1 0 1 1 w	mod 1 1 1 r/m		
AAD ASCII adjust for divide	1 1 0 1 0 1 0 1	0 0 0 0 1 0 1 0		
CBW Convert byte to word	1 0 0 1 1 0 0 0			
CWD Convert word to double word	1 0 0 1 1 0 0 1			

LOGIC

NOT invert	1 1 1 1 0 1 1 w	mod 0 1 0 r/m
SHL/SAL Shift logical/arithmetic left	1 1 0 1 0 0 v w	mod 1 0 0 r/m
SHR Shift logical right	1 1 0 1 0 0 v w	mod 1 0 1 r/m
SAR Shift arithmetic right	1 1 0 1 0 0 v w	mod 1 1 1 r/m
ROL Rotate left	1 1 0 1 0 0 v w	mod 0 0 0 r/m
ROR Rotate right	1 1 0 1 0 0 v w	mod 0 0 1 r/m
RCL Rotate through carry flag left	1 1 0 1 0 0 v w	mod 0 1 0 r/m
RCR Rotate through carry right	1 1 0 1 0 0 v w	mod 0 1 1 r/m

AND And:

Reg/memory and register to either	0 0 1 0 0 0 d w	mod reg r/m		
Immediate to register/memory	1 0 0 0 0 0 0 w	mod 1 0 0 r/m	data	data if w 1
Immediate to accumulator	0 0 1 0 0 1 0 w	data	data if w 1	

TEST And function to flags, no result:

Register/memory and register	1 0 0 0 0 1 0 w	mod reg r/m		
Immediate data and register/memory	1 1 1 1 0 1 1 w	mod 0 0 0 r/m	data	data if w 1
Immediate data and accumulator	1 0 1 0 1 0 0 w	data	data if w 1	

OR Or:

Reg/memory and register to either	0 0 0 0 1 0 d w	mod reg r/m		
Immediate to register/memory	1 0 0 0 0 0 0 w	mod 0 0 1 r/m	data	data if w 1
Immediate to accumulator	0 0 0 0 1 1 0 w	data	data if w 1	

XOR Exclusive or:

Reg/memory and register to either	0 0 1 1 0 0 d w	mod reg r/m		
Immediate to register/memory	1 0 0 0 0 0 0 w	mod 1 1 0 r/m	data	data if w 1
Immediate to accumulator	0 0 1 1 0 1 0 w	data	data if w 1	

STRING MANIPULATION

REP = Repeat	1 1 1 1 0 0 1 z
MOVS = Move byte/word	1 0 1 0 0 1 0 w
CMPS = Compare byte/word	1 0 1 0 0 1 1 w
SCAS = Scan byte/word	1 0 1 0 1 1 1 w
LODS = Load byte/wd to AL/AX	1 0 1 0 1 1 0 w
STDS = Stor byte/wd from AL/A	1 0 1 0 1 0 1 w

Mnemonics ©Intel, 1978

CONTROL TRANSFER

CALL - Call:

	7 6 5 4 3 2 1 0	7 6 5 4 3 2 1 0	7 6 5 4 3 2 1 0
Direct within segment	1 1 1 0 1 0 0 0	disp-low	disp-high
Indirect within segment	1 1 1 1 1 1 1 1	mod 0 1 0 r/m	
Direct intersegment	1 0 0 1 1 0 1 0	offset-low	offset-high
		seg-low	seg-high
Indirect intersegment	1 1 1 1 1 1 1 1	mod 0 1 1 r/m	

JMP - Unconditional Jump:

	7 6 5 4 3 2 1 0	7 6 5 4 3 2 1 0	7 6 5 4 3 2 1 0
Direct within segment	1 1 1 0 1 0 0 1	disp-low	disp-high
Direct within segment-short	1 1 1 0 1 0 1 1	disp	
Indirect within segment	1 1 1 1 1 1 1 1	mod 1 0 0 r/m	
Direct intersegment	1 1 1 0 1 0 1 0	offset-low	offset-high
		seg-low	seg-high
Indirect intersegment	1 1 1 1 1 1 1 1	mod 1 0 1 r/m	

RET - Return from CALL:

	7 6 5 4 3 2 1 0	7 6 5 4 3 2 1 0	7 6 5 4 3 2 1 0
Within segment	1 1 0 0 0 0 1 1		
Within seg. adding immed to SP	1 1 0 0 0 0 1 0	data-low	data-high
Intersegment	1 1 0 0 1 0 1 1		
Intersegment, adding immediate to SP	1 1 0 0 1 0 1 0	data-low	data-high
JE/JZ-Jump on equal/zero	0 1 1 1 0 1 0 0	disp	
JL/JNGE-Jump on less/not greater or equal	0 1 1 1 1 1 0 0	disp	
JLE/JNG-Jump on less or equal/not greater	0 1 1 1 1 1 1 0	disp	
JB/JNAE-Jump on below/not above or equal	0 1 1 1 0 0 1 0	disp	
JBE/JNA-Jump on below or equal/not above	0 1 1 1 0 1 1 0	disp	
JP/JPE-Jump on parity/parity even	0 1 1 1 1 0 1 0	disp	
JO-Jump on overflow	0 1 1 1 0 0 0 0	disp	
JS-Jump on sign	0 1 1 1 1 0 0 0	disp	
JNE/JNZ-Jump on not equal/not zero	0 1 1 1 0 1 0 1	disp	
JNL/JGE-Jump on not less/greater or equal	0 1 1 1 1 1 0 1	disp	
JNLE/JG-Jump on not less or equal/greater	0 1 1 1 1 1 1 1	disp	

	7 6 5 4 3 2 1 0	7 6 5 4 3 2 1 0
JNB/JAE Jump on not below/above or equal	0 1 1 1 0 0 1 1	disp
JNBE/JA Jump on not below or equal/above	0 1 1 1 0 1 1 1	disp
JNP/JPO Jump on not par/par odd	0 1 1 1 1 0 1 1	disp
JNO Jump on not overflow	0 1 1 1 0 0 0 1	disp
JNS Jump on not sign	0 1 1 1 1 0 0 1	disp
LOOP Loop CX times	1 1 1 0 0 0 1 0	disp
LOOPZ/LOOPE Loop while zero/equal	1 1 1 0 0 0 0 1	disp
LOOPNZ/LOOPNE Loop while not zero/equal	1 1 1 0 0 0 0 0	disp
JCXZ Jump on CX zero	1 1 1 0 0 0 1 1	disp

INT Interrupt

	7 6 5 4 3 2 1 0	7 6 5 4 3 2 1 0
Type specified	1 1 0 0 1 1 0 1	type
Type 3	1 1 0 0 1 1 0 0	
INTO Interrupt on overflow	1 1 0 0 1 1 1 0	
IRET Interrupt return	1 1 0 0 1 1 1 1	

PROCESSOR CONTROL

	7 6 5 4 3 2 1 0	7 6 5 4 3 2 1 0
CLC Clear carry	1 1 1 1 1 0 0 0	
CMC Complement carry	1 1 1 1 0 1 0 1	
STC Set carry	1 1 1 1 1 0 0 1	
CLD Clear direction	1 1 1 1 1 1 0 0	
STD Set direction	1 1 1 1 1 1 0 1	
CLI Clear interrupt	1 1 1 1 1 0 1 0	
STI Set interrupt	1 1 1 1 1 0 1 1	
HLT Halt	1 1 1 1 0 1 0 0	
WAIT Wait	1 0 0 1 1 0 1 1	
ESC Escape (to external device)	1 1 0 1 1 x x x	mod x x x r/m
LOCK Bus lock prefix	1 1 1 1 0 0 0 0	

Footnotes:

AL = 8-bit accumulator
AX = 16-bit accumulator
CX = Count register
DS = Data segment
ES = Extra segment
Above/below refers to unsigned value.
Greater = more positive;
Less = less positive (more negative) signed values
if d = 1 then "to" reg; if d = 0 then "from" reg
if w = 1 then word instruction; if w = 0 then byte instruction

if mod = 11 then r/m is treated as a REG field
if mod = 00 then DISP = 0*, disp-low and disp-high are absent
if mod = 01 then DISP = disp-low sign-extended to 16-bits, disp-high is absent
if mod = 10 then DISP = disp-high: disp-low

if r/m = 000 then EA = (BX) + (SI) + DISP
if r/m = 001 then EA = (BX) + (DI) + DISP
if r/m = 010 then EA = (BP) + (SI) + DISP
if r/m = 011 then EA = (BP) + (DI) + DISP
if r/m = 100 then EA = (SI) + DISP
if r/m = 101 then EA = (DI) + DISP
if r/m = 110 then EA = (BP) + DISP*
if r/m = 111 then EA = (BX) + DISP
DISP follows 2nd byte of instruction (before data if required)

*except if mod = 00 and r/m = 110 then EA = disp-high: disp-low.

if s:w = 01 then 16 bits of immediate data form the operand.
if s:w = 11 then an immediate data byte is sign extended to form the 16-bit operand.
if v = 0 then "count" = 1; if v = 1 then "count" in (CL)
x = don't care
z is used for string primitives for comparison with ZF FLAG.

SEGMENT OVERRIDE PREFIX

0 0 1 reg 1 1 0

REG is assigned according to the following table:

16-Bit (w = 1)		8-Bit (w = 0)		Segment	
000	AX	000	AL	00	ES
001	CX	001	CL	01	CS
010	DX	010	DL	10	SS
011	BX	011	BL	11	DS
100	SP	100	AH		
101	BP	101	CH		
110	SI	110	DH		
111	DI	111	BH		

Instructions which reference the flag register file as a 16-bit object use the symbol FLAGS to represent the file:

FLAGS = X:X:X:X:(OF):(DF):(IF):(TF):(SF):(ZF):X:(AF):X:(PF):X:(CF)

Source: From Uffenbeck (Appendix D).

REVISION QUESTIONS

22.1 For this question use the contents of address block A010 to A01B as shown.

(a) MOV AL, 0FH
State contents of AH, AL
(b) MOV BX, 936BH
State contents of BH, BL
(c) MOV CL, [A011H]
State contents of CH, CL
(d) MOV CX, [A014H]
State contents of CH, CL
(e) MOV [A109H], DL
State contents of DH, DL, A019H
(f) MOV AX, [A012H]
MOV [A10AH], AX
State contents of AH, AL, A01AH, A01BH
(g) MOV AX, [A012H]
ADD AX, [A017H]
State contents of AL, AH
(h) MOV AX, [A013H]
AND AX, 0FF3H
State contents of AX
(i) MOV BL, 46H
OR BL, [A019H]
State contents of BL

0B	A010H
31	1H
53	2H
A7	3H
91	4H
44	5H
AA	6H
FF	7H
29	8H
5A	9H
83	AH
90	BH

22.2 For the following instructions state whether or not the S (sign), Z (zero) and C (carry/borrow) flags are set.

(a) MOV AL, 21H
ADD AL, 61H
(b) MOV AL, 21H
SUB AL, 61H
(c) MOV AX, F375H
SUB Al, 04H
(d) MOV BH, 85H
SAR BH, 1
(e) MOV DL, 0
DEC DL
(f) MOV CX, FFFFH
ADD CX, 1
(g) MOV AL, A1H
MOV BL, 56H
AND AL, BL

(h) MOV CL, [A01AH]
    OR CL, 0AH

22.3   Add the numbers in locations 4300H and 4301H and store the result in 4302H.

22.4   Write a program which compares the numbers in locations 8020H and 8021H. If they are equal clear location 8000H, otherwise store their number in 8001H.

22.5   Write a program which adds all numbers 01H to 0AH and stores the result in 150H. Use no more than seven instructions.

22.6   Write a program which checks whether a number in location 120H is odd or even. If it is odd store it in location 120H, if even store it in location 122H. (Hint: if bit 0 is 0 the number is even.)

22.7   Write a program which outputs:

(a) a 10 kHz square wave
(b) a 10 Hz square wave

on bit 2 of PPI port A (address 1F0H).

22.8   Write a program which sorts the three numbers in locns 0459H, 045AH and 045BH into numerical order with the highest number in the first location.

22.9   If the byte on port A of a PPI has even parity (an even number of 1s including no 1s) then a 1 is sent out on port C line 0. Draw up a flow chart and an assembly language program. (Hint: shift or rotate number into the carry bit position, count the number carrys = 1 and if the number ends in 0 it is even.)

22.10   Attempt text Examples 22.7 and 22.8

(a) with a view to economy of instruction;
(b) using Pascal.

Appendix 1

Resistors: Preferred Values and Color Codes

Readers will have noticed a tendency to use resistor values such as 180 Ω, 47 kΩ and so on. Such values occur because manufacturers provide ranges of components known as 'preferred values' which enable users to stock a limited but representative range which will meet most of their requirements. The preferred value ranges are quoted as being of certain percentage tolerances, the most common of which is 10%. Thus a 470 Ω 10% tolerance resistor can have any value between ±10% of 470 Ω. If this tolerance is unacceptable the user must either use a lower tolerance range, which involves stocking more component values, or be prepared to measure each resistor value before using it.

The nominal values have been chosen as follows. Beginning with a nominal value of 1.0 Ω then with 10% tolerance, the highest value of a 1.0 Ω ± 10% will be 1.1 Ω. If we now select 1.2 as the nominal value then its minimum value will be about 1.1 so it is possible to have any value between 1.0 Ω and 1.2 Ω if the actual

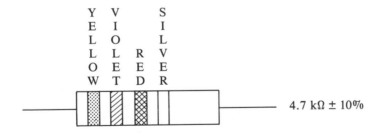

4.7 kΩ ± 10%

Key
Colored bands indicate the resistance in ohms
The first digit is given by the band nearest the edge
The second digit is given by the next band
The number of noughts is given by the third band
The colors represent numbers as follows: 0 black, 1 brown, 2 red, 3 orange, 4 yellow, 5 green, 6 blue, 7 violet, 8 gray, 9 white

The *tolerance* is indicated by the next band, with a color code as follows: 1% brown, 2% red, 5% gold, 10% silver, 20% no band
High stability would be indicated by a fifth pink band

Fig. A.1 Resistor Color Code, An Example

625

value can be within 10% of the nominal value. Carrying on this principle through the decade $1.0\,\Omega$ to $10\,\Omega$, the nominal values, in ohms, are:

1.0, 1.2, 1.5, 1.8, 2.2, 2.7, 3.3, 3.9, 4.7, 5.6, 6.8, 8.2, 10

Other decades generally available are $10\,\Omega$ to $100\,\Omega$, $100\,\Omega$ to $1\,k\Omega$, and so on up into the megohm ranges.

Resistors are usually color coded to indicate their nominal value to two significant figures, their tolerance and sometimes their degree of stability against environmental changes. The color coding system is given in Fig. A.1.

Appendix 2

PSpice Simulation

SPICE is a simulator which was developed at the University of California, at Berkeley. PSpice was derived from SPICE by the Microsim Corporation. It is public domain software and widely used by students of electrical engineering for analog simulation. It can be run on a wide range of computers, including the IBM PC.

We can make a start in stating the rules of PSpice by analysing the network of Fig. A.2.1a.

1. As shown, all nodes have to be numbered, the earth node must be 0.
2. All resistors must be labelled R and voltages V. Suffixes can be numbers or letters. Similar rules apply to all components. Most of the ones used in PSpice are listed in Table A.2.1. E, F, G and H will be explained below. Some of the scaling factor labels are listed in Table A.2.2.

Table A.2.1 PSpice Component Labels

Label	Component
V	Voltage source
I	Current source
R	Resistor
L	Inductor
C	Capacitor
E	VCVS
F	CCCS
G	VCCS
H	CCVS

3. In program statements, component quantities can be expressed either decimally or exponentially, using appropriate scaling factors. Assigning units is optional. For example, these three statements all have the same meaning:

```
V4    0.01
V4    1E-2 Volts
V4    10M
```

627

Table A.2.2   Scaling Factors

Label	Scaling Factor
MEG	10^6
K	10^3
M	10^{-3}
U	10^{-6}
N	10^{-9}
P	10^{-12}

4. PSpice does not distinguish between upper and lower case letters. For example,

$$VCE \equiv vce$$

A PSpice program begins with a circuit description in which each component must,

(a) be labelled;
(b) have the nodes between which it is connected marked;
(c) be given a value.

For Fig. A.2.1a, the program is given in 'Example — Attenuator (1)'. The asterix denotes a comment line which the simulator ignores but which helps readability. All programs must conclude with .END. This is an instruction to execute the program. In the absence of any more instructions the output file, 'Example — Attenuator (1)', will include all node voltages and the current supplied by V_S. The current result is negative because, in PSpice, positive currents flow through the source from positive to negative. In the attenuator circuit of Fig. A.2.1a, V_S is supplying current from its positive terminal around the circuit before reaching the negative terminal, so PSpice calculates the current as a negative quantity. By heading the output file 'Small Signal Bias Solution', the simulator has assumed that only a dc solution was required.

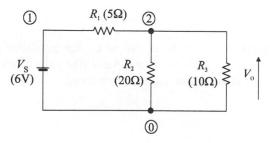

Fig. A.2.1a   Attenuator

```
ATT1.OUT                          1995/12/19      09:40:06      Page: 1


**** 12/19/95 09:40:09 ******** Evaluation PSpice (September 1991) ***********

 Example - Attenuator(1)


  ****      CIRCUIT DESCRIPTION


*******************************************************************************



*Component      Node    Node     Value
R1               1       2         5
R2               2       0        20
R3               2       0        10
VS               1       0        DC   6
.END




ATT1.OUT                          1995/12/19      09:40:06      Page: 2

**** 12/19/95 09:40:09 ******** Evaluation PSpice (September 1991) ***********

 Example - Attenuator(1)

  ****      SMALL SIGNAL BIAS SOLUTION       TEMPERATURE =    27.000 DEG C


*******************************************************************************



 NODE   VOLTAGE      NODE   VOLTAGE      NODE   VOLTAGE      NODE   VOLTAGE


(   1)   6.0000  (   2)   3.4286




      VOLTAGE SOURCE CURRENTS
      NAME           CURRENT

      VS            -5.143E-01

      TOTAL POWER DISSIPATION   3.09E+00   WATTS



         JOB CONCLUDED

         TOTAL JOB TIME           3.68
```

Branch Currents

PSpice only calculates a branch current when there is an independent voltage source in series with the branch. To find the current in, say, R_3 in the attenuator of Fig. A.2.1a, we must modify the circuit by adding an independent voltage source $V_D = 0$ V in series with it, as shown in Fig. A.2.1b. This also requires an extra node. The program is listed in 'Example — Attenuator (2)'.

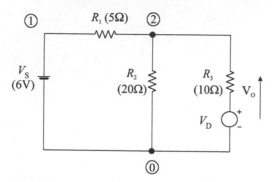

Fig. A.2.1b   Attenuator modified for calculating a branch current

The 'Attenuator (2)' output file shows two currents, the one against V_D being the current in R_3. It is positive because the current flows from positive to negative through the source V_D.

Dependent (Controlled) Sources of Voltage and Current

Many electronic circuits have controlled voltage or current sources, listed as E, F, G and H in Table A.2.1. Application of one of these types of source, the Voltage Controlled Voltage Source (VCVS), whose label must be E (with optional suffix), can be demonstrated with the inverting op amp circuit of Fig. A.2.2a. In the circuit, redrawn for PSpice simulation (Fig. A.2.2b), practical rather than ideal values have been inserted for the op amp. This is essential in the case of R_I because PSpice requires that all nodes have a dc path to earth, so $R_I = \infty$ is not permitted.

The dependent source is E04 20 1E5. This is compiled as follows. E04 indicates that the controlled source appears between nodes 0 and 4. The nodes are in this order because the op amp is connected to invert the voltage between its input terminals. 20 indicates the nodes across which the controlling voltage occurs: it is amplified 100,000 (1E5) times before appearing between nodes 0 and 4. The program and output file are shown as 'Example — Inverting Op-Amp'. The output voltage is V(3).

```
ATT2.OUT                       1995/12/6    11:50:16    Page: 1

**** 12/06/95 11:50:34 ******** Evaluation PSpice (September 1991) ***********

Example - Attenuator(2)


 ****     CIRCUIT DESCRIPTION


*****************************************************************************



*Component     Node    Node    Value
R1              1       2        5
R2              2       0        20
R3              2       3        10
VS              1       0        DC  6
VD              3       0        DC  0
.END



ATT2.OUT                       1995/12/6    11:50:16    Page: 2

**** 12/06/95 11:50:34 ******** Evaluation PSpice (September 1991) ***********

 Example - Attenuator(2)


 ****     SMALL SIGNAL BIAS SOLUTION       TEMPERATURE =   27.000 DEG C


*****************************************************************************


 NODE   VOLTAGE     NODE   VOLTAGE     NODE   VOLTAGE     NODE   VOLTAGE

(    1)   6.0000  (    2)   3.4286  (    3)    0.0000



     VOLTAGE SOURCE CURRENTS
     NAME          CURRENT

     VS           -5.143E-01
     VD            3.429E-01

     TOTAL POWER DISSIPATION   3.09E+00  WATTS


        JOB CONCLUDED

        TOTAL JOB TIME           2.14
```

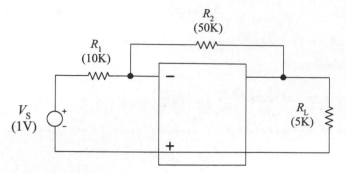

(a) Op Amp Circuit

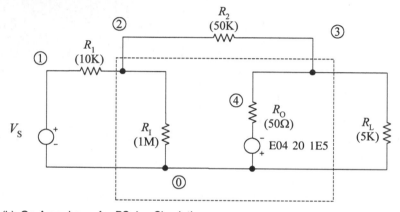

(b) Op Amp drawn for PSpice Simulation

Fig. A.2.2   Inverting Op Amp

Transfer Function Statement

The TF statement facilitates the recording of such quantities as gain. In the last example, if the .END instruction had been preceded by

.TF   V(3)   V1,

then the output file would have included the voltage gain of the circuit.

PRINT Statement

In addition to the node voltages, this statement will produce the voltage across any branch. For the last example, if the voltages across R_1 and R_2 were required, then the program would include the instruction

.PRINT   DC   V(1, 2)   V(2, 3)

```
INVOP.OUT                      1995/12/19    09:12:22    Page: 1


**** 12/19/95 09:12:41 ******** Evaluation PSpice (September 1991) ***********

Example - Inverting Op-Amp


 ****     CIRCUIT DESCRIPTION


*****************************************************************************



R1    1    2    10K
R2    2    3    50K
RI    2    0    1MEG
R0    3    4    50
R3    3    0    5K
VS    1    0    DC 1
E     0    4    2    0    1E5
.END




INVOP.OUT                      1995/12/19    09:12:22    Page: 2

**** 12/19/95 09:12:41 ******** Evaluation PSpice (September 1991) ***********

Example - Inverting Op-Amp


 ****     SMALL SIGNAL BIAS SOLUTION       TEMPERATURE =   27.000 DEG C


*****************************************************************************



NODE   VOLTAGE     NODE   VOLTAGE     NODE   VOLTAGE     NODE   VOLTAGE

(   1)   1.0000  (   2) 50.55E-06  (   3)   -4.9997  (   4)   -5.0547



    VOLTAGE SOURCE CURRENTS
    NAME          CURRENT

    VS         -1.000E-04

    TOTAL POWER DISSIPATION  1.00E-04  WATTS



    JOB CONCLUDED

    TOTAL JOB TIME          3.95
```

AC Analysis

There is a often a need to analyse the performance of circuits containing reactive components over a range of frequencies. PSpice enables this to be done with economy of labour.

In the low pass filter of Fig. A.2.3, it is required to tabulate V_0 over the frequency range 10 Hz to 100 kHz for $V_S = 1$ V. Using an input voltage of 1 V means that the output voltage will also represent gain or loss.

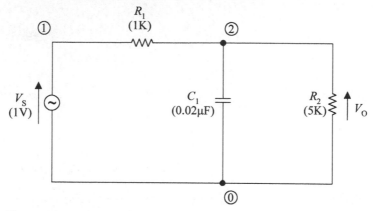

Fig. A.2.3 Low Pass Filter

For the frequency spectrum, PSpice offers a choice between linear (LIN), octave (OCT) and decade (DEC) scaling. It is also necessary to specify the total number of points for a linear sweep or the number of points per octave or decade for the other two sweeps. Using two points/decade for this problem, the instruction specifying the sweep is

.AC DEC 2 10 100K

The program and output file are shown under 'Example — Low Pass Filter'.

Simulation of Small Signal Equivalent Circuits: PROBE Statement

Figure A.2.4a is the h equivalent circuit of a single stage common emitter (CE) amplifier. It is required to tabulate the output voltage V_O over the frequency range 10 Hz to 100 kHz. For the BJT, base current is labelled I_1 and $h_{fe} = 100$. In the same circuit

```
LPF1.OUT                    1995/12/19    09:14:08    Page: 1


**** 12/19/95 09:14:12 ******** Evaluation PSpice (September 1991) ***********

Example - Low Pass Filter


   ****     CIRCUIT DESCRIPTION


*******************************************************************************



R1    1    2    1K
R2    2    0    5K
C1    2    0    0.02U
VS    1    0    AC    1
.AC   DEC  2    10    100K
.PRINT AC V(2)
.END
```

```
LPF1.OUT                    1995/12/19    09:14:08    Page: 3

**** 12/19/95 09:14:12 ******** Evaluation PSpice (September 1991) ***********

Example - Low Pass Filter


   ****     AC ANALYSIS                 TEMPERATURE =   27.000 DEG C


*******************************************************************************



   FREQ         V(2)


   1.000E+01    8.333E-01
   3.162E+01    8.333E-01
   1.000E+02    8.333E-01
   3.162E+02    8.329E-01
   1.000E+03    8.288E-01
   3.162E+03    7.911E-01
   1.000E+04    5.755E-01
   3.162E+04    2.409E-01
   1.000E+05    7.922E-02


        JOB CONCLUDED

        TOTAL JOB TIME        3.51
```

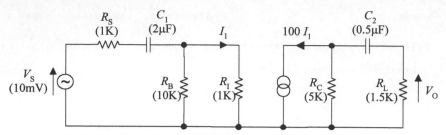

(a) CE Equivalent Circuit

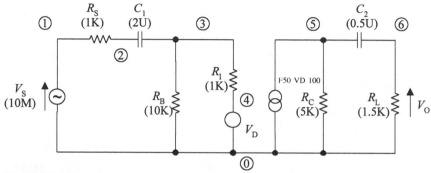

(b) CE Equivalent Circuit re-drawn for PSpice

Fig. A.2.4   CE Amplifier

CEMP1.OUT   **1995/12/19   09:17:09   Page: 1**

**** 12/19/95 09:17:14 ******** Evaluation PSpice (September 1991) **********

Example CE Amp

   ****   CIRCUIT DESCRIPTION

****


```
RS    1    2    1K
RB    3    0    10K
RI    3    4    1K
RC    5    0    5K
RL    6    0    1.5K
C1    2    3    2U
C2    5    6    0.5U
VS    1    0    AC     10M
VD    4    0    0
F     5    0    VD     100
.AC     DEC    4    10    100K
.PRINT AC V(6)
.PROBE  V(6)
.END
```

```
CEMP1.OUT                    1995/12/19    09:17:09    Page: 3


**** 12/19/95 09:17:14 ******** Evaluation PSpice (September 1991) ***********

Example CE Amp


 ****      AC ANALYSIS                    TEMPERATURE =    27.000 DEG C


 ***************************************************************************



     FREQ         V(6)


    1.000E+01    2.565E-02
    1.778E+01    7.359E-02
    3.162E+01    1.801E-01
    5.623E+01    3.329E-01
    1.000E+02    4.555E-01
    1.778E+02    5.158E-01
    3.162E+02    5.383E-01
    5.623E+02    5.459E-01
    1.000E+03    5.483E-01
    1.778E+03    5.491E-01
    3.162E+03    5.493E-01
    5.623E+03    5.494E-01
    1.000E+04    5.494E-01
    1.778E+04    5.494E-01
    3.162E+04    5.495E-01
    5.623E+04    5.495E-01
    1.000E+05    5.495E-01


        JOB CONCLUDED

        TOTAL JOB TIME          3.85
```

prepared for PSpice (Fig. A.2.4b), $V_D = 0$ V must be added in series with R_I to obtain a value for I_1. R_I is used instead of h_{ie} because PSpice requires that R must be the first letter for a resistance. The current controlled source $100I_1$ will, in the PSpice circuit, be F50 VD 100. This can be read: 'the current flowing between nodes 5 and 0 will be equal to 100 times the current flowing through V_D'. Using decade frequency scaling again, but taking 4 frequency points/decade, the program and output file are as shown in 'Example — CE Amp'.

The .PROBE V(6) instruction produces the graphic display of V_6 vs frequency, also labelled 'Example — CE Amp'.

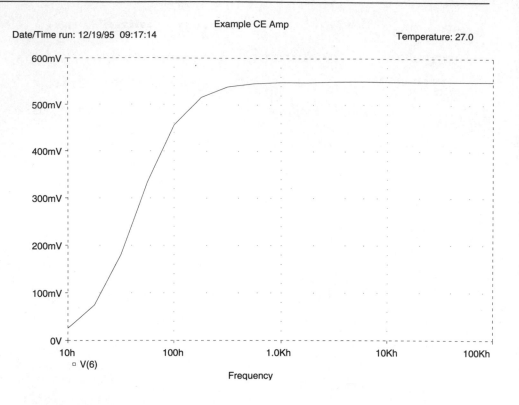

Date/Time run: 12/19/95 09:17:14   Example CE Amp   Temperature: 27.0

Frequency

Revision Questions

To consolidate this work on PSpice it would be a useful exercise to confirm the results of some of the worked examples in chapters 3, 4 and 9 by simulation.

Bibliography and Further Reading

1. Ahmed, H. & Spreadbury, P. J., *Electronics for Engineers*. London: Cambridge University Press, 1973.
2. Alley C. & Attwood, K., *Electronic Engineering*. New York: Wiley, 1973.
3. Almaini, A. E. A., *Electronic Logic Systems*, 2nd Edition. London: Prentice Hall, 1989. Excellent book on modern digital logic systems.
4. Blakeslee. T. R, *Digital Design with Standard MSI and LSI*. New York: Wiley, 1979.
5. Brey, Barry B., *Microprocessor/Hardware Interfacing and Application*. Columbus, Ohio: Merrill, 1984.
   Good on microprocessor hardware.
6. Boylestad R. & Nashelsky, L., *Electronic Devices and Circuit Theory,* 3rd Edition. Englewood Cliffs, NJ: Prentice Hall, 1982.
7. Cirovic, Michael, *Basic Electronics: Devices, Circuits and Systems*. Reston, VA: Reston Publishing Co., 1979.
   Good introduction to circuit update of analog and digital electronics.
8. Clayton, G. B., *Operational Amplifiers,* 2nd Edition. London: Newnes-Butterworths, 1979.
9. Craine, J. F. & Martin, G. R., *Microcomputers in Engineering and Science*. Wokingham, Berks: Addison-Wesley, 1985.
   Orientated towards instrumentation aspects of microcomputers.
10. Dennis, W. H., *Electronic Components and Systems*. Guildford: Butterworths, 1982.
11. Ferguson, John, *Microprocessor Systems Engineering*. Wokingham, Berks: Addison-Wesley, 1985.
    Includes fault finding on principal microprocessor families.
12. Fletcher, William I., *An Engineering Approach to Digital Design*. Englewood Cliffs, NJ: Prentice Hall, 1980.
    An indepth treatment of digital design.
13. Floyd, Thomas L., *Digital Fundamentals,* 2nd Edition. Columbus, Ohio: Merrill, 1982.
    Basic text, well presented.
14. Gosling, W., *A First Course in Applied Electronics*. London: Macmillan, 1975.
15. Hayt, W. H. & Neudeck, G. W., *Electronic Circuit Analysis and Design,* 2nd Edition. Boston, MA: Houghton Mifflin, 1984.

16. Heffer, D. E., King, G. A. & Keith, D., *Basic Principles and Practice of Microprocessors*. London: Edward Arnold, 1981.
Basic treatment of 6800, 8080 and Z80 microprocessor.

17. Leventhal, Lance A., *Introduction to Microprocessors: Software and Hardware Programming*. Englewood Cliffs. NJ: Prentice Hall, 1978.
Comprehensive on microprocessor programming.

18. Lewin, Douglas, *Logical Design of Switching Circuits,* 2nd edition. London: Nelson, 1974.
Clear exposition of digital switching circuit design.

19. Lind, L. F. & Nelson, J. C. C., *Analysis and Design of Digital Systems*. London: Macmillan, 1977.

20. Mano, M., Morris, *Digital Design*. Englewood Cliffs, NJ: Prentice Hall International, 1984.

21. Millman, Jacob, *Microelectronics: digital and analog circuits and systems*. New York: McGraw-Hill, 1979.
Detailed treatment of analog and digital circuits.

22. RS Components Limited, *Data Library*. Corby, Northants, 1986.
Useful set of data sheets on a wide range of electronic devices.

23. Simpson, R. J. & Terrel, T. J., *Introduction to 6800/6802 Microprocessor Systems*. Sevenoaks, Kent: Butterworth, 1982.

24. Tocci, Ronald J., *Digital Systems: principles and applications*. Englewood Cliffs, NJ: Prentice Hall, 1980.

25. Uffenbeck, John, The 8086/8088 Family. Englewood Cliffs, NJ: Prentice Hall, 1987.

26. Kraus, Allan D., *Circuit Analysis*. West Publishing Co., 1991.

27. Dorf, R.C., Introduction to Electronic Circuits, 2nd Edition. New York: Wiley, 1993.

28. Glasford, Glenn M., *Analog Electronic Circuits*. Englewood Cliffs, NJ: Prentice Hall, 1986.

29. Singh, Jasprit, *Semiconductor Devices: An Introduction*. New York: McGraw-Hill, 1994.

30. Tuinenga, Paul W., *SPICE: A Guide to Circuit Simulation and Analysis Using PSpice®* . Englewood Cliffs, NJ: Prentice Hall, 1992.

Index